Process Reactor Design

Process Reactor Design

Ning Hsing Chen
Professor of Chemical Engineering
University of Lowell

ALLYN AND BACON, INC.
Boston London Sydney Toronto

This book is part of the
ALLYN AND BACON SERIES IN ENGINEERING
Consulting Editor: Frank Kreith
University of Colorado

Library of Congress Cataloging in Publication Data

Chen, Ning Hsing.
Process reactor design.

(Allyn and Bacon series in engineering)
Bibliography: p. 534
Includes index.
1. Chemical reactors—Design and construction.
I. Title. II. Series.
TP157.C434 1983 660.2′81 82–22598
ISBN 0–205–07903–2

Printed in the United States of America

10 9 8 7 6 5 4 3 2 1 88 87 86 85 84 83

Contents

Preface

Over the past two decades, chemical kinetics and reactor design have become so important and so attractive that several good textbooks have been published. None, however, contains more than one computer program. Most of the problems in reactor design are long and complicated; and without using a computer, some of the solutions require many hours of computation. In other fields of study, computer programs have already been included in the available texts. It is the author's belief that computer programs in kinetics and reactor design are necessary because their use greatly reduces the time for achieving solutions. In this book, fourteen important, useful, and short programs (CYBER 71) have been included, each successfully tested in the senior-level course "Kinetics and Reactor Design" offered by the author for the past fifteen years in the Department of Chemical Engineering at the University of Lowell.

The conventional method of process stoichiometry is not suitable for writing computer programs. Fortunately a new method, the molar extent of reaction, has been developed. In this book, this method has been renamed, enlarged, and used everywhere in the calculations. In this manner, equations for the total pressure method and the variable volume method, and also generalized equations for treating constant volume rate data, can be developed. This method is reviewed and modified in Chapter 1, is applied to thermodynamics in Chapter 2, and is used extensively in Chapter 3 to analyze the rate data.

Another feature of this book is the development of the generalized mass, energy, and momentum equations in Chapter 4. Incorporated with the divergence theorem, it is shown that these equations can be applied to all cases: batch, semibatch, well-mixed, and plug-flow reactors. Furthermore, three-dimensional generalized mass, energy, and momentum equations are also included in Chapter 9 which can also be used in each case.

In discussing homogeneous reactors, it seems more logical to treat each kind of reactor (batch, semibatch, well-mixed, and plug-flow) separately, with isothermal and nonisothermal conditions all in a single chapter. This has been done in Chapters 5 through 8. Particularly in dealing with the semibatch reactor in Chapter 6, existing textbooks discuss only the first-order reaction. In this book, a comprehensive treatment, including not only the first-order reaction but also second- and higher-order reactions, is described in greater detail, employing a useful program so that all reactions can be handled.

With regard to multiple, complex, and cyclic reactions, using differential equations to solve the simultaneous equations resulting from the formulation of the chemical reactions seems less efficient than using the matrix method. For this reason, the matrix method is introduced to the reader. One illustration is given in Chapter 3. The Laplace transform method is also given in Chapter 3.

Most of the equations generated from the well-mixed reactors are finite-difference equations. In the existing textbooks these equations are solved graphically, but the classical method of solving finite-difference equations is more useful. Hence the inclusion of this method is in Chapter 7. Furthermore, in the sense that the Laplace transform is a useful method for solving differential equations, its equivalent in treating finite-difference equations is the generating function, which is described, for the first time in this kind of textbook, in Chapter 7.

In addition to the above important classical methods of solution, numerical integration and differentiation, as well as the Newton-Raphson method, are also used in Chapters 3, 5, and 7. Numerical methods of solving differential equations, such as the improved Euler method and the Runge-Kutta fourth-order method, are employed in Chapter 8. The numerical finite-difference method is used in Chapter 9.

Because optimization plays an important role in the chemical industry today, a great deal of material has been covered in Chapters 5 through 9 in which the maximum or minimum principle has been recommended.

The inclusion of these sophisticated methods, such as matrices, Laplace transformation, generating functions, finite difference, maximum and minimum principles, and numerical as well as optimization techniques, aims to enhance their applications to chemical engineering science. The choice of the material for class coverage is left to the instructor.

Due to page limitations, heterogeneous reactions and reactors are discussed in a single chapter. Chapter 9 covers heterogeneous reactions and reactors, adsorption and surface reactions, internal diffusion, external mass and heat transfer, catalysts, different models of heterogeneous catalytic reactors, fixed- and fluidized-bed catalytic reactors, slurry reactors, and trickle-bed reactors, and also gas-solid, gas-liquid, and liquid-liquid noncatalytic reactors.

Chapters 10 and 11, describe the nonideal reactor and some of the important aspects in designing reactors, such as size comparison, reactor productivity, reactor selectivity, autothermal operation, and many others.

The author wishes to thank Professor Oliver Ford of the English Department for reviewing the manuscript for consistency of style and expression.

Lastly, it is the author's sincere hope that this book, *Process Reactor Design*, will prove to be useful not only to chemical engineering students but to professional chemical engineers as well.

Process Reactor Design

CHAPTER ONE

Fundamentals

1.1 Introduction

Every chemical engineer should know not only the processes of manufacturing chemicals but also the design and operation of the equipment needed to carry out the processes. This equipment can be divided into two main groups. The first group consists of the equipment used to purify or separate raw materials to the extent necessary to obtain optimum yields. Inasmuch as the properties of the substances after processing remain the same as those of the raw materials, these separations are essentially physical treatment steps; the design and operation of the physical separation equipment are studied in unit operations. The second group consists of chemical reactors in which the processed raw materials react to create products with entirely new physical and chemical properties. This is a chemical treatment step. The design and operation of chemical reactors are the subjects of this book.

When we look at chemical processes, we see that every process behaves as shown in Fig. 1.1-1. That is to say, the raw materials go through a series of physical separation devices and then enter the chemical reactors in which the transformation is carried out. From the reactors, the reaction mixtures which contain the new desired products, side undesired products, and unreacted reactants will be once more purified or separated to obtain products of high quality, to recover the potentially valuable unreacted raw materials, and to destroy or separate the side products. In most cases, recycling the reaction mixtures from the last groups to the first group of the physical separation devices increases the yield of the reaction. In this book, we are primarily concerned with the chemical treatment step. We will discuss the type and size of the reactor needed, provisions for exchange

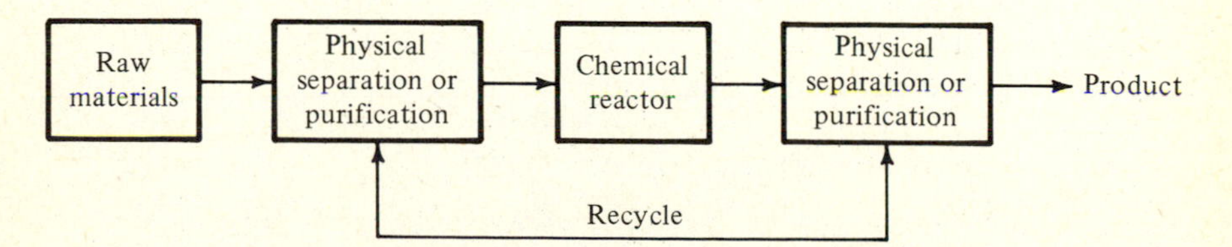

Figure 1.1-1 Flow diagram for manufacture of chemical

of energy with surroundings, and operating conditions such as temperature, pressure, flow rates, and compositions.

Some economic considerations of importance in process reactor systems are the optimum operating conditions, cost analysis and profitability, the stability of the reaction, the control of the reaction, the material construction, and scale-up problems. Whenever necessary, some of these factors will be briefly described.

In the design of a process reactor, a chemical engineer must consider the following:

1. What reaction will occur in the reactor?
2. How fast could the reaction go?
3. What type and size should the reactor be? What operating temperature, pressure, compositions, and flow rates should be selected?
4. Is the production economical?

The first question deals with the thermodynamics from which the equilibrium composition of the reaction mixture can be estimated. The second concerns the process kinetics from which the rate constant of a reaction can be predicted. The third accounts for mass and energy balances in the reaction system. The incorporation of the second and third determines the type and size of the reactor required for certain reactions. The fourth question considers the economics of the process from which the optimum operating conditions can be obtained. In this book, the first problem will be reviewed very briefly, the second and the third will be described in greater detail, the fourth one will be discussed whenever necessary.

In order to fulfill these requirements, we need information, knowledge, and experience from a variety of areas: thermodynamics, chemical kinetics, fluid mechanics, heat transfer, mass transfer, and economics.

1.2 Classification of Reactions

Process reactions may be classified in several ways:

1. On the basis of mechanism
   - (a) Irreversible $\quad A + B \rightarrow C$
   - (b) Reversible $\quad A + B \rightleftarrows C$
   - (c) Parallel, concurrent, or side $\quad A \rightarrow C$, $A \rightarrow B$
   - (d) Series or consecutive $\quad A \rightarrow B \rightarrow C$
   - (e) Complex

$$\begin{array}{ccc} & B & \\ \nearrow\!\!\swarrow & & \nwarrow\!\!\searrow \\ A & \leftrightarrows & C \end{array}$$

2. On the basis of phases
   (a) Homogeneous, gaseous or liquid or solid, catalyzed or uncatalyzed
   (b) Heterogeneous, two or more phases, catalyzed or uncatalyzed.

1.3 Classification of Reactors

In general, the industrial process reactors may be classified into the following types: batch reactors, semibatch reactors, continuous homogeneous reactors, and continuous heterogeneous reactors.

1.3-1 Batch reactors

The batch reactor is the one of the most common types of reactors used in industry (see Fig. 1.3-1). A typical example is the autoclave, in which all the reactants are added before the reaction starts. After the reaction is complete, all the products and unreacted reactants are discharged for further purification. In most cases, this batch reactor is good for liquid reactions. Also included in this type are the closed tank, stirred kettle, kettle with outside recirculation but with no material added or removed, and coil with outside recirculation but with no material added or removed. Batch reactors are used for small-scale or complicated processes that manufacture expensive products, as in the pharmaceutical or dyestuff industries. The principal advantage of this type of reactor is its general versatility combined

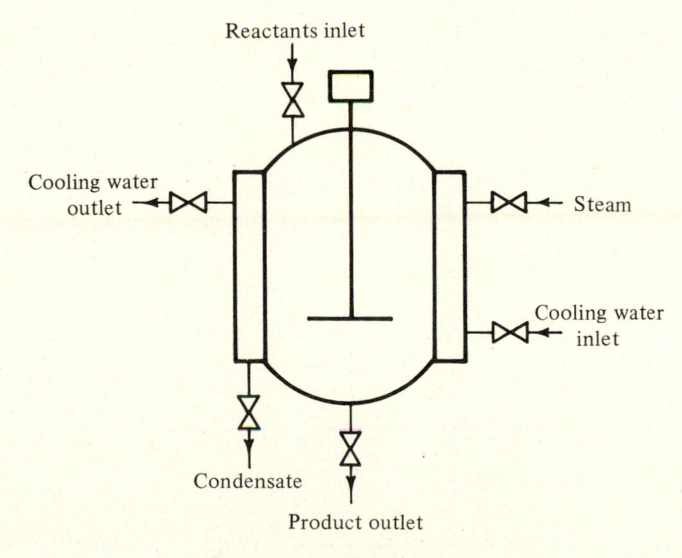

Figure 1.3-1 Batch reactor

with a high yield of the desired product. The disadvantages are small-scale production and high labor costs.

1.3-2 Semibatch reactors

The semibatch reactor is characterized by the feature that one or more reactants are contained in the reactor while the other reactants enter the reactor continuously either with product being removed or remaining in the reactor. There are the following types: (1) batch with continuous removal of products created from gas formation, solid precipitation, or formation of immiscible liquid; (2) batch with continuous addition of one reactant in gas, liquid, or solid phase; (3) batch with combined addition of reactants and removal of one product (Fig. 1.3-2). This type of reactor is often used to carry out homogeneous flow reactions with addition of one product. Its advantages are that the concentration of one reactant is kept low and the temperature can be very well controlled. The disadvantages are that the production is very limited and the labor costs are high.

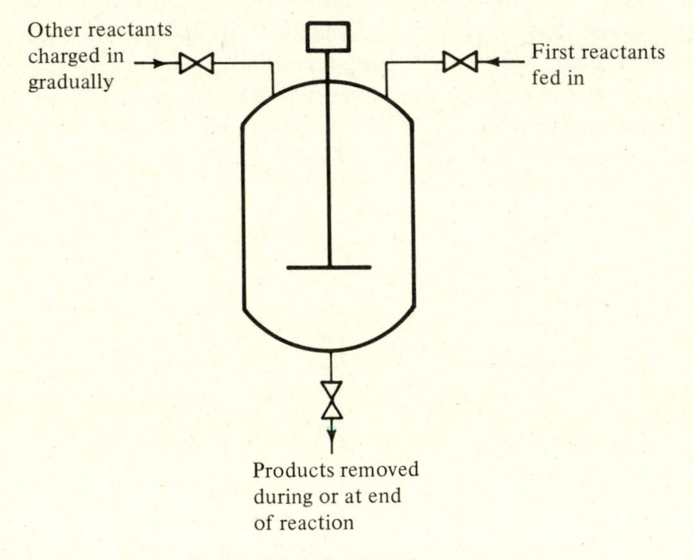

Figure 1.3-2 Semibatch reactor

1.3-3 Continuous homogeneous reactors

There are the following types:

Longitudinal Tubular Reactor (*No Backmixing*) This reactor is used in homogeneous gaseous reaction systems (see Fig. 1.3-3). Its advantages are that high

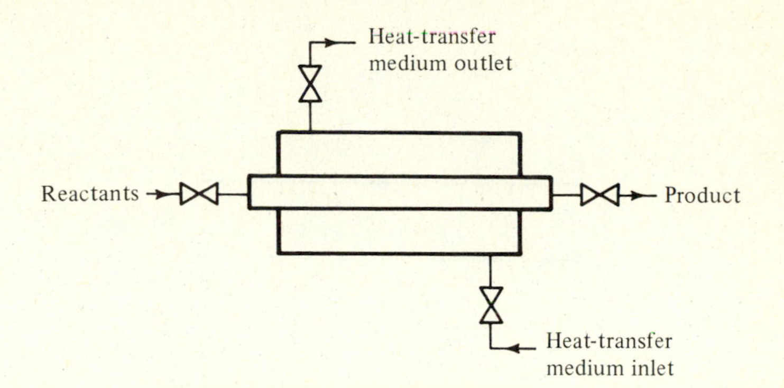

Figure 1.3-3 Longitudinal tubular reactor

yield throughout the reaction can be obtained with little or no backmixing. Its disadvantages are that holding time is fixed for a given throughput and the cost is high for reactions that require long holding time.

Continuous-Stirred-Tank Reactor (*CSTR*) The backmixing is complete in this type of reactor (see Fig. 1.3-4), which is applied in industry where the agitation

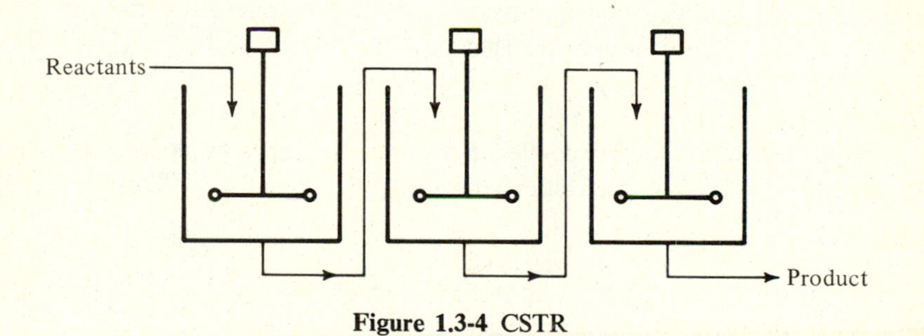

Figure 1.3-4 CSTR

of the reaction mixture is required and where a single tank would be too large to obtain the required degree of conversion. Its advantages are the same as for a batch reactor. Its disadvantages are that the backmixing is sometimes not required and that this series of tanks is more expensive than a single tank.

Baffled-Tank Reactor This reactor is used for homogeneous systems (see Fig. 1.3-5). Its advantages are lower cost per unit volume than in longitudinal reactors and less backmixing than in unbaffled tanks. Its disadvantages are that considerable backmixing is required and the cost is comparatively high.

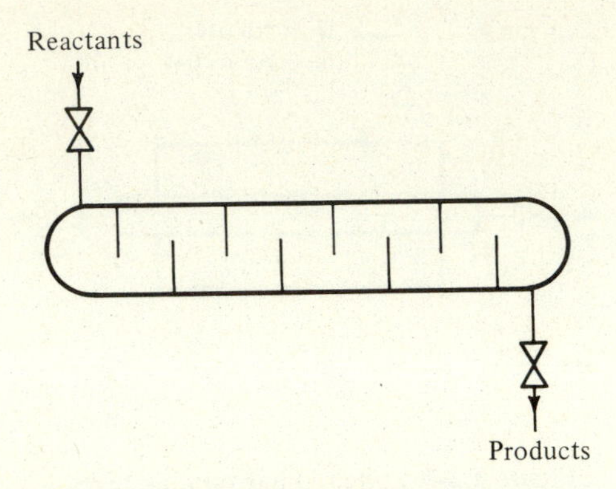

Figure 1.3-5 Baffled-tank reactor

Others These are the tubular tower reactors with some backmixing such as the packed tower or the longitudinal reactor with multiple injections of one reactant.

1.3-4 Continuous heterogeneous reactors

Continuous heterogeneous reactors are used for reactions of gaseous products using a solid or liquid catalyst. Their advantage is that temperature can be closely controlled. Their disadvantages are that capacity is low and the applicability is limited.

Longitudinal Catalytic Fixed-Bed Reactor This reactor is used for vapor-phase reactions with solid catalytic particles (see Fig. 1.3-6). The advantage is

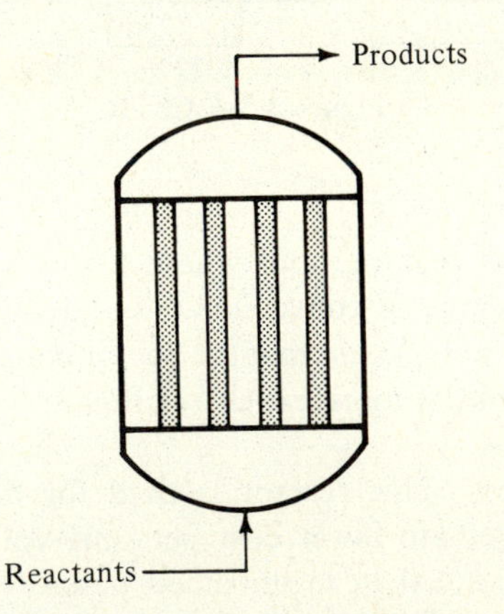

Figure 1.3-6 Fixed-bed reactor

that there is little backmixing. The disadvantage is the possibility of hot spots developing within the bed. Replacing the catalysts is another difficulty.

Moving-Bed Reactor The feed and the products, as well as the fresh catalyst and the unused catalyst, move along the reactor (see Fig. 1.3-7).

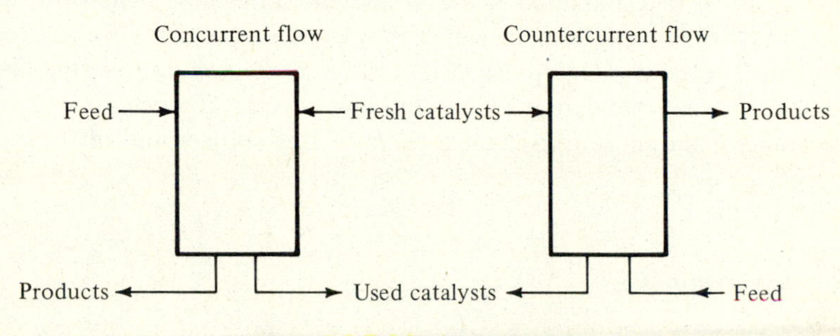

Figure 1.3-7 Moving-bed reactor

Fluidized-Bed Catalytic Reactor Fluidized-bed catalytic reactors are used for vapor-phase reactions requiring a solid catalyst (see Fig. 1.3-8). Its advantages are ease of replacement of the catalyst for regeneration and very even temperature distribution in the catalyst bed. The disadvantages are backmixing and high equipment cost.

Others The packed-tower countercurrent reactors and the distillation-column reactors are other types of continuous homogeneous reactors.

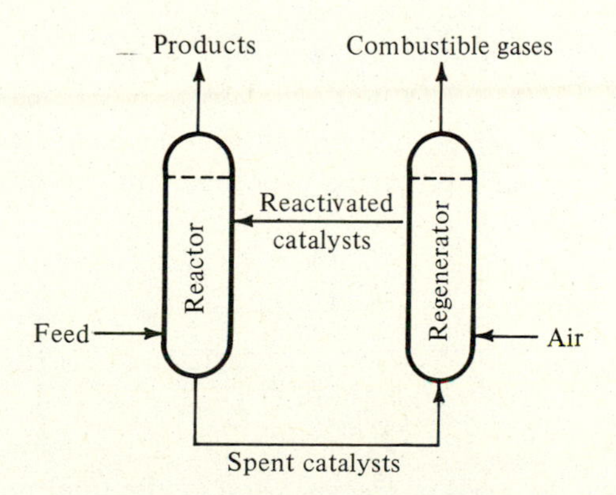

Figure 1.3-8 Fluidized-bed reactor

1.4 Process Stoichiometry

When dealing with a process, the major concern is to determine the necessary amounts of reactants in order to estimate the yield and to evaluate the final composition of the mixture. All of these may be grouped together in a single topic called *process stoichiometry* which is essentially a mass balance of a chemical process. Stoichiometry is the determination of the atomic and molecular weights of elements and compounds, the proportions in which they combine, and the weight relations in any chemical reaction. One must have a clear understanding of stoichiometry or industrial chemical calculation before studying process kinetics, so we are going to review some of the more important aspects of this subject and introduce some new concepts as well.

1.4-1 Process terminology

When a chemical reaction occurs, several species are involved and we need to know how to measure them. In a gaseous system, it is usually more convenient to deal with the number of moles of the gases because the volume of gases changes considerably with the temperature and pressure of the system, whereas for liquids, concentration is used. For each species of the component j, let us denote this number of moles by N_j. If V is the volume of all the species, the molar concentration of component j is

$$C_j = \frac{N_j}{V} \qquad \textbf{(1.4-1)}$$

and the total molar concentration is the sum of the concentration of each species

$$C = \Sigma C_j = \Sigma \frac{N_j}{V} = \frac{\Sigma N_j}{V} = \frac{N}{V} \qquad \textbf{(1.4-2)}$$

where N is the total number of moles of all the chemical species in the system. Similarly, the individual and total mass concentration can be defined as density.

$$\rho_j = \frac{M_j}{V} \qquad \textbf{(1.4-3)}$$

$$\rho = \Sigma \rho_j = \frac{M}{V} \qquad \textbf{(1.4-4)}$$

where M_j and M are the mass of species j and of the total mixture, respectively. Thus the mole fraction x_j and mass fraction g_j become

$$x_j = \frac{N_j}{\Sigma N_j} = \frac{N_j}{N} \tag{1.4-5}$$

$$g_j = \frac{M_j}{\Sigma M_j} = \frac{M_j}{M} \tag{1.4-6}$$

For ideal gas, Dalton's law states that

$$p_j = x_j P \tag{1.4-7}$$

where p_j is the partial pressure of species j, and P is the total pressure of the system. In a closed system, the ideal gaseous mixture becomes

$$p_j V = N_j RT \tag{1.4-8}$$

or

$$\frac{N_j}{V} = \frac{p_j}{RT} \tag{1.4-9}$$

or

$$C_j = \frac{p_j}{RT} = \frac{x_j P}{RT} \tag{1.4-10}$$

where T is the temperature of the system and R is the gas constant.

1.4-2 Process reaction

When a chemical process is carried out, there is a major reaction with or without side reactions. In general, we can classify the major and side reactions into two types: (1) single reactions and (2) multiple reactions

Single Reaction A single reaction is usually denoted by a chemical formula. The typical example is

$$2H_2(g) + O_2(g) \rightarrow 2H_2O(l) \tag{1.4-11}$$

This means that 2 mol gaseous hydrogen react with 1 mol gaseous oxygen to produce 2 mol liquid water. The use of chemical notation is adequate in this simple system in which only a few chemical reactions occur. However, today computer calculation of more complicated systems is facilitated by the use of algebraic notation. For a system having S chemical species, the reaction can be generalized as

$$\sum_{1}^{S} \alpha_j X_j(p) = 0 \tag{1.4-12}$$

in which $X_j(p)$ is the chemical formula for species j in a physical state p. α_j, which is the stoichiometric coefficient of species j, is negative for reactant, positive for product, and zero for inert. For the previous reaction, Eq. (1.4-11) becomes

$$\alpha_1 X_1(p) + \alpha_2 X_2(p) + \alpha_3 X_3(p) = 0 \tag{1.4-13}$$

Here

$$X_1(p) = H_2(g) \qquad \alpha_1 = -2$$
$$X_2(p) = O_2(g) \qquad \alpha_2 = -1$$
$$X_3(p) = H_2O(l) \qquad \alpha_3 = 2$$

Multiple Reactions For multiple reactions, the generalized equations are

$$\sum_{j} \alpha_{1j} X_j(p) = 0 \qquad \text{first reaction} \tag{1.4-14}$$

$$\sum_{j} \alpha_{2j} X_j(p) = 0 \qquad \text{second reaction} \tag{1.4-15}$$

$$\cdots$$

$$\sum_{j} \alpha_{qj} X_j(p) = 0 \qquad q\text{th reaction} \tag{1.4-16}$$

Summing gives

$$\sum_{i=1}^{q} \left(\sum_{j=1}^{S} \alpha_{ij} \right) X_j(p) = 0 \tag{1.4-17}$$

$$\sum_{j=1}^{S} \left(\sum_{i=1}^{q} \alpha_{ij} \right) X_j(p) = 0 \tag{1.4-18}$$

Letting

$$\sum_{i=1}^{q} \alpha_{ij} = \alpha_j \tag{1.4-19}$$

then

$$\sum_{j=1}^{S} \alpha_j X_j(p) = 0 \tag{1.4-20}$$

For example, the water-gas synthesis reactions

$$C(s) + \frac{1}{2} O_2(g) \rightarrow CO(g) \tag{A}$$

$$H_2O(g) \longrightarrow H_2(g) + \frac{1}{2} O_2(g) \tag{B}$$

can be represented by

$$\alpha_{11}X_1(p) + \alpha_{12}X_2(p) + \alpha_{13}X_3(p) = 0 \tag{C}$$

$$\alpha_{24}X_4(p) + \alpha_{25}X_5(p) + \alpha_{22}X_2(p) = 0 \tag{D}$$

where

$$X_1(p) = C(s) \qquad \alpha_{11} = -1$$

$$X_2(p) = O_2(g) \qquad \alpha_{12} = -\frac{1}{2}$$

$$X_3(p) = CO(g) \qquad \alpha_{13} = 1$$

$$X_4(p) = H_2O(g) \qquad \alpha_{24} = -1$$

$$X_5(p) = H_2(g) \qquad \alpha_{25} = 1$$

$$X_2(p) = O_2(g) \qquad \alpha_{22} = \frac{1}{2}$$

Since

$$\alpha_j = \sum_{i=1}^{2} \alpha_{ij}$$

thus

$$\text{For } j = 1 \qquad \alpha_1 = \alpha_{11} + \alpha_{21} = -1 + 0 = -1$$

$$j = 2 \qquad \alpha_2 = \alpha_{12} + \alpha_{22} = -\frac{1}{2} + \frac{1}{2} = 0$$

$$j = 3 \qquad \alpha_3 = \alpha_{13} + \alpha_{23} = 1 + 0 = 1$$

$$j = 4 \qquad \alpha_4 = \alpha_{14} + \alpha_{24} = 0 + (-1) = -1$$

$$j = 5 \qquad \alpha_5 = \alpha_{15} + \alpha_{25} = 0 + 1 = 1$$

Hence

$$-C(s) + CO(g) - H_2O(g) + H_2(g) = 0$$

or

$$C(s) + H_2O(g) = CO(g) + H_2(g)$$

which can be verified by adding Eqs. (A) and (B).

1.4-3 Limiting reactant

In a chemical reaction, the reactant present in the smallest stoichiometric amount is called the *limiting reactant.* For example, the oxidation of sulfur dioxide to sulfur trioxide is represented by

$$SO_2 + \frac{1}{2} O_2 \rightarrow SO_3$$

This stoichiometric equation requires 1 mol or 64 kg SO_2 and 0.5 mol or 16 kg O_2 to produce 1 mol or 80 kg SO_3. Let us consider three cases:

1. If the reacting mixture contains 64 kg SO_2 and 16 kg O_2, then there is no limiting reactant.
2. If the reacting mixture contains 64 kg SO_2 and 20 kg O_2, oxygen is in excess and the limiting reactant is SO_2.
3. If the reacting mixture contains 80 kg SO_2 and 16 kg O_2, sulfur dioxide is in excess and the limiting reactant is O_2.

1.4-4 Yield and selectivity

In a chemical reaction, the amount produced is called the *yield* from the reaction. The yield calculated from the stoichiometric quantities of the chemical reaction is called the *theoretical yield.* The yield actually produced by the reaction is the *actual yield.* The ratio of these two is the *fractional yield.* For example, there are *a* mol alcohol in which *x* mol alcohol decompose to *y* mol aldehyde and *z* mol ethylene as follows:

$$C_2H_5OH \xrightarrow{k_1} CH_3CHO + H_2$$

$$C_2H_5OH \xrightarrow{k_2} C_2H_4 \qquad + H_2O$$

The theoretical yield of CH_3CHO should be *a* mol. Since the actual yield is *y* mol, the fractional yield is therefore y/a on the basis of the feed; since *x* mol react, *x* mol CH_3CHO should be produced, and the fractional yield is therefore y/x. Similarly, the yield for C_2H_4 is z/a and z/x. Hence the meaning of yield depends on the basis we choose. In industry, reactant feed is the basis used most frequently.

The term *selectivity* is used to express the product distribution. Expressing the amount of product formed in terms of the amount of A reacted, the selectivity is defined as

$$\text{Overall selectivity} \quad S_o = \frac{\text{Moles of A transformed into the desired product}}{\text{Moles of A transformed into unwanted products}} \qquad \textbf{(1.4-21)}$$

$$\text{Point selectivity} \quad S_p = \frac{\text{Rate of production of desired product}}{\text{Rate of production of unwanted products}} \qquad \textbf{(1.4-22)}$$

In the example given, assuming first order for both reactions,

$$S_o = \frac{y}{z}$$

$$S_p = \frac{d(\text{aldehyde})/dt}{d(\text{ethylene})/dt} = \frac{k_1(\text{conc. of alcohol})}{k_2(\text{conc. of alcohol})} = \frac{k_1}{k_2}$$

1.4-5 Unit conversion method

After discussing the algebraic representation of chemical reactions, it is appropriate now to study the process stoichiometry of the reactions. First we review the principle of stoichiometry and then indicate the latest scientific method of the use of unit conversion. Let us consider a general chemical reaction

$$\alpha_1 X_1(p) + \alpha_2 X_2(p) \rightarrow \alpha_3 X_3(p) + \alpha_4 X_4(p) \qquad \textbf{(1.4-23)}$$

and choose $X_1(p)$, the limiting reactant, as the basis of calculation. Dividing the reaction by the stoichiometric coefficient of $X_1(p)$ gives, after rearrangement,

$$-X_1(p) - \frac{\alpha_2}{\alpha_1} X_2(p) + \frac{\alpha_3}{\alpha_1} X_3(p) + \frac{\alpha_4}{\alpha_1} X_4(p) = 0 \qquad \textbf{(1.4-24)}$$

The following stoichiometric table can be made:

Chemical species	Initial amount	Final amount	Change
$X_1(g)$	N_{1o}	N_1	$N_1 - N_{1o}$
$X_2(g)$	N_{2o}	N_2	$N_2 - N_{2o}$
$X_3(g)$	N_{3o}	N_3	$N_3 - N_{3o}$
$X_4(g)$	N_{4o}	N_4	$N_4 - N_{4o}$

Since $(\alpha_2/\alpha_1)(N_1 - N_{1o})$ mol X_2 are required for $N_1 - N_{1o}$ mol X_1,

$$\frac{\alpha_2}{\alpha_1}(N_1 - N_{1o}) = N_2 - N_{2o} \tag{1.4-25}$$

Similarly, for X_3 and X_4, we have

$$\frac{\alpha_3}{\alpha_1}(N_1 - \mathrm{N}_{1o}) = N_3 - N_{3o} \tag{1.4-26}$$

$$\frac{\alpha_4}{\alpha_1}(N_1 - N_{1o}) = N_4 - N_{4o} \tag{1.4-27}$$

Equating these three expressions gives

$$\frac{N_1 - N_{1o}}{\alpha_1} = \frac{N_2 - N_{2o}}{\alpha_2} = \frac{N_3 - N_{3o}}{\alpha_3} = \frac{N_4 - N_{4o}}{\alpha_4} = \beta \tag{1.4-28}$$

in which β is defined by some authors as molar extent or degree of advancement of the reaction in moles or a reaction coordinate. Since the difference of the initial moles of X_1 and the final moles of X_1 is the measure of the conversion X_1 and the α_1 is the stoichiometric coefficient of the X_1, it seems more appropriate to define β as the unit conversion. This equation can be expressed in a general form

$$\frac{N_j - N_{jo}}{\alpha_j} = \beta \tag{1.4-29}$$

in which the stoichiometric coefficient α_j is negative for reactants and positive for products or

$$N_j = N_{jo} + \alpha_j \beta \tag{1.4-30}$$

For constant volume V, the equation becomes

$$C_j = C_{jo} + \alpha_j \lambda \tag{1.4-31}$$

where $\lambda = \beta/V$ and is defined as unit conversion per unit volume of reaction mixture. For the limiting reactant X_k, we have

$$N_k = N_{ko} + \alpha_k \beta \tag{1.4-32}$$

Eliminating β yields

$$\frac{N_j - N_{jo}}{N_k - N_{ko}} = \frac{\alpha_j}{\alpha_k} \tag{1.4-33}$$

which can be rearranged as

$$N_j = N_{jo} + \frac{\alpha_j}{\alpha_k}(N_k - N_{ko}) = \left(M - \frac{\alpha_j}{\alpha_k}\right) N_{ko} + \frac{\alpha_j}{\alpha_k} N_k \tag{1.4-34}$$

where $M = N_{jo}/N_{ko}$. Summing Eq. (1.4-30) over the s species gives

$$N = N_o + \alpha\beta \tag{1.4-35}$$

where

$$N = \Sigma N_j,\ N_o = \Sigma N_{jo},\ \alpha = \Sigma \alpha_j,\ j = 1, 2, 3, \cdots, s$$

Dividing Eq. (1.4-30) by Eq. (1.4-35) yields

$$x_j = \frac{N_j}{N} = \frac{N_{jo} + \alpha_j \beta}{N_o + \alpha\beta} \tag{1.4-36}$$

or

$$x_j = \frac{x_{jo} + \alpha_j \delta}{1 + \alpha\delta} \tag{1.4-37}$$

where $\delta = \beta/N_o$. Similarly, summing Eq. (1.4-31) gives

$$C = C_o + \alpha\lambda \tag{1.4-38}$$

where $C = \Sigma C_j$, $C_o = \Sigma C_{jo}$. Dividing Eq. (1.4-31) by Eq. (1.4-38) gives

$$x_j = \frac{C_j}{C} = \frac{C_{jo} + \alpha_j \lambda}{C_o + \alpha\lambda} = \frac{x_{jo} + \alpha_j \gamma}{1 + \alpha\gamma} \tag{1.4-39}$$

where $\gamma = \lambda/C_o$. The fractional conversion f_k is defined as

$$N_k = N_{ko}(1 - f_k) \tag{1.4-40}$$

Combining with Eq. (1.4-30) for the limiting reactant k yields

$$\alpha_k \beta = f_k N_{ko} \tag{1.4-41}$$

in which α_k is always positive whereas α_j can be positive or negative. If the stoichiometric coefficient of the limiting reactant is unity, $\alpha_k = 1$

$$\beta = f_k N_{ko} \tag{1.4-42}$$

Substituting Eq. (1.4-41) into Eq. (1.4-36) gives

$$x_j = \frac{N_{jo} + \alpha_j f_k N_{ko}/\alpha_k}{N_o + \boldsymbol{\alpha} f_k N_{ko}/\alpha_k} \tag{1.4-43}$$

Dividing by N_o yields

$$x_j = \frac{x_{jo} + \alpha_j f_k x_{ko}/\alpha_k}{1 + \boldsymbol{\alpha} f_k x_{ko}/\alpha_k} \tag{1.4-44}$$

Similarly, we may have

$$\lambda = \frac{f_k C_{ko}}{\alpha_k} \tag{1.4-45}$$

and

$$x_j = \frac{C_{jo} + \alpha_j f_k C_{ko}/\alpha_k}{C_o + \boldsymbol{\alpha} f_k C_{ko}/\alpha_k} = \frac{x_{jo} + \alpha_j f_k x_{ko}/\alpha_k}{1 + \boldsymbol{\alpha} f_k x_{ko}/\alpha_k} \tag{1.4-46}$$

For multiple reactions, the number of moles for j component is

$$N_j = N_{jo} + \sum_{i=1}^{q} \alpha_{ij}\beta_i \qquad j = 1, 2, \cdots, s \tag{1.4-47}$$

If $j = k$,

$$N_k = N_{ko} + \sum_{i=1}^{q} \alpha_{ik}\beta_i \tag{1.4-48}$$

That means j represents the species from 1 to s with k being the limiting reactant; i represents multiple reactions from 1 to q. Eliminating β_i gives

$$\frac{N_j - N_{jo}}{N_k - N_{ko}} = \frac{\Sigma\alpha_{ij}}{\Sigma\alpha_{ik}} \tag{1.4-49}$$

Summing Eq. (1.4-47) gives

$$N = N_o + \boldsymbol{\alpha}_i\beta_i \qquad i = 1, 2, \cdots, q \tag{1.4-50}$$

where

$$N = \Sigma N_j, \quad N_o = \Sigma N_{jo}, \quad \boldsymbol{\alpha}_i = \sum_{j=1}^{s} \alpha_{ij}$$

Dividing Eq. (1.4-47) by Eq. (1.4-50) gives

$$x_j = \frac{N_j}{N} = \frac{N_{jo} + \Sigma\alpha_{ij}\beta_i}{N_o + \Sigma\alpha_i\beta_i} = \frac{x_{jo} + \Sigma\alpha_{ij}\delta_i}{1 + \Sigma\alpha_i\delta_i} \tag{1.4-51}$$

where $\delta_i = \beta_i/N_o$.

Example 1.4-1 Stoichiometry by the Conventional and Unit Conversion Methods for Single Reaction

Hydrogen can be manufactured from the carbon monoxide by the water-gas shift reaction

$$CO + H_2O \rightleftarrows CO_2 + H_2$$

2.34 g mol steam per g mol CO is required for the reaction. Find the compositions of the gas mixture for 90 mol percent conversion of CO.

Solution

(1) Conventional method:

Basis of calculation: 1 g mol of CO

Initial gram moles of CO	$= 1$
Initial gram moles of H_2O	$= 2.34$
gram moles of CO after reaction	$= 1 - 0.9 = 0.1$
gram moles of H_2O after reaction	$= 2.34 - 0.9 = 1.44$
gram moles of CO_2 after reaction	$= 0.9$
gram moles of H_2 after reaction	$= 0.9$
Total gram moles of gaseous mixture after reaction	$= 0.1 + 1.44 + 0.9 + 0.9 = 3.34$
Mole fraction of CO after reaction	$= 0.1/3.34 = 0.0299$
Mole fraction of H_2O after reaction	$= 1.44/3.34 = 0.4311$
Mole fraction of CO_2 after reaction	$= 0.9/3.34 = 0.2695$
Mole fraction of H_2 after reaction	$= 0.9/3.34 = 0.2695$
	1.0000

(2) Unit conversion method:

Using Eq. (1.4-43),

$$x_j = \frac{N_{jo} + \alpha_j f_k N_{ko}/\alpha_k}{N_o + \alpha f_k N_{ko}/\alpha_k}$$

$$\alpha_{CO} = -1 \qquad \alpha_{H_2O} = -1 \qquad \alpha_{CO_2} = 1 \qquad \alpha_{H_2} = 1$$

$$\alpha = 1 + 1 - 1 - 1 = 0 \qquad f_k = 0.9 \qquad N_{ko} = 1 \qquad \alpha_k = 1$$

$$\text{For } j = CO \qquad x_{CO} = \frac{1 - 0.9(1)}{3.34 + 0} = \frac{0.1}{3.34} = 0.0299$$

$$j = H_2O \qquad x_{H_2O} = \frac{2.34 - 0.9(1)}{3.34} = \frac{1.44}{3.34} = 0.4311$$

$$j = CO_2 \qquad x_{CO_2} = \frac{0 + 0.9(1)}{3.34} = \frac{0.9}{3.34} = 0.2695$$

$$j = H_2 \qquad x_{H_2} = \frac{0 + 0.9(1)}{3.34} = \frac{0.9}{3.34} = \underline{0.2695}$$

$$1.0000$$

Example 1.4-2 Stoichiometry by the Conventional and Unit Conversion Methods for Multiple Reactions

A mixture of 70 mol percent H_2, 23 mol percent N_2, and 7 mol percent CO_2 reacts until 16 mol percent NH_3 and 4 mol percent H_2O are formed in the following simultaneous reactions. What then are the percentages of H_2 and N_2?

$$3H_2 + N_2 \rightarrow 2NH_3$$

$$H_2 + CO_2 \rightarrow CO + H_2O$$

Solution

1. Conventional method:

   Basis of calculation: 100 g mol of original mixture

$$\text{Initial moles of } H_2 = 70$$

$$\text{Initial moles of } N_2 = 23$$

$$\text{Initial moles of } CO_2 = 7$$

Let y = total gram moles of mixture after reaction
Moles of NH_3 after reaction = $0.16y$
Moles of H_2O after reaction = $0.04y$
To produce $0.16y$ mol of NH_3, the amount of H_2 required is

$$0.16y\ (3/2) = 0.24y$$

To produce $0.04y$ mol of H_2O, the amount of H_2 required is

$$0.04y\ (1/1) = 0.04y$$

Total amount of H_2 required = $0.24y + 0.04y = 0.28y$

Amount of N_2 required = $0.16y(1/2) = 0.08y$

N_2 remaining = $23 - 0.08y$

Amount of CO_2 required = $0.04y(1/1) = 0.04y$

CO_2 remaining = $7 - 0.04y$

Total moles of final mixture = $70 - 0.28y + 23 - 0.08y + 7 - 0.04y + 0.16y + 0.04y + 0.04y = y$

Hence $100 = 1.16y$; therefore $y = 86.21$.

$$x_{H_2} = \frac{70 - 0.28y}{y} = 0.5320 \qquad f_1 = \frac{0.24(86.21)}{70} = 0.2955$$

$$x_{N_2} = \frac{23 - 0.08y}{y} = 0.1868 \qquad f_2 = \frac{0.04(86.21)}{70} = 0.04926$$

2. Unit conversion method:

$$2NH_3 - 3H_2 - N_2 = 0 \quad \text{or} \quad 2X_1 - 3X_2 - X_3 = 0$$

$$CO + H_2O - H_2 - CO_2 = 0 \quad \text{or} \quad X_4 + X_5 - X_2 - X_6 = 0$$

$$\alpha_{11} = 2 \quad \alpha_{12} = -3 \quad \alpha_{13} = -1 \quad \alpha_{14} = 0 \quad \alpha_{15} = 0 \quad \alpha_{16} = 0$$

$$\alpha_{21} = 0 \quad \alpha_{22} = -1 \quad \alpha_{23} = 0 \quad \alpha_{24} = 1 \quad \alpha_{25} = 1 \quad \alpha_{26} = -1$$

$$\alpha_1 = 2 - 3 - 1 = -2 \qquad \alpha_2 = -1 + 1 + 1 - 1 = 0$$

Using Eq. (1.4-51)

$$x_j = \frac{x_{jo} + \alpha_{1j}\delta_1 + \alpha_{2j}\delta_2}{1 + \alpha_1\delta_1 + \alpha_2\delta_2}.$$

$$NH_3\text{: } 0.16 = \frac{0 + \alpha_{11}\delta_1 + \alpha_{21}\delta_2}{1 + (-2)\delta_1 + 0\,\delta_2} = \frac{2\delta_1}{1 - 2\delta_1} \quad \text{hence } \delta_1 = 0.069$$

$$H_2O\text{: } 0.04 = \frac{0 + \alpha_{15}\delta_1 + \alpha_{25}\delta_2}{1 + (-2)\delta_1 + 0\,\delta_2} = \frac{\delta_2}{1 - 2\delta_1} \quad \text{hence } \delta_2 = 0.0344$$

$$H_2\text{: } x_{H_2} = \frac{0.70 - 3(0.069) - 0.0344}{1 - 2(0.069)} = 0.532$$

$$N_2\text{: } x_{N_2} = \frac{0.23 - 0.069}{1 - 2(0.069)} = 0.1868$$

Hence

$$\beta_1 = \delta_1 N_o = 0.069\ (100) = 6.9$$

$$\beta_2 = \delta_2 N_o = 0.0344(100) = 3.44$$

$$f_1 = \frac{\beta_1 \alpha_{1k}}{N_{ko}} = \frac{6.9(3)}{70} = 0.2957$$

$$f_2 = \frac{\beta_2 \alpha_{2k}}{N_{ko}} = \frac{3.44(1)}{70} = 0.049$$

Problems

1. In a bomb calorimeter, 0.03 kmol propane is completely oxidized with 0.72 kmol air that contains 21 mol percent oxygen and 79 mol percent nitrogen to carbon dioxide and water vapor. What is the final composition of the system?

2. Ethylene is oxidized in the presence of a catalyst with air at 200 to 250°C according to the following two reactions:

$$\text{(a)}\ C_2H_4(g) + \frac{1}{2} O_2(g) \rightarrow C_2H_4O(g)$$

$$\text{(b)}\ C_2H_4(g) + 3O_2(g) \rightarrow 2CO_2(g) + 2H_2O(g)$$

   For y mol of air per mole of ethylene in the feed, express the final compositions of the reaction mixture in terms of conversions of ethylene in these two reactions.

3. For the reaction

$$SO_2 + \frac{1}{2} O_2 \rightarrow SO_3$$

   if the entering gas mixture has the composition of 11 mol SO_2, 10 mol percent O_2 and 79 mol percent N_2, calculate the final compositions of the mixture for a fractional conversion of 0.9.

References

(A1). Aris, R. *Elementary Chemical Reactor Analysis.* Englewood Cliffs, N.J.: Prentice-Hall, Inc., 1969.

(H1). Hill, C.G., Jr. *An Introduction to Chemical Engineering Kinetics and Reactor Design.* New York: John Wiley & Sons, Inc., 1977.

(L1). Levenspiel, O. *Chemical Reaction Engineering,* 2nd ed. New York: John Wiley & Sons, Inc., 1972.

(S1). Smith, J.M. *Chemical Engineering Kinetics,* 3rd ed. New York: McGraw-Hill Book Company, 1981.

CHAPTER TWO
Process Thermodynamics

2.1 *Introduction*

As discussed in Chapter 1, the study of process thermodynamics is important because it reveals how far a process reaction can go. Particularly when dealing with the kinetics of a reversible reaction, the equilibrium constant of the reaction is involved. Hence it is necessary to know what it is and how it can be evaluated. When the energy balance of a reaction is made, we need to know the heat of reaction, which is another important topic in process thermodynamics. Although these have been covered in the study of thermodynamics, it is still worthwhile to review some important topics.

2.2 *Heat of Combustion*

When a compound is burned, that is, when a compound reacts with oxygen, heat is evolved or absorbed. This heat is defined as the heat of combustion. For example, when carbon is burned 94,052 cal heat is evolved for 1 g mol carbon. This reaction is expressed as

$$C + O_2 \rightarrow CO_2 \qquad \Delta H_c^\circ = -94{,}052 \text{ cal/g mol}$$

where ΔH_c° is the standard heat of combustion measured at 25°C and 1 atm. The reaction may also be written as

$$C + O_2 \rightarrow CO_2 + 94{,}052$$

or

$$-C - O_2 + CO_2 + 94{,}052 = 0$$

In general, we may represent the equation as

$$\Sigma \alpha_j X_j - (\Delta H_c) = 0 \qquad \textbf{(2.2-1)}$$

where ΔH_c is positive when heat is absorbed and negative when heat is evolved. The values of the heat of combustion for different inorganic and organic compounds can be found in various sources, such as Perry's *Chemical Engineers Handbook*, Lange's *Handbook of Chemistry*, and Weast's *Handbook of Chemistry and Physics*.

2.3 *Heat of Formation*

When a compound is formed from its elements, heat is evolved or absorbed. This heat is defined as the *heat of formation*. When one of the elements is oxygen, the heat of formation is equal to the heat of combustion. For example, the heat of formation of CO_2 at 25°C and 1 atm is $\Delta H_f^\circ = -94{,}052$ cal/g mol, and the heat of formation of water in the gaseous state is

$$H_2 + \frac{1}{2} O_2 \rightarrow H_2O(g) \qquad \Delta H_{298}^\circ = -57{,}800 \text{ cal/g mol}$$

or

$$-H_2 - \frac{1}{2} O_2 + H_2O(g) + 57{,}800 = 0$$

Conversely, the heat of decomposition of water in the gaseous state is

$$H_2O(g) \rightarrow H_2 + \frac{1}{2} O_2 \qquad \Delta H_{298}^\circ = 57{,}800 \text{ cal/g mol}$$

or

$$H_2 + \frac{1}{2} O_2 - H_2O(g) - 57{,}800 = 0$$

In general, we may write

$$\Sigma \alpha_j X_j - (\Delta H_f^\circ) = 0 \tag{2.3-1}$$

Similarly, the values of the heat of formation for inorganic and organic compounds can be found from the sources mentioned above.

2.4 *Heat of Reaction*

For the reaction

$$\alpha_1 X_1 + \alpha_2 X_2 \rightarrow \alpha_3 X_3 + \alpha_4 X_4$$

the heat of reaction ΔH_r is defined as

$$\begin{aligned} \Delta H_r &= \alpha_3 H_3 + \alpha_4 H_4 - \alpha_1 H_1 - \alpha_2 H_2 \\ &= \Sigma \alpha_j H_j \end{aligned} \tag{2.4-1}$$

where H is the enthalpy. This reaction implies that when α_1 mol X_1 reacts with α_2 mol X_2 to produce α_3 mol X_3 and α_4 mol X_4, heat is evolved or absorbed. This is the heat of reaction that can be calculated from the data of standard heat of combustion or standard heat of formation

$$\Delta H_r^\circ = \Sigma \alpha_j \Delta H_{fj}^\circ \tag{2.4-2}$$

$$\Delta H_r^\circ = -\Sigma \alpha_j \Delta H_{cj}^\circ \tag{2.4-3}$$

where ΔH_r° is the standard heat of reaction at 25°C and 1 atm.

Example 2.4-1 Calculation of heat of reaction

Calculate heat of reaction from heat of combustion for the following reaction

$$\underset{(e)}{C_2H_5OH(l)} + \underset{(a)}{CH_3COOH(l)} \rightarrow \underset{(r)}{C_2H_5OOCCH_3(l)} + \underset{(w)}{H_2O(l)}$$

$$\begin{aligned} \Delta H_r^\circ &= -(\Delta H_{c,r}^\circ + \Delta H_{c,w}^\circ - \Delta H_{c,e}^\circ - \Delta H_{c,a}^\circ) \\ &= -[-538.75 + 0 - (-326.70 - 208.34)] = 3.71 \text{ kcal/g mol} \end{aligned}$$

Calculate heat of reaction from heat of formation from the following reaction

$$C_2H_2(g) + H_2O(l) \rightarrow CH_3CHO(g)$$

$$\begin{aligned} \Delta H_r^\circ &= \Delta H_{f,CH_3CHO}^\circ - \Delta H_{f,H_2O}^\circ - \Delta H_{f,C_2H_2}^\circ \\ &= -39{,}760 - (54{,}194 - 68{,}317) = -25{,}637 \text{ cal/g mol} \end{aligned}$$

2.5 *Effect of Temperature on Heat of Reaction*

For the reaction

$$\alpha_1 X_1 + \alpha_2 X_2 + \cdots \rightarrow \alpha_3 X_3 + \alpha_4 X_4 + \cdots$$

the heat of reaction at any temperature is

$$\Delta H_r^T = \alpha_3 H_3 + \alpha_4 H_4 + \cdots - \alpha_1 H_1 - \alpha_2 H_2 - \cdots = \Sigma \alpha_j H_j \tag{2.5-1}$$

where H_j is the enthalpy of the j component and α_j is the stoichiometric coefficient. Since at constant pressure, the partial derivative of enthalpy with respect to temperature is the heat capacity at constant pressure, Eq. (2.5-1) after differentiating becomes

$$\left(\frac{\partial \Delta H_r^T}{\partial T}\right)_P = \alpha_3 C_{P3} + \alpha_4 C_{P4} + \cdots - \alpha_1 C_{P1} - \alpha_2 C_{P2} \cdots = \Sigma \alpha_j C_{Pj} \tag{2.5-2}$$

where $C_{Pj} = (\partial H_j / \partial T)_P$. Let us define

$$\Delta C_{Pj} = \Sigma \alpha_j C_{Pj}$$

Integrating Eq. (2.5-2) gives

$$\int_{\Delta H^{T^\circ}}^{\Delta H^T} d\Delta H_r^T = \int_{T^\circ}^{T} \Delta C_{Pj}\, dT \tag{2.5-3}$$

If

$$C_{Pj} = a_j + b_j T + c_j T^2 + \frac{d_j}{T^2}$$

then

$$\Delta C_{Pj} = \Delta a_j + \Delta b_j T + \Delta C_j T^2 + \frac{\Delta d_j}{T^2}$$

where

$$\Delta a_j = \Sigma \alpha_j a_j \qquad \Delta b_j = \Sigma \alpha_j b_j \qquad \Delta c_j = \Sigma \alpha_j c_j \qquad \Delta d_j = \Sigma \alpha_j d_j$$

Integrating Eq. (2.5-3) gives

$$\Delta H_r^T = \Delta H_r^{T^\circ} + \Delta a_j (T - T^\circ) + \frac{\Delta b_j}{2}(T^2 - T^{\circ 2}) + \frac{\Delta C_j}{3}(T^3 - T^{\circ 3}) + \frac{\Delta d_j}{T^\circ}\left(\frac{\phi - 1}{\phi}\right) \tag{2.5-4}$$

where $\phi = T/T^\circ$ and $\Delta H_r^{T^\circ}$ is the heat of reaction at T°. If the standard condition is taken, $T^\circ = 25°C$. If mean heat capacities are known,

$$\Delta \mathbf{C}_{Pj} = \Sigma \alpha_j \mathbf{C}_{Pj}$$

Eq. (2.5-4) becomes

$$\Delta H_r^T = \Delta H_r^{T^\circ} + \Delta \mathbf{C}_{Pj}(T - T^\circ) \tag{2.5-5}$$

where

$$\Delta \mathbf{C}_{Pj} = \frac{\int_{T^\circ}^{T} \Delta C_{Pj}\, dT}{T - T^\circ}$$

Example 2.5-1 Calculation of Heat of Reaction

Calculate the heat of reaction of the following reaction at 500°C

$$C(s) + H_2O(g) \rightarrow CO(g) + H_2(g)$$

$$\text{Data: } \Delta H^{25}_{f,H_2O} = -57.798 \text{ kcal/g mol}$$

$$\Delta H^{25}_{f,CO} = -26.416 \text{ kcal/g mol}$$

	a, cal/g mol · K	$10^3 b$, cal/g mol · K^2	$10^7 c$, cal/g mol · K^3	$10^{-5} d$, cal · K/g mol
C(s)	4.100	1.020	0	−2.10
H_2O(g)	7.219	2.374	2.67	0
CO(g)	6.342	1.836	−2.80	0
H_2(g)	6.947	−0.200	4.81	0

Solution

$\Delta H_r^{25} = -26{,}416 + 57{,}798 = 31{,}382$ cal/g mol

$\Delta a = (-1)(4.10) + (-1)(7.219) + (1)(6.342) + (1)(6.947) = 1.97$ cal/g mol · K

$\Delta b = [(-1)(1.02) + (-1)(2.374) + (1)(1.836) - (0.2)]10^{-3} = -1.758 \times 10^{-3}$ cal/g mol · K^2

$\Delta c = [-2.67 - 2.8 + 4.81]10^{-7} = -0.66 \times 10^{-7}$ cal/g mol · K^3

$\Delta d = 2.1 \times 10^5$ cal · K/g mol

$\phi = T/T^\circ = 773/298 = 2.5939$

Using Eq. (2.5-4)

$$\Delta H_r^{500} = 31{,}382 + 298(1.97)(2.5939 - 1) + \frac{1}{2}(298)^2(-1.758 \times 10^{-3})(2.5939^2 - 1)$$
$$+ \frac{1}{3}(298)^3(-0.66 \times 10^{-7})(2.5939^3 - 1) + \frac{(2.1 \times 10^5)(2.5939 - 1)}{298(2.5939)}$$
$$= 32{,}294.014 \text{ cal/g mol}$$

2.6 *Effect of Pressure on Heat of Reaction*

Differentiate Eq. (2.5-1) at constant temperature with respect to pressure.

$$\left(\frac{\partial \Delta H}{\partial P}\right)_T = \alpha_3 \frac{\partial H_3}{\partial P} + \alpha_4 \frac{\partial H_4}{\partial P} + \cdots - \alpha_1 \frac{\partial H_1}{\partial P} - \alpha_2 \frac{\partial H_2}{\partial P} \cdots = \Sigma \alpha_j \frac{\partial H_j}{\partial P} \quad \textbf{(2.6-1)}$$

Integrating yields

$$\int_{H^{T^\circ}_{(1)}}^{H^{T^\circ}_{(P)}} d\Delta H^T = \Sigma \alpha_j \int_1^P \frac{\partial H_j}{\partial P}\, dP \quad \textbf{(2.6-2)}$$

where $H^{T^\circ}_{(P)}$ is the enthalpy at temperature T° and pressure P, and $H^{T^\circ}_{(1)}$ is the enthalpy at temperature T° and pressure of 1 atm. From thermodynamics, it is seen that

$$\left(\frac{\partial H_j}{\partial P}\right)_T = \left(V_j - T\frac{\partial V_j}{\partial T}\right)_P \quad \textbf{(2.6-3)}$$

Hence Eq. (2.6-2) becomes

$$(\Delta H_r^{T^\circ})_P = (\Delta H_r^{T^\circ})_1 + \Sigma \alpha_j \int_1^P \left(V_j - T\frac{\partial V_j}{\partial T}\right)_P dP \quad \textbf{(2.6-4)}$$

where $(\Delta H_r^{T^\circ})_1$ is the standard heat of reaction at temperature T° and 1 atm. The integral in Eq. (2.6-4) is almost negligible for liquid systems and gases at low pressure. For high pressure, it can be evaluated by the analytical method, the graphical method, and the generalized Pitzer correlation.

2.6-1 Analytical method

If an equation of state, such as the Beattie-Bridgeman, Van der Waals, or Redlich-Kwong for j component is known, the integral can be evaluated. The discussion of the accuracy of these equations is beyond the scope of this book, but it has been treated very extensively in the literature of thermodynamics. Here only the analytical method of evaluating the integral of Eq. (2.6-4) is illustrated by an example, from which the method can be extended to various cases.

Example 2.6-1 Effect of Pressure on Enthalpy

Calculate the integral of Eq. (2.6-4) for methane at 37.73°C and 102.07 atm by the Beattie-Bridgeman equation.

Solution

The Beattie-Bridgeman equation of state is

$$PV = RT + \frac{\beta}{V} + \frac{\gamma}{V^2} + \frac{\delta}{V^3} \tag{A}$$

where

$$\beta = RTB_o - A_o - \frac{Rc}{T^2}$$

$$\gamma = -RTB_ob + aA_o - \frac{RB_oc}{T^2}$$

$$\delta = \frac{RB_obc}{T^2}$$

The numerical constants are:

$$A_o = 2.2769,\ a = 0.01855,\ B_o = 0.05587,\ b = -0.01587,\ c = 12.83 \times 10^4$$

From thermodynamics, it is known that

$$d(PV) = P\,dV + V\,dP$$

Thus

$$\int_{P_0}^{P} V\,dP = PV - P_oV_o - \int_{V_0}^{V} P\,dV \tag{B}$$

Because the coefficient $(\partial V/\partial T)_P$ is not convenient to work with in the case of an equation not explicit in the volume, it may be transformed by

$$\left(\frac{\partial V}{\partial T}\right)_P = -\left(\frac{\partial P}{\partial T}\right)_V \left(\frac{\partial V}{\partial P}\right)_T \tag{C}$$

At constant T,

$$\left(\frac{\partial V}{\partial T}\right)_P dP = -\left(\frac{\partial P}{\partial T}\right)_V dV \tag{D}$$

Using Eqs. (B) and (D),

$$\int_{P_0}^{P} \left[V - T\left(\frac{\partial V}{\partial T}\right)_P\right] dP = PV - P_oV_o - \int_{V_0}^{V} P\,dV + T\int_{V_0}^{V} \left(\frac{\partial P}{\partial T}\right)_V dV \tag{E}$$

The first and second integrals on the right-hand side of Eq. (E) can be obtained from the Beattie-Bridgeman equation. Substituting the results into Eq. (E) gives

$$\int_{P_o}^{P}\left[V - T\left(\frac{\partial V}{\partial T}\right)_P\right]dP = PV - P_oV_o$$

$$-\left[\alpha_1\left(\frac{1}{V}-\frac{1}{V_o}\right)+\beta_1\left(\frac{1}{V^2}-\frac{1}{V_o^2}\right)-\gamma_1\left(\frac{1}{V^3}-\frac{1}{V_o^3}\right)\right] \quad \text{(F)}$$

where

$$\alpha_1 = A_o + \frac{3Rc}{T^2}$$

$$\beta_1 = \frac{3RcB_o}{2T^2} - \frac{aA_o}{2}$$

$$\gamma_1 = \frac{RB_obc}{T^2}$$

Using trial and error, the value of V can be evaluated from Eq. (A)

$$V = 0.2189 \text{ liter/g mol}$$

$$V_o = \frac{RT}{P_o} = \frac{0.08206(310.93)}{1} = 25.5149 \text{ liters/g mol}$$

The values of α_1, β_1, and γ_1 in Eq. (F) are

$$\alpha_1 = 2.6036 \qquad \beta_1 = -0.01199 \qquad \gamma_1 = -9.6558 \times 10^{-5}$$

Substituting all these numerical values into Eq. (F) gives

$$\int_{P=1}^{P=102.07}\left[V - T\left(\frac{\partial V}{\partial T}\right)_P\right]dP = -14.7227 \text{ liter}\cdot\text{atm/g mol}$$

The value reported in *Chemical Engineering,* February 25, 1980, is −14.764 liter·atm/g mol.

2.6-2 Graphical method

If the *PVT* relationship is tabulated, the integral in Eq. (2.6-4) can be evaluated by the graphical method which is illustrated by the following example.

Example 2.6-2 Effect of Pressure on Enthalpy by the Graphical Method

From the following data, calculate

$$\int_{40}^{229.3}\left[V - T\left(\frac{\partial V}{\partial T}\right)_P\right]dP$$

Specific volumes V, ft³/lb

P, psia	100°F	130°F	160°F	190°F	220°F
40	2.411	2.565	2.715	2.861	3.006
60	1.538	1.651	1.757	1.860	1.962
100		0.9121	0.9888	1.060	1.128
150			0.5983	0.6557	0.7091
200				0.4505	0.4977
229.3				0.3687	0.4137

Solution

The values of the partial derivative of volume with respect to temperature at constant pressure $(\partial V/\partial T)_P$ are first evaluated from the given data.

At 40 psia, using the five-point formula,

$$\left(\frac{\partial V}{\partial T}\right)_{40,190°F} = \frac{1}{12(30)}[-2.411 + 6(2.565) - 18(2.715) + 10(2.861) + 3(3.006)]$$
$$= 0.004825$$

At 60 psia, using the five-point formula,

$$\left(\frac{\partial V}{\partial T}\right)_{60,190°F} = \frac{1}{360}[-1.538 + 6(1.651) - 18(1.757) + 10(1.86) + 3(1.962)]$$
$$= 0.003411$$

At 100 psia, using the four-point formula,

$$\left(\frac{\partial V}{\partial T}\right)_{100,190°F} = \frac{1}{180}(0.9121 - 6(0.9888) + 3(1.06) + 2(1.128)) = 0.0023072$$

At 150 psia, using the three-point formula,

$$\left(\frac{\partial V}{\partial T}\right)_{150,190°F} = \frac{1}{60}(0.7091 - 0.5983) = 0.0018467$$

At 200 psia, using the forward-difference formula,

$$\left(\frac{\partial V}{\partial T}\right)_{200,190°F} = \frac{1}{30}(0.4977 - 0.4505) = 0.0015733$$

At 229.3 psia, using the forward-difference formula,

$$\left(\frac{\partial V}{\partial T}\right)_{229.3,190°F} = \frac{1}{30}(0.4137 - 0.3687) = 0.0015$$

Next the values of $[V - T(\partial V/\partial T)_P]$ are calculated at various pressures:

Pressure, psia	$[V - T(\partial V/\partial T)_P]$
40	2.861 − 650(0.004825) = −0.27525
60	1.860 − 650(0.003411) = −0.35715
100	1.060 − 650(0.002307) = −0.43968
150	0.6557 − 650(0.001847) = −0.54465
200	0.4505 − 650(0.001573) = −0.57215
229.3	0.3687 − 650(0.001500) = −0.60630

Finally, the values of $-[V - T(\partial V/\partial T)_P]$ are plotted versus pressure (see Fig. 2.6-1) and the area under the curve is evaluated by Simpson's rule as −69.67 ft³/lb (psia).

Then

$$\int_{40}^{229.3} [V - T(\partial V/\partial T)_P]\, dP = -69.67(144)/788 = -12.73 \text{ Btu/lb}$$

2.6-3 *Generalized Pitzer correlation*

More generally, the integral in Eq. (2.6-4) can be evaluated by the generalized Pitzer correlation. The second virial coefficient in an expansion involving the acentric factor described by Pitzer and Curl is*

$$\frac{BP_c}{RT_c} = B^0 + B^1\Omega \tag{2.6-5}$$

where

$$B^0 = f^0(T_r) = 0.1445 - \frac{0.330}{T_r} - \frac{0.1385}{T_r^2} - \frac{0.0121}{T_r^3}$$

$$B^1 = f^1(T_r) = 0.073 + \frac{0.46}{T_r} - \frac{0.5}{T_r^2} - \frac{0.097}{T_r^3} - \frac{0.0073}{T_r^8}$$

The real gas equation for 1 mol is

$$Z = \frac{PV}{RT} \tag{2.6-6}$$

* See Reid and Sherwood: *The Properties of Gases and Liquids*, 2nd ed. (New York: McGraw-Hill Book Co., 1966), p. 74.

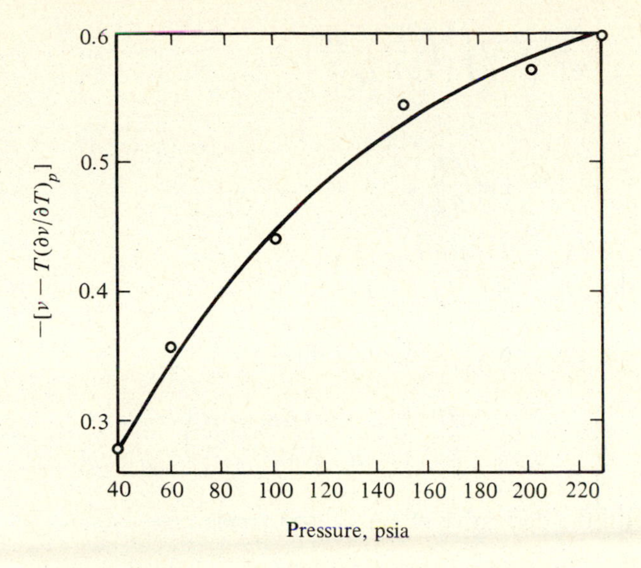

Figure 2.6-1 Graphical method for effect of pressure on enthalpy

The total differential dZ is

$$dZ = \frac{Z}{P}\,dP + \frac{Z}{V}\,dV - \frac{Z}{T}\,dT \tag{2.6-7}$$

Hence at constant pressure

$$\left(\frac{\partial Z}{\partial T}\right)_P = \left(\frac{Z}{V}\right)\left(\frac{\partial V}{\partial T}\right)_P - \frac{Z}{T} = \left(\frac{P}{RT}\right)\left(\frac{\partial V}{\partial T}\right)_P - \frac{Z}{T} \tag{2.6-8}$$

Therefore

$$\left(\frac{\partial V}{\partial T}\right)_P = \left(\frac{RZ}{P}\right) + \left(\frac{RT}{P}\right)\left(\frac{\partial Z}{\partial T}\right)_P \tag{2.6-9}$$

Substituting Eq. (2.6-6) and Eq. (2.6-9) into the integral in Eq. (2.6-4) and defining H' gives

$$H' = \int_1^P \left[\frac{ZRT}{P} - T\left(\frac{RZ}{P} + \frac{RT}{P}\left(\frac{\partial Z}{\partial T}\right)_P\right)\right] dP \tag{2.6-10}$$

or

$$\frac{H'}{RT} = -T\int_1^P \left(\frac{\partial Z}{\partial T}\right)_P \frac{dP}{P} \tag{2.6-11}$$

But

$$Z = 1 + \frac{BP}{RT} \tag{2.6-12}$$

$$\left(\frac{\partial Z}{\partial T}\right)_P = \frac{P}{R}\left(\frac{1}{T}\frac{dB}{dT} - \frac{B}{T^2}\right) \tag{2.6-13}$$

Substituting into Eq. (2.6-11) and integrating gives

$$\frac{H'}{RT} = \frac{1-P}{R}\left(\frac{dB}{dT} - \frac{B}{T}\right) \tag{2.6-14}$$

Combining with the derivative of Eq. (2.6-5) yields

$$\frac{H'}{RT} = \frac{1-P}{P_c}\left[\left(\frac{dB^0}{dT_r} - \frac{B^0}{T_r}\right) + \Omega\left(\frac{dB^1}{dT_r} - \frac{B^1}{T_r}\right)\right] \tag{2.6-15}$$

where B^0, B^1, dB^0/dT_r, and dB^1/dT_r can be obtained from Eq. (2.6-5). After the integrals are evaluated, the heat of reaction at any pressure can be calculated from Eq. (2.6-4) as follows:

$$(\Delta H_r^{T^\circ})_p = (\Delta H_r^{T^\circ})_1 + \Sigma \alpha_j H_j' \tag{2.6-16}$$

Example 2.6-3 Heat of Reaction by Pitzer Correlation

Calculate the heat of reaction at 400°C and 200 atm of

$$\frac{1}{3}N_2 + H_2 \rightarrow \frac{2}{3}NH_3$$

Data: $\Delta H_{r,\,1\,\text{atm}}^{400°\text{C}} = -35{,}450$ J/g mol H_2

	T_c, K	P_c, atm	Ω
NH_3	405.6	112.5	0.25
N_2	126.2	33.5	0.04
H_2	33.2	12.8	0.00

Solution

	NH_3	N_2	H_2
T_r	1.6598	5.3344	20.2771
P_r	1.7777	5.9701	15.6250
$B^0 = 0.1445 - 0.33/T_r - 0.1385/T_r^2 - 0.0121/T_r^3$	−0.1072	0.0777	0.1278
$B^1 = 0.073 + 0.46/T_r - 0.5/T_r^2 - 0.097/T_r^3 - 0.0073/T_r^8$	0.1473	0.1410	0.0945
$dB^0/dT_r = 0.33/T_r^2 + 0.277/T_r^3 + 0.0363/T_r^4$	0.1851	0.0135	8.36 $\times 10^{-4}$
$dB^1/dT_r = -0.46/T_r^2 + 1/T_r^3 + 0.291/T_r^4 + 0.0584/T_r^9$	0.0907	−9.2182 $\times 10^{-3}$	Neg.
H', kJ/kmol	−2,475.5	82.88	448.96

Using Eq. (2.6-16),

$$\Delta H_{r,\,200\text{ atm}}^{400^\circ\text{C}} = 35{,}450 - \tfrac{1}{3}(82.88) - 448.96 - \tfrac{2}{3}(2{,}475.5)$$

$$= -37{,}576.92 \text{ J/g mol } H_2$$

2.7 *Heat of Reaction in a Flow Process*

A macroscopic energy balance of a flow process is

$$\int_{\text{c.s.}} (\hat{H} + \hat{\phi} + \hat{K})\rho(\mathbf{v}\cdot\mathbf{n})dA + \frac{d}{dt}\int_{\text{c.v.}} (\hat{U} + \hat{\phi} + \hat{K})\rho dV = Q - W_s \quad \textbf{(2.7-1)}$$

where $\hat{H}$, $\hat{\phi}$, $\hat{K}$ are enthalpy, potential energy, and kinetic energy per unit mass, respectively. ρ is the density. Q is the heat transferred from the surroundings to the system. W_s is the shaft work done on the surroundings. Expanding Eq. (2.7-1) gives

$$\frac{d}{dt}\int(\hat{U} + \hat{\phi} + \hat{K})dm + [(\hat{H} + \hat{\phi} + \hat{K})(\rho u A)]_2 - [(\hat{H} + \hat{\phi} + \hat{K})(\rho u A)]_1 = Q - W_s \quad \textbf{(2.7-2)}$$

At steady state,

$$\frac{d}{dt}\int(\hat{U} + \hat{\phi} + \hat{K})dm = 0 \quad \textbf{(2.7-3)}$$

Neglecting $\hat{\phi}$, $\hat{K}$, W_s gives

$$(\hat{H}\dot{m})_2 - (\hat{H}\dot{m})_1 = Q$$

or

$$\dot{m}_2\hat{H}_2 - \dot{m}_1\hat{H}_1 = Q \tag{2.7-4}$$

When this equation applies to a reaction, it means that heat transferred from the surroundings is equal to the difference of the total enthalpy of all the products at the exit and the total enthalpy of all the reactants at the entrance. The equation becomes

$$Q = \dot{n}_p H_p - \dot{n}_f H_f \tag{2.7-5}$$

where $\dot{n}_p$ is the total molar product rate, $\dot{n}_f$ is the total molar feed rate, H_p is the molar enthalpy of the products, and H_f is the molar enthalpy of feed. In other words, we may write

$$Q = \dot{n}_p H_p^{Tp} - \dot{n}_f H_f^{Tf} = \dot{n}_p H_p^{Tp} - \dot{n}_p H_p^{298} - (\dot{n}_f H_f^{Tf} - \dot{n}_f H_f^{298}) + (\dot{n}_p H_p^{298} - \dot{n}_f H_f^{298})$$

or

$$Q = \int_{298}^{Tp} (\Sigma \dot{n}_j)_p C_{Pj}\, dT - \int_{298}^{Tf} (\Sigma \dot{n}_j)_f C_{Pj}\, dT - (\Sigma \dot{n}_{jp} - \Sigma \dot{n}_{jf}) H^{298} \tag{2.7-6}$$

Because

$$\dot{n}_{jp} - \dot{n}_{jf} = \alpha_j \dot{\beta}$$

and

$$\Sigma \alpha_j H^{298} = \Delta H^{298}$$

hence

$$Q = \int_{298}^{Tp} (\Sigma \dot{n}_j)_p C_{Pj}\, dT - \int_{298}^{Tf} (\Sigma \dot{n}_j)_f C_{Pj}\, dT + \Sigma \alpha_j \dot{\beta} H^{298}$$

$$= \int_{298}^{Tp} \Sigma(\dot{n}_j C_{Pj})_p\, dT - \int_{298}^{Tf} \Sigma(\dot{n}_j C_{Pj})_f\, dT + \dot{\beta} \Delta H^{298} \tag{2.7-7}$$

Example 2.7-1 Heat Removal from a Reactor

A gaseous mixture of 11 mol percent SO_2, 10 mol percent O_2, and 79 mol percent N_2 at 400°C enters a converter and leaves at 500°C with an equilibrium fractional conversion of 0.9. Calculate the heat removal from the converter. ΔH^{298} is −23,490 cal/g mol.

Solution

$$SO_2 + \frac{1}{2} O_2 \rightarrow SO_3$$
$$(1) \qquad (2) \qquad (3)$$

Basis of calculation: 100 g mol/h of the entering gaseous mixture

$$\dot{n}_{1f} = 11 \text{ g mol/h} \qquad \dot{n}_{2f} = 10 \text{ g mol/h} \qquad \dot{n}_{3f} = 0; \ \dot{n}_{4f} = 79 \text{ g mol/h}$$

$$\alpha_{11} = -1 \qquad \alpha_{12} = -\frac{1}{2} \qquad \alpha_{13} = 1 \qquad \alpha_{14} = 0$$

$$\dot{n}_j = \dot{n}_{jo} + \frac{\alpha_j f_k \dot{n}_{ko}}{\alpha_k}$$

$j = 1$ for SO_2 $\quad \dot{n}_1 = 11 - 1(0.9)(11)/(1) = 1.1$ g mol/h

$j = 2$ for O_2 $\quad \dot{n}_2 = 10 - \frac{1}{2}(0.9)(11)/(1) = 5.05$ g mol/h

$j = 3$ for SO_3 $\quad \dot{n}_3 = 0 + 1(0.9)(11)/(1) = 9.9$ g mol/h

$j = 4$ for N_2 $\quad \dot{n}_4 = 79 + 0 = 79$ g mol/h

Hence

$$\dot{\beta} = \frac{\dot{n}_{1P} - \dot{n}_{1f}}{\alpha_{11}} = \frac{1.1 - 11}{-1} = 9.9 \text{ g mol/h}$$

or

$$\dot{\beta} = \frac{f_k \dot{n}_{ko}}{\alpha_k} = \frac{0.9(11)}{1} = 9.9 \quad \text{check}$$

$$\begin{aligned}\int_{298}^{500+273} (\Sigma \dot{n}_j C_{Pj})_p \, dT = {} & 1.1[7.116(773 - 298) + 4.756(773^2 - 298^2)10^{-3} \\ & + 1.17(10)^{-6}(777^3 - 298^3)] + 5.05[6.148(475) \\ & + 1.551(509)10^{-3} - 0.308(437.5)10^{-6}] + 79[6.449(475) \\ & + 0.706(509)10^{-3} - 0.0269(437.5)10^{-6}] + 9.9[6.077(475) \\ & + 11.768(509)10^{-3} - 0.229(437.5)10^{-6}] \\ = {} & 292{,}354.96 \text{ cal/h}\end{aligned}$$

$$\begin{aligned}\int_{298}^{400+273} (\Sigma \dot{n}_j C_{Pj})_f \, dT = {} & 11[7.116(673 - 298) + 4.7556(673^2 - 298^2)10^{-3} \\ & + 1.1703(673^3 - 298^3)10^{-6}] + 10[6.148(375) \\ & + 1.551(364.5)10^{-3} - 0.3077(278.5)10^{-6}] + 79[6.449(375) \\ & + 0.706(364.5)10^{-3} - 0.0269(278.5)10^{-6}] \\ = {} & 266{,}117.45 \text{ cal/h}\end{aligned}$$

Hence

$$Q = 292{,}354.97 - 266{,}117.45 + 9.9(-23{,}490) = -206{,}313.50 \text{ cal/h}$$

If the reaction is carried out adiabatically, the temperature of the reaction attained in the converter is called *adiabatic flame temperature,* which can also be calculated by Eq. (2.7-7) using $Q = 0$.

2.8 Heat of Reaction in a Nonflow Process

In this case, there is no mass flow, $\int_{c.s.}(\hat{H} + \hat{\phi} + \hat{K})\rho(\mathbf{v} \cdot \mathbf{n})dA = 0$. By neglecting potential and kinetic energies and shaft work, Eq. (2.7-1) becomes

$$\frac{d}{dt}\int_{c.v.} \rho \hat{U}\, dV = Q \quad \text{or} \quad \frac{d}{dt}(\hat{U}m) = Q \tag{2.8-1}$$

That is,

$$\frac{dU}{dt} = Q \tag{2.8-2}$$

Define

$$Q = \frac{dq}{dt}$$

Then

$$\frac{dU}{dt} = \frac{dq}{dt}$$

or

$$U_2 - U_1 = q \tag{2.8-3}$$

Standard heats of reaction are thus reported and tabulated on a nonflow constant pressure basis where $q = \Delta H$. Nevertheless, many heats of reaction are determined by running the reaction in a closed constant-volume bomb calorimeter where $q = \Delta U$. It is often necessary to correct these measurements to the constant-pressure basis. For example

$$H_2(g) + \frac{1}{2} O_2(g) \rightarrow H_2O(l)$$

The calorimeter determination is -67.43 kcal/g mol

$$\Delta H^{25} = \Delta U + P\Delta V \quad \text{because of constant pressure}$$

$$= -67.43 + -1.5[1.987(298)(1/1{,}000)]$$

$$= -68.32 \text{ kcal/g mol}$$

Hence heat is evolved in this reaction.

2.9 *Chemical Equilibrium*

A criterion of chemical equilibrium is the free-energy change of a system at constant temperature and constant pressure is zero.

$$dG_{T,P} = 0 \tag{2.9-1}$$

For a reversible reaction

$$\alpha_1 X_1 + \alpha_2 X_2 = \alpha_3 X_3 + \alpha_4 X_4$$

the change in free energy due to the change of composition is

$$\begin{aligned} dG_{T,P} &= \left(\frac{\partial G}{\partial N_3} dN_3 + \frac{\partial G}{\partial N_4} dN_4\right) - \left(\frac{\partial G}{\partial N_1} dN_1 + \frac{\partial G}{\partial N_2} dN_2\right) \\ &= (\mathbf{G}_3\, dN_3 + \mathbf{G}_4\, dN_4) - (\mathbf{G}_1\, dN_1 + \mathbf{G}_2\, dN_2) \end{aligned} \tag{2.9-2}$$

where $\mathbf{G}_1$, $\mathbf{G}_2$, $\mathbf{G}_3$, $\mathbf{G}_4$ are partial molal free energies. The relationship of the final composition and the initial composition of each species is

$$N_j = N_{jo} + \sum_{j=1}^{s} \alpha_{ij}\beta_i \qquad i = 1, \cdots, q \tag{2.9-3}$$

Differentiating Eq. (2.9-3) gives

$$dN_1 = -\alpha_1\, d\beta \qquad dN_2 = -\alpha_2\, d\beta \qquad dN_3 = \alpha_3\, d\beta \qquad dN_4 = \alpha_4\, d\beta$$

Substituting into Eq. (2.9-2) yields

$$dG_{T,P} = (\mathbf{G}_3\alpha_3 + \mathbf{G}_4\alpha_4 - \mathbf{G}_1\alpha_1 - \mathbf{G}_2\alpha_2)d\beta \tag{2.9-4}$$

The free-energy change for component j is

$$\mathbf{G}_j - G_j^\circ = RT \ln a \tag{2.9-5}$$

where a is the activity. Hence Eq. (2.9-4) becomes

$$\begin{aligned} dG_{T,P} = \alpha_3(G_3^\circ + RT \ln a_3) + \alpha_4(G_4^\circ + RT \ln a_4) - \alpha_1(G_1^\circ + RT \ln a_1) \\ - \alpha_2(G_2^\circ + RT \ln a_2)d\beta \end{aligned} \tag{2.9-6}$$

Define

$$\Delta G_r^\circ = \alpha_3 G_3^\circ + \alpha_4 G_4^\circ - \alpha_1 G_1^\circ - \alpha_2 G_2^\circ \tag{2.9-7}$$

And because

$$dG_{T,P} = 0$$

thus

$$-\frac{\Delta G_r}{RT} = \ln \frac{(a_3)^{\alpha_3}(a_4)^{\alpha_4}}{(a_1)^{\alpha_1}(a_2)^{\alpha_2}} = \ln K \tag{2.9-8}$$

This ΔG_r° is defined as the standard free-energy change of the reaction at 25°C. K is defined as the equilibrium constant which is evaluated for the following cases:

2.9-1 *Perfect gas*

For this case, the activity is equal to the partial pressure $a = p$. Thus

$$K = \frac{(p_3)^{\alpha_3}(p_4)^{\alpha_4}}{(p_1)^{\alpha_1}(p_2)^{\alpha_2}} = K_p \tag{2.9-9}$$

Because

$$p_i = y_i P \tag{2.9-10}$$

where y_i is the gas composition and P is the total pressure of the system, Eq. (2.9-9) can be expressed in terms of mole fraction as

$$K = \frac{(y_3P)^{\alpha_3}(y_4P)^{\alpha_4}}{(y_1P)^{\alpha_1}(y_2P)^{\alpha_2}} = \frac{y_3^{\alpha_3} y_4^{\alpha_4}}{y_1^{\alpha_1} y_2^{\alpha_2}} P^{\Delta\alpha} \tag{2.9-11}$$

$$= K_y P^{\Delta\alpha}$$

where

$$\Delta\alpha = \alpha_3 + \alpha_4 - \alpha_1 - \alpha_2$$

Because

$$p_i = C_i RT \tag{2.9-12}$$

where C_i is the concentration. Equation (2.9-9) can be expressed in terms of concentration as

$$K = \frac{(C_3RT)^{\alpha_3}(C_4RT)^{\alpha_4}}{(C_1RT)^{\alpha_1}(C_2RT)^{\alpha_2}} = \frac{C_3^{\alpha_3} C_4^{\alpha_4}}{C_1^{\alpha_1} C_2^{\alpha_2}} (RT)^{\Delta\alpha} \tag{2.9-13}$$

$$= K_C (RT)^{\Delta\alpha}$$

2.9-2 Imperfect gas

Thermodynamics indicates that

$$\nu_A = \frac{f_A}{p_A} \tag{2.9-14}$$

where f_A is the fugacity of component A, p_A is the partial pressure of A, ν_A is the fugacity coefficient of component A and is equal to

$$\nu_A = \frac{f'_A}{P} \tag{2.9-15}$$

where f'_A is the fugacity of pure A and P is the total pressure of the system. Hence

$$K = \frac{p_3^{\alpha_3} p_4^{\alpha_4} \nu_3^{\alpha_3} \nu_4^{\alpha_4}}{p_1^{\alpha_1} p_2^{\alpha_2} \nu_1^{\alpha_1} \nu_2^{\alpha_2}} = K_p K_\nu = \frac{y_3^{\alpha_3} y_4^{\alpha_4}}{y_1^{\alpha_1} y_2^{\alpha_2}} P^{\Delta\alpha} K_\nu = K_y P^{\Delta\alpha} K_\nu = K_C (RT)^{\Delta\alpha} K_\nu \tag{2.9-16}$$

Example 2.9-1 Fugacity Coefficient by Analytical Method

Calculate the fugacity coefficient of methane at 37.73°C and 102.07 atm by the Beattie-Bridgeman equation.

Solution

From thermodynamics, the free-energy change is

$$dG = -S\,dT + V\,dP \tag{A}$$

where S is the entropy. At constant temperature,

$$dG = V\,dP \tag{B}$$

For ideal gas

$$V = \frac{RT}{P} \tag{C}$$

Then

$$dG = RT\,d\ln P \tag{D}$$

For nonideal gas

$$dG = RT\,d\ln f = V\,dP \tag{E}$$

Integrating gives

$$G - G^\circ = RT \ln \frac{f}{f^\circ} = \int_{P^\circ}^{P} V\,dP \tag{F}$$

The Beattie-Bridgeman equation is

$$P = \frac{RT}{V} + \frac{\beta}{V^2} + \frac{\gamma}{V^3} + \frac{\delta}{V^4} \tag{G}$$

where

$$\beta = RTB_o - A_o - \frac{Rc}{T^2}$$

$$\gamma = -RTB_o b + aA_o - \frac{RB_o c}{T^2}$$

$$\delta = RB_o \frac{bc}{T^2}$$

Differentiating Eq. (G) gives

$$dP = -\left(\frac{RT}{V^2} + \frac{2\beta}{V^3} + \frac{3\gamma}{V^4} + \frac{4\delta}{V^5}\right) dV \tag{H}$$

Hence

$$\int_{P^\circ}^{P} V\,dP = \int_{V_\infty}^{V} -\left(\frac{RT}{V} + \frac{2\beta}{V^2} + \frac{3\gamma}{V^3} + \frac{4\delta}{V^4}\right) dV \tag{I}$$

Eq. (F) becomes

$$RT \ln \frac{f}{f^\circ} = \left[-RT \ln V + \frac{2\beta}{V} + \frac{3\gamma}{2V^2} + \frac{4\delta}{3V^3}\right]_{V_\infty}^{V} \tag{J}$$

Since

$$f^\circ = P^\circ = \frac{RT}{V} \tag{K}$$

Eq. (J) is reduced to

$$\ln f = \ln \frac{RT}{V} + \frac{2\beta}{RTV} + \frac{3\gamma}{2RTV^2} + \frac{4\delta}{3RTV^3} \tag{L}$$

The numerical values of the empirical constants for methane in the Beattie-Bridgeman equation are

$A_o = 2.2769$ $\quad a = 0.01855$ $\quad B_o = 0.05587$ $\quad b = -0.01587$ $\quad c = 12.83 \times 10^4$

Then the values of β, γ, and δ can be calculated as

$$\beta = -9.6028 \times 10^{-1} \qquad \gamma = 5.8775 \times 10^{-2} \qquad \delta = -9.6558 \times 10^{-5}$$

From Example 2.6-1, $V = 0.2189$ liter/g mol. Substituting these values into Eq. (L) gives f = 88.7779 atm and then f/P = 88.7779/102.07 = 0.8698 versus the 0.8714 reported in *Chemical Engineering,* February 25, 1980.

Example 2.9-2 Fugacity Coefficient by the Graphical Method

Calculate the fugacity coefficient of methane at 37.73°C and 102.07 atm from the following *PVT* relationship

Pressure P, atm	1	6.80	13.61	27.22	40.83	54.44	68.05	102.07
Volume V, liter/ g mol	25.46	3.71	1.836	0.9007	0.5891	0.4338	0.3410	0.2191

Solution

The equation for the fugacity has been derived as

$$\ln \frac{f}{f^{\circ}} = \frac{1}{RT} \int_{P^{\circ}}^{P} V \, dP \tag{A}$$

Define a residual volume

$$\alpha = \frac{RT}{P} - V \tag{B}$$

Hence

$$V = \frac{RT}{P} - \alpha \tag{C}$$

Substituting Eq. (C) into Eq. (A) and integrating

$$\ln \frac{f}{f^{\circ}} = \ln \frac{P}{P^{\circ}} - \frac{1}{RT} \int_{0}^{P} \alpha \, dP \tag{D}$$

Since $f^{\circ} = P^{\circ}$,

$$\ln \frac{f}{P} = -\frac{1}{RT} \int_{0}^{P} \alpha \, dP \tag{E}$$

From the given data, the values of α can be calculated at various pressure as:

Pressure P, atm	1	6.804	13.61	27.22	40.83	54.44	68.05	102.07
α	0.0549	0.0410	0.0389	0.0367	0.0358	0.0349	0.0340	0.0309

A plot of α versus pressure can be made in Figure 2.9-1. The area under the curve from $P = 0$ to $P = 102.07$ is calculated by Simpson's rule as 3.6373.

Thus

$$\ln \frac{f}{P} = -\frac{1}{RT}(3.6373) = -\frac{3.6373}{0.08206(310.93)} = -0.1425$$

or $f/P = 0.8672$ versus the 0.8714 reported in *Chemical Engineering,* February 25, 1980.

Example 2.9-3 Fugacity Coefficient by Pitzer Correlation

A gaseous mixture of nitrogen and hydrogen in stoichiometric amounts reacts at 773 K and 600 atm at which the equilibrium constant K is 0.0034 atm. Calculate the equilibrium fractional conversion for the following reaction:

$$\frac{1}{2} N_2 + \frac{3}{2} H_2 = NH_3$$

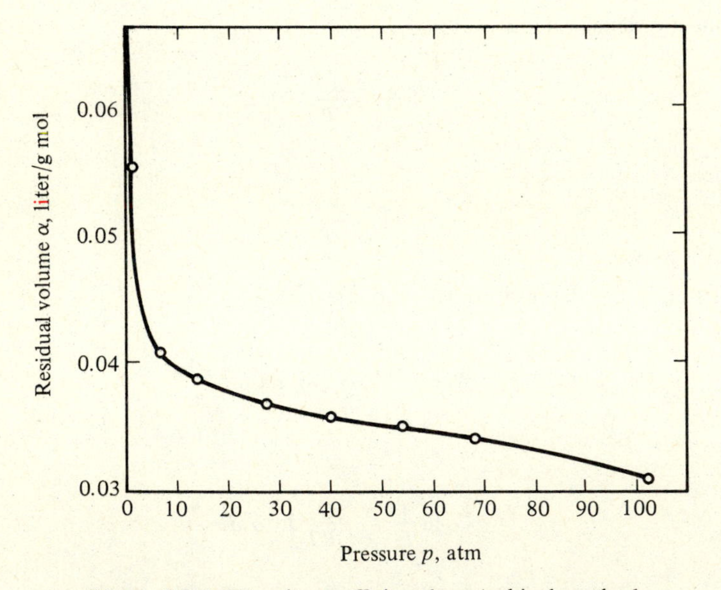

Figure 2.9-1 Fugacity coefficient by graphical method

Solution

Since the reaction is carried out at high pressure, the ideal-gas behavior cannot be assumed. The fugacity coefficient used in the calculation can be evaluated by the generalized Pitzer correlation as follows. Thermodynamics indicates that, under constant temperature and constant composition, the fugacity coefficient relates to the total pressure of the system and the compressibility factor as

$$d(\ln \nu) = (Z - 1)\frac{dP}{P} \tag{A}$$

Eliminating z in Eq. (A) by

$$Z = 1 + \frac{BP}{RT} \tag{B}$$

and integrating from 0 to P gives

$$\ln \nu = \frac{BP}{RT} \tag{C}$$

where

$$B = \frac{RT_c}{P_c}(B^0 + \Omega B^1) \tag{D}$$

Substituting Eq. (D) into Eq. (C) gives

$$\ln \nu = \frac{P_r}{T_r}(B^0 + \Omega B^1) \tag{E}$$

where B^0 and B^1 can be found in Eq. (2.6-5). The calculation can be tabulated as follows:

	N_2	H_2	NH_3
T_r	6.1	23.3	1.91
P_r	17.9	46.9	5.33
Ω	0.04	0	0.25
$B^0 = 0.1445 - 0.33/T_r - 0.1385/T_r^2 - 0.0121/T_r^3$	0.087	0.1301	−0.0680
$B^1 = 0.073 + 0.46/T_r - 0.5/T_r^2 - 0.097/T_r^3 - 0.0073/T_r^8$	0.1345	0.0918	0.1628
$\nu = e^{P_r(B^0+\Omega B^1)/T_r}$	1.311	1.299	0.9267
α_j	$-\frac{1}{2}$	$-\frac{3}{2}$	1
N_{jo}	0.5	1.5	0
$y_j = \dfrac{N_{jo} + \alpha_j\beta}{N_o + \alpha\beta}$	$\dfrac{\frac{1}{2} - \frac{1}{2}\beta}{2-\beta}$	$\dfrac{\frac{3}{2} - \frac{3}{2}\beta}{2-\beta}$	$\dfrac{\beta}{2-\beta}$

Substituting into Eq. (2.9-16) gives

$$\frac{\beta/(2-\beta)}{[0.5(1-\beta)/(2-\beta)]^{1/2}[1.5(1-\beta)/(2-\beta)]^{3/2}}600^{1-(1/2)-(3/2)}\frac{0.9267}{1.311^{1/2}\,1.299^{3/2}}=0.0034$$

Solving gives $\beta = 0.5864$. Hence the composition of NH_3 in the product

$$y_{NH_3}=\frac{\beta}{2-\beta}=\frac{0.5864}{2-0.5864}=0.4148 \text{ vs experimental value of } 0.4215$$

2.9-3 *Ideal solution*

$$K=\frac{x_3^{\alpha 3}x_4^{\alpha 4}}{x_1^{\alpha 1}x_2^{\alpha 2}} \tag{2.9-17}$$

Example 2.9-4 Equilibrium Constant for the Ideal Solution

378.5 liters of an aqueous solution containing 100 kg acetic acid, 186 kg ethyl alcohol, and the same concentration of HC1 as a catalyst are reacted in a reactor. Assuming that the density is constant at 1042.5 g/liter and the equilibrium constant K is 2.91, what is the equilibrium conversion of the acetic acid?

Solution

$$\underset{(1)}{CH_3COOH}+\underset{(2)}{C_2H_5OH}\rightleftarrows\underset{(3)}{CH_3COOC_2H_5}+\underset{(4)}{H_2O}$$

$$C_1=\frac{100(1{,}000)}{(378.5)(60)}=4.4 \text{ g mol/liter}$$

$$C_2=\frac{186(1{,}000)}{(378.5)(46)}=10.68 \text{ g mol/liter}$$

$$C_3=0$$

$$C_4=\frac{(1042.5)(378.5)-(100+186)(1{,}000)}{(378.5)(18)}=15.94 \text{ g mol/liter}$$

C_o = initial total concentration = 4.4 + 10.68 + 15.94 = 31.02 g mol/liter

Substituting into Eq. (1.4-39) gives

$$x_1=\frac{4.4-\lambda}{31.02}\qquad x_2=\frac{10.68-\lambda}{31.02}\qquad x_3=\frac{\lambda}{31.02}\qquad x_4=\frac{15.94+\lambda}{31.02}$$

Substituting into Eq. (2.9-17) yields

$$\frac{\lambda(15.94+\lambda)}{(4.4-\lambda)(10.68-\lambda)}=2.91$$

Solving for λ

$$\lambda = 2.4827$$

Hence equilibrium conversion of the acetic acid is $2.4827/4.4 = 0.5643$

2.9-4 Nonideal solution

$$K=\frac{(\gamma_3 x_3)^{\alpha_3}(\gamma_4 x_4)^{\alpha_4}}{(\gamma_1 x_1)^{\alpha_1}(\gamma_2 x_2)^{\alpha_2}} \quad \textbf{(2.9-18)}$$

where γ is the activity coefficient, which can be evaluated by the following two constant Margules equations for a binary mixture

$$\ln \gamma_1 = A_{12}x_2^2\left[1+2x_1\left(\frac{A_{21}}{A_{12}}-1\right)\right] \quad \textbf{(2.9-19)}$$

$$\ln \gamma_2 = A_{21}x_1^2\left[1+2x_2\left(\frac{A_{12}}{A_{21}}-1\right)\right] \quad \textbf{(2.9-20)}$$

or the following van Laar equations

$$\ln \gamma_1 = \frac{A'_{12}x_2^2}{\{1+[(A'_{12}/A'_{21})-1]x_1\}^2} \quad \textbf{(2.9-21)}$$

$$\ln \gamma_2 = \frac{A'_{21}x_1^2}{\{1+[(A'_{21}/A'_{12})-1]x_2\}^2} \quad \textbf{(2.9-22)}$$

The use of these two kinds of equations, Margules or van Laar, can be found in any thermodynamics textbook.

2.9-5 Heterogeneous mixtures

The fugacity of a pure liquid or a solid in a reaction mixture can be taken as a unity. The corresponding terms do not appear in the product. For example, when carbon dioxide is reduced by carbon to carbon monoxide

$$2CO - C - CO_2 = 0$$

the equilibrium relationship is

$$K = \frac{f_{CO}^2}{f_{CO_2}}$$

where f's are the fugacities. If a solid decomposes into another solid and a gas, such as the dissociation of calcium carbonate into calcium oxide and carbon dioxide, the reaction will proceed until the pressure of the gas equals the equilibrium pressure. The pressure is known as the *dissociation pressure.* The equilibrium relationship is simply

$$K = P_{CO_2}$$

where P is the pressure of the carbon dioxide.

Example 2.9-5 Equilibrium in a Heterogeneous System

At 1200°C, the following two reactions occur:

$$CaCO_3(s) = CaO(s) + CO_2(g) \qquad K_1 = 31.4 \text{ atm}$$

$$CO_2(g) = CO(g) + \frac{1}{2} O_2(g) \qquad K_2 = 10^{-5} \text{ atm}^{1/2}$$

What is the partial pressure of oxygen at equilibrium?

Solution

Using Eq. (1.4-39), we calculate $\boldsymbol{\alpha} = 1 + \frac{1}{2} - 1 = \frac{1}{2}$

$$x_{CO_2} = \frac{1-\gamma}{1+\frac{1}{2}\gamma}$$

$$x_{CO} = \frac{\gamma}{1+\frac{1}{2}\gamma}$$

$$x_{O_2} = \frac{\frac{1}{2}\gamma}{1+\frac{1}{2}2}$$

Hence

$$P\left(\frac{1-\gamma}{1+\frac{1}{2}\gamma}\right) = 31.4$$

and

$$\frac{\left[\frac{1}{2}\gamma P/\left(1+\frac{1}{2}\gamma\right)\right]^{1/2}\left[\frac{\gamma P}{\left(1+\frac{1}{2}\gamma\right)}\right]}{\left(\frac{1-\gamma}{1+\frac{1}{2}\gamma}\right)P} = 10^{-5}$$

where P is the total pressure. Solving these two equations, substituting back to get the mole fraction of oxygen, and then calculating the partial pressure of oxygen gives

$$p_{O_2} = 2.9092 \times 10^{-3} \text{ atm}$$

2.9-6 Simultaneous reactions

For example, it is desired to determine the equilibrium compositions of the following reactions:

$$-X_1 + X_2 = 0 \qquad \text{and} \qquad -X_2 + X_3 = 0$$

The equilibrium relationships of these two reactions are

$$\frac{C_2}{C_1} = K_{C_1} \qquad \text{and} \qquad \frac{C_3}{C_2} = K_{C_2}$$

Example 2.9-6 Equilibrium for Simultaneous Reactions

If 5 mol steam and 1 mol methane react in a converter at 1 atm, the following reactions take place

(1) $CH_4 + H_2O = CO + 3H_2$ methane-steam reaction

(2) $CO + H_2O = CO_2 + H_2$ water-gas shift reaction

What are the equilibrium compositions at 600°C at which $K_1 = 0.54$ atm² and $K_2 = 2.49$ atm?

Solution

Expand Eq. (1.4-51) for these two reactions $i = 2$ as

$$x_j = \frac{N_{jo} + \alpha_{1j}\beta_1 + \alpha_{2j}\beta_2}{N_o + \alpha_1\beta_1 + \alpha_2\beta_2}$$

where

$$N_o = 1 \text{ mol } CH_4 + 5 \text{ mol } H_2O = 6 \text{ mol}$$

$$\alpha_1 = 3 + 1 - 1 - 1 = 2$$

$$\alpha_2 = 1 + 1 - 1 - 1 = 0$$

$$j = 1 \text{ for } CH_4 \qquad x_1 = \frac{1 - \beta_1}{6 + 2\beta_1}$$

$$j = 2 \text{ for } H_2O \qquad x_2 = \frac{5 - \beta_1 - \beta_2}{6 + 2\beta_1}$$

$$j = 3 \text{ for } CO \qquad x_3 = \frac{\beta_1 - \beta_2}{6 + 2\beta_1}$$

$$j = 4 \text{ for } H_2 \qquad x_4 = \frac{3\beta_1 + \beta_2}{6 + 2\beta_1}$$

$$j = 5 \text{ for } CO_2 \qquad x_5 = \frac{\beta_2}{6 + 2\beta_1}$$

Substituting into Eq. (2.9-11) and simplifying

$$\frac{(\beta_1 - \beta_2)(3\beta_1 + \beta_2)^3}{(6 + 2\beta_1)^2(1 - \beta_1)(5 - \beta_1 - \beta_2)} = 0.54 \tag{A}$$

$$\frac{\beta_2(3\beta_1 + \beta_2)}{(\beta_1 - \beta_2)(5 - \beta_1 - \beta_2)} = 2.49 \tag{B}$$

Solving Eq. (A) and Eq. (B) simultaneously gives

$$\beta_1 = 0.911 \qquad \text{and} \qquad \beta_2 = 0.653$$

Hence the equilibrium compositions are

$$x_1 = 0.01135 \qquad x_2 = 0.4393 \qquad x_3 = 0.03298 \qquad x_4 = 0.4328 \qquad x_5 = 0.08346$$

2.10 Effect of Temperature on Equilibrium Constant

From the previous treatment, it is seen that once the free-energy change is determined, the equilibrium constant can be evaluated because

$$-\Delta G_r^\circ RT \ln K \tag{2.10-1}$$

If the effect of temperature on the free-energy change is calculated, its effect on the equilibrium constant can be obtained. There are three methods to determine ΔG_r°.

1. From standard heat of reaction ΔH_r°
Thermodynamics indicates that, at constant pressure,

$$\frac{d}{dT}\left(\frac{\Delta G_T^\circ}{T}\right)_P = -\frac{\Delta H_T^\circ}{T^2} \tag{2.10-2}$$

where ΔH_T° is the standard heat of reaction at temperature T. Substituting the corresponding equation of Eq. (2.5-4) and integrating gives

$$\Delta G_T^\circ = \Delta H_{T^\circ}^\circ + (\Delta G_{T^\circ}^\circ - \Delta H_{T^\circ}^\circ)\phi + T^\circ(\Delta a)(\phi - 1 - \phi \ln \phi)$$
$$-\frac{T^{\circ 2}\Delta b}{2}(\phi^2 - 2\phi + 1) - \frac{T^{\circ 3}\Delta c}{6}(\phi^3 - 3\phi + 2) - \frac{\Delta d}{2T^\circ}\left(\frac{\phi^2 - 2\phi + 1}{\phi}\right) \tag{2.10-3}$$

where $\phi = T/T^\circ$.

Using Eq. (2.10-1), we obtain the variation of K with temperature as

$$\ln K = -\frac{1}{RT}\left[\Delta H_{T^\circ}^\circ + (\Delta G_{T^\circ}^\circ - \Delta H_{T^\circ}^\circ)\phi + T^\circ(\Delta a)(\phi - 1 - \phi \ln \phi)\right.$$
$$\left. -\frac{T^{\circ 2}\Delta b}{2}(\phi^2 - 2\phi + 1) - \frac{T^{\circ 3}\Delta c}{6}(\phi^3 - 3\phi + 2) - \frac{\Delta d}{2T^\circ}\left(\frac{\phi^2 - 2\phi + 1}{\phi}\right)\right] \tag{2.10-4}$$

2. From standard heat of reaction ΔH° and standard entropy ΔS°
The third law of thermodynamics is

$$\Delta G_T^\circ = \Delta H_T^\circ - T\Delta S_T^\circ \tag{2.10-5}$$

Neglecting the Δd term, we obtain

$$\Delta H_T^\circ = \Delta H_{T^\circ}^\circ + \Delta a(T - T^\circ) + \frac{\Delta b}{2}(T^2 - T^{\circ 2}) + \frac{\Delta c}{3}(T^3 - T^{\circ 3}) \tag{2.10-6}$$

$$\Delta S_T^\circ = \Delta S_{T^\circ}^\circ + \int_{T^\circ}^{T}\left(\frac{\Delta a}{T} + \Delta b + T\Delta c\right)dT$$
$$= \Delta S_{T^\circ}^\circ + \Delta a \ln\frac{T}{T^\circ} + \Delta b(T - T^\circ) + \frac{\Delta c}{2}(T^2 - T^{\circ 2}) \tag{2.10-7}$$

3. From spectroscopic data

$$\frac{\Delta G_T^\circ}{T} = \Sigma\left(\frac{G_T^\circ - H_{T^\circ}^\circ}{T} + \frac{\Delta H_f}{T}\right)_P - \Sigma\left(\frac{G_T^\circ - H_{T^\circ}^\circ}{T} + \frac{\Delta H_f}{T}\right)_R \tag{2.10-8}$$

where $(G_T^\circ - H_{T^\circ}^\circ)/T$ is defined as free-energy function.

Example 2.10-1 Equilibrium Composition from Free-Energy Change

Assuming ΔH_T° to be independent of temperature, calculate $\Delta G^\circ_{250^\circ C}$, $\Delta G^\circ_{500^\circ C}$ and hence equilibrium gas composition at these temperatures and 1 atm pressure for the reaction

$$UO_2(s) + 4HF(g) = UF_4(s) + 2H_2O(g)$$

with $\Delta H^\circ_{298} = -43.2$ kcal/g mol UO_2 and $\Delta G^\circ_{298} = -31.2$ kcal/g mol UO_2.

Solution

Combining Eqs. (2.10-1) and (2.10-2) gives

$$\frac{d(\ln K)}{dT} = \frac{\Delta H_T^\circ}{RT^2} \tag{A}$$

Assuming ΔH_T° to be independent of temperature, integrating yields

$$\ln \frac{K_2}{K_1} = \frac{\Delta H^\circ}{R}\left(\frac{1}{T_1} - \frac{1}{T_2}\right) \tag{B}$$

The same result can be obtained by integrating Eq. (2.10-2) to give

$$\frac{\Delta G_T^\circ}{\mathrm{T}} = \frac{\Delta G^\circ_{298}}{298} + \Delta H_T^\circ\left(\frac{1}{T} - \frac{1}{298}\right) \tag{C}$$

and combining with Eq. (2.10-1). Substituting the numerical values into Eq. (C) gives the free-energy change at any temperature as

$$\frac{\Delta G^\circ}{T} = -\frac{31{,}200}{298} - 43{,}200\left(\frac{1}{T} - \frac{1}{298}\right)$$
$$= 40.27 - \frac{43{,}200}{T}$$

At $T = 250^\circ C$

$$\Delta G^\circ_{523} = -22{,}138.79$$

Thus

$$\ln K = -\frac{-22{,}138.79}{1.987(523)} = 21.3$$

or

$$\ln \frac{p^2_{H_2O}}{p^4_{HF}} = 21.3$$

because

$$p_{H_2O} + p_{HF} = 1$$

Solving these two equations gives $p_{HF} = 0.005$ atm.

Example 2.10-2 Calculation of Free-Energy Change

Evaluate ΔG°_{298} and develop a function ΔG°_T for the reaction

$$\frac{1}{2} N_2(g) + \frac{3}{2} H_2(g) = NH_3(g)$$

$$\Delta H^\circ_{298} = -11.04 \text{ kcal/g mol}; \qquad \Delta S^\circ_{298} = -23.69 \text{ e.u.}$$

$$C_P, NH_3 = 6.086 + 8.812 \times 10^{-3}T - 1.506 \times 10^{-6}T^2 \text{ cal/g mol} \cdot \text{K}$$

$$C_P, N_2 = 6.449 + 1.412 \times 10^{-3}T - 0.0807 \times 10^{-6}T^2 \text{ cal/g mol} \cdot \text{K}$$

$$C_P, H_2 = 6.947 - 0.200 \times 10^{-3}T + 0.481 \times 10^{-6}T^2 \text{ cal/g mol} \cdot \text{K}$$

Solution

(a) Calculation of ΔG°_T at 298 K

Using Eq. (2.10-5)

$$\Delta G^\circ_{298} = -11{,}040 - [298(-23.69)] = -3{,}980 \text{ cal/g mol}$$

From the equations of the C_p, we obtain

$$\Delta a = 6.086 - \frac{1}{2}(6.449) - \frac{3}{2}(6.947) = -7.559$$
$$\Delta b = (8.812 - 0.706 + 0.3)10^{-3} = 8.406 \times 10^{-3}$$
$$\Delta c = (-1.506 + 0.0404 - 0.7215)10^{-6} = -2.187 \times 10^{-6}$$

(b) Calculation of ΔG°_T

$$\Delta H^\circ_T = \Delta H^\circ_{298} + \int_{298}^{T} (-7.559 + 8.406 \times 10^{-3}T - 2.187 \times 10^{-6}T^2)dT$$

$$= -9{,}141 - 7.559T + 4.203 \times 10^{-3}T^2 - 0.729 \times 10^{-6}T^3$$

$$\Delta S^\circ_T = \Delta S^\circ_{298} + \int_{298}^{T} \left(-\frac{7.559}{T} + 8.406 \times 10^{-3} - 2.187 \times 10^{-6}T\right)dT$$

$$= 16.96 - 7.559 \ln T + 8.406 \times 10^{-3}T - 1.0935 \times 10^{-6}T^2$$

Substituting into Eq. (2.10-5)

$$\Delta G_T^\circ = -9{,}141 - 24.52T - 4.2 \times 10^{-3}T^2 + 0.36 \times 10^{-6}T^3 + 7.56T \ln T$$

Example 2.10-3 Equilibrium Constant from Free-Energy Function

At 800 K, the following reaction takes place

$$C_3H_8(g) = C_3H_6(g) + H_2(g)$$

Calculate the equilibrium constant at this temperature from the following data:

	C_3H_6	H_2	C_3H_8
$-\dfrac{G_T^\circ - H_0^\circ}{T}$ cal/°C	67.04	31.19	68.74
$(\Delta H_f^\circ)_0$ kcal	8.47	0	−19.48

Solution

Using Eq. (2.10-8),

$$\frac{\Delta G_T^\circ}{T} = \left(-67.04 - 31.19 + \frac{8.47(1000)}{800}\right)_{\text{products}} - \left(-68.74 - \frac{19.48(1000)}{800}\right)_{\text{reactants}}$$

$$= 5.44$$

Hence $\ln K = -5.44/1.987 = -2.745$ and $K = 0.0643$.

Problems

1. The standard heats of combustion at 25°C for CS_2, C, and S are −265, −94 and −71 kcal/g mol, respectively. Calculate the standard heat of formation of CS_2.

2. The standard heats of formation for CO_2, H_2O(g), and H_2O(l) are −94.052, −57.798, and −68.317 kcal, respectively. Evaluate the standard heat of formation of propane from its heat of combustion of 530.605 kcal.

3. For the reaction

$$C_3H_8(g) + 5O_2(g) \longrightarrow 3CO_2(g) + 4H_2O \text{ (g or l)}$$

with the values of the standard heat of formation from Problem (2) and that of C_3H_8 being −24.82 kcal/g mol, calculate the internal energy and enthalpy changes at 25°C for the above two cases, water vapor or liquid water.

4. Calculate the heat of combustion of CO at 827°C from the following data:
   (a) Standard heats of formation of CO and CO_2 at 25°C are −26.416 and −94.052 kcal/g mol, respectively.
   (b) Specific heats C_P at T K are:

$$CO: 6.726 + 0.04001T \times 10^{-2} + 0.1283T^2 \times 10^{-5} - 0.5307T^3 \times 10^{-9} \text{ cal/g mol} \cdot °C$$

$$CO_2: 5.316 + 1.42850T \times 10^{-2} - 0.8362T^2 \times 10^{-5} + 1.7840T^3 \times 10^{-9} \text{ cal/g mol} \cdot °C$$

$$O_2: 6.085 + 0.36310T \times 10^{-2} - 0.1709T^2 \times 10^{-5} + 0.3133T^3 \times 10^{-9} \text{ cal/g mol} \cdot °C$$

5. Ethylene is oxidized to ethylene oxide by the reaction

$$C_2H_4(g) + \frac{1}{2}O_2(g) \rightarrow C_2H_4O(g) \qquad \Delta H^\circ_{18} = -28.611 \text{ kcal/g mol}$$

5,000 kg/day of the oxide is produced by feeding oxygen at 150°C with 10 percent in excess and ethylene at 190°C. The products leave at 275°C. Calculate the hourly heat removal from the reactor for converting 80 percent of the ethylene feed.

Data: Specific heats C_p at T K are:

$$C_2H_4: 0.944 + 3.735 \times 10^{-2}T - 1.993 \times 10^{-5}T^2 + 4.22 \times 10^{-9}T^3 \text{ cal/g mol} \cdot K$$

$$C_2H_4O: -1.12 + 4.925 \times 10^{-2}T - 2.389 \times 10^{-5}T^2 + 3.149 \times 10^{-9}T^3 \text{ cal/g mol} \cdot K$$

$$O_2: 6.732 + 0.1505 \times 10^{-2}T - 0.01791 \times 10^{-5}T^2 \text{ cal/g mol} \cdot K$$

6. Calculate the final gas composition and the heat removal for the following reaction

$$H_2O(g) + CO(g) = CO_2(g) + H_2(g) \qquad \Delta H^\circ_{298} = -9840 \text{ cal/g mol}$$

The reaction is carried out by two steps: (a) first CO is produced by passing air over coke for complete oxygen conversion. Then the CO produced is mixed with N_2 stream at 150°C. (b) Then the above CO–N_2 stream is mixed with 2 mol steam per mole of CO at 270°C in a reactor in which 98 percent CO conversion is obtained. The products leave at 480°C.

7. Ammonia is produced by passing a gaseous mixture of 60 mol percent H_2, 20 mol percent N_2, and 20 mol percent inerts over a catalyst at 50 atm pressure. For the reaction

$$N_2 + 3H_2 = 2NH_3$$

if the K_p is 0.0125 and the system behaves ideally, what is the maximum conversion? If the feed is nonstoichiometric, say 70 mol percent H_2, 10 mol percent N_2, and 20 mol percent inerts, what is the equilibrium conversion?

8. For the following reaction

$$2CO_2 + 4H_2 = C_2H_5OH + H_2O$$

a mol H_2 per mole of CO_2 is fed to a reactor at 20 atm pressure and 573 K at which $\Delta G° = 2{,}126$ cal/g mol. Find an expression for the fractional conversion of carbon dioxide. When CO_2 is in large excess, develop the corresponding expression for the fractional conversion of hydrogen.

9. For the reactions

$$CO_2 + 3H_2 = CH_3OH + H_2O \qquad K_p = 2.82 \times 10^{-6}$$

$$CO + 2H_2 = CH_3OH \qquad K_p = 2.00 \times 10^{-5}$$

a gaseous mixture of 25 mol percent CO_2 and 75 mol percent CO, after mixing with the stoichiometric amount of H_2, passes over a catalyst at a pressure of 300 atm. Assuming ideal behavior, calculate the fractional conversion for each reaction and then the volume composition of the mixture at equilibrium.

10. For the following reaction

$$C_2H_6 = C_2H_4 + H_2$$

the data are:

	C_2H_6	C_2H_4
Standard heat of formation $\Delta H°$, cal/g mol	−20,236	+12,496
Standard free energy of formation $\Delta G°$, cal/g mol	−7,860	+16,282

Assuming that $\Delta H°$ is independent of temperature, develop an expression for $\Delta G°$ at any temperature and hence estimate the fractional conversion at 500°C and 2.0 atm pressure for a feed of 85 percent C_2H_6 and 10 percent H_2.

11. For the reaction

$$C_2H_4(g) + H_2O(g) = C_2H_5OH(g)$$

$\Delta G° = 1685$ cal/g mol and $\Delta H° = -10{,}400$ cal/g mol at 418°K. A feed of 2.5 mol H_2O per mole C_2H_4 produces 300 kg C_2H_5OH per hour. Assuming that the feed, products, and the reaction are isothermal at 418 K and 2 atm, find the fractional conversion and the heat removal.

12. For the oxidation of ammonia

$$4NH_3 + 5O_2 \rightarrow 4NO + 6H_2O(g)$$

Calculate ΔG° and ΔH° at 500 K with the variation of specific heat with temperature taking into account.

13. For the reaction

$$N_2(g) + C_2H_2(g) \rightarrow 2HCN(g) \qquad \Delta G^\circ_{300^\circ C} = 7{,}190 \text{ cal/g mol}$$

estimate the maximum mole fraction of HCN (T_c = 456.7 K, P_c = 48.9 atm, and acentric factor = 0.4) in the product stream at 300°C if the reactor pressure is (a) 1 atm (b) 200 atm.

14. For the reaction

$$C_2H_4(g) + H_2O(g) \rightarrow C_2H_5OH(g) \qquad \Delta G^\circ_{125^\circ C} = 1{,}082 \text{ cal/ g mol}$$

a gaseous mixture containing 50 mol percent ethylene and 50 mol percent steam enters the converter at 1 atm. Estimate the product composition at 125°C.

References

(A1). Abbott, M.M., and Van Ness, H.C. *Theory and Problems of Thermodynamics.* New York: McGraw-Hill Book Company, 1972.

(D1). Dodge, B.F. *Chemical Engineering Thermodynamics.* New York: McGraw-Hill Book Company, 1944.

(P1). Perry, R.H., and Chilton, C.H. *Chemical Engineer's Handbook,* 5th ed. New York: McGraw-Hill Book Company, 1973.

(S1). Smith, J.M. *Introduction to Chemical Engineering Thermodynamics.* New York: McGraw-Hill Book Company, 1949.

(W1). Weber, J.H. Calculate Equation-of-State Variables. *Chemical Engineering,* Feb. 25, 1980, pp. 93–100.

CHAPTER THREE
Process Kinetics

3.1 *Introduction*

In Chapter 2 our review of thermodynamics revealed how far a reaction can go. A chemical engineer designing or operating a chemical reactor also should know how fast the process can take place, that is, he or she should be familiar with the *process kinetics.* The rate of the reaction should be formulated, hence the variables that affect the reaction rate must be studied. At present, because rates cannot be predicted, they must be interpreted from the data measured in a reactor. Such data normally involve concentrations of reactants and products. By measuring these concentrations we can evaluate the rate constant that will be used to calculate the reaction rate and then to determine the reactor size. In general, there are two main types of systems: homogeneous and heterogeneous. In this chapter, the chemical rate theories for a homogeneous system are first briefly mentioned, and then the analysis of rate data for different types of homogeneous reactions follows. The heterogeneous kinetics and reactors will be described in Chapter 9.

3.2 *Overall Rates of Reaction*

When a homogeneous reaction takes place isothermally in a batch tank with perfect mixing or in a well-mixed flow reactor, the principle of mass action to reaction rate and chemical equilibria states that the rate of reaction is proportional to the product of the concentration of the reactants raised to powers equal to their respective stoichiometric numbers. This reaction is called an *elementary reaction.* For instance, α_1 mol X_1 reacts with α_2 mol X_2 to produce α_3 mol X_3 and α_4 mol X_4 for the forward reaction. Conversely, in the reverse reaction, α_3 mol X_3 reacts with α_4 mol X_4 to produce α_1 mol X_1 and α_2 mol X_2. This reversible reaction is represented by

$$\alpha_1 X_1 + \alpha_2 X_2 \underset{k'}{\overset{k}{\rightleftharpoons}} \alpha_3 X_3 + \alpha_4 X_4 \tag{3.2-1}$$

Then the forward rate is

$$r_{\text{forward}} = kC_1^{\underline{\alpha}_1} C_2^{\underline{\alpha}_2} \tag{3.2-2}$$

and the reverse rate is

$$r_{\text{reverse}} = k' C_3^{\underline{\alpha}_3} C_4^{\underline{\alpha}_4} \tag{3.2-3}$$

Thus the overall rate of reaction becomes

$$r = r_{\text{forward}} - r_{\text{reverse}} \tag{3.2-4}$$

$$= kC_1^{\underline{\alpha}_1} C_2^{\underline{\alpha}_2} - k' C_3^{\underline{\alpha}_3} C_4^{\underline{\alpha}_4} \tag{3.2-5}$$

In general

$$r = k \prod_{\text{reactants}} C_j^{\underline{\alpha}_j} - k' \prod_{\text{products}} C_j^{\underline{\alpha}'_j} \tag{3.2-6}$$

where Π is the product notation. At equilibrium, $r = 0$. If we define an equilibrium constant K_c as

$$K_c = \frac{k}{k'} \tag{3.2-7}$$

where k and k' are the specific rate constants for the forward and the reverse reactions, respectively (based on concentration unit). C_j is the concentration of j component. Then Eq. (3.2-6) becomes

$$r = k'(K_c \Pi\, C_j^{\underline{\alpha}_j} - \Pi\, C_j^{\underline{\alpha}'_j}) \tag{3.2-8}$$

For multiple reactions, the rates are

$$r_i - k_i \Pi\, C_j^{\underline{\alpha}_j} - k'_i \Pi\, C_j^{\underline{\alpha}'_j} \tag{3.2-9}$$

The exponents $\underline{\alpha}_1$, $\underline{\alpha}_2$, etc., are defined as the order of the reaction with respect to X_1, X_2, etc., respectively. An underlined $\underline{\alpha}$ is shown to distinguish the order of the reaction from the stoichiometric coefficient α. If these exponents are the same as the stoichiometric coefficients in the reaction, $\underline{\alpha} = \alpha$, the reaction is elementary; otherwise it is nonelementary and the exponents should be determined experimentally or by mechanism. The total order of the reaction is the sum of these two exponents $\underline{\alpha}_1$ and $\underline{\alpha}_2$. If only one molecule is involved, the reaction is unimolecular. When two molecules interact, the reaction is bimolecular. For example, the decomposition of nitrous oxide is bimolecular with respect to NO_2. Sometimes this r is also called the *intrinsic rate* of the reaction, expressed in moles per hour per cubic foot, or, in general, in moles per unit time per unit volume. For an nth-order reaction

$$A \longrightarrow \text{product}$$

the reaction rates can be expressed as

$$r_A \frac{\text{lb} \cdot \text{mol}}{\text{h} \cdot \text{ft}^3} = k_c \left[\frac{1}{\text{h}}\left(\frac{\text{ft}^3}{\text{lb} \cdot \text{mol}}\right)^{n-1}\right] C_A^n \left(\frac{\text{lb} \cdot \text{mol}}{\text{ft}^3}\right)^n \tag{3.2-10}$$

$$= k_p \left[\frac{\text{lb} \cdot \text{mol}}{\text{h} \cdot \text{ft}^3 \cdot \text{atm}^n}\right] P_A^n \, (\text{atm}^n) \tag{3.2-11}$$

But

$$P_A = C_A RT \tag{3.2-12}$$

Thus

$$k_c C_A^n = k_p (C_A RT)^n$$

or

$$k_c = k_p (RT)^n \tag{3.2-13}$$

where n is the order of the reaction, k_c is the rate constant based on concentration, and k_p is that based on partial pressure.

3.3 *Component Rate of Reaction*

In a batch reactor, the concentration of j component is

$$C_j = C_{jo} + \alpha_j \lambda$$

It is seen that the overall or intrinsic rate of the reaction r is the derivative of λ with respect to time

$$r = \frac{d\lambda}{dt} = \frac{1}{V}\frac{d\beta}{dt} \tag{3.3-1}$$

Let us define the component rate of reaction r_j by

$$r_j = \frac{dC_j}{dt} \tag{3.3-2}$$

Thus

$$\frac{dC_j}{\alpha_j dt} = \frac{d\lambda}{dt} = r = \frac{r_j}{\alpha_j} \tag{3.3-3}$$

Similarly, for multiple reactions

$$C_j = C_{jo} + \Sigma \alpha_{ij} \lambda_i \quad i = 1, \cdots, q \tag{3.3-4}$$

$$r_j = \frac{dC_j}{dt} = \Sigma \alpha_{ij} r_i \quad i = 1, \cdots, q \tag{3.3-5}$$

3.4 Mechanism

From a kinetic point of view, the formation of the final products from the original reactants usually occurs in a series of relatively simple steps. Each step has a different rate. The slowest of these steps governs the rate of the overall reaction. The sequence of steps that describes how the final products are formed from the original reactants is called the *mechanism* of the reaction. Usually the formulation of this mechanism is a difficult job and it is an important subject for a chemist. Once the mechanism is known, it is possible to derive the rate equation. For example, the formation of phosgene from carbon monoxide and chlorine could be theoretically represented by

$$CO + Cl_2 \underset{k'}{\overset{k}{\rightleftharpoons}} COCl_2$$

$$r = r_f - r_b = k(CO)(Cl_2) - k'(COCl_2) \tag{3.4-1}$$

if it were an elementary reaction. As a matter of fact, the reaction rate is far from equilibrium and is found to be

$$r = k(CO)(Cl_2)^{3/2} - k'(COCl_2)(Cl_2)^{1/2} \tag{3.4-2}$$

The last equation is the overall reaction of the following elementary postulated steps which are combined to form the so-called mechanism of the reaction:

1. $Cl_2 = 2Cl \qquad K_1 = \dfrac{(Cl)^2}{(Cl_2)}$ or $(Cl) = [K_1(Cl_2)]^{1/2}$

2. $Cl + CO = COCl \qquad K_2 = \dfrac{(COCl)}{(CO)(Cl)}$ or $(COCl) = K_2(CO)(Cl)$

$$= K_2[K_1(Cl_2)]^{1/2}(CO)$$

3. $COCl + Cl_2 \underset{k_3'}{\overset{k_3}{\rightleftarrows}} COCl_2 + Cl$

$$r = r_f - r_b$$

$$r = k_3(COCl)(Cl_2) - k_3'(COCl_2)(Cl)$$

$$= k_3\,[K_1^{1/2}K_2(Cl_2)^{1/2}](CO)(Cl_2) - k_3'(COCl_2)(K_1Cl_2)^{1/2}$$

$$= k_3K_1^{1/2}K_2(CO)(Cl_2)^{3/2} - k_3'K_1^{1/2}(COCl_2)(Cl_2)^{1/2}$$

$$= k(CO)(Cl_2)^{3/2} - k'(COCl_2)(Cl_2)^{1/2}$$

where

$$k = k_3K_1^{1/2}K_2 \quad \text{and} \quad k' = k_3'K_1^{1/2}$$

Another example is the formation of hydrogen bromide, HBr. As we know, its stoichiometric equation is

$$H_2 + Br_2 \xrightarrow{k} 2HBr \tag{3.4-3}$$

If this reaction were elementary, the rate equation would be

$$r = k\,(H_2)(Br_2) \tag{3.4-4}$$

However, the actual facts are much more complex. At about 300°C, this reaction apparently takes place in the following steps, according to Bodenstein and Lind (B1). In other words, the mechanism of the formation of HBr is:

$$Br_2 \xrightarrow{k_1} 2Br \tag{3.4-5}$$

$$Br + H_2 \xrightarrow{k_2} HBr + H \tag{3.4-6}$$

$$H + Br_2 \xrightarrow{k_3} HBr + Br \tag{3.4-7}$$

$$H + HBr \xrightarrow{k_4} H_2 + Br \tag{3.4-8}$$

$$2Br \xrightarrow{k_5} Br_2 \tag{3.4-9}$$

Overall

$$H_2 + Br_2 \xrightarrow{k} 2HBr \tag{3.4-3}$$

Because the equilibrium dissociation of Br_2 represented by Eqs. (3.4-5) and (3.4-9) is fast,

$$\frac{d\,(\mathrm{Br})}{dt} = -k_1\,(\mathrm{Br_2}) + k_5\,(\mathrm{Br})^2 = 0 \tag{3.4-10}$$

Experiment indicates the presence of small amount of atomic hydrogen at any time. This means that the net rate of its formation is zero

$$\frac{d\,(\mathrm{H})}{dt} = k_2(\mathrm{Br_2})(\mathrm{H_2}) - k_3(\mathrm{H})(\mathrm{Br_2}) - k_4(\mathrm{H})(\mathrm{HBr}) = 0 \tag{3.4-11}$$

The net rate of formation of HBr can be obtained from Eqs. (3.4-6), (3.4-7), and (3.4-8)

$$\frac{d\,(\mathrm{HBr})}{dt} = k_2(\mathrm{Br})(\mathrm{H_2}) + k_3(\mathrm{H})(\mathrm{Br_2}) - k_4(\mathrm{H})(\mathrm{HBr}) \tag{3.4-12}$$

Solving Eq. (3.4-10) for Br and Eq. (3.4-11) for H and substituting into Eq. (3.4-12) yields

$$\frac{d\,(\mathrm{HBr})}{dt} = \frac{K'(\mathrm{H_2})(\mathrm{Br_2})^{1/2}}{k'' + (\mathrm{HBr})/(\mathrm{Br_2})} \tag{3.4-12}$$

where

$$k' = \frac{2k_2k_3(k_1/k_5)^{1/2}}{k_4}$$

$$k'' = \frac{k_3}{k_4}$$

Equation (3.4-12) is different from Eq. (3.4-4). Therefore the reaction is nonelementary.

3.5 *Effect of Temperature on Rate Constant*

There are three laws that deal with the effect of temperature on rate constant: Arrhenius's law, collision theory, and transition or activated-state theory.

3.5-1 *Arrhenius's law*

$$k = A_1e^{-E_1/RT} \quad \text{or} \quad \ln k = \ln A_1 - \frac{E_1}{RT} \tag{3.5-1}$$

3.5-2 *Collision theory*

$$k = A_2 T^{1/2} e^{-E_2/RT} \quad \text{or} \quad \ln \frac{k}{T^{1/2}} = \ln A_2 - \frac{E_2}{RT} \tag{3.5-2}$$

3.5-3 *Transition or activated-state theory*

$$k = A_3 T e^{-E_3/RT} \quad \text{or} \quad \ln \left(\frac{k}{T}\right) = \ln A_3 - \frac{E_3}{RT} \tag{3.5-3}$$

where A's and E's are defined as the frequency factors and the activation energies, respectively; T is the absolute temperature. Hence the values of A's and E's can be obtained graphically by plotting the experimental values on a semilog graph paper.

4.6 *Evaluation of A* and *E*

Suppose we have the experimental data points x_i and y_i, and we want to find the best values of a and b in the equation of a straight line. The well-known method of *least squares* is employed.

$$\mathbf{y}_i = a + bx_i \tag{3.6-1}$$

Here $\mathbf{y}_i$, the value of $\mathbf{y}_i$ calculated from the experimental value of x_i by Eq. (3.6-1), is not the same as the experimental value of y_i. The first step is to form the sum of squares (SS) as

$$\text{SS} = \sum_i^N (\mathbf{y}_i - y_i)^2 = \sum_i^N [(a + bx_i) - y_i]^2 \tag{3.6-2}$$

Differentiating the SS with respect to a and b gives

$$\frac{\partial(\text{SS})}{\partial a} = 2\,\Sigma[(a + bx_i) - y_i] \tag{3.6-3}$$

$$\frac{\partial(\text{SS})}{\partial b} = 2\,\Sigma[(a + bx_i) - y_i]x_i \tag{3.6-4}$$

Equating both to zero, expanding, and solving by Cramer's rule yields

$$a = \frac{\begin{vmatrix} \Sigma y_i & \Sigma x_i \\ \Sigma x_i y_i & \Sigma x_i^2 \end{vmatrix}}{\begin{vmatrix} N & \Sigma x_i \\ \Sigma x_i & \Sigma x_i^2 \end{vmatrix}} \qquad b = \frac{\begin{vmatrix} N & \Sigma y_i \\ \Sigma x_i & \Sigma x_i y_i \end{vmatrix}}{\begin{vmatrix} N & \Sigma x_i \\ \Sigma x_i & \Sigma x_i^2 \end{vmatrix}} \tag{3.6-5}$$

For the kinetic rate equation,

$$k = Ae^{-E/RT} \quad \text{or} \quad \ln k = \ln A - \frac{E}{RT}$$

the intercept is $a = \ln A$ or $A = e^a$, and the slope is $b = -E/R$ or $E = -bR$

Example 3.6-1

The reaction of ethanol and acetic acid catalyzed by a cation exchange resin was studied by Saletan and White (S1). Evaluate the frequency factor A and the energy of activation E.

k, liter/g mol · h	0.5	1.1	2.2	4.0	6.0
Temperature, °C	30	40	50	60	70

Solution

Before Eq. (3.6-5) is applied, it is good to tabulate the data as follows:

T, °C	T, K	$1/T$ x	k	ln k y	$\ln k/T$ xy	$(1/T)^2$ xx
30	303	3.3003×10^{-3}	0.5	−0.6931	-2.2876×10^{-3}	1.0892×10^{-5}
40	313	3.1949×10^{-3}	1.1	0.0953	3.0450×10^{-4}	1.0207×10^{-5}
50	323	3.0959×10^{-3}	2.2	0.7885	2.4410×10^{-3}	9.5851×10^{-6}
60	333	3.0030×10^{-3}	4.0	1.3863	4.1630×10^{-3}	9.0180×10^{-6}
70	343	2.9155×10^{-3}	6.0	1.7918	5.2238×10^{-3}	8.4999×10^{-6}
		15.5096×10^{-3}		3.3687	9.8448×10^{-3}	4.8202×10^{-5}

Substituting into Eq. (3.6-5) and solving

$$a = 20.9108 \quad \text{or} \quad A = e^{20.9108} = 1.2063 \times 10^9 h^{-1}$$

$$b = -6{,}524.0407 \quad \text{or} \quad E = -(-6{,}524.0407)1.987 = 12{,}963.269 \text{ cal/g mol}$$

```
00100 PROGRAM ARRIUS(INPUT,OUTPUT)
00110CSK=REACTION RATE CONSTANT,LITER/GMOLE-HR
00120CT=TEMPERATURE,DEGC
00130 DIMENSION X(33),Y(33),SK(33),T(33)
00140 DO 1 I=1,5
00150 READ*,SK(I),T(I)
00160 X(I)=1./(T(I)+273.)
00170 1 Y(I)=ALOG(SK(I))
00180 X2=0.
00190 X1=0.
```

```
00200 XY=0.
00210 YY=0.
00220 DO 2 I=1,5
00230 X2=X2+X(I)*X(I)
00240 X1=X1+X(I)
00250 XY=XY+X(I)*Y(I)
00260 2 YY=YY+Y(I)
00270 DENOM=X2*5.-X1*X1
00280 ZEMX1=XY*5.-X1*YY
00290 XB=ZEMX1/DENOM
00300 ZEMX2=X2*YY-XY*X1
00310 XA=ZEMX2/DENOM
00320 BIGA=EXP(XA)
00330 BIGE=-XB*1.987
00340 PRINT*,XB,XA,BIGA,BIGE
00350 STOP
00360 END
READY.
RNH

? 0.5,30.
? 1.1,40.
? 2.2,50.
? 4.0,60.
? 6.0,70.
 -6529.385471616 20.92742934903 1.226498684414E+9 12973.8889321
```

3.7 *Total-Pressure Method*

In order to determine the rate constant, two kinds of experimentation are usually employed. The first one is the so-called *total-pressure method* in which a reaction takes place at constant volume and constant temperature. The reaction

$$\alpha_1 X_1 + \alpha_2 X_2 \rightarrow \alpha_3 X_3 + \alpha_4 X_4$$

can be rewritten as

$$X_1 + \frac{\alpha_2}{\alpha_1} X_2 \rightarrow \frac{\alpha_3}{\alpha_1} X_3 + \frac{\alpha_4}{\alpha_1} X_4$$

The number of moles of j component during the reaction is

$$N_j = N_{jo} + \alpha_j \beta$$

Summing for all j components gives

$$N = N_o + \alpha\beta$$

Eliminating β from these two equations yields

$$N_j = N_{jo} + \frac{\alpha_j}{\alpha}(N - N_o) \tag{3.7-1}$$

At constant temperature and volume, from ideal gas law, we obtain

$$p_j = p_{jo} + \frac{\alpha_j}{\alpha}(P - P_o) \tag{3.7-2}$$

where p and P are the partial and total pressures, respectively. $\alpha = \Sigma\alpha_j$. Because

$$C_j = \frac{p_j}{RT} \tag{3.7-3}$$

$$\frac{dC_j}{dt} = \frac{d}{dt}\left(\frac{p_j}{RT}\right) = \frac{1}{RT}\left[p_{jo} + \frac{\alpha_j}{\alpha}(P - P_o)\right] = \frac{1}{RT}\frac{\alpha_j}{\alpha}\frac{dP}{dt} \tag{3.7-4}$$

But

$$\frac{d\lambda}{dt} = \frac{1}{\alpha_j}\frac{dC_j}{dt} = k\Pi C_j^{\alpha j} - k'\Pi C_j^{\alpha j} \tag{3.7-5}$$

Substituting Eq. (3.7-4) into Eq. (3.7-5) gives

$$\frac{1}{RT\alpha}\frac{dP}{dt} = k\Pi\left[p_{jo} + \frac{\alpha_j}{\alpha}(P - P_o)\right]^{\alpha j}\left(\frac{1}{RT}\right)^{\alpha j}$$
$$- k'\Pi\left[p_{jo} + \frac{\alpha_j}{\alpha}(P - P_o)\right]^{\alpha j'}\left(\frac{1}{RT}\right)^{\alpha j'} \tag{3.7-6}$$

Example 3.7-1 Total-Pressure Method

It is found that the gas-phase dissociation of disulfuryl chloride, SO_2Cl_2, into chlorine and sulfur dioxide occurs at 279.2°C. The total-pressure method was employed to follow the course of the reaction. Under constant-volume conditions, the results were (S1)

Time, min	3.4	28.1	54.5	82.4	96.3
Total pressure, mm	325	345	365	385	395

What reaction mechanism do these data suggest? The conversion is 100 percent at infinite time.

Solution

$$SO_2Cl_2 \rightarrow Cl_2 + SO_2$$

or

$$Cl_2 + SO_2 - SO_2Cl_2 = 0$$

$$\alpha = 1 + 1 - 1 = 1$$

$$\alpha_{SO_2Cl_2} = -1$$

Assuming first-order, the exponent is 1, and Eq. (3.7-6) becomes

$$\frac{dP}{dt} = k[P_o - (P - P_o)]$$

or

$$\frac{dP}{2P_o - P} = k\,dt$$

The initial total pressure can be obtained by extending the data to zero time. The result is $P_o = 322.5$ mm. Substituting this value and integrating, we obtain

$$645 - P = 322.5\,e^{-kt}$$

The value of k at different temperatures can be calculated from the experimental data as given in the problem.

Time, min	3.4	28.1	54.5	82.4	96.3
Total pressure P, mm	325	345	365	385	395
$645 - P$	320	300	280	260	250
$k \times 10^3$	−2.288	−2.573	−2.593	−2.614	−2.644

Since the values are fairly close, the assumption of first-order is correct and the average value of k is -2.542×10^{-3} min^{-1} including the first data or $k = -2.606 \times 10^{-3}$ min^{-1}, excluding the first data. Graphically the same result can be obtained by plotting ln $(645 - P)$ versus time t

$$\ln(645 - P) = \ln 322.5 - kt$$

The slope is $-k$ as found in the Figure 3.7-1:

$$k = \text{slope} = \frac{\ln(0.003/0.0025)}{28 - 98} = -0.0026\ \text{min}^{-1}$$

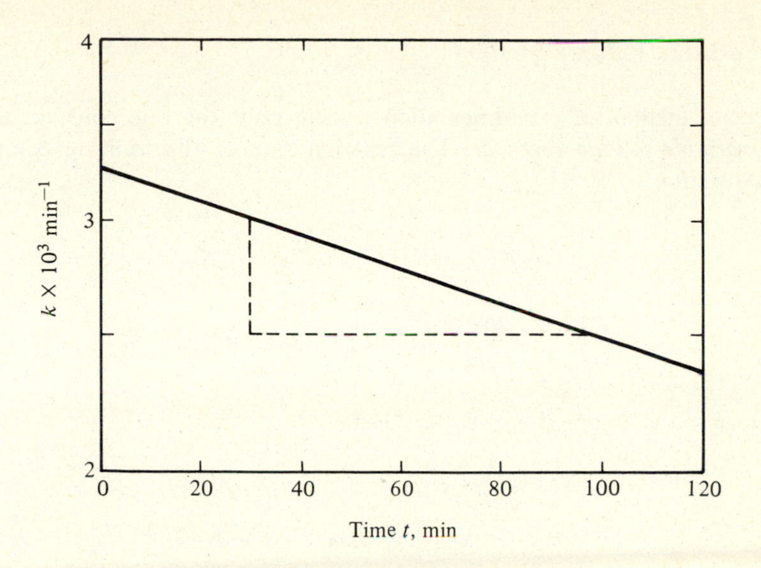

Figure 3.7-1 Rate constant by total-pressure method

Example 3.7-2 Total-Pressure Method

Initially pure ethylene oxide is kept in a vessel at 2 atm and 450°C. After 30 min, what is the total pressure in the vessel? The rate constant for the gas-phase decomposition of ethylene oxide to methane and carbon dioxide at 450°C is 0.0212 min^{-1}. Ideal-gas behavior and constant temperature are assumed.

Solution

$$\underset{\diagdown O \diagup}{CH_2{-}CH_2}(g) \xrightarrow{k} CH_4(g) + CO(g)$$

Using Eq. (3.7-6),

$$k' = 0 \qquad p_{jo} = P_o = 2 \qquad \underline{\alpha} = 1 \qquad \boldsymbol{\alpha} = 1 + 1 - 1 = 1$$

Hence

$$\frac{dP}{dt} = k[p_{jo} - (P - P_o)] = k(2P_o - P)$$

$$= 0.0212(4 - P)$$

Integrating from $P = 2$ at $t = 0$ to $P = P$ at $t = 30$ gives

$$P = 2.94 \text{ atm}$$

3.8 Variable-Volume Method

The second method of experimentation to determine the rate constant is the so-called *variable-volume method.* The reaction rate of the limiting component l for the reaction

$$\alpha_k X_1 + \alpha_2 X_2 \underset{k'}{\overset{k}{\rightleftharpoons}} \alpha_3 X_3 + \alpha_4 X_4$$

is

$$-\frac{dN_l}{V\,dt} = kC_1^{\alpha_k} C_2^{\alpha_2} - k' C_3^{\alpha_3} C_4^{\alpha_4} \tag{3.8-1}$$

if the reaction is assumed elementary. But

$$N_j = N_{jo} + \alpha_j \beta = N_{jo} + \alpha_j N_{ko} f_k / \alpha_k$$

$$\begin{aligned}
&\text{For } j = k \quad N_k &&= N_{ko} - \alpha_k N_{ko} f_k/\alpha_k = N_{ko}(1 - f_k)\\
&\text{For } j = 2 \quad N_2 &&= N_{2o} - \alpha_2 N_{ko} f_k/\alpha_k\\
&\text{For } j = 3 \quad N_3 &&= N_{3o} + \alpha_3 N_{ko} f_k/\alpha_k\\
&\text{For } j = 4 \quad N_4 &&= N_{4o} + \alpha_4 N_{ko} f_k/\alpha_k
\end{aligned}$$

Summing

$$\begin{aligned}
N &= N_o + \boldsymbol{\alpha} N_{ko} f_k/\alpha_k\\
&= N_o(1 + \epsilon_k f_k)
\end{aligned}$$

where

$$\epsilon_k = \frac{\boldsymbol{\alpha} N_{ko}}{\alpha_k N_o} \tag{3.8-2}$$

By ideal gas law

$$V = (1 + \epsilon_k f_k) V_o \frac{P_o T}{P\,T_o} \tag{3.8-3}$$

If the reaction is carried out under constant pressure and constant temperature,

$$V = (1 + \epsilon_k f_k) V_o \tag{3.8-4}$$

Substituting all these equations into Eq. (3.8-1) and simplifying yields

$$\frac{N_{ko}\,df_k}{V\,dt} = k\left[\frac{N_{ko}(1 - f_k)}{V}\right]^{\alpha_k}\left[\frac{N_{2o} - \alpha_2 N_{ko} f_k/\alpha_k}{V}\right]^{\alpha_2} - k'\left[\frac{N_{3o} + \alpha_3 N_{ko} f_k/\alpha_k}{V}\right]^{\alpha_3}\left[\frac{N_{4o} + \alpha_4 N_{ko} f_k/\alpha_k}{V}\right]^{\alpha_4}$$

or in general

$$\frac{N_{ko}\,df_k}{V_o(1+\epsilon_k f_k)dt} = k\Pi\left[\frac{N_{jo}+\alpha_j N_{ko} f_k/\alpha_k}{V_o(1+\epsilon_k f_k)}\right]^{\alpha_j} - k'\Pi\left[\frac{N_{jo}+\alpha_j N_{ko} f_k/\alpha_k}{V_o(1+\epsilon_k f_k)}\right]^{\alpha_j'} \quad (3.8\text{-}5)$$

Example 3.8-1 Variable-Volume Method

Develop an expression to relate the fractional conversion of NO with temperature, pressure, and time for the following elementary reaction:

$$2NO + O_2 \rightarrow 2NO_2$$

Solution

Since NO is chosen as the limiting reactant, the reaction is rearranged to

$$NO + \frac{1}{2}O_2 \rightarrow NO_2$$

Therefore for the original reaction

$$N_{ko} = 2 \text{ mol NO};\ N_o = 2 \text{ mol NO} + 1 \text{ mol } O_2 = 3$$

$$\alpha = 2 - 1 - 2 = -1$$

$$\epsilon = \frac{\alpha N_{ko}}{\alpha_k N_o} = -\frac{1}{2}\left(\frac{2}{3}\right) = -\frac{1}{3}$$

Substituting into Eq. (3.8-4)

$$V = V_o\left(1 - \frac{1}{3}f_k\right) \quad \text{(A)}$$

$$\text{For } j = k \qquad N_{jo} + \frac{\alpha_j N_{ko} f_k}{\alpha_k} = N_{ko} - N_{ko} f_k = 2(1 - f_k) \quad \text{for NO} \quad \text{(B)}$$

$$\text{For } j = j \qquad N_{jo} + \frac{\alpha_j N_{ko} f_k}{\alpha_k} = 1 - \frac{1}{2}(2)f_k = 1 - f_k \quad \text{for } O_2 \quad \text{(C)}$$

Since it is an irreversible reaction, $k' = 0$. Since it is an elementary reaction, the exponent α_j is 2 with respect to NO and 1 with respect to O_2. Hence substituting Eqs. (A), (B), and (C) into Eq. (3.8-5) yields

$$\frac{2\,df_k}{V_o(1 - \frac{1}{3}f_k)dt} = k\left[\frac{2(1 - f_k)}{V_o(1 - \frac{1}{3}f_k)}\right]^2\left[\frac{1 - f_k}{V_o(1 - \frac{1}{3}f_k)}\right] \quad \text{(D)}$$

Simplifying and using $V_o = 3RT/P$, we get

$$\frac{df_k}{dt}=2k\left\{\frac{(1-f_k)^3}{[3(RT/P)(1-f_k/3)]^2}\right\} \tag{E}$$

Integrating between $t = 0$, $f_k = 0$, and $t = t$, $f_k = f_k$ and simplifying

$$\frac{2kP^2t}{(3RT)^2}=\frac{1}{9}\left[\frac{2(4f_k-3f_k^2)}{(1-f_k)^2}+\ln\frac{1}{1-f_k}\right] \tag{F}$$

Example 3.8-2 Variable-Volume Method

A gaseous mixture of 0.35 mol A, 0.35 mol B, 0.21 mol C, and 0.09 mol inerts reacts at 400°C in an isothermal reactor that will be maintained at a constant pressure of 1 atm. If the reaction

$$A+B\xrightarrow{k}C$$

is first-order in A and first-order in B and if $k = 2$ liters/g mol · min, how long will it take before the mole fraction of C reaches 0.3?

Solution

Basis of calculation: 1 g mol feed $= N_o$. Thus

Initial gram mole of A	= 0.35
Initial gram mole of B	= 0.35
Initial gram mole of C	= 0.21
Initial gram mole of inert	= 0.09
Total initial gram mole N_o	= 1.00

$$\alpha=1-1-1=-1$$

$$\epsilon_k=\frac{\alpha\, N_{ko}}{\alpha_k N_0}=\frac{(-1)(0.35)}{1}=-0.35$$

$$V_o=\frac{N_oRT}{P}=\frac{1(0.082)(673)}{1}=55.186 \text{ liters}$$

Substituting into Eq. (3.8-5)

$$\frac{df_k}{dt}=\frac{2(0.35-0.35f_k)^2}{0.35(55.186)(1-0.35f_k)} \tag{A}$$

Integrating by partial fractions gives

$$\frac{0.65}{1-f_k}=1.2684\times10^{-2}t+0.65+0.35\ln(1-f_k) \tag{B}$$

Using

$$N_j = N_{jo} + \alpha_j \beta = N_{jo} + \alpha_j N_{ko} f_k / \alpha_k$$

$$\begin{aligned} N_A &= 0.35 - 0.35 f_k \\ N_B &= 0.35 - 0.35 f_k \\ N_C &= 0.21 + 0.35 f_k \\ N_I &= 0.09 \\ \hline N &= 1.00 - 0.35 f_k \end{aligned}$$

Hence the mole fraction of C

$$x_C = \frac{0.21 + 0.35 f_k}{1 - 0.35 f_k} = 0.3$$

Thus

$$f_k = 0.1978$$

Substituting into Eq. (B) gives

$$t = 18.72 \text{ min}$$

3.9 *Generalized Equations for Chemical Reactions*

The relationship of the process variables in a generalized chemical reaction

$$\alpha_1 X_1 + \alpha_2 X_2 + \cdots = \alpha_3 X_3 + \alpha_4 X_4 + \cdots \tag{3.9-1}$$

or

$$X_1 + \frac{\alpha_2}{\alpha_1} X_2 + \cdots = \frac{\alpha_3}{\alpha_1} X_3 + \frac{\alpha_4}{\alpha_1} X_4 + \cdots \tag{3.9-2}$$

can be calculated by the following equations:

1. Stoichiometric equation:

For a single reaction

$$\Sigma \alpha_j X_j = 0 \qquad C_j = C_{jo} + \lambda \alpha_j \tag{3.9-3}$$

For multiple reactions

$$\Sigma \alpha_{ij} X_j = 0 \qquad C_j = C_{jo} + \Sigma \alpha_{ij} \lambda_i \tag{3.9-4}$$

2. Rate equations:

(A) Overall

Single reaction

$$\frac{d\lambda}{dt} = r = k\Pi C_j^{\alpha_j} - k'\Pi C_j^{\alpha_j'} \tag{3.9-5}$$

Multiple reactions

$$\frac{d\lambda_i}{dt} = r_i = k_i\Pi C_j^{\alpha_j} - k_i'\Pi C_j^{\alpha_j'} \tag{3.9-6}$$

(B) Component

Single reaction

$$\frac{dC_j}{\alpha_j dt} = r \quad \text{or} \quad \frac{dC_j}{dt} = \alpha_j r \tag{3.9-7}$$

Multiple reactions

$$\frac{dC_j}{dt} = \Sigma\alpha_{ij}r_i \tag{3.9-8}$$

3. Fractional conversion:

$$\lambda = \frac{f_k C_{ko}}{\alpha_k} \tag{3.9-9}$$

3.10 Analysis of Constant-Volume Batch Data

In order to determine the rate constant of different kinds of chemical reaction, the previous generalized Eq. (3.9-1) to (3.9-9) can be employed under constant volume and temperature conditions.

3.10-1 Zero-order reaction

$$X_1 + \frac{\alpha_2}{\alpha_1}X_2 \xrightarrow{k} \text{product}$$

1. Stoichiometric equation:

$$C_1 = C_{1o} - \lambda \tag{3.10-1}$$

2. Overall rate equation:

$$\frac{d\lambda}{dt} = r = k \tag{3.10-2}$$

3. Component rate equation:

$$\frac{dC_1}{dt} = -r \tag{3.10-3}$$

Combining Eq. (3.10-2) with Eq. (3.10-3) gives

$$\frac{dC_1}{dt} = -k \tag{3.10-4}$$

Integrating yields

$$C_1 - C_{1o} = -kt \tag{3.10-5}$$

Hence a plot of C_1 versus t gives slope equal to $-k$ (see Fig. 3.10-1).

Half-life is defined as the time when the concentration reduces to half of the initial concentration, i.e., at $t = t_{1/2}$, $C_1 = \frac{1}{2} C_{1o}$. Thus

$$\frac{1}{2} C_{1o} - C_{1o} = -kt_{1/2} \quad \text{or} \quad k = \frac{1}{2} \frac{C_{1o}}{t_{1/2}} \tag{3.10-6}$$

The unit of the specific rate constant for zero-order reaction is concentration per unit time [(conc)(time)$^{-1}$] in gram moles per liter per second.

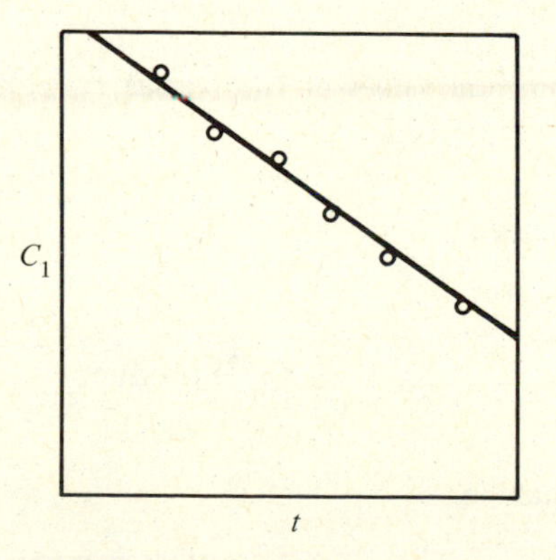

Figure 3.10-1 Zero-order rate constant

Example 3.10-1 Zero-Order Rate Constant

Carbon tetrachloride reacts with water vapor in an ultrasonic field. The experimental data are (C2)

$$H_2O + CCl_4 \longrightarrow Cl_2 + \text{others}$$

Time, min	0	1	2	3	4
Conc. of Cl_2, meq/liter	0.0	0.09	0.17	0.265	0.385

Assuming zero-order, determine the rate constant.

Solution

$$\frac{d\lambda}{dt} = k \text{ for zero-order reaction}$$

But

$$C_{Cl_2} = (C_{Cl_2})_o + \lambda = 0 + \lambda = \lambda$$

Therefore

$$\frac{dC_{Cl_2}}{dt} = k$$

Integrating gives

$$C_{Cl_2} = kt$$

Plotting the experimental values of C_{Cl_2} versus t yields the value of k = 0.0945 meq/liter²/min (see Fig 3.10-2).

3.10-2 First-order reaction

$$X_1 + \frac{\alpha_2}{\alpha_1} X_2 \xrightarrow{k} \text{product}$$

1. Stoichiometric equation:

$$C_1 = C_{1o} - \lambda \tag{3.10-7}$$

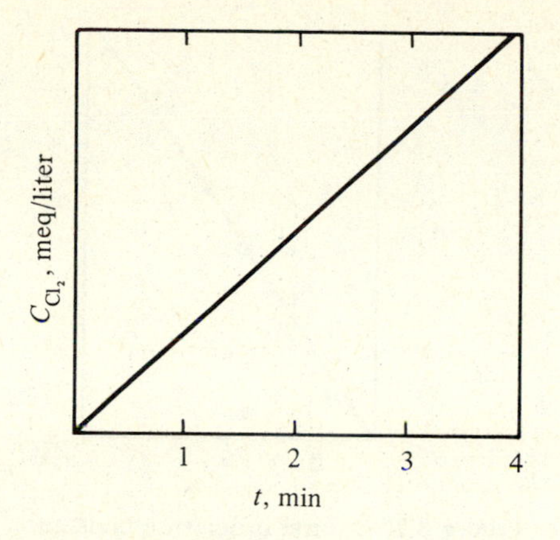

Figure 3.10-2 Solution of Example 3.10-1

2. Overall rate equation:

$$\frac{d\lambda}{dt} = r = kC_1 \tag{3.10-8}$$

3. Component rate equation:

$$\frac{dC_1}{dt} = -r \tag{3.10-9}$$

Combining Eqs. (3.10-8) and (3.10-9) gives

$$\frac{dC_1}{dt} = -kC_1 \tag{3.10-10}$$

Integrating yields

$$\ln \frac{C_{1o}}{C_1} = kt \tag{3.10-11}$$

Hence plotting $-\ln (C_1/C_{1o})$ versus t produces slope = k (see Fig 3.10-3). In terms of half-time,

$$k = \frac{\ln 2}{t_{1/2}}$$

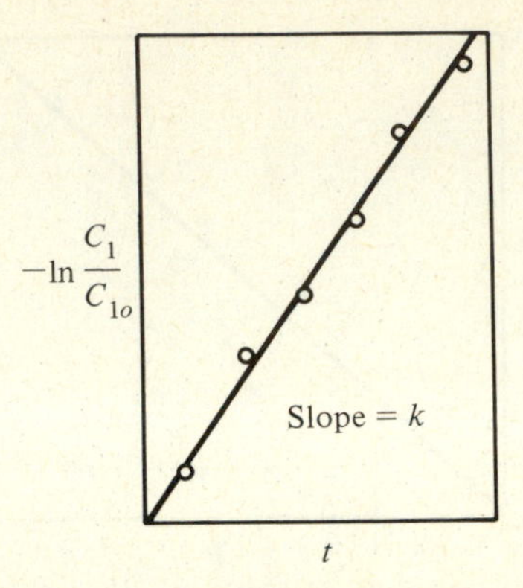

Figure 3.10-3 First-order rate constant

Example 3.10-2 First-Order Rate Constant

The experimental data for the decomposition of bibromosuccinic acid [2,3-dibromobutanedionic acid, $C_2H_2Br_2(COOH)_2$] in a batch reactor are:

Time, min	0	10	20	30	40	50
Acid remaining, g	5.11	3.77	2.74	2.02	1.48	1.08

Assuming first-order, find the specific rate constant (R1).

Solution

Since m_1 = grams of acid remaining = C_1 g/liter (V liters) substituting into Eq. (3.10-10) and simplifying gives

$$-\frac{dm_1}{dt} = km_1$$

Integrating yields

$$\ln \frac{m_1}{m_{1o}} = -kt$$

t	0	10	20	30	40	50
$-\ln \frac{m_1}{m_{1o}}$	0	0.3041	0.6232	0.9281	1.2392	1.5542
k		0.03041	0.03116	0.03094	0.03098	0.03108

$$k_{av} = 0.03092$$

3.10-3 Pseudo first-order

Sometimes one of the reactants is in large excess. Its presence does not affect the rate of the reaction. For example, the reaction

$$X_1 + X_2 \xrightarrow{k} nX_3$$

can be carried out with C_2 much larger than C_1. This would occur when X_2 is the solvent in which X_1 is dissolved. Then

1. Stoichiometric equation:

$$C_1 = C_{1o} - \lambda, \; C_2 = C_{2o} - \lambda \tag{3.10-12}$$

2. Overall rate equation:

$$\frac{d\lambda}{dt} = r = kC_1C_2 \tag{3.10-13}$$

3. Component rate equation:

$$\frac{dC_1}{dt} = -r \tag{3.10-14}$$

From Eq. (3.10-12), eliminating λ gives

$$C_{1o} - C_1 = C_{2o} - C_2$$

or

$$C_2 = C_{2o} - C_{1o} + C_1 \tag{3.10-15}$$

Since C_{2o} is much greater than C_{1o}, then $C_2 \approx C_{2o}$. With this approximation, the combination of Eqs. (3.10-13) and (3.10-14) yields

$$\frac{dC_1}{dt} = -kC_1C_{2o} = -k'C_1 \tag{3.10-16}$$

Integrating gives

$$\ln \frac{C_1}{C_{1o}} = -k't \tag{3.10-17}$$

which resembles first-order. However, originally it is a second-order reaction. Due to the presence of large amount of X_2, the reaction behaves as a first-order.

This kind of reaction is called a *pseudo first-order reaction.* For instance, the hydrolysis of acetic anhydride in excess water to form acetic acid has been studied by Eldridge and Piret as a pseudo first-order reaction with a rate constant of 0.0567 min^{-1} at $-15°\text{C}$ (E1). There is another manner in which a reaction in solute might appear to be first-order. For example, a catalytic reaction

$$A + B \rightarrow nD + B$$

in which B reacts as a catalyst which is not used up in the reaction. Hence its concentration is then unchanged throughout the reaction.

Example 3.10-3 Pseudo First-Order Rate Constant

The catalyzed reaction of the cleavage of diacetone alcohol at 25°C is

$$\text{Diacetone alcohol} + OH^- \rightarrow 2\ \text{acetone} + OH^-$$

The experimentally determined rate of reaction of diacetone alcohol in the presence of NaOH is (F2)

$$r = -0.47 C_A C_B$$

Since C_B is constant in any experiment, the rate becomes

$$r = -k' C_A$$

where $k' = 0.47 C_B$

3.10-4 Second-order

Type I:

$$\alpha_1 X_1 + \alpha_2 X_2 \xrightarrow{k} \text{product}$$

or

$$X_1 + \frac{\alpha_2}{\alpha_1} X_2 \xrightarrow{k} \text{product}$$

1. Stoichiometric equation:

$$C_1 = C_{1o} - \lambda \qquad C_2 = C_{2o} - \frac{\alpha_2}{\alpha_1} \lambda \tag{3.10-18}$$

2. Overall rate equation:

$$\frac{d\lambda}{dt} = r = kC_1C_2 \tag{3.10-19}$$

3. Component rate equation:

$$\frac{dC_1}{dt} = -r \tag{3.10-20}$$

Eliminating λ in Eq. (3.10-18) gives

$$C_2 = C_{2o} - \frac{\alpha_2}{\alpha_1}(C_{1o} - C_1) \tag{3.10-21}$$

Substituting into the combination of Eq. (3.10-19) and Eq. (3.10-20)

$$\frac{dC_1}{dt} = -kC_1\left[C_{2o} - \frac{\alpha_2}{\alpha_1}(C_{1o} - C_1)\right] \tag{3.10-22}$$

By partial fraction, this equation can be integrated. However, there is another easier method through the use of λ.

$$\frac{d\lambda}{dt} = k(C_{1o} - \lambda)\left(C_{2o} - \frac{\alpha_2}{\alpha_1}\lambda\right) \tag{3.10-23}$$

Integrating by partial fraction yields

$$\ln\frac{C_{2o} - (\alpha_2/\alpha_1)\lambda}{C_{2o}} - \ln\frac{C_{1o} - \lambda}{C_{1o}} = \left(C_{2o} - \frac{\alpha_2}{\alpha_1}C_{1o}\right)kt$$

or

$$\ln\frac{C_2C_{1o}}{C_{2o}C_1} = \left(C_{2o} - \frac{\alpha_2}{\alpha_1}C_{1o}\right)kt \tag{3.10-24}$$

See Fig. 3.10-4 for a graph.

Type II:

$$2X_1 \xrightarrow{k} \text{product}$$

1. Stoichiometric equation:

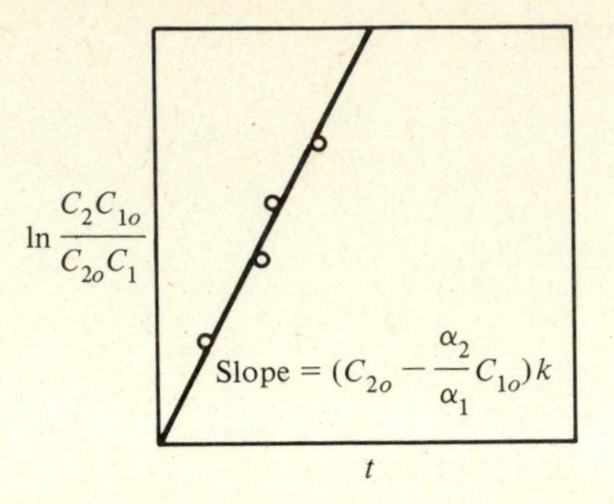

Figure 3.10-4 Second-order type I rate constant

$$C_1 = C_{1o} - 2\lambda \tag{3.10-25}$$

2. Overall rate equation:

$$\frac{d\lambda}{dt} = r = kC_1^2 \tag{3.10-26}$$

3. Component rate equation:

$$\frac{dC_1}{dt} = -2r \tag{3.10-27}$$

Combining Eq. (3.10-26) and Eq. (3.10-27) and integrating yields

$$\frac{1}{C_1} = \frac{1}{C_{1o}} + 2kt \tag{3.10-28}$$

Plotting $1/C_1$ versus t gives the slope as $2k$ (see Fig. 3.10-5), at half-time, $2k = 1/t_{1/2}C_{1o}$.

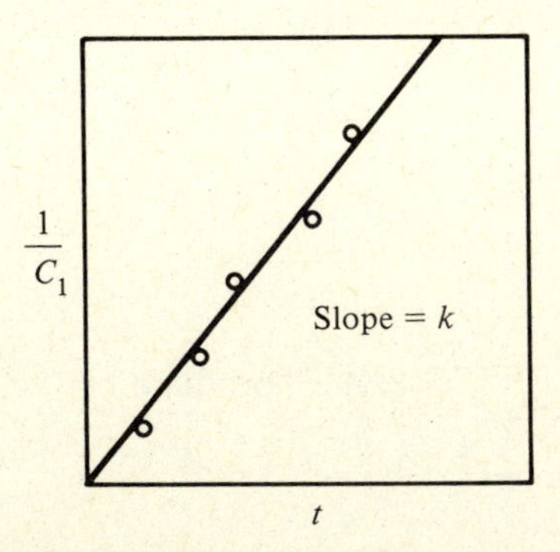

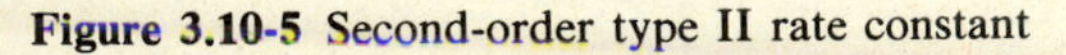

Figure 3.10-5 Second-order type II rate constant

Example 3.10-4 Second-Order Type I Rate Constant

A liquid mixture of ethyl acetate and sodium hydroxide reacts at 25°C. Initially 100 ml of the mixture requires 68.2 ml of 0.05 *N* HCl for neutralization. After 2.28 min of reaction, it requires 49.7 ml of the same strength of acid. At the end of the reaction, it requires 15.6 ml. Determine the specific rate constant.

Solution

$$CH_3COOC_2H_5 + NaOH \rightarrow CH_3COONa + C_2H_5OH$$

The initial concentration of NaOH = 0.05(68.2/100) = 0.034 g mol/liter = C_{2o}. At 2.28 min the concentration of NaOH = 0.05(49.7/100) = 0.0249 g mol/liter = C_2. At completion the concentration of NaOH = 0.05(15.6/100) = 0.0078 g mol/liter = $C_{2\infty}$. From Eq. (3.10-18)

$$C_{2\infty} = C_{2o} - \lambda_\infty \quad \text{or} \quad \lambda_\infty = C_{2o} - C_{2\infty} = 0.034 - 0.0078 = 0.0262 \text{ g mol/liter}$$

$$C_{1\infty} = C_{1o} - \lambda_\infty \quad \text{or} \quad C_{1o} = C_{1\infty} + \lambda_\infty = 0.0262 \text{ g mol/liter}$$

At 2.28 min

$$C_2 = C_{2o} - \lambda \quad \text{or} \quad \lambda = C_{2o} - C_2 = 0.034 - 0.0249 = 0.0091 \text{ g mol/liter}$$

$$C_1 = C_{1o} - \lambda = 0.0262 - 0.0091 = 0.0171 \text{ g mol/liter}$$

Substituting into Eq. (3.10-24)

$$\ln \frac{(0.0249)(0.0262)}{(0.0340)(0.0171)} = (0.0340 - 0.0262)k(2.28)$$

Hence

$$k = 6.48 \text{ liter/g mol} \cdot \text{min}$$

Example 3.10-5 Second-Order Type II Rate Constant

The following are the experimental data for the dimerization of butadiene at 326°C and constant volume (V1). Assuming second-order, determine the rate constant.

Time t, min	0	20	40	60	80
Total pressure, mmHg	632	558.2	514	484	463

Solution

$$2 \text{ butadiene} \xrightarrow{k} (\text{butadiene})_2$$

$$C_{10} = P_0/RT$$

$$C_1 = P_1/RT = \frac{1}{RT}\left[P_{10} + \frac{\alpha_j}{\alpha}(P - P_0)\right]$$

$$= \frac{1}{RT}\left[P_0 + \frac{-2}{-1}(P - P_0)\right] = \frac{1}{RT}(2P - P_0)$$

Using Eq. (3.10-28),

$$\frac{1}{2P - P_o} - \frac{1}{P_o} = \frac{2k_c t}{RT} = \frac{k_c' t}{RT} = k_p t$$

Time t, min	0	20	40	60	80
Total pressure P, mmHg	632	558.2	514	484	463
$k_p \times 10^5$, $\text{mmHg}^{-1}\ \text{min}^{-1}$		2.41	2.36	2.32	2.27

$$\text{ave. } k_p = 2.34 \times 10^{-5}\ \text{mmHg}^{-1}\ \text{min}^{-1}$$

$$k_c' = 2.34 \times 10^{-5}(62.4)(599) = 0.8746\ \text{liter/g mol} \cdot \text{min}$$

3.10-5 Third-order reaction

Type I:

$$X_1 + \frac{\alpha_2}{\alpha_1} X_2 + \frac{\alpha_3}{\alpha_1} X_3 \rightarrow \text{products}$$

1. Stoichiometric equations:

$$C_1 = C_{1o} - \lambda \qquad C_2 = C_{2o} - \frac{\alpha_2}{\alpha_1}\lambda \qquad C_3 = C_{3o} - \frac{\alpha_3}{\alpha_1}\lambda \tag{3.10-29}$$

2. Overall rate equation:

$$\frac{d\lambda}{dt} = r = kC_1C_2C_3 \tag{3.10-30}$$

3. Component rate equation:

$$\frac{dC_1}{dt} = -r \tag{3.10-31}$$

Combining Eqs. (3.10-29) and (3.10-30) gives

$$\frac{d\lambda}{dt} = k(C_{1o} - \lambda)\left(C_{2o} - \frac{\alpha_2}{\alpha_1}\lambda\right)\left(C_{3o} - \frac{\alpha_3}{\alpha_1}\lambda\right) \tag{3.10-32}$$

which can be integrated by partial fractions. For simplicity, let $\alpha_1 = \alpha_2 = \alpha_3 = 1$ and combination of Eqs. (3.10-29), (3.10-30), and (3.10-31) yields

$$-\frac{dC_1}{dt} = kC_1(C_{2o} - C_{1o} + C_1)(C_{3o} - C_{1o} + C_1) \tag{3.10-33}$$

Integrating by partial fractions results in

$$\frac{1}{(C_{2o} - C_{1o})(C_{3o} - C_{1o})} \ln \frac{C_{1o}}{C_1} + \frac{1}{(C_{1o} - C_{2o})(C_{3o} - C_{2o})} \ln \frac{C_{2o}}{C_2} + \frac{1}{(C_{1o} - C_{3o})(C_{2o} - C_{3o})} \ln \frac{C_{3o}}{C_3} = kt$$

or

$$\frac{\ln [(C_1/C_{1o})^{C_{2o} - C_{3o}} (C_2/C_{2o})^{C_{3o} - C_{1o}} (C_3/C_{3o})^{C_{1o} - C_{2o}}]}{(C_{1o} - C_{2o})(C_{2o} - C_{3o})(C_{3o} - C_{1o})} = kt \tag{3.10-34}$$

Type II:

(A) $X_1 + X_2 \longrightarrow$ products

1. Stoichiometric equations:

$$C_1 = C_{1o} - \lambda, \; C_2 = C_{2o} - \lambda \tag{3.10-35}$$

2. Overall rate equation:

$$\frac{d\lambda}{dt} = r = kC_1^2 C_2 \tag{3.10-36}$$

3. Component rate equation:

$$\frac{dC_1}{dt} = -r \tag{3.10-37}$$

Eliminating λ in Eq. (3.10-35) gives

$$C_2 = C_{2o} - C_{1o} + C_1 \tag{3.10-38}$$

Combining Eqs. (3.10-36), (3.10-37), and (3.10-38) yields

$$-\frac{dC_1}{dt} = kC_1^2(C_{2o} - C_{1o} + C_1) \tag{3.10-39}$$

Integrating results in

$$\frac{1}{(C_{2o} - C_{1o})}\left(\frac{1}{C_1} - \frac{1}{C_{1o}}\right) + \frac{1}{(C_{2o} - C_{1o})^2}\ln\frac{C_{2o}C_1}{C_{1o}C_2} = kt \tag{3.10-40}$$

Type II:

(B) $2X_1 + X_2 \rightarrow \text{product}$

1. Stoichiometric equations:

$$C_1 = C_{1o} - 2\lambda, \; C_2 = C_{2o} - \lambda \tag{3.10-41}$$

2. Overall rate equation:

$$\frac{d\lambda}{dt} = kC_1^2C_2 = r \tag{3.10-42}$$

3. Component rate equation:

$$-\frac{dC_1}{2dt} = r \tag{3.10-43}$$

Eliminating λ in Eq. (3.10-41) gives

$$C_2 = C_{2o} - \frac{1}{2}C_{1o} + \frac{1}{2}C_1 \tag{3.10-44}$$

Combining Eqs. (3.10-42) and (3.10-43) gives

$$-\frac{dC_1}{dt} = 2kC_1^2C_2 = k_1C_1^2C_2 \tag{3.10-45}$$

where $k_1 = 2k$. Eliminating C_2 in Eq. (3.10-45) by Eq. (3.10-44) yields

$$\frac{dC_1}{dt} = k_1C_1^2\left(C_{2o} - \frac{C_{1o}}{2} + \frac{C_1}{2}\right) \tag{3.10-46}$$

Integrating results in

$$\frac{2}{(2C_{2o} - C_{1o})}\left(\frac{1}{C_1} - \frac{1}{C_{1o}}\right) + \frac{2}{(2C_{2o} - C_{1o})^2} \ln \frac{C_{2o}C_1}{C_{1o}C_2} = k_1 t \quad \textbf{(3.10-47)}$$

Example 3.10-6 Third-Order Type IIB Rate Constant

Initially 0.054 mol/liter methanol reacts with 0.106 mol/liter triphenyl methyl (trityl) chloride in dry benzene solution.

$$CH_3OH + (C_6H_5)_3CCl \rightarrow (C_6H_5)_3COCH_3 + HCl$$
$$(1) \qquad (2)$$

The reaction is second-order with respect to CH_3OH and first-order with respect to $(C_6H_5)_3CCl$. Find the rate constant from the following experimental data (S4).

Solution

t, min	426	1150	1660	3120
x reacted, mol/liter	0.0189	0.0318	0.0354	0.0416
$C_1 = 0.054 - x$	0.0351	0.0222	0.0186	0.0124
$C_2 = 0.106 - x$	0.0871	0.0742	0.0706	0.0644
k Eq. (3.10-47)	0.2466	0.2724	0.2616	0.2676

$$\text{Average } k = 0.262 \text{ (g mol/liter)}^{-2}\text{(min)}^{-1}$$

Type III: $X \rightarrow$ products

1. Stoichiometric equation:

$$C_1 = C_{1o} - \lambda \quad \textbf{(3.10-48)}$$

2. Overall rate equation:

$$\frac{d\lambda}{dt} = r = kC_1^3 \quad \textbf{(3.10-49)}$$

3. Component rate equation:

$$\frac{dC_1}{dt} = -r \quad \textbf{(3.10-50)}$$

Combining Eqs. (3.10-49) and (3.10-5) gives

$$-\frac{dC_1}{dt} = kC_1^3 \tag{3.10-51}$$

Integrating results in

$$\frac{1}{C_1^2} - \frac{1}{C_{1o}^2} = -2kt \tag{3.10-52}$$

Therefore, plotting $1/C^2$ versus t yields slope $= -2k$.

3.10-6 Fractional or n*th-order (except first)*

$$X \longrightarrow \text{products}$$

1. Stoichiometric equation:

$$C_1 = C_{1o} - \lambda \tag{3.10-53}$$

2. Overall rate equation:

$$\frac{d\lambda}{dt} = r = kC_1^n \tag{3.10-54}$$

3. Component rate equation:

$$\frac{dC_1}{dt} = -r \tag{3.10-55}$$

Combining Eqs. (3.10-54) and (3.10-55) gives

$$-\frac{dC_1}{dt} = kC_1^n \tag{3.10-56}$$

Integrating results in

$$\frac{1}{n-1}\left(\frac{1}{C_1^{n-1}} - \frac{1}{C_{1o}^{n-1}}\right) = kt \tag{3.10-57}$$

3.11 Determination of Orders

From experimental data of chemical reactions, the following methods can be employed to determine the order of the reaction.

3.11-1 Integration method

This method has been described in previous sections for different orders. The result is summarized as follows:

Order	Specific rate constant k	
0	$tk = (C_{1o} - C_1)$	(3.10-5)
1	$tk = \ln\ (C_{1o}/C_1)$	(3.10-11)
Pseudo 1	$tk' = \ln\ (C_{1o}/C_1)$	(3.10-17)
2 Type I	$tk = \ln\ (C_2 C_{1o}/C_{2o}C_1)/(C_{2o} - C_{1o})$	(3.10-24)
2 Type II	$tk = (1/C_1 - 1/C_{1o})/2$	(3.10-28)
3 Type I	$tk = \dfrac{\ln\,[(C_1/C_{1o})^{C_{2o}-C_{3o}}(C_2/C_{2o})^{C_{3o}-C_{1o}}(C_3/C_{3o})^{C_{1o}-C_{2o}}]}{(C_{1o} - C_{2o})(C_{2o} - C_{3o})(C_{3o} - C_{1o})}$	(3.10-34)
3 Type IIA	$tk = \dfrac{1}{(C_{2o} - C_{1o})}\left(\dfrac{1}{C_1} - \dfrac{1}{C_{1o}}\right) + \dfrac{1}{(C_{2o} - {}_{1o})^2}\ln\dfrac{C_{2o}C_1}{C_{1o}C_2}$	(3.10-40)
3 Type IIB	$tk = \dfrac{2}{2C_{2o} - C_{1o}}\left(\dfrac{1}{C_1} - \dfrac{1}{C_{1o}}\right) + \dfrac{2}{(2C_{2o} - C_{1o})^2}\ln\dfrac{C_{2o}C_1}{C_{1o}C_2}$	(3.10-47)
3 Type III	$tk = \dfrac{1}{2}(1/C_{1o}^2 - 1/C_1^2)$	(3.10-52)
n	$tk = \dfrac{1}{n-1}\left(\dfrac{1}{C_1^{n-1}} - \dfrac{1}{C_{1o}^{n-1}}\right)$	(3.10-57)

Example 3.11-1 Determination of Order

Calculate the rate constants for zero-, first-, second-orders from the experimental data for $C_{1o} = 0.01$.

t, min	f_1	$C_1 = C_{1o}(1 - f)$	k_0 Eq. (3.10-5)	k_1 Eq. (3.10-11)	k_2 Eq. (3.10-28)
1	0.049	0.00951	4.9×10^{-4}	5.1×10^{-2}	5.15
2	0.094	0.00906	4.6×10^{-4}	4.9×10^{-2}	5.07
5	0.206	0.00794	4.1×10^{-4}	4.6×10^{-2}	5.19
10	0.342	0.00658	3.4×10^{-4}	4.2×10^{-2}	5.20
15	0.438	0.00562	2.9×10^{-4}	3.8×10^{-2}	5.19
25	0.565	0.00435	2.3×10^{-4}	3.3×10^{-2}	5.20

Since k_2 is fairly constant, the order is second.

3.11-2 Half-life method

Sometimes the experimental data are more conveniently determined by half-life, defined as the time for the initial concentration of the limiting reactant reducing to half, i.e., at $t = t_{1/2}$, $C_k = \frac{1}{2}C_{ko}$.

Order	Half-life	
0	$t_{1/2} = C_{ko}/2k$	(3.11-1)
1	$t_{1/2} = (\ln 2)/k$	(3.11-2)
pseudo 1	$t_{1/2} = (\ln 2)/k'$	(3.11-3)
2 Type I	$t_{1/2} = \ln (2C_2/C_{2o})/k(C_{2o} - C_{ko})$	(3.11-4)
2 Type II	$t_{1/2} = 1/(2kC_{ko})$	(3.11-5)
3 Type I	$t_{1/2} = \dfrac{\ln [(1/2)^{C_{2o} - C_{3o}}(C_2/C_{2o})^{C_{3o} - C_{ko}}(C_3/C_{3o})^{C_{ko} - C_{2o}}]}{k(C_{ko} - C_{2o})(C_{2o} - C_{3o})(C_{3o} - C_{ko})}$	(3.11-6)
3 Type IIA	$t_{1/2} = \{1/C_{ko}(1/C_{2o} - C_{ko}) + [\ln (C_{2o}/2C_2)]/(C_{2o} - C_{ko})^2\}/k$	(3.11-7)
3 Type IIB	$t_{1/2} = \{2/C_{ko}(1/2C_{2o} - C_{ko}) + [2 \ln (C_{2o}/2C_2)]/(2C_{2o} - C_{ko})^2\}/k$	(3.11-8)
3 Type III	$t_{1/2} = -(3/2)/C_{ko}^2 k$	(3.11-9)
n	$t_{1/2} = (2^{n-1} - 1)/kC_{ko}^{n-1}(n - 1)$	(3.11-10)

From Eq. (3.11-10), it is seen that $t_{1/2}$ is proportional to C_{ko}^{1-n}. Hence plotting $t_{1/2}$ versus C_{ko} gives the slope $= 1 - n$.

3.11-3 Differential method

$$X_1 \rightarrow \text{product}$$

$$\frac{d\lambda}{dt} = k(C_{1o} - \lambda)^n \qquad \textbf{(3.11-11)}$$

or

$$\frac{-dC_1}{dt} = kC_1^n$$

Taking the logarithm gives

$$\log\left(-\frac{dC_1}{dt}\right) = n \log C_1 + \log k \tag{3.11-12}$$

Thus plotting log $(-dC_1/dt)$ versus log C_1 gives slope equal to n. The value of dC_1/dt can be approximated by $\Delta C_1/\Delta t$.

Example 3.11-2 Determination of Order by Differential Method

Determine the order of the reaction

$$-\frac{dC_1}{dt} = kC_1^m$$

from the following experimental data: (N1)

Time t, min	0	8.00	15.60	30.00	47.00	57.00	63.00
C_1, mol/liter	0.3335	0.2255	0.1632	0.1053	0.0678	0.0553	0.0482

where C_1 is the concentration of Br_2.

Solution

Calculate $-\Delta C_1/\Delta t$ and C_1 from the data:

Time t min	C_1	$-\frac{\Delta C_1}{\Delta t} \times 10^3$	C_1
0	0.3335		
		13.50	0.2795
8.00	0.2255		
		8.20	0.1944
15.60	0.1632		
		4.02	0.1343
30.00	0.1053		
		2.21	0.0866
47.00	0.0678		
		1.25	0.0616
57.00	0.0553		
		1.18	0.0518
63.00	0.0482		

From Fig. 3.11-1,

$$m = \frac{\log 11.6 - \log 1.74}{\log 0.25 - \log 0.07} = 1.4903$$

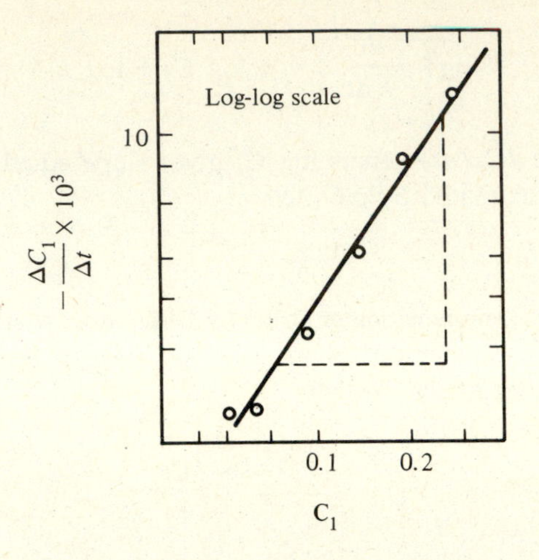

Figure 3.11-1 Order by differential method

Therefore the order of the reaction is 1.5. Sometimes the data can be analyzed by the so-called three-point, five-point, or seven-point formulas. A program for the five-point formula is included in Fig 3.11-2.

```
00100 PROGRAM FIVEPT(INPUT,OUTPUT)
00110CX=VALUE OF ORDINATE
00120CT=VALUE OF ABSCISSA
00130 READ*,T1,T2,T3,T4,T5,X1,X2,X3,X4,X5
00140 H=0.5
00150 DXDT1=(-25.*X1+48.*X2-36.*X3+16.*X4-3.*X5)/12./H
00160 DXDT2=(-3.*X1-10.*X2+18.*X3-6.*X4+X5)/12./H
00170 DXDT3=(X1-8.*X2+8.*X4-X5)/12./H
00180 DXDT4=(-X1+6.*X2-18.*X3+10.*X4+3.*X5)/12./H
00190 DXDT5=(3.*X1-16.*X2+36.*X3-48.*X4+25.*X5)/12./H
00200 PRINT*,DXDT1,DXDT2,DXDT3,DXDT4,DXDT5
00210 STOP
00220 END
READY.
RNH

? 2.,2.5,3.,3.5,4.,2.3,3.062,4.2,5.788,7.9
 1.197333333333 1.875333333333 2.701333333333 3.675333333333
 4.797333333333
```

Figure 3.11-2 Program for the five-point formula

3.11-4 Initial slope method

For a general forward reaction, the rate equation is

$$r_o = kC_{1o}^{\alpha_1}C_{2o}^{\alpha_2}\cdots = k\Pi C_{io}^{\alpha_i} \tag{3.11-13}$$

The order of reaction can be determined by making a rate measurement at two different initial concentrations of that species to which the order is determined while holding all other concentrations constant between the two runs. Let the other run be r'_o, then

$$r_o = (kC_{2o}^{\alpha_2}C_{3o}^{\alpha_3}\cdots)C_{1o}^{\alpha_1} \tag{3.11-14}$$

$$r'_o = (kC_{2o}^{\alpha_2}C_{3o}^{\alpha_3}\cdots)C'^{\alpha_1}_{1o} \tag{3.11-15}$$

Dividing and solving gives

$$\alpha_1 = \frac{\log(r_o/r'_o)}{\log(C_{1o}/C'_{1o})} \tag{3.11-16}$$

The order of reaction with respect to other species can be determined in the same manner.

Example 3.11-3 Determination of Order by Initial Slope Method

The data for the reaction (F1)

$$4HCl + CaMg(CO_3)_2 \rightarrow Mg^{2+} + Ca^{2+} + 4Cl^- + 2CO_2 + 2H_2O$$

are as follows:

Volume of the solution = 675 ml; area of dolomite surface = 20.2 cm².

Run I: 1 *N* HCl

Time t, min	0	2	4	6	8
C_{HCl}, g mol/liter	4.0000	3.9993	3.9986	3.9979	3.9968

Run II: 1 *N* HCl

Time t, min	0	2	4	6	8
C_{HCl}, g mol/liter	1.0000	0.9996	0.9991	0.9986	0.9980

Solution

The initial rates can be calculated as:

Run I:

$$r_o = \frac{3.9968 - 4.0000}{8}\left(\frac{0.6751}{20.2\ \text{cm}^2}\right)\left(\frac{1\ \text{min}}{60\ \text{s}}\right) = -2.228 \times 10^{-7}\ \text{g mol/s}\cdot\text{cm}^2$$

Run II:

$$r'_o = \frac{0.9980 - 1.0000}{8}\left(\frac{0.6751}{20.2\ \text{cm}^2}\right)\left(\frac{1\ \text{min}}{60\ \text{s}}\right) = -1.392 \times 10^{-7}\ \text{g mol/s}\cdot\text{cm}^2$$

Hence

$$\underline{\alpha}_1 = \frac{\log(-2.228 \times 10^{-7}/-1.392 \times 10^{-7})}{\log(4/1)} = -0.3393^*$$

3.11-5 Isolation method

By using large excess concentrations, the general forward reaction equation

$$-\frac{dC_1}{dt} = k_n C_1^{\underline{\alpha}_1} C_2^{\underline{\alpha}_2} \cdots C_n^{\underline{\alpha}_n} \tag{3.11-17}$$

can be reduced to

$$-\frac{dC_1}{dt} = k_s C_1^{\underline{\alpha}_1} \tag{3.11-18}$$

where

$$k_s = k_n C_2^{\underline{\alpha}_2} C_3^{\underline{\alpha}_3} \cdots C_n^{\underline{\alpha}_n}$$

Hence the order of reaction $\underline{\alpha}_1$ and the rate constant k_s can be determined.

3.11-6 Reference curve method

The rate equation for a general nth-order reaction is

$$\frac{1}{C_t^{n-1}} - \frac{1}{C_o^{n-1}} = (n-1)k_n t \tag{3.11-19}$$

---

* This calculation is based on the assumption that all the data points fall on straight lines. If a more accurate result is desired, the values of r_o and r'_o should be calculated more accurately by some other means.

At 80 percent conversion, Eq. (3.11-19) becomes

$$\frac{1}{C_{0.8}^{n-1}} - \frac{1}{C_o^{n-1}} = (n-1)k_n t_{o\,0.8} \tag{3.11-20}$$

Dividing Eq. (3.11-19) by Eq. (3.11-20) gives

$$\frac{t}{t_{0.8}} = \frac{(C_o/C_t)^{n-1} - 1}{(C_o/C_{0.8})^{n-1} - 1} \tag{3.11-21}$$

Using n as parameters, a plot of percent conversion versus $t/t_{0.8}$ can be made because

$$\frac{C_{0.8}}{C_o} = \frac{1}{5} \quad \text{and} \quad \frac{C_t}{C_o} = 1 - f$$

Eq. (3.11-21) reduces to

$$\frac{t}{t_{0.8}} = \frac{[1/(1-f)]^{n-1} - 1}{5^{n-1} - 1} \quad \text{for } n \neq 1 \tag{3.11-22}$$

and

$$\frac{t}{t_{0.8}} = \frac{\ln [1/(1-f)]}{\ln 5} \quad \text{for } n = 1 \tag{3.11-23}$$

where f is the fractional conversion or percent conversion. For the experimental data using the same basis, another plot can be made on the same figure. Then the order of the reaction can be read from Fig. 3.11-3.

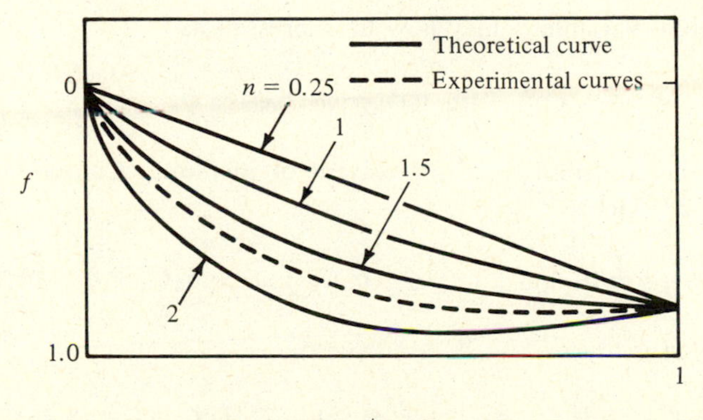

Figure 3.11-3 Reference curve

3.11-7 Physical property method

Sometimes it is not easy to measure the concentration of the reactants and therefore the rate of the reaction can hardly be calculated. However, for some of the reaction, measuring one of the physical properties of some of the reactants during the course of the reaction is quite feasible. If the relationship of the concentration with the physical property is known, the order of the reaction can be determined. Let us denote the physical property by ϕ, which is measured experimentally and is the sum of the contribution of all the ith species

$$\phi = \sum_i \phi_i \tag{3.11-24}$$

Assume this individual physical property to be linear with the concentration, i.e.,

$$\phi_i = a_i + b_i C_i \tag{3.11-25}$$

where a_i and b_i are empirical constants characteristic of species i. Let us consider the following general reaction

$$\alpha_1 X_1 + \alpha_2 X_2 \rightarrow \alpha_3 X_3 + \alpha_4 X_4$$

or

$$\sum_i \alpha_i X_i = 0$$

The concentration of species i at any time is

$$C_i = C_{io} + \alpha_i \lambda \tag{3.11-26}$$

The experimental variable ϕ may now be expressed as

$$\phi = \phi_M + \phi_1 + \phi_2 + \phi_3 + \phi_4 \tag{3.11-27}$$

where ϕ_M is the contribution of the solvent or medium. Combining Eqs. (3.11-25) and (3.11-27) yields

$$\phi = \phi_M + a_1 + b_1 C_1 + a_2 + b_2 C_2 + a_3 + b_3 C_3 + a_4 + b_4 C_4 \tag{3.11-28}$$

The initial value is

$$\phi = \phi_M + a_1 + b_1 C_{1o} + a_2 + b_2 C_{2o} + a_3 + b_3 C_{3o} + a_4 + b_4 C_{4o} \tag{3.11-29}$$

Subtracting Eq. (3.11-29) from Eq. (3.11-28) gives

$$\phi - \phi_o = b_1(C_1 - C_{1o}) + b_2(C_2 - C_{2o}) + b_3(C_3 - C_{3o}) + b_4(C_4 - C_{4o}) \quad \textbf{(3.11-30)}$$

Using Eq. (3.11-26) gives

$$\phi - \phi_o = b_1\alpha_1\lambda + b_2\alpha_2\lambda + b_3\alpha_3\lambda + b_4\alpha_4\lambda = \lambda \sum_i \alpha_i b_i \quad \textbf{(3.11-31)}$$

At infinity

$$\phi_\infty - \phi_o = \lambda_\infty \sum_i \alpha_i b_i \quad \textbf{(3.11-32)}$$

The ratio of Eqs. (3.11-31) and (3.11-32) is then

$$\frac{\lambda}{\lambda_\infty} = \frac{\phi - \phi_o}{\phi_\infty - \phi_o} \quad \textbf{(3.11-33)}$$

Combining Eqs. (3.11-33) with (3.11-26) yields

$$C_i = C_{io} + \alpha_i \lambda_\infty \left(\frac{\phi - \phi_o}{\phi_\infty - \phi_o}\right) \quad \textbf{(3.11-34)}$$

These concentrations may now be used in the various integral or differential formula.

Example 3.11-4 Determination of Order by Physical Property Method

The data for the reaction (B1)

$$C_6H_5COCl + CH_3OH \rightarrow C_6H_5COOCH_3 + HCl$$

with

$$r = k' C_{C_6H_5COCl}^{\alpha_1}$$

have been obtained as follows:

Time t, s	0	55	79	100	120	10,800
Solution conductivity, $(10^4)\,(\Omega^{-1}\text{cm}^{-1})$	negligible	0.732	0.969	1.21	1.40	3.50

Solution

Since the reaction is pseudo first-order, Eq. (3.10-17) applies. Substituting

$$C_1 = C_{1o} - \lambda$$

into Eq. (3.10-17) gives

$$\ln\left(1 - \frac{\lambda}{C_{1o}}\right) = -k't$$

But

$$\frac{\mathrm{H} - \mathrm{H}_o}{\mathrm{H}_\infty - \mathrm{H}_o} = \frac{\lambda}{\lambda_\infty} = \frac{\lambda}{C_{1o}} \tag{A}$$

where H is the conductivity and H_∞ is taken to be the reading at 10,800 s.

Then

$$\ln\left(1 - \frac{\mathrm{H} - \mathrm{H}_o}{\mathrm{H}_\infty - \mathrm{H}_o}\right) = -k't$$

or

$$\ln\frac{\mathrm{H}_\infty - \mathrm{H}}{\mathrm{H}_\infty - \mathrm{H}_o} = -k't$$

$$\ln(\mathrm{H}_\infty - \mathrm{H}) = \ln(\mathrm{H}_\infty - \mathrm{H}_o) - k't$$

Therefore a plot of $\ln(\mathrm{H}_\infty - \mathrm{H})$ versus t gives slope equal to $-k'$ (Fig. 3.11-4). It is found to be $4.3 \times 10^{-3}\ \mathrm{s}^{-1}$.

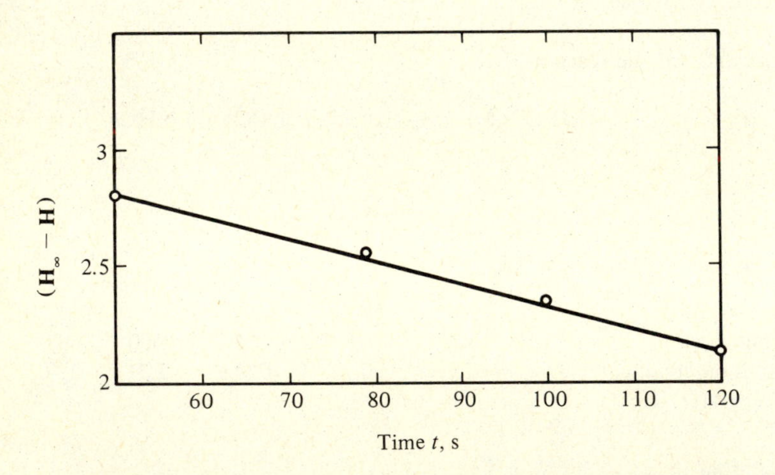

Figure 3.11-4 Determination of order by physical property method

3.12 *Reversible Reactions*

We are going to discuss first-, second-, and mixed-order reversible reactions.

3.12-1 *First-order reactions*

$$X_1 \underset{k'}{\overset{k}{\rightleftarrows}} \frac{\alpha_2}{\alpha_1} X_2$$

1. Stoichiometric equation:

$$C_1 = C_{1o} - \lambda, \; C_2 = C_{2o} + \frac{\alpha_2}{\alpha_1}\lambda \tag{3.12-1}$$

2. Overall rate equation:

$$\frac{d\lambda}{dt} = r = kC_1 - k'C_2 \tag{3.12-2}$$

3. Component rate equation:

$$\frac{dC_1}{dt} = -r \tag{3.12-3}$$

Substituting Eq. (3.12-1) into Eq. (3.12-2) yields

$$\frac{d\lambda}{dt} = k(C_{1o} - \lambda) - k'\left(C_{2o} + \frac{\alpha_2}{\alpha_1}\lambda\right) \tag{3.12-4}$$

At equilibrium

$$\frac{d\lambda}{dt} = 0 = k(C_{1o} - \lambda_e) - k'\left(C_{2o} + \frac{\alpha_2}{\alpha_1}\lambda_e\right) \tag{3.12-5}$$

and

$$K = \frac{k}{k'} \qquad C_{1e} = C_{1o} - \lambda_e \qquad C_{2e} = C_{2o} + \frac{\alpha_2}{\alpha_1}\lambda_e$$

Subtracting Eq. (3.12-5) from Eq. (3.12-4) gives

$$\frac{d\lambda}{dt} = (k + \frac{\alpha_2}{\alpha_1}k')(\lambda_e - \lambda) \tag{3.12-6}$$

Integrating from $t = 0$, $\lambda = 0$ to $t = t$, $\lambda = \lambda$ results in

$$\ln \frac{\lambda_e - \lambda}{\lambda_e} = -\left(k + \frac{\alpha_2}{\alpha_1} k'\right)t \tag{3.12-7}$$

Converting back to concentrations gives

$$\ln \frac{C_{1o} - C_{1e}}{C_1 - C_{1e}} = \left(k + \frac{\alpha_2}{\alpha_1} k'\right)t \tag{3.12-8}$$

Hence a plot of $\ln (C_1 - C_{1e})$ versus t yields a slope equal to $-[k + (\alpha_2/\alpha_1) k']$.

Example 3.12-1 First-Order Reversible Reaction

The following result was obtained for a first-order reversible reaction:

Time, min	0	180	360	540	900	∞
Polarimeter reading	115.6°	102.0°	90.5°	80.6°	65.1°	23.5°

Find the forward and reverse rate constants (C3).

Solution

Substituting $\lambda_\infty = C_{1o} - C_{1e}$ into Eq. (3.11-34) yields

$$\frac{C_1 - C_{1e}}{C_{1o} - C_{1e}} = \frac{\phi - \phi_\infty}{\phi_o - \phi_\infty} \tag{A}$$

With $\alpha_2 = \alpha_1 = 1$, substituting Eq. (A) into Eq. (3.12-8) yields

$$\ln \frac{\phi - \phi_\infty}{\phi_o - \phi_\infty} = -(k + k')t$$

A plot of $\ln(\phi - \phi_\infty)$ versus t yields slope $= -(k + k') = -8.83 \times 10^{-4}$ (see Fig. 3.12-1).

t	0	180	360	540	900	∞
ϕ	115.6	102.0	90.5	80.6	65.1	23.5
$\phi - \phi_\infty$	92.1	78.5	67.0	57.1	41.6	0
f	0	0.1176	0.2171	0.3027	0.4369	0.7969

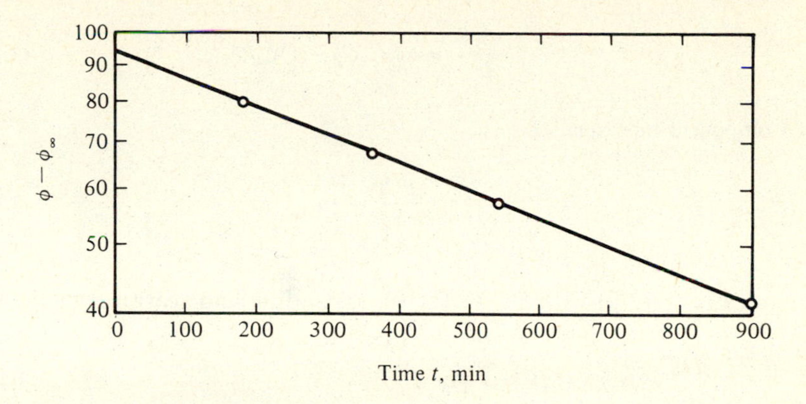

Figure 3.12-1 First-order reversible reaction

From these data, it is seen that the fractional conversion is

$$f = \frac{\phi_o - \phi}{\phi_o} \tag{B}$$

which has been calculated and included in the above table. At equilibrium, Eq. (3.12-5) can be simplified to

$$K = \frac{k}{k'} = \frac{C_{2o} + \lambda_e}{C_{1o} - \lambda_e} = \frac{C_{1o} f_e}{C_{1o} - C_{1o} f_e} = \frac{f_e}{1 - f_e} = \frac{0.7967}{1 - 0.7967} = 3.918$$

Solving with

$$k + k' = 8.83 \times 10^{-4}$$

results in

$$k = 0.000705 \text{ min}^{-1} \quad \text{and} \quad k' = 0.00018 \text{ min}^{-1}$$

3.12-2 Second-order reactions

Type I:

$$X_1 + \frac{\alpha_2}{\alpha_1} X_2 \underset{k'}{\overset{k}{\rightleftarrows}} \frac{\alpha_3}{\alpha_1} X_3 + \frac{\alpha_4}{\alpha_1} X_4$$

1. Stoichiometric equations:

$$C_1 = C_{1o} - \lambda \quad C_2 = C_{2o} - \frac{\alpha_2}{\alpha_1}\lambda \quad C_3 = C_{3o} + \frac{\alpha_3}{\alpha_1}\lambda \quad C_4 = C_{4o} + \frac{\alpha_4}{\alpha_1}\lambda \tag{3.12-9}$$

2. Overall rate equation:

$$\frac{d\lambda}{dt} = r = kC_1C_2 - k'C_3C_4 \tag{3.12-10}$$

3. Component rate equation:

$$\frac{dC_1}{dt} = -r \tag{3.12-11}$$

Combining Eq. (3.12-9) with Eq. (3.12-10), expanding and rearranging yields

$$\frac{d\lambda}{dt} = D_1\lambda^2 + D_2\lambda + D_3 \tag{3.12-12}$$

where

$$D_1 = k\frac{\alpha_2}{\alpha_1} - \frac{\alpha_3\alpha_4}{\alpha_1^2}k'$$

$$D_2 = -\left[k\left(C_{1o}\frac{\alpha_2}{\alpha_1} + C_{2o}\right) + k'\left(C_{3o}\frac{\alpha_4}{\alpha_1} + C_{4o}\frac{\alpha_3}{\alpha_1}\right)\right]$$

$$D_3 = kC_{1o}C_{2o} - k'C_{3o}C_{4o}$$

At equilibrium

$$\frac{d\lambda}{dt} = 0 = D_1\lambda_e^2 + D_2\lambda_e + D_3 \tag{3.12-13}$$

Subtracting Eq. (3.12-13) from Eq. (3.12-12) gives

$$\frac{d\lambda}{dt} = D_1(\lambda^2 - \lambda_e^2) + D_2(\lambda - \lambda_e) \tag{3.12-14}$$

Factoring

$$\frac{d\lambda}{dt} = D_1\left(\lambda + \lambda_e + \frac{D_2}{D_1}\right)(\lambda - \lambda_e) \tag{3.12-15}$$

Letting

$$\sigma = \lambda_e + \frac{D_2}{D_1}$$

and substituting

$$\frac{d\lambda}{dt} = D_1(\lambda + \sigma)(\lambda - \lambda_e) \tag{3.12-16}$$

and then integrating by partial fractions yield

$$\ln \frac{\sigma(\lambda_e - \lambda)}{\lambda_e(\sigma + \lambda)} = (\lambda_e + \sigma)D_1 t \tag{3.12-17}$$

Example 3.12-2 Second-Order Reversible Reaction

1.15 g mol/liter methyl acetate and 48.76 g mol/liter water are hydrolyzed by the catalyst HCl acid at 25°C. It was found that 34.92 percent was hydrolyzed in 1 h. Find the forward and reverse rate constants if K is 0.22.

Solution

$$\underset{(1)}{CH_3COOCH_3} + \underset{(2)}{H_2O} \underset{k'}{\overset{k}{\rightleftharpoons}} \underset{(3)}{CH_3COOH} + \underset{(4)}{CH_3OH}$$

$$C_{1o} = 1.15 \text{ g mol/liter} \qquad C_{2o} = 48.76 \text{ g mol/liter}$$

$$f = 0.3492 \qquad K = 0.22 \qquad \lambda = C_{1o}f = 1.15(0.3492) = 0.4016$$

$$D_1 = k - k' = 0.22k' - k' = -0.78k'$$

$$D_2 = -[k(1.15 + 48.76)] = -49.91k = -10.98k'$$

$$D_3 = k(1.15)(48.76) = 56.074k = 12.34k'$$

Substituting into Eq. (3.12-12) gives

$$\frac{d\lambda}{dt} = -0.78k'\lambda^2 - 10.98k'\lambda + 12.34k'$$

At equilibrium

$$-0.78\lambda_e^2 - 10.98\lambda_e + 12.34 = 0$$

Solving

$$\lambda_e = 1.046 \quad \text{(the other value is rejected)}$$

$$\sigma = \lambda_e + \frac{D_2}{D_1} = 1.046 + \frac{(-10.98k')}{(-0.78k')} = 15.1229$$

Substituting into Eq. (3.12-17) yields

$$\ln \frac{15.1229(1.046 - 0.4016)}{1.046(15.1229 + 0.4016)} = (1.046 + 15.1229)(-0.78k')(60)$$

Solving for k'

$$k' = 6.7477 \times 10^{-4} \text{ liter/g mol} \cdot \text{min}$$

Then

$$k = 0.22k' = 1.4845 \times 10^{-4} \text{ liter/g mol} \cdot \text{min}$$

Type II:

$$X_1 + \frac{\alpha_2}{\alpha_1} X_2 \underset{k'}{\overset{k}{\rightleftharpoons}} 2X_3$$

1. Stoichiometric equations (assuming $\alpha_2/\alpha_1 = 1$):

$$C_1 = C_{1o} - \lambda, \quad C_2 = C_{2o} - \lambda, \quad C_3 = C_{3o} + 2\lambda \tag{3.12-18}$$

2. Overall rate equation:

$$\frac{d\lambda}{dt} = r = kC_1C_2 - k'C_3^2 \tag{3.12-19}$$

3. Component rate equation:

$$\frac{dC_1}{dt} = -r \tag{3.12-20}$$

Combining Eq. (3.12-18) with Eq. (3.12-19), expanding, and rearranging yields

$$\frac{d\lambda}{dt} = D_1\lambda^2 + D_2\lambda + D_3 \tag{3.12-21}$$

where

$$D_1 = k - 4k'$$

$$D_2 = -[(C_{1o} + C_{2o})k + 4C_{3o}k']$$

$$D_3 = C_{1o}C_{2o}k - k'C_{3o}^2$$

The final result is the same as Eq. (3.12-17) with these new D values.

Example 3.12-3 Second-Order Type II Reversible Reaction

5.5 kmol/m³ of sulfuric acid reacts with 5.5 kmol/m³ of diethyl sulfate at 22.9°C. The following data were obtained:

$$H_2SO_4 + (C_2H_5)_2SO_4 \underset{k'}{\overset{k}{\rightleftharpoons}} 2C_2H_5SO_4H$$

(1) (2) (3)

Time t, s	0	5,760	10,800	19,080	22,740	24,600	∞
C_3, kmol/m³	0	2.75	4.11	5.15	5.35	5.42	5.80

Find the forward and reverse rate constants (H1).

Solution

$$C_{1o} = C_{2o} = 5.5 \text{ kmol/m}^3 \qquad C_{3e} = 5.8 \text{ kmol/m}^3$$

Because

$$C_{3e} = C_{3o} + 2\lambda_e$$

$$\lambda_e = \frac{1}{2}(C_{3e} - C_{3o}) = \frac{1}{2}(5.8 - 0) = 2.9 \text{ kmol/m}^3$$

$$C_{1e} = C_{1o} - \lambda_e = 5.5 - 2.9 = 2.6 \text{ kmol/m}^3$$

$$C_{2e} = C_{2o} - \lambda_e = 5.5 - 2.9 = 2.6 \text{ kmol/m}^3$$

$$K = \frac{k}{k'} = \frac{C_{3e}^2}{(C_{1e})(C_{2e})} = \frac{5.8^2}{2.6(2.6)} = 4.9763$$

Then

$$k = 4.9763k'$$

$$D_1 = k - 4k' = 0.9763k'$$

$$D_2 = -(5.5 + 5.5)(4.9763)k' = -54.7393k'$$

$$D_3 = 5.5(5.5)(4.9763k') = 150.533k'$$

$$\sigma = 2.9 + \frac{-54.7393k'}{0.9763k'} = -53.1681$$

Substituting into Eq. (3.12-17) gives

$$\ln \frac{-53.1681(2.9 - \lambda)}{2.9(\lambda - 53.1681)} = -50.2681 D_1 t = -50.2681(0.9763k')t$$

For each value of t, the experimental value of C_3 is given. Hence λ can be calculated ($= C_3/2$) in this case. Thus k' can be evaluated at each time. The average value of k' for the given five runs is 2.235×10^{-6} m³/kmol · s and the average $k = 1.11 \times 10^{-5}$ m³/kmol · s.

3.12-3 Mixed-order reactions

Some reversible reactions may be first-order in the forward reaction and second-order in the reverse reaction or vice versa. The treatment is the same. For example, let us consider the following elementary reaction:

$$X_1 + X_2 \underset{k'}{\overset{k}{\rightleftharpoons}} X_3$$

1. Stoichiometric equations:

$$C_1 = C_{1o} - \lambda, \ C_2 = C_{2o} - \lambda, \ C_3 = C_{3o} + \lambda \qquad \textbf{(3.12-22)}$$

2. Overall rate equation:

$$\frac{d\lambda}{dt} = r = kC_1C_2 - k'C_3 \qquad \textbf{(3.12-23)}$$

3. Component rate equation:

$$\frac{dC_1}{dt} = -r \qquad \textbf{(3.12-24)}$$

Combining Eq. (3.12-22) with Eq. (3.12-23), expanding, and rearranging yields

$$\frac{d\lambda}{dt} = D_1\lambda^2 + D_2\lambda + D_3$$

where

$$D_1 = k$$

$$D_2 = -[k(C_{1o} + C_{2o}) + k'C_{3o}]$$

$$D_3 = kC_{1o}C_{2o} - k'C_{3o}$$

With these new D values, the final result can also be obtained by Eq. (3.12-17).

3.13 Multiple Reactions

3.13-1 Irreversible concurrent, parallel, and side simultaneous reactions

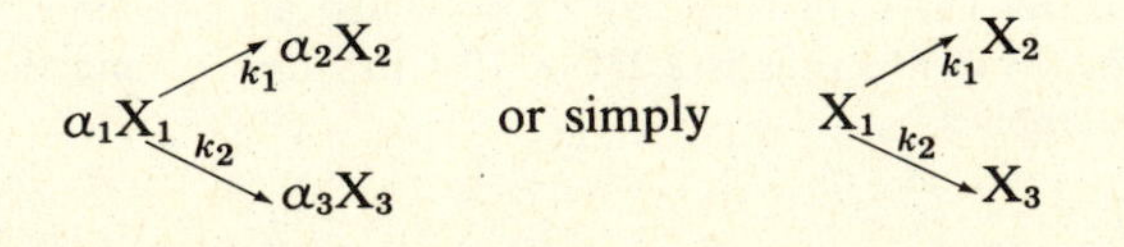

1. Stoichiometric equation:

$$C_j = C_{jo} + \Sigma\alpha_{ij}\lambda_i \tag{3.13-1}$$

$$C_1 = C_{1o} + \alpha_{11}\lambda_1 + \alpha_{21}\lambda_2 = C_{1o} - \lambda_1 - \lambda_2 \tag{3.13-2}$$

$$C_2 = C_{2o} + \alpha_{12}\lambda_1 = C_{2o} + \lambda_1 \tag{3.13-3}$$

$$C_3 = C_{3o} + \alpha_{23}\lambda_2 = C_{3o} + \lambda_2 \tag{3.13-4}$$

2. Overall rate equations:

$$\frac{d\lambda_i}{dt} = \mathrm{r}_i = k_i C_j \tag{3.13-5}$$

$$\frac{d\lambda_1}{dt} = r_1 = k_1 C_1 \tag{3.13-6}$$

$$\frac{d\lambda_2}{dt} = r_2 = k_2 C_1 \tag{3.13-7}$$

3. Component rate equations:

$$\frac{dC_j}{dt} = \Sigma\alpha_{ij} r_i = \Sigma\alpha_{ij}\frac{d\lambda_i}{dt} \tag{3.13-8}$$

$$\frac{dC_1}{dt} = \alpha_{11}\frac{d\lambda_1}{dt} + \alpha_{21}\frac{d\lambda_2}{dt} = \alpha_{11}r_1 + \alpha_{21}r_2$$

$$= -\frac{d\lambda_1}{dt} - \frac{d\lambda_2}{dt} = -r_1 - r_2 = -k_1C_1 - k_2C_1 \tag{3.13-9}$$

$$\frac{dC_2}{dt} = \alpha_{12}\frac{d\lambda_1}{dt} + \alpha_{22}\frac{d\lambda_2}{dt} = \alpha_{12}r_1 + \alpha_{22}r_2$$

$$= \frac{d\lambda_1}{dt} + \frac{d\lambda_2}{dt} = r_1 + 0 = k_1C_1 \tag{3.13-10}$$

$$\frac{dC_3}{dt} = \alpha_{13}\frac{d\lambda_1}{dt} + \alpha_{23}\frac{d\lambda_2}{dt} = \alpha_{13}r_1 + \alpha_{23}r_2$$

$$= \frac{d\lambda_1}{dt} + \frac{d\lambda_2}{dt} = +r_2 = k_2C_1 \tag{3.13-11}$$

Integrating Eq. (3.13-9) gives

$$\ln\frac{C_1}{C_{1o}} = -(k_1 + k_2)t \quad \text{or} \quad \frac{C_1}{C_{1o}} = \exp(-k_1 + k_2)t \tag{3.13-12}$$

Substituting Eq. (3.13-12) into Eq. (3.13-10) and rearranging give

$$\frac{dC_2}{dt} = k_1 C_{1o} \exp[-(k_1 + k_2)t] \tag{3.13-13}$$

Integrating yields

$$C_2 = C_{2o} + \frac{k_1 C_{1o}}{k_1 + k_2} \{1 - \exp[-(k_1 + k_2)t]\} \tag{3.13-14}$$

Substituting Eq. (3.13-12) into Eq. (3.13-11) and rearranging give

$$\frac{dC_3}{dt} = k_2 C_{1o} \exp[-(k_1 + k_2)t] \tag{3.13-15}$$

Integrating yields

$$C_3 = C_{3o} + \frac{k_2 C_{1o}}{k_1 + k_2} (1 - e^{(-k_1 + k_2)t}) \tag{3.13-16}$$

Example 3.13-1 Irreversible Concurrent Reactions

3 kmol/m^3 of nitric acid reacted with 1 kmol/m^3 of mononitrobenzene to produce *o*-, *m*-, and *p*-dinitrobenzene. After 20 min, the mononitrobenzene was half used up and the dinitrobenzenes were present in the proportions of 6.4, 93.5, and 0.1, respectively. Find the specific rate constants for all these three second-order parallel reactions (W1).

$$HNO_3 + C_6H_5NO_2 \overset{k1}{\rightleftharpoons} o\text{-}C_6H_4(NO_2)_2 + H_2O$$

$$HNO_3 + C_6H_5NO_2 \overset{k2}{\rightleftharpoons} m\text{-}C_6H_4(NO_2)_2 + H_2O$$

$$HNO_3 + C_6H_5NO_2 \overset{k3}{\rightleftharpoons} p\text{-}C_6H_4(NO_2)_2 + H_2O$$

Solution

1. Stoichiometric equations:

$$HNO_3: \quad C_1 = C_{1o} - \lambda_1 - \lambda_2 - \lambda_3 \tag{A}$$

$$C_6H_5NO_2: \quad C_2 = C_{2o} - \lambda_1 - \lambda_2 - \lambda_3 \tag{B}$$

$$o\text{-}C_6H_4(NO_2)_2: \quad C_3 = C_{3o} + \lambda_1 \tag{C}$$

$$m\text{-}C_6H_4(NO_2)_2: \quad C_4 = C_{4o} + \lambda_2 \tag{D}$$

$$p\text{-}C_6H_4(NO_2)_2: \quad C_5 = C_{5o} + \lambda_3 \tag{E}$$

$$H_2O: \quad C_6 = C_{6o} + \lambda_1 + \lambda_2 + \lambda_3 \tag{F}$$

2. Overall rate equations:

$$\frac{d\lambda_1}{dt} = r_1 = k_1 C_1 C_2 \tag{G}$$

$$\frac{d\lambda_2}{dt} = r_2 = k_2 C_1 C_2 \tag{H}$$

$$\frac{d\lambda_3}{dt} = r_3 = k_3 C_1 C_2 \tag{I}$$

3. Component rate equation:

$$\frac{dC_1}{dt} = -(k_1 + k_2 + k_3) C_1 C_2 \tag{J}$$

$$\frac{dC_2}{dt} = -(k_1 + k_2 + k_3) C_1 C_2 \tag{K}$$

$$\frac{dC_3}{dt} = k_1 C_1 C_2 \tag{L}$$

$$\frac{dC_4}{dt} = k_2 C_1 C_2 \tag{M}$$

$$\frac{dC_5}{dt} = k_3 C_1 C_2 \tag{N}$$

$$\frac{dC_6}{dt} = (k_1 + k_2 + k_3) C_1 C_2 \tag{O}$$

From Eqs. (A) and (B), we obtain

$$C_2 = C_1 - C_{1o} + C_{2o} = C_1 - 3 + 1 = C_1 - 2 \tag{P}$$

Substituting into Eq. (J) and integrating from $t = 0$, $C_1 = C_{1o} = 3$ by partial fractions gives

$$\ln \frac{3(C_1 - 2)}{C_1} = -2(k_1 + k_2 + k_3)t \tag{Q}$$

From Eqs. (L), (M), (N), it is seen that

$$\frac{dC_3}{k_1} = \frac{dC_4}{k_2} = \frac{dC_5}{k_3} \tag{R}$$

Integrating from $C_{3o} = C_{4o} = C_{5o} = 0$ and inserting the given data yields

$$k_2 = \frac{93.5}{6.4} k_1 \qquad \text{and} \qquad k_3 = \frac{0.1}{6.4} k_1 \tag{S}$$

At t = 20 min, half of the nitrobenzene is reacted. Hence

$$\lambda = \frac{f}{C_{2o}} = \frac{0.5}{1} = 0.5 \tag{T}$$

$$C_2 = C_{2o} - \lambda = 1 - 0.5 = 0.5 \tag{U}$$

$$C_1 = C_{1o} - \lambda = 3 - 0.5 = 2.5 \tag{V}$$

Solving Eqs. (Q), (S), and (V) for t = 20 min results in

$$k_1 = 0.000817 \text{ m}^3/\text{kmol}\cdot\text{min} \qquad k_2 = 0.01194 \text{ m}^3/\text{kmol}\cdot\text{min}$$
$$k_3 = 1.27 \times 10^{-5} \text{ m}^3/\text{kmol}\cdot\text{min}$$

3.13.2 Irreversible consecutive, series simultaneous reaction

Type I: First-Order for Both Reactions

$$\alpha_1 X_1 \xrightarrow{k_1} \alpha_2 X_2 \xrightarrow{k_2} \alpha_3 X_3$$

or simply

$$X_1 \xrightarrow{k_2} X_2 \xrightarrow{k_2} X_3$$

1. Stoichiometric equations:

$$C_j = C_{jo} + \lambda_i \alpha_{ij}$$

$$C_1 = C_{1o} + \alpha_{11}\lambda_1 + \alpha_{21}\lambda_2 = C_{1o} - \lambda_1 \tag{3.13-17}$$

$$C_2 = C_{2o} + \alpha_{12}\lambda_1 + \alpha_{22}\lambda_2 = C_{2o} + \lambda_1 - \lambda_2 \tag{3.13-18}$$

$$C_3 = C_{3o} + \alpha_{13}\lambda_1 + \alpha_{23}\lambda_2 = C_{3o} \qquad + \lambda_2 \tag{3.13-19}$$

2. Overall rate equations:

$$\frac{d\lambda_1}{dt} = r_1 = k_1 C_1 \tag{3.13-20}$$

$$\frac{d\lambda_2}{dt} = r_2 = k_2 C_2 \tag{3.13-21}$$

3. Component rate equations:

$$\frac{dC_1}{dt} = \alpha_{11}\frac{d\lambda_1}{dt} = \alpha_{11}r_1 = -r_1 = -\ k_1C_1 \tag{3.13-22}$$

$$\frac{dC_2}{dt} = \alpha_{12}\frac{d\lambda_1}{dt} + \alpha_{22}\frac{d\lambda_2}{dt} = \alpha_{12}r_1 + \alpha_{22}r_2 = r_1 - r_2 = k_1C_1 - k_2C_2 \tag{3.13.23}$$

$$\frac{dC_3}{dt} = \alpha_{23}\frac{d\lambda_2}{dt} = \alpha_{23}r_2 = r_2 = k_2C_2 \tag{3.13-24}$$

Taking the Laplace transform of Eq. (3.13-22) with $C_1(0) = C_{1o}$ yields

$$s\mathbf{C}_1 - C_{1o} = -k_1\mathbf{C}_1 \quad \text{or} \quad \mathbf{C}_1 = \frac{C_{1o}}{s + k_1} \tag{3.13-25}$$

Inverting gives

$$C_1 = C_{1o}e^{-k_1 t} \tag{3.13-26}$$

Taking the Laplace transform of Eq. (3.13-23) with $C_2(0) = 0$ yields

$$s\mathbf{C}_2 = k_1\mathbf{C}_1 - k_2\mathbf{C}_2 \quad \text{or} \quad (s + k_2)\mathbf{C}_2 = k_1\mathbf{C}_1$$

or

$$\mathbf{C}_2 = \frac{k_1C_{1o}}{(s + k_2)(s + k_1)} \tag{3.13-27}$$

Inverting Eq. (3.13-27) gives

$$C_2 = C_{1o}k_1\frac{e^{-k_1 t} - e^{-k_2 t}}{k_2 - k_1} \tag{3.13-28}$$

Taking the Laplace transform of Eq. (3.13-24) with $C_3(0) = 0$ yields

$$s\mathbf{C}_3 = k_2\mathbf{C}_2 \quad \text{or} \quad \mathbf{C}_3 = \frac{k_2\mathbf{C}_2}{s} = \frac{k_1k_2C_{1o}}{s(s + k_1)(s + k_2)} \tag{3.13-29}$$

Inverting Eq. (3.13-29) gives

$$C_3 = \frac{C_{1o}}{1/k_1 - 1/k_2}\left(\frac{1 - e^{-k_1 t}}{k_1} - \frac{1 - e^{-k_1 t}}{k_2}\right) \tag{3.13-30}$$

Example 3.13-2 Irreversible Consecutive Reaction

One kmol per cubic meter of isopropylbenzene undergoes the following first-order consecutive series reactions: (A2)

$$\text{isopropyldibenzene} \xrightarrow{k_1} \text{isopropyl-}sec\text{-butylbenzene} \xrightarrow{k_2} \text{isopropyldi-}sec\text{-butylbenzene}$$

(1) (2) (3)

At 1.5 min, the concentrations are $C_1 = 0.2231$ kmol/m³ and $C_2 = 0.6769$ kmol/m³. Calculate the rate constants k_1 and k_2. What is then the concentration of 3? What maximum concentration of 2 can be obtained if the reaction is allowed to continue? What is t_{max}?

Solution

k_1, k_2, and C_3 can be calculated from Eq. (3.13-26), Eq. (3.13-28) by trial and error, and Eq. (3.13-30), respectively, to yield

$$k_1 = 1 \text{ min}^{-1} \qquad k_2 = 0.15 \text{ min}^{-1} \qquad C_3 = 0.0999 \text{ kmol/m}^3$$

Since there is a maximum for C_2, differentiating C_2 in Eq. (3.13-28) with respect to time t, equating to zero, and solving for t yields

$$t_{max} = \frac{\ln (k_1/k_2)}{k_1 - k_2} = 2.23 \text{ min} \tag{3.13-31}$$

Eliminating t in Eq. (3.13-28) by Eq. (3.13-31) gives

$$C_{2,max} = C_{1o}\left(\frac{k_1}{k_2}\right)^{k_2/(k_2-k_1)} = 1\left(\frac{1}{0.15}\right)^{0.15/(0.15-1)} = 0.7155 \tag{3.13-32}$$

Type II: Second-Order in First Reaction, First-Order in Second Reaction

$$X_1 \xrightarrow{k_1} X_2 \xrightarrow{k_2} X_3$$

1. Stoichiometric equations:

$$C_j = C_{jo} + \Sigma \alpha_{ij} \lambda_i$$

$$C_1 = C_{1o} + \alpha_{11}\lambda_1 + \alpha_{21}\lambda_2 = C_{1o} - \lambda_1 \tag{3.13-33}$$

$$C_2 = C_{2o} + \alpha_{12}\lambda_1 + \alpha_{22}\lambda_2 = C_{2o} + \lambda_1 - \lambda_2 \tag{3.13-34}$$

$$C_3 = C_{3o} + \alpha_{13}\lambda_1 + \alpha_{23}\lambda_2 = C_{3o} \qquad + \lambda_2 \tag{3.13-35}$$

2. Overall rate equations:

$$\frac{d\lambda_1}{dt} = r_1 = k_1 C_1^2 \tag{3.13-36}$$

$$\frac{d\lambda_2}{dt} = r_2 = k_2 C_2 \tag{3.13-37}$$

3. Component rate equations:

$$\frac{dC_1}{dt} = \alpha_{11}\frac{d\lambda_1}{dt} = \alpha_{11}r_1 = -r_1 = -k_1C_1^2 \qquad \textbf{(3.13-38)}$$

$$\frac{dC_2}{dt} = \alpha_{12}\frac{d\lambda_1}{dt} + \alpha_{22}\frac{d\lambda_2}{dt} = \alpha_{12}r_1 + \alpha_{22}r_2 = r_1 - r_2 = k_1C_1^2 - k_2C_2 \qquad \textbf{(3.13-39)}$$

$$\frac{dC_3}{dt} = \alpha_{23}\frac{d\lambda_2}{dt} = \alpha_{23}r_2 = r_2 = k_2C_2 \qquad \textbf{(3.13-40)}$$

Integrating Eq. (3.13-38) from $t = 0$, $C_1 = C_{1o}$ to $t = t$, $C_1 = C_1$ gives

$$C_1 = \frac{C_{1o}}{1 + k_1C_{1o}t} \qquad \textbf{(3.13-41)}$$

Substituting Eq. (3.13-41) into Eq. (3.13-39) yields

$$\frac{dC_2}{dt} + k_2C_2 = k_1\left(\frac{C_{1o}}{1 + k_1C_{1o}t}\right)^2 \qquad \textbf{(3.13-42)}$$

Multiplying Eq. (3.13-42) by e^{k_2t} and integrating

$$e^{k_2t}C_2 = \int_0^t e^{k_2t}k_1\left(\frac{C_{1o}}{1 + C_{1o}k_1t}\right)^2 dt \qquad \textbf{(3.13-43)}$$

Letting

$$1 + C_{1o}k_1t = z \quad \text{or} \quad t = \frac{z-1}{(C_{1o}k_1)}$$

then

$$dt - \frac{1}{C_{1o}k_1}dz \qquad \textbf{(3.13-44)}$$

Substituting into Eq. (3.13-43) and rearranging yields

$$C_2 = e^{-k_2t}\int_1^z e^{k_2(z-1)/k_1C_{10}}C_{10}\frac{dz}{z^2} \qquad \textbf{(3.13-45)}$$

$$= C_{1o}e^{-k_2t}e^{-(k_2/k_1C_{10})}\int_1^z e^{k_2z/k_1C_{10}}\frac{dz}{z^2} \qquad \textbf{(3.13-46)}$$

Letting

$$w = \frac{k_2z}{k_1C_{1o}}$$

then Eq. (3.13-46) can be reduced to

$$C_2 = C_{10}e^{-k_2 t}e^{-(k_2/k_1C_{10})}\int_{k_2/k_1C_{10}}^{w} \frac{e^w k_1 C_{10}}{(k_1C_{10}w/k_2)^2 k_2} \tag{3.13-47}$$

$$= C_{10}e^{-k_2 t}e^{-(k_2/k_1C_{10})}\left(\frac{k_2}{k_1C_{10}}\right)\int_{k_2/k_1C_{10}}^{w} \frac{e^w\,dw}{w^2} \tag{3.13-48}$$

Integrating by parts

$$C_2 = \frac{-e^w}{w} + \frac{k_1C_{10}}{k_2}e^{(k_2/k_1C_{10})} + \int_{-\infty}^{w}\frac{e^w\,dw}{w} - \int_{-\infty}^{k_2/k_1C_{10}}\frac{e^w\,dw}{w}$$

$$= \frac{-e^w}{w} + \frac{k_1C_{10}}{k_2}e^{k_2/k_1C_{10}} + Ei(w) - Ei\left(\frac{k_2}{k_1C_{10}}\right) \tag{3.13-49}$$

where

$$Ei(w) = \int_{-\infty}^{w}\frac{e^w}{w}\,dw$$

Substituting back for the original variable t yields

$$C_2 = \frac{-C_{10}}{1 + C_{10}k_1 t} + C_{10}e^{-k_2 t}$$
$$+ \frac{k_2}{k_1}e^{-k_2 t-(k_2/k_1C_{10})}\left[Ei\left(\frac{k_2}{k_1C_{10}} + k_2 t\right) - Ei\left(\frac{k_2}{k_1C_{10}}\right)\right] \tag{3.13-50}$$

$$C_3 = C_{10} - C_1 - C_2 = C_{10}\left\{1 - e^{-k_2 t} + \frac{k_2}{k_1C_{10}}e^{(-k_2/k_1C_{10})-k_2 t}\right.$$
$$\left.\left[Ei\left(k_2 t + \frac{k_2}{k_1C_{10}}\right) - Ei\left(\frac{k_2}{k_1C_{10}}\right)\right]\right\} \tag{3.13-51}$$

3.14 Complex Reactions

3.14-1 Reversible parallel or consecutive reaction

$$X_1 \underset{k_1'}{\overset{k_1}{\rightleftharpoons}} X_2 \underset{k_2'}{\overset{k_2}{\rightleftharpoons}} X_3$$

1. Stoichiometric equations:

$$C_1 = C_{1o} - \lambda_1 \tag{3.14-1}$$

$$C_2 = C_{2o} + \lambda_1 - \lambda_2 \tag{3.14-2}$$

$$C_3 = C_{3o} \qquad + \lambda_2 \tag{3.14-3}$$

2. Overall rate equations:

$$\frac{d\lambda_1}{dt} = r_1 = k_1 C_1 - k_1' C_2 \tag{3.14-4}$$

$$\frac{d\lambda_2}{dt} = r_2 = k_2 C_2 - k_2' C_3 \tag{3.14-5}$$

3. Component rate equations:

$$\frac{dC_1}{dt} = -r_1 = -(k_1 C_1 - k_1' C_2) \tag{3.14-6}$$

$$\frac{dC_2}{dt} = r_1 - r_2 = (k_1 C_1 - k_1' C_1) - (k_2 C_2 - k_2' C_3) \tag{3.14-7}$$

$$\frac{dC_3}{dt} = r_2 = k_2 C_2 - k_2' C_3 \tag{3.14-8}$$

The solution of Eqs. (3.14-6) to (3.14-8) with the initial conditions $C_1(0) = C_{1o}$, $C_2(0) = C_3(0) = 0$ can be accomplished by the use of matrices, Laplace transformation, differential operator, etc. The complete solution, which is left as an exercise because it is too complicated, is as follows:

$$C_1 = C_{1o}\left\{\frac{k_2' k_1'}{\alpha\beta} + \frac{k_1 k_1'}{(\alpha-\beta)}\left[\frac{(\alpha + k_2')e^{\alpha t}}{\alpha(\alpha + k_1)} - \frac{(\beta + k_2')e^{\beta t}}{\beta(\beta + k_1)}\right]\right\} \tag{3.14-9}$$

$$C_2 = C_{1o}\left\{\frac{k_1 k_2'}{\alpha\beta} + \frac{k_1}{(\alpha-\beta)}\left[\frac{(\alpha + k_2')e^{\alpha t}}{\alpha} - \frac{(\beta + k_2')e^{\beta t}}{\beta}\right]\right\} \tag{3.14-10}$$

$$C_3 = C_{1o}\left\{\frac{k_1 k_2}{\alpha\beta} + \frac{k_1 k_2}{(\alpha-\beta)}\left[\frac{e^{\alpha t}}{\alpha} - \frac{e^{\beta t}}{\beta}\right]\right\} \tag{3.14-11}$$

where α and β are determined from the following equations:

$$\alpha \times \beta = k_1 k_2' + k_1 k_2 + k_1' k_2'$$

$$\alpha + \beta = -(k_1 + k_1' + k_2 + k_2')$$

Example 3.14-1 Parallel Reversible Reaction

Calculate the concentrations of the species for the reaction in Section 3.14-1 if $k_1 = 1\ \text{min}^{-1}$, $k_1' = 2\ \text{min}^{-1}$, $k_2 = 3\ \text{min}^{-1}$, $k_2' = 4\ \text{min}^{-1}$ and $C_{1o} = 1\ \text{kmol/m}^3$

Solution

$$\alpha \times \beta = (1)(4) + (1)(3) + (2)(4) = 15$$

$$\alpha + \beta = -(1 + 2 + 3 + 4) = -10$$

Hence

$$\alpha = -8.1622 \quad \text{and} \quad \beta = -1.8377$$

$$C_1 = \frac{8}{15} + 0.0225e^{-8.1622t} + 0.444e^{-1.8377t}$$

$$C_2 = \frac{4}{15} - 0.0806e^{-8.1622t} - 0.1865e^{-1.8377t}$$

$$C_3 = \frac{3}{15} + 0.0581e^{-8.1622t} - 0.2580e^{-1.8377t}$$

Substituting the value of the time t into these equations gives the answer.

3.14-2 System or cyclic reaction

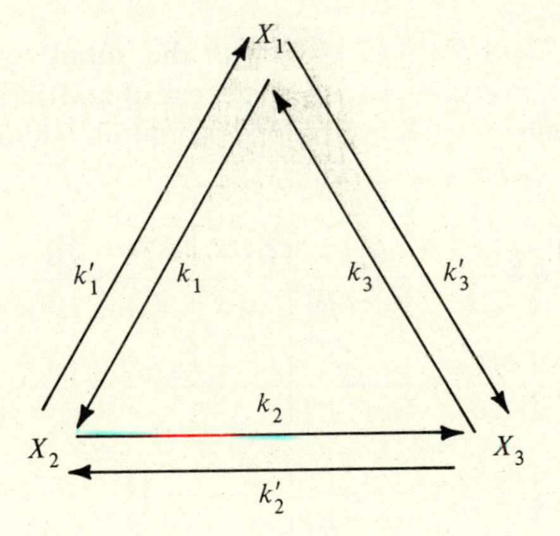

1. Stoichiometric equations:

$$C_1 = C_{1o} - \lambda_1 + \lambda_3 \tag{3.14-12}$$

$$C_2 = C_{2o} + \lambda_1 - \lambda_2 \tag{3.14-13}$$

$$C_3 = C_{3o} + \lambda_2 - \lambda_3 \tag{3.14-14}$$

2. Overall rate equations:

$$\frac{d\lambda_1}{dt} = r_1 = k_1 C_1 - k'_1 C_2 \tag{3.14-15}$$

$$\frac{d\lambda_2}{dt} = r_2 = k_2C_2 - k_2'C_3 \tag{3.14-16}$$

$$\frac{d\lambda_3}{dt} = r_3 = k_3C_3 - k_3'C_1 \tag{3.14-17}$$

3. Component rate equations:

$$-\frac{dC_1}{dt} = (k_1C_1 - k_1'C_2) - (k_3C_3 - k_3'C_1) \tag{3.14-18}$$

$$-\frac{dC_2}{dt} = (k_2C_2 - k_2'C_3) - (k_1C_1 - k_1'C_2) \tag{3.14-19}$$

$$-\frac{dC_3}{dt} = (k_3C_3 - k_3'C_3) - (k_2C_2 - k_2'C_3) \tag{3.14-20}$$

The solution of Eqs. (3.14-18) to (3.14-20) with the initial conditions $C_1(0) = C_{1o}$, $C_2(0) = C_3(0) = 0$ can be done by the use of matrix as follows:

$$[dC] = [A]\,[C] \tag{3.14-21}$$

where

$$[A] = \begin{bmatrix} -(k_1 + k_3') & k_1' & k_3 \\ k_1 & -(k_2 + k_1') & k_2' \\ k_3' & k_2 & -(k_3 + k_2') \end{bmatrix}$$

The final solution is

$$[C] = e^{[A]t}[C_o] \tag{3.14-22}$$

where $e^{[A]t}$ should be evaluated by Sylvester's theorem. For further detail, see Appendix 6.

Example 3.14-2 Cyclic Reaction

Find the concentrations of the reaction in 3.14-2 as functions of time if $k_1 = 1$ min^{-1}, $k_1' = 3$ min^{-1}, $k_2 = 2$ min^{-1}, $k_3 = 3$ min^{-1}, $k_2' = 2$ min^{-1}, $k_3' = 1$ min^{-1} and $C_{1o} = C_{2o} = C_{3o} = 5$ kmol/m^3.

Solution

Substituting the numerical values into Eqs. (3.14-18) to (3.14-20) gives

$$\frac{dC_1}{dt} = -2C_1 + 3C_2 + 3C_3$$

$$\frac{dC_2}{dt} = C_1 - 5C_2 + 2C_3$$

$$\frac{dC_3}{dt} = C_1 + 2C_2 - 5C_3$$

Hence

$$[A] = \begin{bmatrix} -2 & 3 & 3 \\ 1 & -5 & 2 \\ 1 & 2 & -5 \end{bmatrix}$$

The characteristic equation is

$$\det(\phi I - A) = \begin{vmatrix} \phi + 2 & -3 & -3 \\ -1 & \phi + 5 & -2 \\ -1 & -2 & \phi + 5 \end{vmatrix} = 0$$

Expanding

$$\phi^3 + 12\phi^2 + 35\phi = 0$$

Solving for ϕ

$$\phi_1 = 0 \qquad \phi_2 = -5 \qquad \phi_3 = -7$$

Then

$$\det'(0) = 35 \qquad \det'(-5) = -10 \qquad \det'(-7) = 14$$

$$\operatorname{adj}(\phi I - A) = \begin{bmatrix} \phi^2 + 10\phi + 21 & 3\phi + 21 & 3\phi + 21 \\ \phi + 7 & \phi^2 + 7\phi + 7 & 2\phi + 7 \\ \phi + 7 & 2\phi + 7 & \phi^2 + 7\phi + 7 \end{bmatrix}$$

$$e^{[A]t} = \frac{1}{35}\begin{bmatrix} 21 & 21 & 21 \\ 7 & 7 & 7 \\ 7 & 7 & 7 \end{bmatrix} + \frac{e^{-5t}}{-10}\begin{bmatrix} -4 & 6 & 6 \\ 2 & -3 & -3 \\ 2 & -3 & -3 \end{bmatrix} + \frac{e^{-7t}}{14}\begin{bmatrix} 0 & 0 & 0 \\ 0 & 7 & -7 \\ 0 & -7 & 7 \end{bmatrix}$$

$$= \begin{bmatrix} \frac{2}{5}e^{-5t} + \frac{3}{5} & \frac{-3}{5}e^{-5t} + \frac{3}{5} & \frac{-3}{5}e^{-5t} + \frac{3}{5} \\ \frac{-1}{5}e^{-5t} + \frac{1}{5} & \frac{1}{2}e^{-7t} + \frac{3}{10}e^{-5t} + \frac{1}{5} & \frac{3}{10}e^{-5t} - \frac{1}{2}e^{-7t} + \frac{1}{5} \\ \frac{-1}{5}e^{-5t} + \frac{1}{5} & \frac{3}{10}e^{-5t} - \frac{1}{2}e^{-7t} + \frac{1}{5} & \frac{3}{10}e^{-5t} + \frac{1}{2}e^{-7t} + \frac{1}{5} \end{bmatrix}$$

$$[C] = \begin{bmatrix} C_1 \\ C_2 \\ C_3 \end{bmatrix} = e^{[A]t}[C(0)] = e^{[A]t}\begin{bmatrix} 5 \\ 5 \\ 5 \end{bmatrix} = \begin{bmatrix} -4e^{-5t} + 9 \\ 2e^{-5t} + 3 \\ 2e^{-5t} + 3 \end{bmatrix}$$

3.15 *Miscellaneous Reactions*

In addition to the above reactions, there are some other reactions which may occur occasionally and are important in some industries. The reactions are:

3.15-1 *Chain reactions*

The mechanism of chain reactions can be indicated as

Initiation	$\ldots \rightarrow mR + \ldots$
Propagation	$R + \ldots \rightarrow R + \ldots$
Breaking	$R + R + \ldots \rightarrow \ldots$

where R represents a chain carrier, typically an atom or radical. The rate equations can then be written for each step. The overall rate equation or controlling rate equation can be derived as the case may be.

3.15-2 *Branching chain reactions*

The mechanism for this case is

Initiation	$\ldots \rightarrow mR + \ldots$
Propagation	$R + \ldots \rightarrow R + \ldots$
Branching	$R + \ldots \rightarrow 2R + \ldots$
Breaking	$R + \ldots \rightarrow \ldots$

3.15-3 *Organic molecule decomposition reactions*

The mechanism is

$$M_1 \rightarrow R_1 + M_2$$
$$R_1 + M_1 \rightarrow R_1H + R_2$$
$$R_2 \rightarrow R_1 + M_3$$
$$R_1 + R_2 \rightarrow M_4$$
$$2R_1 \rightarrow M_5$$
$$2R_2 \rightarrow M_6$$

3.15-4 *Photochemical reaction*

Smith (S3) indicates that the rate of consumption of Cl_2 in the photochlorination of propane at 25°C and 1 atm pressure is independent of propane, second-order in chlorine, and first-order in light intensity. If h is Planck's constant and ν is the frequency of radiation, the initiation reaction for a photochemical activation may be written as

$$Cl_2 + h\nu \rightarrow 2Cl$$

Using steady-state approximation, the following rate equation can be derived in Smith (S4)

$$-\frac{dC_{Cl_2}}{dt} = \frac{2k_3\phi_i}{k_7}\alpha C_{Cl_2}^2 I$$

where k_3 and k_7 are rate constants in the intermediate steps, ϕ_i is the quantum yield of the initiation step, I is the intensity of radiation, and α is the absorptivity.

3.15-5 Polymerization

The mechanism is

Initiation

$$M \rightarrow P_1$$

Propagation

$$M + P_1 \rightarrow P_2$$
$$M + P_2 \rightarrow P_3$$
$$M + P_{r-1} \rightarrow P_r$$

Termination

$$P_r + P_n \rightarrow M_{n+r}$$
$$P_{r-1} + P_n \rightarrow M_{r-1+n}$$

3.15-5 Catalytic reactions

There are two general types of catalytic reactions: homogeneous and heterogeneous. The latter will be described in Chapter 9. Homogeneous catalytic reactions in the presence of catalysts occur in either gas or liquid phases. Chamber oxidation of sulfur dioxide catalyzed by nitric oxide NO in the gas phase is an example. The reactions are

$$NO + \frac{1}{2}O_2 \rightarrow NO_2$$

$$NO_2 + SO_2 \rightarrow SO_3 + NO$$

Here NO, the catalyst, reacts with oxygen to produce NO_2 which, in turn, reacts with SO_2 to yield SO_3 and to regenerate the catalyst NO.

Either acids or bases can be employed as the catalysts for homogeneous liquid phase catalytic reactions. The hydrolysis of an ester in the presence of an acid is an example. The mechanism is

$$\text{Ester} + H^+ \underset{k_2}{\overset{k_1}{\rightleftharpoons}} EH^+$$

$$EH^+ + H_2O \xrightarrow{k_3} \text{acid} + \text{alcohol} + H^+$$

Sometimes a reaction is catalyzed by one of its products and is then defined as an autocatalytic reaction.

Problems

1. For the esterification of acetic acid with ethyl alcohol in the presence of HCl, the rate constants for forward and reverse reactions at 100°C are 4.76×10^{-4} and 1.63×10^{-4} liter/g mol · min, respectively. An equal mass of 85 wt. percent aqueous solution of acid and 90 wt. percent solution of ethanol is allowed to react at constant volume. Calculate the conversion of acid at various times. Assuming complete miscibility, estimate the equilibrium conversion. Density of original mixture is 0.848 g/cm^3.

2. A pure ethylene oxide is allowed to decompose to methane and carbon monoxide at 2.0 atm and 450°C isothermally in a constant-volume reactor. Assuming ideal behavior, determine the total pressure in the reactor after 52 min. The rate constant for the gas phase decomposition of the ethylene oxide at 450°C is 0.0212 min^{-1}.

3. The half-life data for thermal decomposition of nitrous oxide (N_2O) in the gas phase at 1030 K in a constant volume vessel at various initial pressures of N_2O are as follows: (H3)

p_0, mmHg	52.5	139	290	360
$t_{1/2}$, s	860.0	470	255	212

Determine the rate constant.

4. The following reaction is of first order

$$C_6H_5N_2Cl(aq) \rightarrow C_6H_5Cl(aq) + N_2(g)$$

In one experiment, at 50°C, the initial concentration of $C_6H_5N_2Cl$ is 10 g/liter and the following amounts of N_2 will be liberated: (C1)

Time, min	6	12	18	22	26	30
N_2 liberated, cm^3 at 50°C, 1 atm	19.3	32.6	41.3	45.0	48.4	50.3

Complete decomposition of the diazo salt liberates 58.3 cm^3 N_2. Calculate the rate constant.

5. The experimental data for decomposing dimethyl ether in the gas phase in a batch reactor at 504°C are (H2)

$$(CH_3)_2O \rightarrow CH_4 + H_2 + CO$$

Time, s	0	390	777	1,195	3,155	∞
Total pressure, mmHg	312	408	488	562	779	931

Determine the rate constant.

6. The reactions for the decomposition of nitrogen pentoxide are (D1)

$$\text{(a)}\quad 2N_2O_5 \rightarrow 2N_2O_4 + O_2$$

$$\text{(b)}\quad N_2O_4 = 2NO_2$$

Since the second reaction is in equilibrium, K_p at 25°C = 97.5 mmHg, only the first reaction need be considered. Initially only N_2O_5 is present. Determine the rate constant from the following data:

Time, min	0	40	80	120	∞
Total pressure, mmHg	268.7	302.2	318.9	332.3	473.0

7. The experimental data for decomposing nitrogen dioxide are as follows:

Temperature, K	592	603	627	651.5	656
k, $cm^3/gmol \cdot s$	522	755	1700	4020	5030

Assuming a second-order reaction, determine the energy of activation E_c. If the reaction is written as (F3)

$$2NO_2 \rightarrow 2NO + O_2$$

Also compute the activation energy E_p at 600 K. (Note the E_c and E_p)

8. The experimental data for the polymerization of styrene in the solvent of benzene by the use of a butyllithium initiator have been reported by Worsford and Bywater (W1) as:

Temperature, T°C	10	15	20	25	30.3
k_c, (liter/g mol)$^{1/2} \cdot$ min^{-1}	0.155	0.265	0.387	0.563	0.929

Determine the activation energy E_A and the frequency factor A in the Arrhenius equation.

9. Show that the reaction order is 1.5 and hence evaluate the rate constant for the following data:

Time, min	0	15	25	35	45
Concentration, g/liter	16.0	12.09	10.24	8.78	7.61

10. The half-life of an irreversible decomposition of A in solution catalyzed by B in a constant-volume reactor is determined by varying the initial concentration of A and B as follows:

$t_{1/2}$, min	15.87	11.20	9.15	7.93
A_o, gmol/liter	1.00	1.00	2.00	2.00
B_o, gmol/liter	0.001	0.002	0.003	0.004

Estimate the orders of reaction with respect to A and B and hence calculate the rate constant.

11. Determine the values of A and E in the Arrhenius equation from the following data, assuming first order reaction:

$t_{1/2}$, min	945	232	62	18	5.68
Temperature, °C	15	25	35	45	55

12. Calculate the values of A and E in the Arrhenius equation from the following data for the decomposition of nitrogen penoxide at different temperatures, assuming first-order reaction: (M1)

Temperature, °C	15	25	40	50	65
k (10^5) s^{-1}	1.04	3.38	24.7	75.9	487

13. At 25°C, an aqueous solution of methyl acetate (1.15 g mol/liter ester and 48.76 g mol/liter water) is hydrolyzed by means of hydrochloric acid. If the rate constants are those shown in Example 3.12-2, what is the half-life?

14. Andrews (A1) studied the alcoholysis of cinnamal chloride

$$C_6H_5CH{=}CH{-}CHCl_2 + C_2H_5OH \rightarrow C_6H_5{-}\underset{\displaystyle OC_2H_5}{\underset{|}{CH}}{-}CH{=}CHCl + HCl$$

by measuring the optical density of the reaction mixture. The cinnamal chloride, having a double bond conjugated with the benzene ring, absorbs strongly at 2,600 Å, whereas the product being unconjugated does not absorb until 2,100 Å. Hence a measurement of the optical density at 2,500 Å as a function of time permits the rate of the reaction to be measured.

Time, min	0	10	74	178	1,200
Optical density at 2,600 Å	0.406	0.382	0.255	0.143	0.01

Assuming first-order, determine the rate constant with the following data. At 22.6°C,

$$\text{Concentration of cinnamal chloride} = 2.11 \times 10^{-5} \text{ g mol/liter}$$

$$\text{Concentration of sodium ethoxide} = 0.547 \text{ g mol/liter}$$

15. For the following first-order reactions

$$X \xrightarrow{k_1} Y \xrightarrow{k_2} Z$$

express the concentrations of each component as functions of time. Initially only X is present at C_{Xo}.

16. For the following reactions with $C_{Xo} = C_{Yo}$ at constant density

$$X = Y \xrightarrow{k_1} Z$$

$$X + Z \xrightarrow{k_2} W$$

Assuming second-order reaction, derive expressions for the concentrations of Z and W in terms of C_{Xo} and C_{Yo} when k_1 is not equal to k_2.

17. A first-order homogeneous reaction $X \rightarrow 3Y$ is carried out in a constant-pressure batch reactor. It is found that, starting with pure A, the volume after 12 min is increased by 70 percent at a pressure of 1.8 atm. If the same reaction is to be carried out in a constant-volume reactor and the initial pressure is 1.8 atm., estimate the time required to bring the pressure to 2.5 atm.

18. Meissner (M1) obtained the following experimental data for the formation of hexamethylenetetramine at 0°C from aqueous ammonia and formaldehyde in a 1 liter reactor:

NH_3 present, wt %	0.8	0.6	0.41	0.20	0.016
NH_3 converted, mol/min	0.025	0.020	0.010	0.002	0.00002

Determine the order of the reaction with respect to NH_3.

19. The solution of the rate equation

$$\frac{dy}{dt} = 1.5k_1 e^{-k_1 t} - k_2 y$$

is

$$y = \frac{1.5k_1}{k_2 - k_1}(e^{-k_1 t} - e^{(-k_2 t)})$$

From the following data, evaluate k_1 and k_2.

t, min	20	30	40	50
y, moles converted	0.0878	0.0196	0.0044	0.0009

20. Find the concentrations of A, B, C, D, E as functions of time for the Denbigh's reaction system

$$\begin{array}{ccccc} A & \xrightarrow{k_1} & B & \xrightarrow{k_2} & C \\ \downarrow k^3 & & \downarrow k^4 & & \\ D & & E & & \end{array}$$

All are first-order irreversible reactions.

21. For the reaction

$$A \xrightarrow{k_1} B \xrightarrow{k_2} C$$

if the first reaction is of second-order, find the concentrations of A and B for $k_1 = 2\ (\text{kmol/m}^3)^{-1} \cdot \text{min}^{-1}$, $k_2 = 1\ \text{min}^{-1}$; $C_{Ao} = 1\ \text{kmol/m}^3$, $C_{Bo} = 0$.

22. The reaction

$$aA \rightarrow bB$$

takes place in a constant volume reactor in which initially contains A at a partial pressure p_{Ao}, B at a partial pressure p_{Bo} and inerts at a partial pressure of p_I, show that when the total pressure reaches P, the partial pressure of A will be:

$$\frac{bp_{Ao} + a(p_{Bo} + p_I - P)}{b - a}$$

23. The following reaction in gas phase

$$X + Y \xrightarrow{k_1} Z$$

takes place isothermally at 300°C under constant pressure of 1 atm in a reactor in which there are initially 0.35 mol percent X, 0.45 mol percent Y, 0.15 mol percent Z, and 0.05 mol percent inerts. Assuming that the reaction is first-order to both X and Y, how long will it take before the mole fraction of Z reaches 0.25 for $k_1 = 2$ liters/mol · min?

References

(A1). Andrews, I.J. *J. Am. Chem. Soc.,* 69:3062 (1947).

(A2). Aris, R. *Elementary Chemical Reactor Analysis.* Englewood Cliffs, N.J.: Prentice-Hall, Inc., 1969.

(B1). Biordi, J.C. *J. Chem. Eng. Data,* 15:166 (1970).

(C1). Cain, J.C., and Nicoll, F. *Proc. Chem. Soc.,* 24:282 (1909).

(C2). Chendke, P.K., Fogler, H.S., *Fourth International UMA Symposium* (n.p. 1970), or Fogler, H.S., *The Elements of Chemical Kinetics and Reactor Calculations.* Englewood Cliffs, N.J.: Prentice-Hall, Inc., 1974.

(C3). Cooper, A.R., and Jeffreys, G.V. *Chemical Kinetics and Reactor Design.* Englewood Cliffs, N.J.: Prentice-Hall, Inc., 1973.

(D1). Daniels, F., and Johnson, E.H. *J. Am. Chem. Soc.,* 43:53 (1921).

(E1). Eldridge, J.W. and Piret, E.L. *Chem. Eng. Prog.,* 46:290 (1950).

(F1). Fogler, H.S. *The Elements of Chemical Kinetics and Reactor Calculations.* Englewood Cliffs, N.J.: Prentice-Hall, Inc., 1974.

(F2). French, C.C. *J. Am. Chem. Soc.,* 51:3215 (1929).

(F3). Frost, A.A., and Pearson, R.G. *Kinetics and Mechanism.* New York: John Wiley & Sons, Inc., 1953.

(H1). Hellin, M. and Jungers, J.C. *Bull. Soc. Chim. France,* p. 386 (1957).

(H2). Hinshelwood, C.N., and Askey, P.J. *Proc. Roy. Soc.* (London), A115:215 (1927).

(H3). Hinshelwood, C.N., and Burk, R.E. *Proc. Roy. Soc.,* 106A:284 (1924).

(M1). Moelwyn-Hughes, E.A. *Physical Chemistry.* New York: Pergamon Press, 1957, p. 1109.

(N1). Neyens, A. Unpublished results cited on page 88 of "Cinetique Chimique Appliquee" by J.C. Jungers, J.C. Balaceanu, F. Coussemant, F. Eschard, A. Giraud in C.G. Hill, *An Introduction to Chemical Engineering Kinetics and Reactor Design.* New York: John Wiley & Sons, Inc., 1977.

(R1). Russell, T.W.F., and Denn, M.M. *Introduction to Chemical Engineering Analysis.* New York: John Wiley & Sons, Inc., 1972.

(S1). Saleton, D.I., and White, R.R. *Chem. Eng. Prog.,* 4:59 (1952).

(S2). Smith, D.F. *J. Am. Chem. Soc.,* 47:1862 (1925).

(S3). Smith, J.N. *Chemical Engineering Kinetics,* 3rd ed. New York: McGraw Hill Book Company, 1981.

(S4). Swain, C.G. *J. Am. Chem. Soc.,* 70:1119 (1948).

(V1). Vanghan, W.E. *J. Am. Chem. Soc.,* 54:3863 (1932).

(W1). Worsford, S., and Bywater, S. *Can. J. Chem.,* 38:1891 (1960).

CHAPTER FOUR

Mass, Energy, and Momentum Balances

4.1 *Introduction*

To design a chemical reactor, we need to know how far a chemical reaction can go, as we examined in Chapter 2 on thermodynamics, and how fast a chemical reaction can occur, as we examined in Chapter 3 on kinetics. In addition, we must also understand mass, energy, and momentum balances. Mass balance accounts for the changes of the masses of the reactants and the products during the course of the reaction; energy balance deals mainly with the change of the temperature of the reactants and products in the reaction; and momentum balance concerns the pressure drop of the feed during the reaction. All three balances are essential to the design of a chemical reactor. In this chapter, the general equations for each balance are derived. The applications of these three balances to each type of reactor are discussed in Chapter 5 for the ideal batch reactor, in Chapter 6 for the ideal semibatch reactor, in Chapter 7 for the continuous stirred-tank reactor, and finally in Chapter 8 for the ideal flow reactor. Chapter 9 will discuss heterogeneous kinetics and heterogeneous reactor design. Chapter 10 will address the nonideal reactors. In Chapter 4, we will also examine the similarity of the interphase transfer term in these balances.

4.2 *Mass Balance*

We separate the mass balance into two kinds: overall and component.

4.2-1 *Overall mass balance*

When a stream of fluid flows in and out of a vessel, the mass should be conserved. That is, for an increment volume of the vessel, or more precisely for an increment of length in a tube,

$$\text{Mass flow in} - \text{Mass flow out} = \text{Mass accumulated} \tag{4.2-1}$$

If the fluid with density ρ flows through the increment length Δx during the time Δt at a volumetric rate q, assuming that the compositions are uniform at a cross section, the balance becomes (see Fig 4.2-1)

$$\Delta t\,(q_x \rho_x - q_{x+\Delta x}\rho_{x+\Delta x}) = \rho A\,\Delta x \tag{4.2-2}$$

where A is the cross-sectional area of the tube. The subscripts indicate the length dependence. Dividing through by Δt and taking limits as Δt approaches zero yields

$$q_x \rho_x - q_{x+\Delta x}\rho_{x+\Delta x} = \rho A \lim_{\Delta t \to 0} \frac{\Delta x}{\Delta t} \tag{4.2-3}$$

Using the definition of a derivative, Eq. (4.2-3) becomes

$$q_x \rho_x - q_{x+\Delta x}\rho_{x+\Delta x} = \rho A \frac{dx}{dt} \tag{4.2-4}$$

Because the mass m is

$$m = \rho V = \rho A x \tag{4.2-5}$$

and the mass rate $\dot{m}$ is

$$\dot{m} = \rho q \tag{4.2-6}$$

Equation (4.2-4) reduces to

$$\dot{m}_1 - \dot{m}_2 = \frac{dm}{dt} \tag{4.2-7}$$

where subscripts 1 and 2 refer to inlet and outlet, respectively, and m is the mass of the fluid in any position x at any time t. If the density ρ is independent of position x, Eq. (4.2-4) becomes

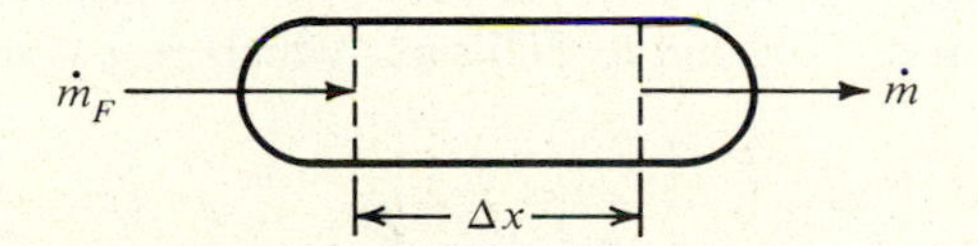

Figure 4.2-1 Flow in a tube

$$q_1 - q_2 = A\frac{dx}{dt} = \frac{dV}{dt} \qquad \textbf{(4.2-8)}$$

where V is the volume through which the fluid is passing.

4.2-2 Component mass balance

The law of conservation of mass is applied to the increment length of a tube as indicated above.

$$\begin{bmatrix}\text{Accumulation}\\ \text{of mass}\end{bmatrix} = \begin{bmatrix}\text{Mass in}\\ \text{by bulk}\\ \text{flow and}\\ \text{diffusion}\end{bmatrix} - \begin{bmatrix}\text{Mass out}\\ \text{by bulk}\\ \text{flow and}\\ \text{diffusion}\end{bmatrix} + \text{Generation} - \begin{bmatrix}\text{Interphase}\\ \text{transfer}\end{bmatrix} \qquad \textbf{(4.2-9)}$$

(I) (II) (III) (IV) (V)

I. Accumulation of mass

$$= C_j\,(\text{g mol/m}^3)A(\text{m}^2)\,\Delta x(\text{m})\,|_{t+\Delta t} - C_jA\,\Delta x\,|_t \qquad \textbf{(4.2-10)}$$

II. Mass in by bulk flow and diffusion

$$= u(\text{m/h})C_j(\text{g mol/m}^3)A\,(\text{m}^2) + J_A(\text{g mol/h}\cdot\text{m}^2)A\,(\text{m}^2)\,|_x\Delta t \qquad \textbf{(4.2-11)}$$

III. Mass out by bulk flow and diffusion

$$= uC_jA + J_A\,A\,|_{x+\Delta x}\Delta t \qquad \textbf{(4.2-12)}$$

IV. Generation by reaction

$$= \sum_{j=1,i=1}^{j=s,i=q} \alpha_{ij}r_i(\text{g mol/h}\cdot\text{m}^3)A\,(\text{m}^2)\,\Delta x\,(\text{m})\,\Delta t(\text{h}) \qquad \textbf{(4.2-13)}$$

V. Interphase transfer

$$= k_m[\text{g mol/h}\cdot\text{m}^2\,(\text{g mol/m}^3)]P(\text{m})\Delta x(m)[C_j(\text{g mol/m}^3) - \text{C*}]\,\Delta t\,(\text{h}) \qquad \textbf{(4.2-14)}$$

where k_m is the mass transfer coefficient, P is the perimeter of the tube = (π) (diameter of the tube), C^* is the concentration of the j component in the surrounding, and J_A is the mass flux due to diffusion, which is, according to Fick's law, equal to

$$J_A = -D\frac{dC_j}{dx} \qquad \textbf{(4.2-15)}$$

where D is the diffusion coefficient. Substituting Eqs. (4.2-10) to (4.2-15) into Eq. (4.2-9) and defining

$$a\left(\frac{\text{m of length}}{\text{m}^2 \text{ of area}}\right) = \frac{P}{A} = \frac{\pi D}{\pi D^2/4} = \frac{4}{D} \tag{4.2-16}$$

result in

$$\frac{\partial C_j}{\partial t} + \frac{\partial (u C_j)}{\partial x} = D \frac{\partial^2 C_j}{\partial x^2} + \Sigma \alpha_{ij} r_i - k_m a (C_j - C_j^*) \tag{4.2-17}$$

Equation (4.2-17) is the generalized equation for mass balance with chemical reaction occurring in a tube. One should note in particular the portion of the equation for interphase transfer. For the homogeneous phase this term is unnecessary, but for two or three phases this term is essential. When we discuss gas-liquid reactors in Chapter 9, this term will be evident.

4.3 Energy Balance

Similarly, an energy balance for the fluid flowing through an increment length of the tube in Figure 4.2-1 can be made as follows:

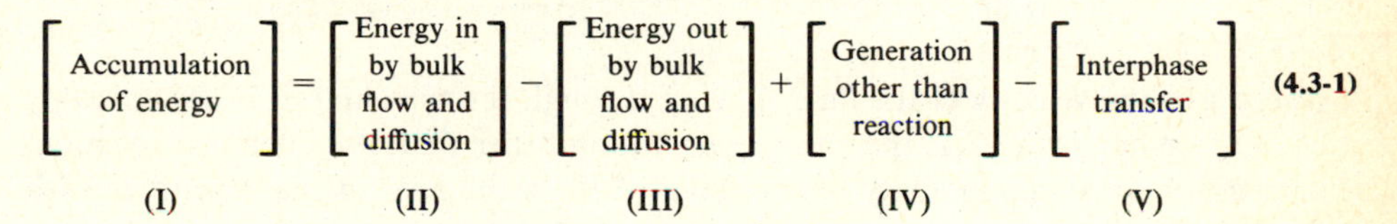

I. Accumulation of energy

$$= \Sigma C_j \,(\text{g mol/m}^3) h_j \,(\text{J/g mol}) A \,(\text{m}^2)\, \Delta x \,(\text{m}) \,|_{t+\Delta t} - \Sigma C_j h_j A \,\Delta x \,|_t \tag{4.3-2}$$

II. Energy in by bulk flow and diffusion

$$= u(\text{m/h}) A \,(\text{m}^2) \Sigma [C_j \,(\text{g mole/m}^3) h_j \,(\text{J/g mol})] + Q\,(\text{J/h}\cdot\text{m}^2) A\,(\text{m}^2) \,|_x \,\Delta t(\text{h}) \tag{4.3-3}$$

III. Energy out by bulk flow and diffusion

$$= uA \,\Sigma (C_j h_j) + QA \,|_{x+\Delta x} \Delta t \tag{4.3-4}$$

IV. Generation

$$= G(\text{J/h}\cdot\text{m}^3) A\,(\text{m}^2)\, \Delta x\,(\text{m})\, \Delta t(\text{h}) \tag{4.3-5}$$

V. Interphase transfer

$$= U(\text{J/h}\cdot\text{m}^2\cdot{}^\circ\text{C})\, P(\text{m})\, \Delta x(\text{m})(T - T^*)({}^\circ\text{C})\, \Delta t(\text{h}) \qquad \textbf{(4.3-6)}$$

Substituting Eqs. (4.3-2) to (4.3-6) into Eq. (4.3-1) and remembering that the heat flux $Q = -k(dT/dx)$ by Fourier's law, dividing by $A\ \Delta x\ \Delta t$ and taking the limit result in

$$\frac{\partial \Sigma\,(C_j h_j)}{\partial t} + \frac{\partial[u\Sigma(C_j h_j)]}{\partial x} = k\frac{\partial^2 T}{\partial x^2} + G - Ua(T - T^*) \qquad \textbf{(4.3-7)}$$

Expanding and using Eq. (4.2-17) by neglecting the diffusion and the mass interphase terms and also using

$$h_j = C_{pj}\,(T - T_f) \quad \text{or} \quad \frac{\partial h_j}{\partial t} = C_{pj}\frac{\partial T}{\partial x} \quad \text{and} \quad \frac{\partial h_j}{\partial x} = C_{pj}\frac{\partial T}{\partial x}$$

and

$$\Sigma h_j \alpha_{ij} = \Delta H_i$$

we obtain

$$\Sigma C_j C_{pj}\left(\frac{\partial T}{\partial t} + u\frac{\partial T}{\partial x}\right) + \Sigma \Delta H_i r_i = k\frac{\partial^2 T}{\partial x^2} + G - Ua\,(T - T^*) \qquad \textbf{(4.3-8)}$$

where u is the velocity of the fluid flowing through the tube and h_j is the enthalpy of the j component. G is the energy generation other than the chemical reaction, k is the thermal conductivity of the fluid. ΔH_i is the heat of reaction of the ith reaction.

4.4 Momentum Balance

As a fluid passes through a tube, the pressure will drop. Hence a momentum balance should be made as follows:

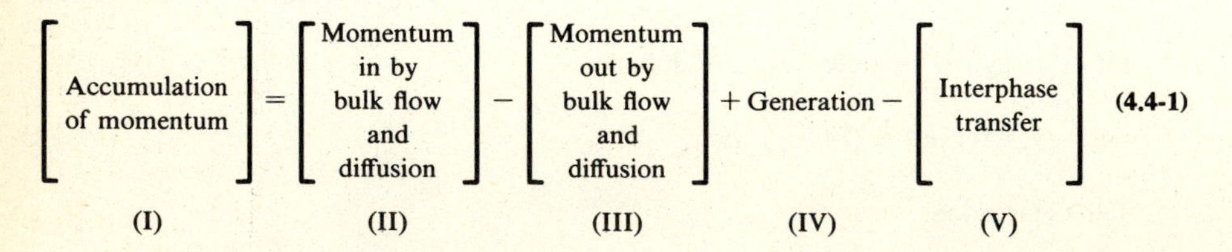

I. Accumulation of momentum

$$= w(\text{kg/s})\, \Delta x\,(\text{m})\,\big|_{t+\Delta t} - w\, \Delta x\,\big|_{t} \qquad \textbf{(4.4-2)}$$

II. Momentum in by bulk flow and diffusion

$$= u(\text{m/s})w\,(\text{kg/s}) + \tau(\text{kg/m}\cdot\text{s}^2)A\,(\text{m}^2)\,|_x\ \Delta t\,(\text{s}) \tag{4.4-3}$$

III. Momentum out by bulk flow and diffusion

$$= uw + \tau A\ |_{x+\Delta x}\ \Delta t(\text{s}) \tag{4.4-4}$$

IV. Generation

$$= -\Delta P(\text{N/m}^2)A\,(\text{m}^2)\,\Delta t(\text{s})\ + g(\text{m/s}^2)\rho(\text{kg/m}^3)\,A\,(\text{m}^2)\,\Delta x\,(\text{m})\,\Delta t(\text{s}) \tag{4.4-5}$$

V. Interphase transfer

$$= \frac{f_f}{2}\,u(\text{m/s})u\rho(\text{kg/s}\cdot\text{m}^2)\pi D(\text{m})\,\Delta x\,(\text{m})\,\Delta t(\text{s}) \tag{4.4-6}$$

Substituting Eqs. (4.4-2) to (4.4-6) into Eq. (4.4-1), dividing by $\Delta t\ \Delta x$, remembering that the momentum flux τ by Newton's law, and taking limit

$$\tau = -\mu\frac{du}{dx} \qquad \text{and} \qquad w = u\rho A$$

$$\rho\frac{\partial u}{\partial t} + u\rho\frac{\partial u}{\partial x} = \mu\frac{\partial^2 u}{\partial x^2} - \frac{\partial P}{\partial x} + g\rho - \frac{f_f u^2 \rho\pi D}{2A} \tag{4.4-7}$$

Eq. (4.4-7) is the generalized momentum equation for a fluid flowing through an increment length of the tube in the previous figure.

4.5 *Interphase Transfer*

It is interesting to note that the interphase transfer term is similar to those in the mass balance and the energy balance. The similarity can be seen from below:

	Flux			
Mass	$\frac{N}{A} = k_m(C_j - C_j^*) = -D_{AB}\frac{dC_j}{dr}\Big\vert_{r=r_i}$	Fick's law	(4.5-1)
Energy	$\frac{Q}{A} = U(T - T^*) = \frac{-k\,dT}{dr}\Big\vert_{r=r_i}$	Fourier's law	(4.5-2)
Momentum	$\frac{\tau}{A} = \frac{\rho u f_f}{2}(u - 0) = \frac{-\mu\,du}{dr}\Big\vert_{r=r_i}$	Newton's law	(4.5-3)

where w is the mass rate of flow, μ is the viscosity of the fluid, f_f is the Fanning friction factor, g is the gravity force, and ΔP is the pressure drop.

CHAPTER FIVE

Ideal Batch Reactor

5.1 Introduction

The characteristic of a batch reactor is that all the reactants are placed in a vessel in which the reaction takes place. After the reaction is complete, the products and the unreacted reactants are withdrawn for purification. Usually an agitator is employed to ensure that the reaction mixture is sufficiently agitated to obtain uniform composition and temperature throughout. This kind of reactor is often used for liquid-phase reactions, particularly when the required production is small. The operation, control, and maintenance of a batch reactor are simple and inexpensive when the production capacities are low. The disadvantages for batch reactors are the high labor cost for filling, emptying, and cleaning the reactor. In this chapter, the design of a batch reactor under isothermal conditions is first described, followed by a description under nonisothermal conditions, and concluding with a discussion of the optimization of the batch reactor.

5.2 Isothermal Batch Reactor

Start from the generalized mass balance equation [Eq. (4.2-17)]. For a batch reactor, there is no flow, and $u = 0$. Neglecting diffusion and interphase transfer, $D = 0$ and $k_m = 0$. Hence Eq. (4.2-17) becomes

$$\frac{\partial C_j}{\partial t} = \Sigma \alpha_{ij} r_i \tag{5.2-1}$$

The stoichiometric equation gives, for $j = k$,

$$C_j = C_{jo} + \alpha_j \lambda = C_{ko}\,(1 - f) \tag{5.2-2}$$

Differentiating Eq. (5.2-2) with respect to t gives

$$\frac{dC_j}{dt} = -C_{ko}\frac{df}{dt} \tag{5.2-3}$$

Equating with Eq. (5.2-1)

$$-C_{ko}\frac{df}{dt} = +\Sigma r_i \alpha_{ij} \tag{5.2-4}$$

If j is the limiting reactant

$$-C_{ko}\frac{df}{dt} = -\Sigma r_i \tag{5.2-5}$$

For $i = 1$

$$-C_{ko}\frac{df}{dt} = -r_1 \tag{5.2-6}$$

Separating variables and integrating

$$t = C_{ko}\int_{f_o}^{f} \frac{df}{r} \tag{5.2-7}$$

But

$$C_{ko} = \frac{m_k}{V} \tag{5.2-8}$$

where m_k is the original mass of the limiting reactant k in the reactor and V is the volume of the mixture. Then

$$t = \frac{m_k}{V}\int_{f_o}^{f} \frac{df}{r} \tag{5.2-9}$$

Example 5.2-1 Isothermal Batch Reactor

The problem is to produce 12,000 kg/day of ethyl acetate in an isothermal batch reactor by the following reaction

$$\underset{(1)}{CH_3COOH} + \underset{(2)}{C_2H_5OH} \underset{k_2}{\overset{k1}{\rightleftharpoons}} \underset{(3)}{CH_3COOC_2H_5} + \underset{(4)}{H_2O}$$

where $k_1 = 8.0 \times 10^{-6}$ and $k_2 = 2.7 \times 10^{-6}$ m^3/kmol·s. The initial charge to the reactor contains an aqueous solution of ethanol, 550 kg/m^3, and acetic acid, 300 kg/m^3, and a small quantity of HCl to act as a catalyst. The time between batches for discharging, cleaning, and recharging is 45 min. Assuming that the density of the reaction mixture is constant at 1,145 kg/m³, determine the volume of the reactor required for 40 percent conversion.

Solution

Basis of calculation: 1 m³ of reactor

$$C_{1o} = \frac{300}{60} = 5.0 \text{ kmol/m}^3$$

$$C_{2o} = \frac{550}{46} = 11.96 \text{ kmol/m}^3$$

$$C_{3o} = 0$$

$$C_{4o} = \frac{1{,}145 - 300 - 550}{18} = 16.39 \text{ kmol/m}^3$$

and

$$r = 8.0 \times 10^{-6} C_1 C_2 - 2.7 \times 10^{-6} C_3 C_4$$

where

$$C_1 = C_{1o}(1 - f_1)$$
$$C_2 = C_{2o} - C_{1o} f_1$$
$$C_3 = C_{3o} + C_{1o} f_1$$
$$C_4 = C_{4o} + C_{1o} f_1$$

Substituting all of these into

$$t = C_{1o} \int_0^{0.4} \frac{df_1}{r}$$

and integrating gives

$$t = 5{,}979 \text{ s.}$$

Hence

$$t_{\text{batch}} = t_{\text{reaction}} + t_{\text{cleaning}}$$
$$= 5{,}979 + 2{,}700 = 8{,}679 \text{ s.}$$

Acetic acid converted = ethyl acetate produced = 5.0(0.4) = 2.00 kmol

$$\text{Average production rate of ethyl acetate} = 2.0(88)\frac{(3{,}600)(24)}{8{,}679}$$
$$= 1{,}752.09 \text{ kg/day}\cdot\text{m}^3$$

Hence

$$\text{Volume of reactor} = \frac{10{,}000 \text{ kg/day}}{1{,}752.09 \text{ kg/day}\cdot\text{m}^3} = 6.85 \text{ m}^3$$

5.3 *Nonisothermal Batch Reactor*

The energy balance for a flow reactor is Eq. (4.3-7). For a batch reactor, the accumulation of energy is $\partial \Sigma C_j E_j/\partial t$ where E_j is the internal energy of the component. For liquids, the internal energy is approximately equal to enthalpy. The treatment for gases will be dealt with in the next section. For a batch reactor, under adiabatic conditions, assuming that the diffusion and generation are negligible, Eq. (4.3-7) becomes

$$\frac{\partial \Sigma C_j h_j}{\partial t} = 0 \tag{5.3-1}$$

Expanding Eq. (5.3-1) gives

$$\Sigma C_j \frac{dh_j}{dt} + \Sigma h_j \frac{dC_j}{dt} = 0 \tag{5.3-2}$$

From the mass balance equation, we obtain

$$\frac{dC_j}{dt} = \Sigma \alpha_{ij} r_i \tag{5.3-3}$$

The definition of the enthalpy is

$$h_j = C_{pj}(T - T_f) \tag{5.3-4}$$

where C_{pj} is the heat capacity of the component j at constant pressure and T_f is the reference temperature. Differentiating Eq. (5.3-4) yields

$$\frac{dh_j}{dt} = C_{pj} \frac{dT}{dt} \tag{5.3-5}$$

Substituting Eqs. (5.3-3) and (5.3-5) into Eq. (5.3-2) gives

$$\Sigma C_j C_{pj} \frac{dT}{dt} + \Sigma h_j \Sigma \alpha_{ij} r_i = 0 \tag{5.3-6}$$

Since

$$\Sigma h_j \,\alpha_{ij} = \Delta H_i$$

Eq. (5.3-6) reduces to

$$\Sigma C_j C_{pj} \frac{dT}{dt} + \Sigma \Delta H_i r_i = 0 \tag{5.3-7}$$

Because the rate of the reaction is

$$r_i = \frac{dC_j}{\alpha\, dt} = C_{ko} \frac{df}{dt} \tag{5.3-8}$$

Thus

$$\Sigma C_j C_{pj} \frac{dT}{dt} + \Sigma \Delta H_i \left(C_{ko} \frac{df}{dt} \right) = 0 \tag{5.3-9}$$

Now letting

$$\Sigma C_j C_{pj} = \mathfrak{z}_p \ (\mathrm{J/m^3 \cdot {}^\circ C}) \tag{5.3-10}$$

we get

$$\mathfrak{z}_p \frac{dT}{dt} = - \ \Sigma \Delta H_i C_{ko} \frac{df}{dt} \tag{5.3-11}$$

Hence assuming $\mathfrak{z}_p$ and $\Sigma \Delta H_i$ constant

$$\int_{T_o}^{T} \frac{dT}{\Sigma \Delta H_i} = - \frac{C_{ko}}{\mathfrak{z}_p} \int_0^f df \tag{5.3-12}$$

or

$$T = T_o - \frac{C_{ko} \Sigma \Delta H_i f}{\mathfrak{z}_p} = T_o - \mathfrak{z} f \tag{5.3-13}$$

where

$$\mathfrak{z} = \left(\frac{C_{ko} \Sigma \Delta H_i}{\mathfrak{z}_p} \right) \tag{5.3-14}$$

Since

$$\begin{aligned} r &= k \Pi C_j^{\alpha_j} - k' \Pi C_j^{\alpha'_j} \\ &= A \phi(f) e^{-E/RT} \\ &= A \phi(f) e^{-E/R(T_o - \mathfrak{z} f)} \end{aligned} \tag{5.3-15}$$

thus

$$t = \frac{C_{ko}}{A}\int_0^f \frac{df}{\phi(f)e^{-E/R(T_o - 3f)}} \tag{5.3-16}$$

where A and E are the Arrhenius constants.

Example 5.3-1 Nonisothermal Batch Reactor

A batch reactor has a 1,070-kg charge of a solution of acetic anhydride at 15°C containing acetic anhydride at a concentration of 0.32 kmol/m³. The solution density is 1,070 kg/m³ and its specific heat 3.8 kJ/kg · K. The first-order reaction has $\Delta H = -210{,}000$ kJ/kmol anhydride hydrolyzed. The variation of the rate constant with temperature is given in the following table. What is the time required for 80 percent conversion under adiabatic conditions?

Temperature, °C	15	20	25	30
Rate constant, s^{-1}	0.00134	0.00188	0.00263	0.00351

Solution

$T_o = 15°C.$ $\quad C_{ko} = 0.32\ \text{kmol/m}^3$ $\quad \Delta H = -210{,}000\ \text{kJ/kmol}$

$3_p = 3.8(1{,}070)\ \text{kJ/m}^3 \cdot \text{K}$

Substituting into Eq. (5.3-13) gives

$$T = 15 + 16.52f \quad (\text{in °C}) \tag{A}$$

Since

$$r = kC_A = kC_{Ao}(1-f) = C_{Ao}\frac{df}{dt}$$

hence

$$t = \int_0^{0.8} \frac{df}{k(1-f)} \tag{B}$$

Thus, for each value of f, T can be calculated from Eq. (A), then k can be estimated by interpolation from the table, and finally the value of $1/[k(1-f)]$ can be plotted against f. The area under the curve from $f = 0$ to $f = 0.8$ can be found graphically and is equal to t. Numerically, Simpson's rule is employed to solve the problem as follows:

Conversion, f	0	0.1	0.2	0.3	0.4	0.5	0.6	0.7	0.8
Temperature, °C	15.00	16.65	18.30	19.96	21.61	23.26	24.91	26.56	28.22
Rate constant, $k(10)^3$	1.346	1.505	1.680	1.873	2.087	2.322	2.580	2.863	3.174
$1/k(1-f)$	742.8	738.3	743.9	762.4	798.6	861.4	969.1	1164.	1575.

By Simpson's rule

$$t = \frac{0.1}{3}[1(742.8) + 4(738.3 + 762.4 + 861.4 + 1{,}164) + 2(743.9 + 798.6 + 969.1) + 1(1{,}575)] = 714.88 \text{ s} = 11.91 \text{ min}$$

A computer program for first calculating the rate constant from the table by the method of least squares and then evaluating the integral numerically by Simpson's rule is given in Fig. 5.3-1.

For gases, the accumulation of energy is $\partial C_j E_j/\partial t$ where E_j is the internal energy of the j component. In dealing with gases, it is easier to use number of moles rather than C_j. Thus Eq. (4.3-7) should be modified for a batch reactor as

$$\frac{\partial \Sigma C_j E_j}{\partial t} = \frac{\partial \Sigma n_j E_j}{V\,\partial t} = 0 \tag{5.3-17}$$

or

$$\frac{\partial \Sigma n_j E_j}{\partial t(V)} = \frac{\partial \Sigma n_j h_j}{V\,\partial t} - \frac{\partial \Sigma n_j PV}{V\,\partial t} = 0 \tag{5.3-18}$$

Expanding and using $PV = RT$ gives

$$\frac{\Sigma n_j\,\partial h_j}{V\,\partial t} + \frac{h_j}{V}\frac{\partial \Sigma n_j}{\partial t} - \frac{\Sigma n_j R}{V}\frac{\partial T}{\partial t} - \frac{RT\,\partial \Sigma n_j}{V\,\partial t} = 0 \tag{5.3-19}$$

From the definition of the enthalpy and the mass balance equation, we get

$$\frac{\Sigma n_j C_{pj}}{V}\frac{\partial T}{\partial t} - \frac{\Sigma n_j R}{V}\frac{\partial T}{\partial t} + \frac{\Sigma h_j \alpha_{ij} r_i V}{V} - \frac{RT\Sigma \alpha_{ij} r_i V}{V} = 0 \tag{5.3-20}$$

Using the definition of heat of reaction, we obtain

$$\frac{\Sigma n_j C_{vj}}{V}\frac{\partial T}{\partial t} + (\Delta H_i^T - \alpha RT) r_i = 0 \tag{5.3-21}$$

where C_{vj} is the heat capacity of j component at constant volume, ΔH_i^T is the heat of reaction i at temperature T, and $\boldsymbol{\alpha} = \Sigma\alpha_j$. Equation (5.3-21) can be also written as for single reaction

$$\Omega \frac{\partial T}{\partial t} + (\Delta H^T - \boldsymbol{\alpha} RT)rV = 0 \tag{5.3-22}$$

where

$$\Omega = \Sigma n_j C_{vj}$$

5.4 *Nonadiabatic Operation*

By neglecting the diffusion and the generation in a batch reactor for no flow, Eq. (4.2-17) becomes

$$\frac{\partial \Sigma(C_j h_j)}{\partial t} = -Ua(T - T^*) \tag{5.4-1}$$

Expanding Eq. (5.4-1) as before by the use of the definition of enthalpy and the mass balance equation yields

$$\Sigma C_j C_{pj} \frac{dT}{dt} + \Sigma \Delta H_i^T C_{ko} \frac{df}{dt} = -U\frac{P}{A}(T - T^*) \tag{5.4-2}$$

Multiplying by V and grouping result in

$$U\pi DL(T^* - T)dt = M_k \Delta H_i^T df + M_T \mathbf{C}_p \, dT \tag{5.4-3}$$

where

$$M_k = C_{ko} V$$

$$M_T = \text{total moles} = \Sigma C_j V = \Sigma n_j$$

$$\mathbf{C}_p = \text{mean specific heat of mixtures}$$

$$D = \text{diameter of the tube}$$

$$L = \text{length of the tube}$$

Example 5.4-1 Nonadiabatic Batch Reactor

Two hundred and fifty kilograms of acetylated castor oil is hydrolyzed in a batch reactor by the reaction:

```
00100 PROGRAM BATCH2(INPUT,OUTPUT)
00110CX=TEMPERATURE,DEGC
00120CR=RATE CONSTANT,SEC**(-1)
00130CF=CONVERSION
00140CA=FREQUENCY FACTOR,SEC**(-1)
00150CE=ACTIVATION ENERGY,CAL/GMOLE
00160 DIMENSIONRK(10),F(10),FCT(10),X(10),R(10)
00170 DATA(X(I),I=1,4)/1.5E+1,2.0E+1,2.5E+1,3.0E+1/
00180 DATA(R(I),I=1,4)/1.34E-3,1.88E-3,2.63E-3,3.51E-3/
00190 DATA(F(I),I=1,9)/0.,0.1,0.2,0.3,0.4,0.5,0.6,0.7,0.8/
00200 CALL LEAS(X,R,A,E)
00210 DO 5 I=1,9
00220 T=15.+16.52*F(I)
00230 T=T+273.
00240 5 RK(I)=EXP(A-E/(1.987*T))
00250 DO 6 I=1,9
00260 6 FCT(I)=1./(RK(I)*(1.-F(I)))
00270 C=2.*(FCT(3)+FCT(5)+FCT(7))
00280 D=4.*(FCT(2)+FCT(4)+FCT(6)+FCT(8))
00290 TIME=(0.1/3.)*(FCT(1)+FCT(9)+C+D)/60.
00300 PRINT 3,TIME
00310 3 FORMAT(1X,"TIME REQUIRED FOR 80% CONVERSION IS ",F6.2,"MINUTES")
00320 STOP
00330 END
00340 SUBROUTINE LEAS(X,R,A,E)
00350 DIMENSION X(10),R(10),DX(10),DY(10)
00360 DO 10 I=1,4
00370 X(I)=1./(273.+X(I))
00380 10 R(I)=ALOG(R(I))
00390 SX=0.
00400 SY=0.
00410 DO 30 I=1,4
00420 SX=SX+X(I)
00430 30 SY=SY+R(I)
00440 XBAR=SX/4.
00450 YBAR=SY/4.
```

```
00460 SDXDY=0.
00470 SDXDX=0.
00480 DO 40 I=1,4
00490 DX(I)=X(I)-XBAR
00500 DY(I)=R(I)-YBAR
00510 SDXDY=SDXDY+DX(I)*DY(I)
00520 40 SDXDX=SDXDX+DX(I)*DX(I)
00530 SMB=SDXDY/SDXDX
00540 SMA=YBAR-SMB*XBAR
00550 A=SMA
00560 E=-1.987*SMB
00570 RETURN
00580 END
READY.
RNH

 TIME REQUIRED FOR 80% CONVERSION IS  11.91MINUTES
```

Figure 5.3-1

$$\text{Acetylated castor oil(l)} \rightarrow CH_3COOH(g) + \text{drying oil (l)}$$

The initial temperature is 613 K. Complete hydrolysis yields 0.156 kg acetic acid per kilogram of ester. The specific heat of the solution is constant at 0.6 cal/g·K. The heat of reaction ΔH is 15,000 cal/g mol of acid produced. Heat is added at a rate of 756,000 cal/min. The first-order rate constant is (Gl)

$$\ln k = 35.2 - \frac{44{,}500}{RT}$$

where k is the rate constant in min^{-1} and T is the temperature in degrees Kelvin. Calculate the time required for 80 percent conversion.

Solution

Using Eq. (5.4-3) gives

$$\int_0^t 756{,}000\, dt = \int_0^f \frac{250{,}000(0.156)}{60}(15{,}000)df + \int_{613}^{T} 250{,}000(0.6)\, dT$$

or

$$T = 613 + 5.04t - 65f \tag{A}$$

The mass balance equation yields

$$\frac{df}{dt} = (1-f)e^{35.2-(22{,}395/T)} \tag{B}$$

Instead of solving Eqs. (A) and (B) by Simpson's rule as in Example 5.3-1, another numerical method, called the *improved Euler method,* can be employed. This method will be described in Appendix 5. A computer program to solve the example by this method, with results, is given in Fig. 5.4-1. The time required for 80 percent conversion as indicated in the printout of the program is 9.77 min.

5.5 *Analytical Solution of Irreversible Reactions (D1)*

For a single irreversible reaction, the combination of Eqs. (3.9-3), (3.9-5), and (3.5-1) gives

$$r(\lambda, T) = k\Pi(C_{jo} + \alpha_j\lambda)^{\underline{\alpha}_j} = Ae^{-E/RT}\Pi(C_{jo} + \alpha_j\lambda)^{\underline{\alpha}_j} \tag{5.5-1}$$

The temperature T is related to λ by Eq. (5.3-7) as

```
 81/08/03. 15.04.16.
PROGRAM    MEULR

00100 PROGRAM MEULR(INPUT,OUTPUT)
00110 READ*,F,T,H,FLAST
00120CF=FRACTIONAL CONVERSION
00130CT=TIME,MIN
00140CH=TIME INCREMENT,MIN
00150CFLAST=LAST FRACTIONAL CONVERSION
00160 PRINT 10,F,T,H
00170 10 FORMAT(3E15.6)
00180 20 TP=DEQF(F,T)
00190 T=T+H/2.*(TP+DEQF(F+H,T+H*TP))
00200 F=F+H
00210 PRINT 11,F,T
00220 11 FORMAT(2F10.6)
00230 IF(F-FLAST) 20,21,21
00240 21 STOP
00250 END
00260 FUNCTION DEQF(F,T)
00270 DEQF=1./((1.-F)*(EXP(35.2-22395./(613.0+5.04*T-65.*F))))
00280 RETURN
00290 END
READY.
RNH

? 0.,0.,.05,.8
     0.              0.                  .500000E-01
   .050000   .209382
   .100000   .459631
   .150000   .755885
   .200000  1.102784
   .250000  1.504114
   .300000  1.962513
   .350000  2.479349
   .400000  3.054780
   .450000  3.688052
   .500000  4.377960
   .550000  5.123463
   .600000  5.924398
   .650000  6.782350
   .700000  7.701848
   .750000  8.692319
   .800000  9.771984
   .850000 10.977290

SRU       0.951 UNTS.
```

Figure 5.4-1

$$\frac{dT}{dt} = -\frac{\Delta Hr}{\Sigma C_j C_{pj}} = Jr \tag{5.5-2}$$

where $J = -\Delta H/(\Sigma C_j C_{pj})$. But $r = d\lambda/dt$, Eq. (5.5-2) becomes

$$\frac{dT}{dt} = J\frac{d\lambda}{dt} \tag{5.5-3}$$

or

$$dT = J\,d\lambda$$

Integrating yields

$$T = T_o + J\lambda \tag{5.5-4}$$

Substituting into Eq. (5.5-1) gives

$$r(\lambda, T) = \frac{d\lambda}{dt} = Ae^{-E/R(T_o + J\lambda)}\Pi(C_{jo} + \alpha_j\lambda)^{\alpha j} \tag{5.5-5}$$

Separating the variables, we obtain

$$\int_0^t dt = \int_0^\lambda \frac{d\lambda}{Ae^{-E/R(T_o+J\lambda)}\Pi(C_{jo} + \alpha_j\lambda)^{\alpha j}} \tag{5.5-6}$$

which can be integrated numerically or analytically by the use of exponential integral, which is defined as

$$-Ei(-x) = \int_x^\infty \frac{e^{-t}}{t}\,dt \tag{5.5-7}$$

or

$$Ei(x) = \int_{-\infty}^x \frac{e^y}{y}\,dy \tag{5.5-8}$$

For first-order reaction, A → B, Eq. (5.5-6) becomes

$$\int_0^t dt = \int_0^\lambda \frac{d\lambda}{Ae^{(-E/R(T_o+J\lambda)}(C_{Ao} - \lambda)} \tag{5.5-9}$$

Letting

$$\frac{E}{R(T_o + J\lambda)} = y$$

we get

$$\lambda = \frac{E}{RJy} - \frac{T_o}{J} \tag{5.5-10}$$

Differentiating gives

$$d\lambda = -\frac{E}{RJ}\frac{dy}{y^2} \tag{5.5-11}$$

and

$$\begin{aligned} C_{Ao} - \lambda &= \frac{T_o + JC_{Ao}}{J} - \frac{E}{RJy} \\ &= \frac{T_a}{J}\left(\frac{y - E/RT_a}{y}\right) \end{aligned} \tag{5.5-12}$$

where

$$T_a = T_o + JC_{Ao}$$

Hence Eq. (5.5-9) becomes

$$t(\lambda,\ T_o) = \frac{E}{RT_aA}\int_{E/RT}^{E/RT_o} \frac{e^y\,dy}{y(y - E/RT_a)} \tag{5.5-13}$$

Integrating by partial fractions and using the definition of exponential integral, we obtain

$$\begin{aligned} At = e^{E/RT_a}\left[Ei\left(\frac{E}{RT_o} - \frac{E}{RT_a}\right) - Ei\left(\frac{E}{RT} - \frac{E}{RT_a}\right)\right] \\ - \left[Ei\left(\frac{E}{RT_o}\right) - Ei\left(\frac{E}{RT}\right)\right] \end{aligned} \tag{5.5-14}$$

Tables of $Ei(x)$ may be found in standard compilation.

For second-order reaction A + B $\longrightarrow$ product, the solution for $C_{Ao} = C_{Bo} = C_o$ is

$$AC_o t = \frac{(T_a - T_o)Te^{E/RT}}{T_a(T_a - T)} - \frac{T_o e^{E/RT_o}}{T_a} - \frac{E(T_a - T_o)e^{E/RT_a}}{RT_a^2}\left[Ei\left(\frac{E}{RT_o} - \frac{E}{RT_a}\right) - Ei\left(\frac{E}{RT} - \frac{E}{RT_a}\right)\right] \tag{5.5-15}$$

Example 5.5-1 Solution by Exponential Integral

Pure A decomposes to B irreversibly with first-order kinetics

$$A \rightarrow B$$

The rate constant is

$$k = 2.6352 \times 10^{14} e^{-28{,}960/1.987T}$$

where T is in K and k in h^{-1}. Determine the time required for 97 percent conversion of A. The data are as follows:

1. Heat of reaction = −83 cal/g of A
2. Initial concentration of A = 0.9 g/cm³
3. Density of mixture = 0.9 g/cm³
4. Heat capacity of mixture = 0.5 cal/g·°C
5. Initial temperature = 163°C

Solution

Using Eq. (5.5-9)

$$t = \int_0^{0.873} \frac{d\lambda}{2.6352 \times 10^{14}(0.9 - \lambda)e^{-14{,}574.74/T}}$$

Since

$$\lambda = f_k C_{ko}/\alpha_k \quad \text{from Eq. (1.4-45)}$$

and

$$C_{ko} = 0.9 \qquad \alpha_k = 1$$

therefore

$$\lambda = 0.9 f_k$$

and

$$t = \int_0^{0.97} \frac{df_k}{2.6352 \times 10^{14}(1 - f_k)e^{-14{,}574.74/T}}$$

Applying Eq. (5.5-14)

$$JC_{Ao} = -\frac{0.9(-83)}{0.9(0.5)} = +166 \qquad T_a = 163 + 166 = 329°C$$

$$A = 2.635 \times 10^{14} \qquad E = 28{,}960 \qquad T = 436 + 166(0.97) = 597.02$$

$$2.6352 \times 10^{14} t = e^{28{,}960/1.987(602)} \left\{ Ei\left[\frac{28{,}960}{1.987(436)} - \frac{28{,}960}{1.987(602)}\right] - Ei\left[\frac{28{,}960}{1.987(597.02)} - \frac{28{,}960}{1.987(602)}\right]\right\} - \left\{Ei\left[\frac{28{,}960}{1.987(436)}\right] - Ei\left[\frac{28{,}960}{1.987(597.02)}\right]\right\}$$

Solving

$$t = 0.1159 \text{ h}$$

5.6 *Optimization*

The operation of a batch reactor frequently poses the following problems:

1. What production time should be selected for obtaining maximum production with the specified reactor?
2. What optimum load should be required to obtain minimum production costs?
3. What optimum temperature should be maintained in the reactor?

We shall discuss each in the following.

5.6-1 *Maximum production*

There are two cases: maximum production based on total batch time and maximum yield based on reaction or operating time.

Maximum Production Based on Total Batch Time Let N_p be the moles of product X_3 created in a batch reactor for the following reaction

$$X_1 + \frac{\alpha_2}{\alpha_1} X_2 \underset{k_2}{\overset{k_1}{\rightleftharpoons}} \frac{\alpha_3}{\alpha_1} X_3 + \frac{\alpha_4}{\alpha_1} X_4 \qquad \textbf{(5.6-1)}$$

If t_c is the cleaning, charging, and discharging time, then the total batch time t_b is

$$t_b = t + t_c \qquad \textbf{(5.6-2)}$$

where t is the reaction time. The production per batch P is

$$P = \frac{N_P}{t_b} = \frac{N_P}{t + t_c} \tag{5.6-3}$$

where N_P is the moles of product produced which is related to the concentration of product C_P by

$$N_P = C_P V \tag{5.6-4}$$

where V is the volume of the reaction mixture in the reactor. In a batch reactor, the volume of the mixture is constant. The concentration of product C_P is related to λ, the unit conversion per unit volume, by

$$C_P = C_{Po} + \frac{\alpha_3}{\alpha_1}\lambda = \alpha'\lambda + C_{Po} \tag{5.6-5}$$

Initially there is no product in the reactor. Hence

$$C_P = \alpha'\lambda \tag{5.6-6}$$

Substituting Eqs. (5.6-4) and (5.6-5) into Eq. (5.6-3) gives

$$P = \frac{\alpha'\lambda V}{t + t_c} \tag{5.6-7}$$

Note that this λ is also a function of the reaction time t. This λ for various types of reaction has already been treated in detail in Chapter 3. Here in order to have maximum production, we differentiate Eq. (5.6-7) with respect to time t, equate to zero and solving for t yields

$$\frac{d\lambda^*}{dt} = \frac{\lambda^*}{t^* + t_c} \tag{5.6-8}$$

Then t^*, the optimum reaction time, can be solved by trial and error. Substituting into Eq. (5.6-2) gives the optimum batch time, and substituting into Eq. (5.6-7) yields the maximum production. If the expression for λ is too complicated, Eq. (5.6-8) can be solved graphically (see Fig. 5.6-1) by drawing the curve $\lambda = f(t)$, laying a distance t_c on the other side of the abscissa t, and finally drawing a tangent to the curve. The point of contact gives the optimum t^* and λ^*.

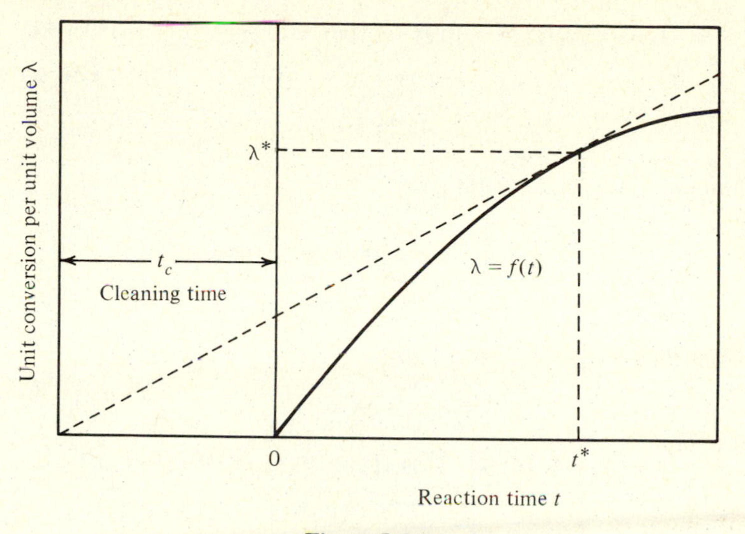

Figure 5.6-1

Example 5.6-1 Optimum Batch Time

Ethyl acetate is produced in a batch reactor which is charged with 3.91 kmol/m³ acetic acid, 10.20 kmol/m³ ethyl alcohol, and 17.56 kmol/m³ water initially. The forward rate constant k and the equilibrium constant K at 100°C for the following reaction

$$CH_3COOH + C_2H_5OH \underset{k'}{\overset{k}{\rightleftharpoons}} CH_3COOC_2H_5 + H_2O$$

are $k = 7.93 \times 10^{-6}$ m³/kmol·s, $K = k/k' = 2.93$. If the cleaning time is 1 h, find the optimum batch time to obtain maximum production.

Solution

The time dependence of the λ for the given reaction has been described in Chapter 3, Section 3.12-2 as

$$\ln \frac{\sigma(\lambda_e - \lambda)}{\lambda_e(\sigma + \lambda)} = (\lambda_e + \sigma)D_1 t \qquad \textbf{(A)}$$

where

$$D_1 = k - k' = 7.93 \times 10^{-6} - 2.7065 \times 10^{-6} = 5.2235 \times 10^{-6}$$

$$D_2 = -[k(C_{1o} + C_{2o}) + k'(C_{3o} + C_{4o})]$$

$$= -[7.93 \times 10^{-6}(3.91 + 10.2) + 2.7065 \times 10^{-6}(17.65)] = -159.4184 \times 10^{-6}$$

$$D_3 = kC_{1o}C_{2o} - k'C_{3o}C_{4o} = 7.93 \times 10^{-6}(3.91)(10.2) - 2.7065 \times 10^{-6}(17.65)(0)$$

$$= 316.26 \times 10^{-6}$$

Substituting into Eq. (3.12-12) gives

$$\frac{d\lambda}{dt} = 5.2235 \times 10^{-6}\lambda^2 - 159.4184 \times 10^{-6}\lambda + 316.26 \times 10^{-6} \qquad \textbf{(B)}$$

At equilibrium, $\dfrac{d\lambda}{dt} = 0.$

Therefore

$$5.2235 \times 10^{-6}\lambda_e^2 - 159.4184 \times 10^{-6}\lambda_e + 316.26 \times 10^{-6} = 0$$

Solving

$$\lambda_e = 2.1329 \qquad \text{the other value is rejected}$$

Then

$$\sigma = \lambda_e + \frac{D_2}{D_1} = 2.1329 - \frac{159.4184 \times 10^{-6}}{5.2235 \times 10^{-6}} = -28.3865 \qquad \textbf{(C)}$$

Substituting into Eq. (A) yields

$$\ln \frac{-28.3865(2.1329 - \lambda)}{2.1329(28.3865 + \lambda)} = (2.1329 - 28.3865)(5.2235 \times 10^{-6}t)$$

Solving for λ

$$\lambda = \frac{60.5455(1 - e^{-137.35\times10^{-6}t})}{28.3865 - 2.1329(e^{-137.135\times10^{-6}t})} \qquad \textbf{(D)}$$

where t is in seconds or

$$\lambda = \frac{60.5455(1 - e^{-0.4936\,t'})}{28.3865 - 2.1329(e^{-0.4936\,t'})} \qquad \textbf{(E)}$$

where t' is in hours.

Differentiating Eq. (E) with respect to t' gives

$$\frac{d\lambda}{dt'} = 60.5455(0.4936e^{-0.4936\,t'})$$

$$\left[\frac{(28.3865 - 2.1329e^{-0.4936\,t'}) - (1 - e^{-0.4936\,t'})2.1329}{(28.3865 - 2.1329e^{-0.4936\,t'})^2}\right] \qquad \textbf{(F)}$$

Substituting Eqs. (E) and (F) into Eq. (5.6-8) and simplifying yields

$$\frac{1}{e^{0.4936t'^*}-1}+\frac{1}{1-13.3088e^{0.4936t'^*}}=\frac{2.0259}{t'^*+1} \tag{G}$$

Solving by trial and error yields

$$t'^* = 1.64 \text{ h}$$

(See Fig. 5.6-2.) Substituting into Eq. (E) gives

$$\lambda^* = 1.2245 \text{ kmol/m}^3$$

Then the fractional conversion

$$f = \frac{1.2245}{3.91} = 0.3132$$

$$\text{Production rate} = \frac{1.2245}{1+1.64} = 0.4638 \text{ kmol/m}^3\cdot\text{h}$$

$$= 0.4638\ (24)(88) = 979.6 \text{ kg/day}\cdot\text{m}^3$$

If the volume of the mixture is 52 m^3, then the production rate = 50,939 kg/day.

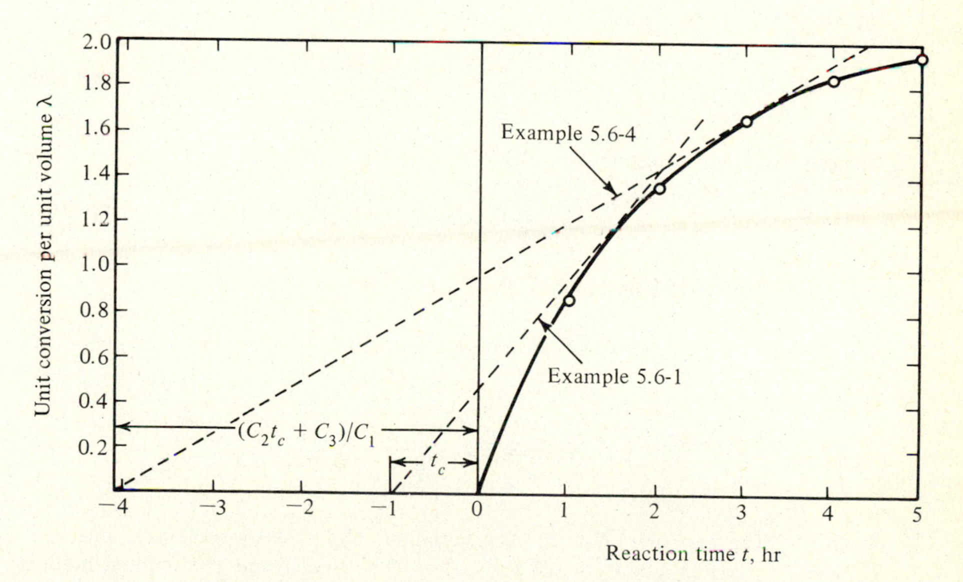

Figure 5.6-2 Optimum batch time and optimum load

Maximum Yield Based on Reaction Time In Chapter 3 the variation of concentrations of each component in different types of reaction has been fully described. In order to obtain the maximum yield based on reaction time, we simply differentiate the function with respect to the reaction time, equate to zero and solve for the time.

Example 5.6-2 Optimum Holding Time for a Series Reaction

For the following set of first-order series reactions

$$X_1 \xrightarrow{k_1} X_2 \xrightarrow{k_2} X_3$$

determine the optimum holding time to maximize the concentration of X_2.

Solution

In Chapter 3, Section 3.13-2, the governing equations are Eqs. (3.13-22), (3.13-23), and (3.13-24). The relationship of the concentration of component X_2 with time is given by Eq. (3.13-28) as

$$C_2 = C_{1o} k_1 \frac{e^{-k_1 t} - e^{-k_2 t}}{k_2 - k_1} \qquad \textbf{(3.13-28)}$$

Differentiating C_2 in Eq. (3.13-28) with respect to t, equating to zero, and solving for t gives

$$t_{\text{max}} = \frac{\ln k_1/k_2}{k_1 - k_2} = \frac{1}{k_{\text{logmean}}} \qquad \textbf{(3.13-31)}$$

and the maximum concentration of X_2 is

$$C_{2,\text{max}} = C_{1o}(k_1/k_2)^{k_2/(k_2 - k_1)} \qquad \textbf{(3.13-32)}$$

Example 5.6-3 Optimum Holding Time for a Complex Reaction

For the following set of first-order complex reactions

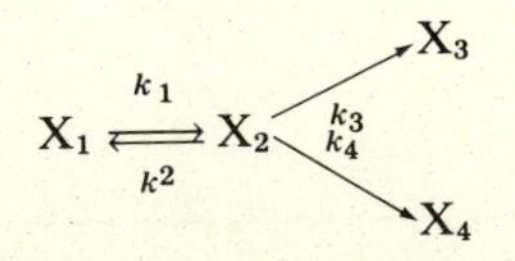

if C_{1o} = 50 g mol/liter, C_{2o} = 5.0 g mol/liter, C_{4o} = 3.0 g mol/liter, and k_1 = 1.75 s^{-1}, k_2 = 1.25 s^{-1}, k_3 = 0.2 s^{-1}, k_4 = 0.8 s^{-1}, determine the optimal holding time in a batch reactor and also the maximum concentration of X_2.

Solution

1. Stoichiometric equations:

$$C_j = C_{jo} + \Sigma \alpha_{ij} \lambda_i$$

$$C_1 = C_{1o} + \alpha_{11}\lambda_1 + \alpha_{21}\lambda_2 = C_{1o} - \lambda_1 + \lambda_2 \tag{A}$$

$$C_2 = C_{2o} + \alpha_{12}\lambda_1 + \alpha_{22}\lambda_2 + \alpha_{32}\lambda_3 + \alpha_{42}\lambda_4 = C_{2o} + \lambda_1 - (\lambda_2 + \lambda_3 + \lambda_4) \tag{B}$$

$$C_3 = C_{3o} + \alpha_{33}\lambda_3 = C_{3o} + \lambda_3 \tag{C}$$

$$C_4 = C_{4o} + \alpha_{34}\lambda_4 = C_{4o} + \lambda_4 \tag{D}$$

2. Overall rate equations:

$$\frac{d\lambda_I}{dt} = r_i = k_i \Pi C_j^{\alpha j} - k_i' \Pi C_j^{\alpha' j}$$

$$\frac{d\lambda_1}{dt} = r_1 = k_1 C_1 \tag{E}$$

$$\frac{d\lambda_2}{dt} = r_2 = k_2 C_2 \tag{F}$$

$$\frac{d\lambda_3}{dt} = r_3 = k_3 C_2 \tag{G}$$

$$\frac{d\lambda_4}{dt} = r_4 = k_4 C_2 \tag{H}$$

3. Component rate equations:

$$\frac{dC_j}{dt} = \Sigma \alpha_{ij} r_i$$

$$\frac{dC_1}{dt} = -r_1 + r_2 = -k_1 C_1 + k_2 C_2 \tag{I}$$

$$\frac{dC_2}{dt} = r_1 - r_2 - r_3 - r_4 = k_1 C_1 - (k_2 + k_3 + k_4) C_2 \tag{J}$$

$$\frac{dC_3}{dt} = r_3 = k_3 C_2 \tag{K}$$

$$\frac{dC_4}{dt} = r_4 = k_4 C_2 \tag{L}$$

Eliminating the λ's in Eqs. (A), (B), (C), and (D), we get

$$C_1 + C_2 + C_3 + C_4 = C_{1o} + C_{2o} + C_{3o} + C_{4o} \tag{M}$$

Now the four concentrations C_1, C_2, C_3, C_4 at any time can be determined from Eq. (I), (J), (K), and (L) with the given initial concentrations C_{1o} = 50 g mol/liter, C_{2o} = 5.0 g mol/liter, C_{4o} = 3.0 g mol/liter. First solving C_1 from Eq. (J) gives

$$C_1 = \frac{1}{1.75}\left(\frac{dC_2}{dt} + 2.25\right) \tag{N}$$

Differentiating Eq. (N) with respect to t yields

$$\frac{dC_1}{dt} = \frac{1}{1.75}\left(\frac{d^2C_2}{dt^2} + 2.25\frac{dC_2}{dt}\right) \tag{O}$$

Substituting Eq. (N) and Eq. (O) into Eq. (I) and simplifying yields

$$\frac{d^2C_2}{dt^2} + 4\frac{dC_2}{dt} + 1.75C_2 = 0 \tag{P}$$

which is a second-order linear differential equation. With the initial conditions C_{1o} = 50 g mol/liter, and C_{2o} = 5.0 g mol/liter, the solution of Eq. (P) can be determined as

$$C_2 = 31.25e^{-(1/2)t} - 26.25e^{-(7/2)t} \tag{Q}$$

Substituting back into Eq. (N) gives

$$C_1 = 31.25e^{-(1/2)t} + 18.75e^{-(7/2)t} \tag{R}$$

Integrating Eq. (L) with C_2 from Eq. (Q) and C_{4o} = 3.0 g mol/liter yields

$$C_4 = 47 + 6e^{-(7/2)t} - 50e^{-(1/2)t} \tag{S}$$

Finally C_3 can be determined from Eq. (M) by assuming $C_{3o} = 0$ as

$$C_3 = 11 - 12.5e^{-(1/2)t} + 1.5e^{-(7/2)t} \tag{T}$$

In order to obtain the optimal holding time for maximum yield of C_2, we can differentiate Eq. (Q) with respect to the time t, equate to zero, and solve for t, which will become the optimal holding time. The result is $t^* = 0.59$ s. Substituting back into Eq. (Q) gives the maximum yield of X_2 as C_2^* = 19.94 g mol/liter.

5.6-2 *Minimum production cost*

If c_1 is the unit operating cost per unit time, then the operating cost during production is c_1t where t is the reaction time. If c_2 is the unit cost per unit time for charging, cleaning, and discharging, then the cost of cleaning etc. is c_2t_c where

t_c is the cleaning time, and if c_3 is the cost for overhead and other expenses, then the total production cost is

$$\text{Total production cost} = c_1 t + c_2 t_c + c_3 \tag{5.6-9}$$

During this period, $t + t_c$, N_P moles of product are made. Therefore

$$\text{Total production cost per mole of product} = \frac{c_1 t + c_2 t_c + c_3}{N_P} \tag{5.6-10}$$

In order to have this total production cost per mole of product minimum, we differentiate Eq. (5.6-10) with respect to the operating time t.

$$\frac{d[(c_1 t + c_2 t_c + c_3)/N_P]}{dt} = \frac{c_1}{N_P} - \frac{c_1 t + c_2 t_c + c_3}{N_P^2} \frac{dN_P}{dt} \tag{5.6-11}$$

Equating to zero gives

$$\frac{dN_P^*}{dt} = \frac{N_P^*}{t + [(c_2 t_c + c_3)/c_1]} \tag{5.6-12}$$

But the combination of Eq. (5.6-4) and Eq. (5.6-5) gives

$$N_P = \alpha' \lambda V \tag{5.6-13}$$

Differentiating Eq. (5.6-13) and substituting into Eq. (5.6-12) gives the final design equation as

$$\frac{d\lambda^*}{dt} = \frac{\lambda^*}{t + [(c_2 t_c + c_3)/c_1]} \tag{5.6-14}$$

As indicated in Section 5.6-1, the time dependence of λ for different kinds of reactions has been described in Chapter 3. Substituting λ and its derivative into Eq. (5.6-14), we can employ the trial-and-error method to evaluate the optimal time t^*. If $\lambda = f(t)$ is too complicated, the graphical method which is similar to that described in Section 5.6-1 can be used with the replacement of t_c by $(c_2 t_c + c_3)/c_1$ (see Fig. 5.6-3). After t^* is determined, λ^* and N_P^* can be evaluated.

Example 5.6-4 Minimum Total Cost

If c_1 = \$27.60 h^{-1}, c_2 = \$8.40 h^{-1}, and c_3 = \$104.00, determine the minimum total cost for operating the reaction as shown in Example 5.6-1.

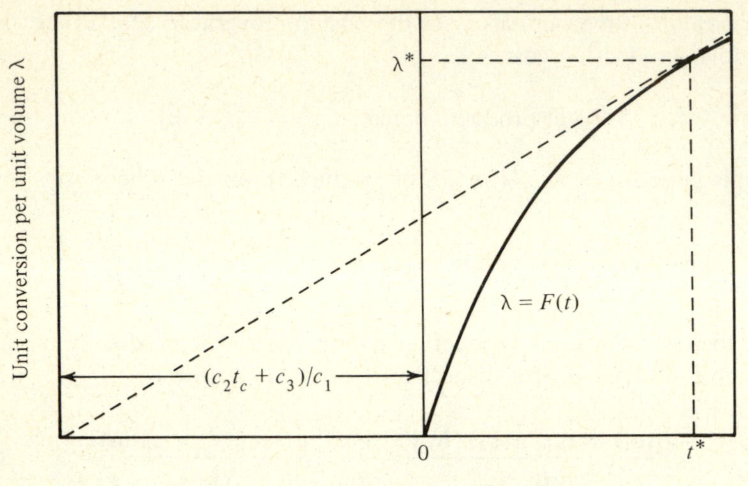

Figure 5.6-3 Minimum total cost

Solution

The calculation is exactly the same as Example 5.6-1 except that the t_c is replaced by

$$\frac{c_2 t_c + c_3}{c_1} = \frac{8.40(1) + 104.00}{27.60} = 4.0724$$

Hence Eq. (G) in Example 5.6-1 is replaced by

$$\frac{1}{e^{0.4936\,t'^*} - 1} + \frac{1}{1 - 13.3088 e^{0.4936\,t'^*}} = \frac{2.0259}{t'^* + 4.0724}$$

Solving by trial and error gives $t'^* = 2.89$ h. (See Fig. 5.6-2.) Substituting into Eq. (E) in Example 5.6-1 yields $\lambda^* = 1.65$ kmol/m^3. Therefore

$$f = \frac{1.65}{3.91} = 0.42$$

Assuming $V = 52$ m^3

$$N_P^* = 1.65(52) = 85.8 \text{ kmol} = 7550 \text{ kg}$$

$$\text{Total production cost} = \frac{27.6(2.89) + 8.40(1) + 104}{85.8} = \$2.24 \text{ kmol}^{-1} = \$0.025 \text{ kg}^{-1}$$

5.6-3 Optimal Operating Temperature

In operating a batch reactor, we want to have the time for the reaction as short as possible. From the design equation Eq. (5.2-7)

$$t = C_{ko} \int_{f_0}^{f} \frac{df}{r} \tag{5.2-7}$$

we see that the minimization of the time is equivalent to maximize the rate of reaction. Because the reaction rate is a function of concentrations of the constituents as well as the rate constants, which in turn is dependent on the temperature by the well-known Arrhenius's law, we can anticipate an optimum operating temperature to carry the reaction to the maximum if such a maximum exists. For irreversible reactions, the reaction rate increases with temperature at all levels of fractional conversion. There exists no such maximum. Thus the operating temperature is governed by the material of the reactor and possibly by some other factors. The maximum temperature which the material of the vessel can endure is usually taken as the operating temperature to give the maximum rate of the reaction. Then the performance of the reactor can be assessed by Eq. (5.2-7) for a fixed value of k. For reversible reactions

$$\alpha_1 X_1 + \alpha_2 X_2 \underset{k_2}{\overset{k_1}{\rightleftharpoons}} \alpha_3 X_3 + \alpha_4 X_4$$

the reaction rate equation is

$$r = k_1 C_1^{\alpha_1} C_2^{\alpha_2} - k_2 C_3^{\alpha_3} C_4^{\alpha_4} \tag{3.2-5}$$

or in general

$$r = k_1 \Pi C_j^{\alpha_j} - k_2 \Pi C_j^{\alpha'_j} = k_1 \left(\Pi C_j^{\alpha_j} - \frac{\Pi C_j^{\alpha'_j}}{K} \right) \tag{3.2-6}$$

The equilibrium ratio K is defined as the ratio of the forward rate constant to the reverse rate constant

$$K = \frac{k_1}{k_2} = \frac{k_{10} e^{-E_1/RT}}{k_{20} e^{-E_2/RT}} = \frac{k_{10}}{k_{20}} e^{(E_2 - E_1)/RT} \tag{5.6-15}$$

However, the effect of temperature of the equilibrium ratio can be indicated by combining Eqs. (2.10-1) and (2.10-2) to yield

$$\frac{d(\ln K)}{dT} = \frac{\Delta H_T^\circ}{RT^2} \tag{5.6-16}$$

By assuming ΔH_T°, which is the heat of reaction at temperature T, constant in the short interval of temperatures, Eq. (5.6-16) can be integrated to

$$K = Ae^{-\Delta H_T^\circ / RT} \tag{5.6-17}$$

where A is an integrating constant. Equating Eq. (5.6-15) with Eq. (5.6-17) gives

$$A = \frac{k_{1o}}{k_{2o}} \tag{5.6-18}$$

and

$$E_2 - E_1 = -\Delta H_T^\circ \tag{5.6-19}$$

Therefore, if $E_2 < E_1$, $-\Delta H_T^\circ$ is negative, the reaction is endothermic. Equation (5.6-17) indicates that K increases with increasing T. Then the terms within the parentheses in Eq. (3.2-6) will increase with increasing temperature. Since the rate constant k_1 also increases with temperature, the rate of the reaction will also increase. Thus the maximum temperature the material of the reactor can endure is selected as the operating temperature for having the maximum rate of reaction. Therefore, if $E_2 > E_1$, $-\Delta H_T^\circ$ is positive, the reaction is exothermic. Equation (5.6-17) indicates that K decreases with increasing T. Then the terms within the parentheses in Eq. (3.2-6) will decrease with increasing temperature, whereas the forward rate constant k_1 will increase with temperature. Hence for each possible composition, there is a temperature at which the rate of the reaction is maximum.

Now start with

$$r = k_1(\Pi C_j^{\alpha_j} - \Pi C_j^{\alpha'_j}/K) \tag{3.2-6}$$

Substituting the Arrhenius equation for k_1 by Eq. (3.5-1) and the temperature dependence for K by Eq. (5.6-17) gives

$$r = k_{1o}e^{-E_1/RT}\left[\Pi C_j^{\alpha_j} - \frac{\Pi C_j^{\alpha'_j}}{Ae^{-\Delta H/RT}}\right] \tag{5.6-20}$$

The derivative of r with respect to temperature T is

$$\frac{dr}{dT} = k_{1o}e^{-E_1/RT}\left[\frac{-\Pi C_j^{\alpha'_j}}{A} e^{+\Delta H/RT}\left(\frac{-\Delta H}{RT^2}\right)\right] + \left[\Pi C_j^{\alpha_j} - \frac{\Pi C_j^{\alpha'_j}}{Ae^{-\Delta H/RT}}\right]k_{1o}e^{-E_1/RT}\left(\frac{E_1}{RT^2}\right) \tag{5.6-21}$$

Setting to zero and simplifying gives

$$K = \frac{\Pi C_j^{\alpha'_j} E_2}{\Pi C_j^{\alpha_j} E_1} = \frac{k_1}{k_2} \tag{5.6-22}$$

Therefore the optimal temperature can be found to satisfy Eq. (5.6-22). Substituting the T_{opt} into Eq. (5.6-20) yields the maximum rate of the reaction. Finally integrating Eq. (5.2-7) gives the minimum time for the reaction.

Example 5.6-5 Maximum Rate of Reaction

Find the optimal temperature sequence to yield maximum rate of reaction for the following reversible exothermic reaction

$$A \underset{k_2}{\overset{k_1}{\rightleftharpoons}} B$$

where $k_1 = 2.51 \times 10^5 e^{-10{,}000/RT}$ h^{-1}, $k_2 = 1.995 \times 10^7 e^{-20{,}000/RT}$ h^{-1}. The maximum allowable temperature for the reactor is 1,000°R. Initially no product is present.

Solution

The rate of the reaction is

$$r = k_1 C_A - k_2 C_B \tag{A}$$

But

$$C_A = C_{Ao}(1 - f)$$

and no B is present at start

$$C_B = C_{Ao} f$$

thus

$$r = k_1 C_{Ao}(1 - f) - k_2 C_{Ao} f \tag{B}$$

Therefore

$$\Pi C_j^{\alpha'_j} = C_{Ao} f \tag{C}$$

$$\Pi C_j^{\alpha_j} = C_{Ao}(1 - f) \tag{D}$$

Substituting into Eq. (5.6-22) gives

$$\frac{C_{Ao} f E_2}{C_{Ao}(1 - f) E_1} = \frac{k_1}{k_2} \tag{E}$$

With the numerical values from the example, Eq. (E) becomes

$$\frac{20{,}000f}{10{,}000(1-f)}=\frac{2.51\times 10^5 e^{-10{,}000/RT}}{1.995\times 10^7 e^{-20{,}000/RT}} \tag{F}$$

Hence for each f, a value of the optimum temperature T^* can be calculated to satisfy Eq. (F). Then k_1 and k_2 can be evaluated and the rate of the reaction can be computed from Eq. (A) r_m. Finally the minimum time t_m^* required for the reaction can be evaluated by Eq. (5.2-7). The calculation is tabulated below:

f	T^*	$k_1 \times 10^{-3}$	$k_2 \times 10^{-2}$	$r_m \times 10^{-3}$	$\frac{1}{r_m} \times 10^4$	$t^* \times 10^5$ isothermal
0	1000.0	1.6368	8.4837	1.6368	6.1090	
0.1	(1752.6)	(14.209)	(639.33)			(1.0217)
	1000.00	1.6368	8.4837	1.3882	7.2031	
0.2	(1366.7)	(6.3159)	(126.32)			(4.8359)
	(1000.)	1.6368	8.4837	1.1398	8.7737	
0.3	(1192.2)	(3.6843)	(42.984)			(13.150)
	1000.0	1.6368	8.4837	89.125	11.220	
0.4	(1079.2)	(2.3684)	(17.764)			(29.049)
	1000.0	1.6368	8.4837	64.273	15.558	
0.5	992.91	1.5789	7.8950	39.470	25.336	58.530
0.6	919.36	1.0526	3.5088	21.051	47.503	114.67
0.7	850.70	0.67671	1.4501	10.150	98.516	230.87
0.8	779.66	0.39473	0.49339	394.75	253.33	518.50
0.9	692.65	0.17544	0.97471	8771.6	1140.0	1617.0

The minimum time required for $f = 0.8$ is evaluated from Eq. (5.2-7) by Simpson's rule as

$$t_m^* = C_{Ao}\int_0^{0.8}\frac{df}{r_m}$$

$$=\frac{0.1}{3}\times 10^{-4}\,[6.1090+4(7.2031+11.220+25.336+98.516)$$
$$+2(8.7737+15.558+47.503)+253.33]$$

$$=3.24\times 10^{-3}=0.00324\text{ h}=11.66\text{s}$$

The last column in the table is calculated by integrating Eq. (B) as follows:

$$r=-\frac{dC_A}{dt}=k_1C_{Ao}(1-f)-k_2C_{Ao}f$$

But

$$C_A=C_{Ao}(1-f)$$

therefore

$$\frac{dC_A}{dt} = -C_{Ao}\frac{df}{dt}$$

and Eq. (B) becomes

$$\frac{df}{dt} = k_1(1-f) - k_2 f \tag{G}$$

Separating variables and integrating

$$t^* = \frac{-1}{(k_1+k_2)} \ln\left[1 - \left(\frac{k_1+k_2}{k_1}\right)f\right] \tag{H}$$

where t^* may be defined as the isothermal optimal operating time. It is seen from the table that for the same conversion, this isothermal optimal operating time is 5.185×10^{-3} h or 18.66 s, which is greater than the time for maximum rate of reaction, 11.66 s. Figure 5.6-4 is the plot of $1/r$ versus f for the calculation of the optimal operating time to obtain the maximum rate of reaction. From the table, it is seen that the optimal temperatures in the initial period are extremely high.

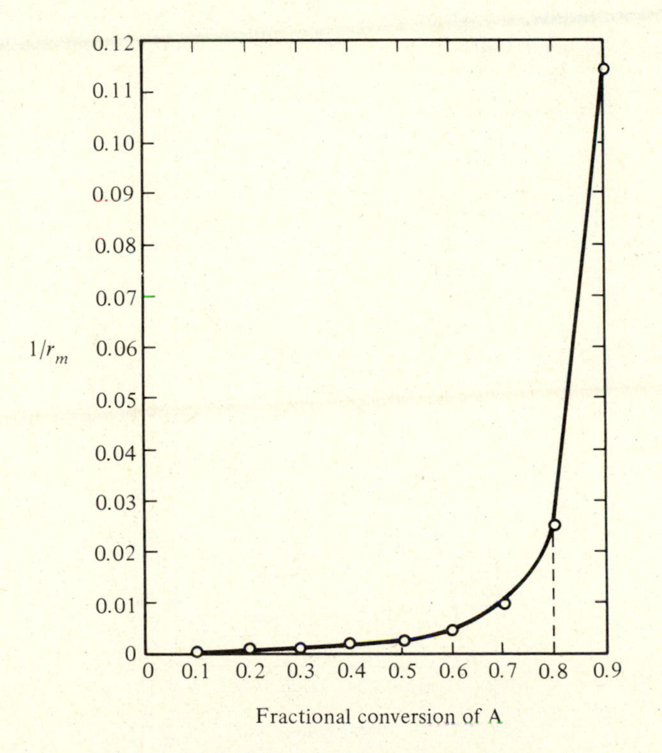

Figure 5.6-4 Optimal operating temperature

Since the allowable temperature is 1,000°R, we use it for the calculation. In the later part of the reaction, it is preferable to use lower optimal operating temperatures. The temperature-conversion time history for isothermal conditions can be obtained from Eq. (H) by assuming a fixed value of t^* and calculating the fractional conversion at various temperatures. The results are tabulated below:

T, °R	1,100	1,050	1,000	950	900	850	800	775	750	700
t^*	Fractional conversion, f									
0.001	0.5447	0.5837	0.6038	0.5917	0.5420					
0.002			0.6541	0.6941	0.7032	0.6634	0.5718			
0.004				0.7148	0.7653	0.7928	0.7685	0.6734		
0.006					0.7709	0.8181	0.8362	0.8231	0.7898	0.6605
0.008					0.7714	0.8230	0.8594	0.8629	0.8494	0.7540
0.01						0.8240	0.8674	0.8780	0.8798	0.8166

The fractional conversion at equilibrium f_e can be shown very easily as

$$f_e = \frac{k_1}{k_1 + k_2}$$

From these results, the temperature-conversion time history is plotted in Fig. 5.6-5. Figure 5.6-6 is a plot of fractional conversion versus temperature using time as parameters.

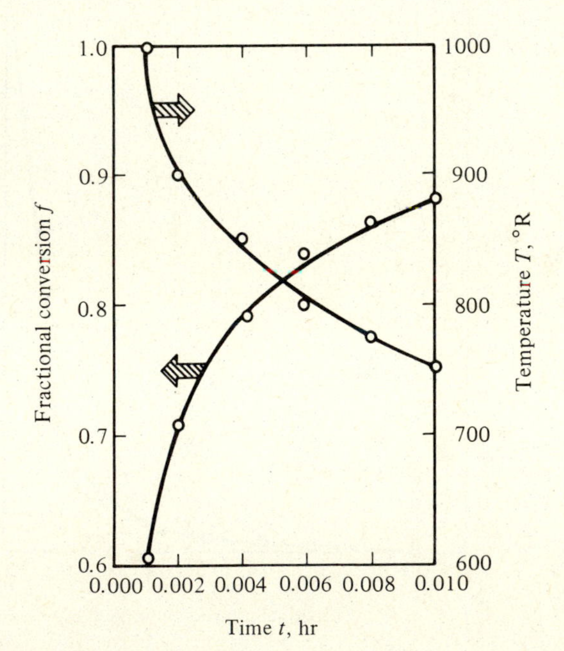

Figure 5.6-5 Isothermal optimum temperature-conversion-time history

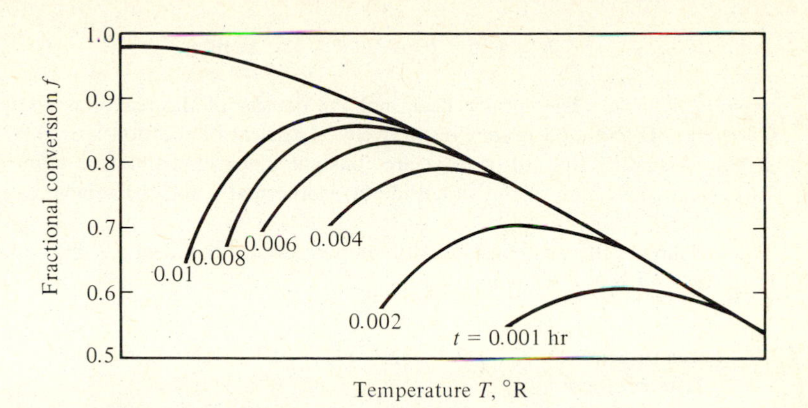

Figure 5.6-6 Isothermal conversion vs. temperature at various times

Problems

1. The rate constant for the decomposition of nitrous oxide at 895°C

$$2N_2O \longrightarrow 2N_2 + O_2$$

is $k_c = 977$ cm³/g mol · s. Calculate the time required to decompose 90 percent of the nitrous oxide at 895°C

(a) at constant volume with initial pressure of 1 atm.

(b) at constant pressure of 1 atm.

2. The rate constants for the acidification of an aqueous solution of sodium propionate

$$C_2H_5COONa + HCl \underset{k'}{\overset{k}{\rightleftharpoons}} C_2H_5COOH + NaCl$$

has been studied by Stevens (S1) and determined as $k = 0.0236$ liter/g mol · min. and $k' = 0.00148$ liter/g mol · min. The initial charge to a batch reactor contains 453.59 kg C_2H_5COONa and 177 kg HCl per 1,514 liters. Assuming that the reaction mixture has a constant density of 1186.3 g/liter, determine the volume of the reactor required for 75 percent conversion of the salt to give an average production of 450 kg/h of the propionic acid if the cleaning and heating times are totally 30 min.

3. A batch reactor for carrying out the following pseudo first-order irreversible reaction

$$(CH_3CO)_2O + H_2) \xrightarrow{k} 2CH_3COOH$$

is charged with an anhydride solution of 0.25 kmol/m³ at 15°C. The rate constant is

$$k = 4.15 \times 10^5 e^{-11,200/RT} \quad \text{in min}^{-1}$$

where T is in K. The specific heat, and the density of the reaction mixture are 0.9 cal/g · °C and 1.05 g/cm³, respectively. The heat of reaction is −50,000 cal/g mol. Assuming that all of these are constant, determine the time required for 70 percent conversion of the anhydride (a) isothermally and (b) adiabatically.

4. The variation of the rate constant with temperature for the ideal gas decomposition reaction A → B + C is as follows:

Temperature T, K	333	338	344	350	355	361	366
Rate constant, $k \times 10^4$, s^{-1}	3.33	4.66	6.41	9.11	12.81	20.00	26.14

Pure A enters a reactor of capacity of 250 liters at 333 K and 5 atm. The specific heat of A is 30, and those of B and C are each 25 cal/g mol · K. The heat of reaction is −1,389 cal/g mol at 333 K. Determine the relationship between conversion and the time under adiabatic and isothermal conditions.

5. The rate constant for the first order irreversible isomerization in liquid solution

$$X(l) \xrightarrow{k} Y(l)$$

is $k = 2.7 \times 10^{14} e^{-15,000/T}$ h⁻¹ where T is in K. The other data are:

Heat of reaction = −85 cal/g at 160°C

Specific heat of X and Y = 0.48 cal/g · °C

Density of X and Y = 0.85 g/cm³

Total time of cleaning = 30 min

Molecular weight of X and Y = 240

Determine the reactor volume required for 90 percent conversion to produce 1 million kg Y in 6,000 h (a) under isothermal conditions at 160°C. (b) under adiabatic conditions assuming the heat of reaction being constant (c) with heat input of 700,000 cal/min.

6. For the hydrolysis of methyl acetate in a batch reactor

$$CH_3COOCH_3 + H_2O \underset{k'}{\overset{k}{\rightleftharpoons}} CH_3COOH + CH_3OH$$

the experimental data for initial concentration of the ester 0.05 kmol/m³ are (C1)

Time, s	1,350	3,060	5,340	7,740	∞
Fractional conversion	0.21	0.43	0.60	0.73	0.90

Prove that the reaction is of pseudo first order in the forward direction and of second order in the reverse. Evaluate these two constants.

7. The reaction A + B $\longrightarrow$ product has the following properties:

   Heat of reaction = −20,000 cal/g mol A

   Activation energy = 18,500 cal/g mol

   Frequency factor = 6.24×10^7 liter/g mol · min

   Density = 1 g/cm^3

   Specific heat = 1 cal/g · °C

   Initial conditions are: $C_{Ao} = C_{Bo} = 8.0$ g mol/liter, $T_o = 96$°C. Estimate the temperature and concentration variation with time for adiabatic operation.

8. Butyl acetate is to be produced batchwise at 100°C. A stream containing 5.00 mol butanol per mole of acetic acid is fed in. Estimate the time required for 55 percent conversion of the acid. If an average production of the ester is 100 kg/h and the cleaning and recharging time is 0.75 h, determine the volume of the reactor for the following data:

$$r = kC_A^2$$

   where C_A is the acetic acid concentration in gram-moles per cubic centimer, k = 17.4 cm^3/g mole · min, densities of acetic acid, butanol, water, and the mixture are 0.958, 0.742, 0.796, and 0.75 g/cm^3, respectively.

9. The chemical reaction A $\xrightarrow{k_1}$ B $\xrightarrow{k_2}$ C is carried out in an isothermal batch reactor. If the initial concentrations of A, B, and C are A_o, 0, and 0, respectively, and the values per unit mole of the species are c_A, c_B, and c_C, find the operating t which maximizes the profit of the manufacture

$$V = c_A[A(t) - A_o] + c_B B(t) + c_C C(t)$$

10. For the system in Prob. 9, suppose that k_1 and k_2 depend on the temperature T in Arrhenius form

$$k_i = k_{io} e^{-E/RT} \qquad i = 1, 2$$

   For fixed total operating time t find the optimal constant temperature. Note the difference in results for the two cases $E_1 < E_2$ (exothermic) and $E_2 < E_1$ (endothermic).

11. Pure phosphine at 672°C and 1 atm decomposes in a batch reactor by the following reaction

$$4PH_3(g) \longrightarrow P_4(g) + 6H_2(g)$$

Determine the time for 20 percent conversion under both isothermal and adiabatic conditions.

1. $\log k = -(18{,}963/T) + 2 \log T + 12.13$ where k in s^{-1}, T in K
2. Heat of reaction at 18°C = 5,665 cal/g mol PH_3
3. Specific heats (cal/g mol · K):
   $P_4: C_p = 5.9 + 0.0096T$
   $PH_3: C_p = 6.7 + 0.0063T$
   $H_2: C_p = 6.88 + 0.066 \times 10^{-3}T + 0.279 \times 10^{-6}T^2$

12. Dimethyl ether decomposes to methane, carbon monoxide, and hydrogen

$$CH_3OCH_3 \xrightarrow{k} CH_4 + CO + H_2$$

with the rate constant

$$k_c = 1.55 \times 10^{13} e^{-58{,}500/RT}\ s^{-1}$$

100 g ether are heated at 500°C at constant-volume conditions in a closed vessel of 125 liters for 30 min. Calculate the percentage conversion of the ether and the final pressure in the closed vessel.

13. Verify Eq. (5.5-15).

14. If C_{Ao} is not equal to C_{Co}, verify the following result:

$$At = \frac{1}{C_{Ao} - C_{Co}} \left\{ e^{E/RT_c} \left[Ei\left(\frac{E}{RT_o} - \frac{E}{RT_c}\right) - Ei\left(\frac{E}{RT} - \frac{E}{RT_c}\right) \right] - e^{E/RT_a} \left[Ei\left(\frac{E}{RT_o} - \frac{E}{RT_a}\right) - Ei\left(\frac{E}{RT} - \frac{E}{RT_a}\right) \right] \right\}$$

15. For zero-order reaction, verify the following result:

$$JAt = Te^{E/RT} - T_o e^{E/RT_o} - \frac{E}{R}\left[Ei\left(\frac{E}{RT}\right) - Ei\left(\frac{E}{RT_o}\right) \right]$$

16. Solve Example 5.3-1 by the exponential integral.

References

(C1). Coulson, J.M., and Richardson, J.F. *Chemical Engineering, Volume III.* New York: Pergamon Press, 1971.

(D1). Douglas, J.M., and Eagleton, L.C. *Ind. Eng. Chem. Fund.*, 1:116 (1962).

(G1). Grummitt, O., and Fleming, H. *Ind. Eng. Chem.*, 37:485 (1945).

(S1). Stevens, W.F. "An Undergraduate Course in Homogeneous Reaction Kinetics," presented at Fourth Summer School for Chemical Engineering Teachers, Pennsylvania State University, June 27, 1955.

CHAPTER SIX

Ideal Semibatch Reactors

6.1 Introduction

A semibatch reactor is a reactor that is initially charged with reactants and then the same or other reactants are added continuously to the vessel. Because the composition of the reaction mixture in the reactor changes with time, the operation is batch; because another stream is added continuously to the vessel, the operation is continuous. Hence the procedure is called a *semibatch operation.*

A semibatch reactor is employed when the reaction evolves a great deal of heat which would cause excessive temperatures in normal batch operation. For example, the reaction of ammonium hydroxide and formaldehyde in the manufacture of hexamethylenetetramine is highly exothermic. If it is carried out in a batch reactor, a large increase in temperature will result and the yield will be

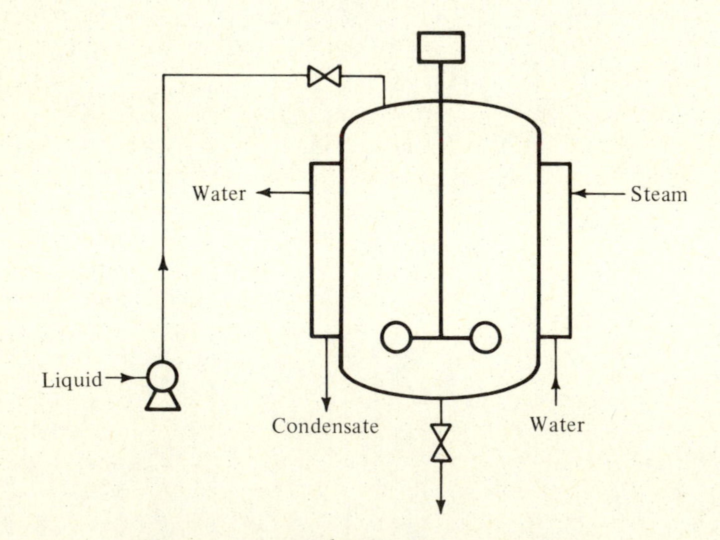

Figure 6.1-1 Liquid-liquid semibatch reactor

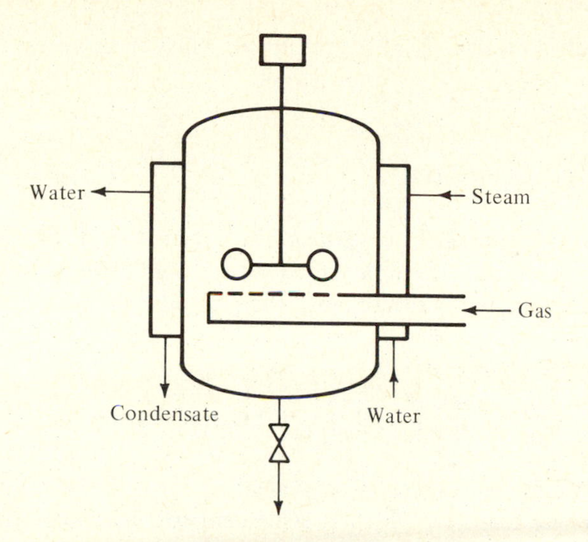

Figure 6.1-2 Gas-liquid semibatch reactor

decreased. However, if the ammonium hydroxide solution is added continuously and gradually at a controlled rate to a reactor containing formalin, (see Fig. 6.1-1), the temperature rise can be reduced and thus can be controlled.

Sometimes semibatch operation is used to prevent the formation of by-products. For example, in the chlorination of benzene, the by-products di- and trichlorobenzene will result if the concentration of chlorine is great. To reduce the yield of the by-products, the benzene may be chlorinated in a stirred-tank reactor by first charging the tank with liquid benzene and the catalyst, and then continuously bubbling chlorine gas through the benzene until the required ratio of chlorine to benzene is obtained.

Another application of the semibatch reactor is to increase the yield of a reversible reaction by removing one of the products. For example, acetic anhydride is hydrolyzed by initially charging the tank with the anhydride solution, then heating to a certain temperature at which the anhydride solution is added continuously to the reactor while the product is withdrawn at the same rate. Another example is the esterification of acetic acid with ethyl alcohol. Initially pure alcohol is charged to a reactor; then an aqueous solution of acetic acid is added continuously. After all the acetic acid is added, the entire amount of the reaction mixture is withdrawn.

The equipment used in a semibatch operation is the same as in a batch operation except that the added stream is pumped to the batch reactor at a controlled rate. The batch reactor should be equipped with a stirrer, a heating and cooling jacket, and a control valve on the outlet so that the rate of the outlet stream can be controlled if necessary. For a gas-liquid system (see Fig. 6.1-2), a distributor for the entering gas stream is immersed in the liquid. Sometimes a vertical condenser should be installed on the top of the reactor to reflux the condensate of the volatile vapor.

6.2 Isothermal Condition

We have used the generalized equations in Chapter 4 to design the batch reactor in Chapter 5. Similarly, the same equations are applicable here to the semibatch reactor. At constant temperature, only the generalized mass equation (4.2-17) is used

$$\frac{\partial C_j}{\partial t} + \frac{\partial(\mathbf{u}C_j)}{\partial x} = D\frac{\partial^2 C_j}{\partial x^2} + \Sigma\alpha_{ij}r_i - k_m a(C_j - C_j^*) \qquad \textbf{(4.2-17)}$$

Neglecting the diffusion, which is equivalent to zero diffusion coefficient $D = 0$, and the interphase transfer, which is equivalent to zero mass transfer coefficient, Eq. (4.2-17) becomes

$$\frac{\partial C_j}{\partial t} + \frac{\partial(\mathbf{u}C_j)}{\partial x} = \Sigma\alpha_{ij}r_i \qquad \textbf{(6.2-1)}$$

Multiplying by dV and integrating from $V = 0$ to $V = V$,

$$\frac{\partial(C_jV)}{\partial t} + \int_0^V \frac{\partial}{\partial x}(\mathbf{u}C_j)dV = \Sigma\alpha_{ij}r_iV \qquad \textbf{(6.2-2)}$$

From vector analysis, the divergence or Gauss theorem indicates that the integral of the divergence of **A** taken over the volume enclosed by the surface is equal to the surface integral of the normal component of a vector **A** taken over a closed surface

$$\iiint_V \nabla\mathbf{A}\, dV = \iint_S \mathbf{A}\cdot\mathbf{n}\, dS \qquad \textbf{(6.2-3)}$$

Using this theorem, the second term of Eq. (6.2-2) becomes

$$\iiint_V \frac{\partial}{\partial x}(\mathbf{u}C_j)dV = \iint_S (\mathbf{u}C_j)\cdot\mathbf{n}\, dS \qquad \textbf{(6.2-4)}$$

Expanding Eq. (6.2-4) and substituting into Eq. (6.2-2)

$$\frac{d}{dt}(C_jV) + (uC_jS)_2 - (uC_jS)_1 = \Sigma\alpha_{ij}r_iV \qquad \textbf{(6.2-5)}$$

in which subscripts 1 and 2 denote inlet and outlet conditions respectively. u is the linear velocity of the fluid. C_j is the concentration of the j component. S is the cross-sectional area of the vessel. Inasmuch as volumetric flow rate q is the product of the linear velocity of the fluid and the cross-sectional area of the vessel through which the fluid is flowing, i.e., $q = uS$, Eq. (6.2-5) after expanding yields

$$V\frac{dC_j}{dt} + C_j\frac{dV}{dt} + (qC_j)_2 - (qC_j)_1 = \Sigma\alpha_{ij}r_iV \tag{6.2-6}$$

An overall balance (Fig. 6.2-1) of the streams around the reactor is

$$\frac{dV}{dt} = q_1 - q_2 \tag{6.2-7}$$

Because of vigorous stirring,

$$C_j = C_{j2} \tag{6.2-8}$$

Combining Eqs. (6.2-7) and (6.2-8) with Eq. (6.2-6)

$$\frac{dC_{j2}}{dt} + \frac{q_1(C_{j2} - C_{j1})}{V} = \Sigma\alpha_{ij}r_i \tag{6.2-9}$$

For a single reaction with $j = k$ and using

$$C_{k2} = C_{k1}(1 - f) \tag{6.2-10}$$

Eq. (6.2-9) becomes

$$\frac{df}{dt} + \frac{f}{V}q_1 = \frac{r}{C_{k1}} \tag{6.2-11}$$

where f is the fractional conversion of the limiting component k.

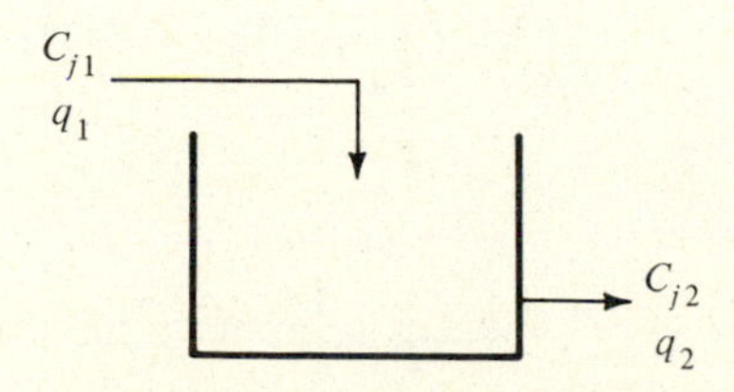

Figure 6.2-1 Overall balance

6.2-1 Irreversible reactions

First-Order Reactions

$$\mathrm{A} \xrightarrow{k} \text{product}$$

The rate of the reaction is

$$r = kC_{\mathrm{A}} \tag{6.2-12}$$

Substituting into Eq. (6.2-9)

$$\frac{dC_{\mathrm{A2}}}{dt} + \frac{q_1(C_{\mathrm{A2}} - C_{\mathrm{A1}})}{V} = -\,kC_{\mathrm{A2}} \tag{6.2-13}$$

Integrating Eq. (6.2-7) gives

$$V = V_o + (q_1 - q_2)t \tag{6.2-14}$$

Substituting into Eq. (6.2-13) and rearranging with $C_{\mathrm{A}} = C_{\mathrm{A2}}$

$$\frac{dC_{\mathrm{A}}}{dt} + \left(\frac{q_1}{V_o + (q_1 - q_2)t} + k\right)C_{\mathrm{A}} = \frac{q_1 C_{\mathrm{A1}}}{V_o + (q_1 - q_2)t} \tag{6.2-15}$$

which is a first-order linear differential equation. There are two simple cases.

1. No product withdrawal, $q_2 = 0$. Then Eq. (6.2-15) reduces to

$$\frac{dC_{\mathrm{A}}}{dt} + \left(\frac{q_1}{V_o + q_1 t} + k\right) C_{\mathrm{A}} = \frac{q_1 C_{\mathrm{A1}}}{V_o + q_1 t} \tag{6.2-16}$$

   Its solution is

$$C_A = \frac{e^{-kt}}{(V_o + q_1 t)} \left\{ C_{\mathrm{A}o} V_o + \left[\frac{q_1 C_{\mathrm{A1}}}{k} (e^{kt} - 1) \right] \right\} \tag{6.2-17}$$

2. Product withdrawal, $q_1 = q_2$. In this case, Eq. (6.2-15) becomes

$$\frac{dC_{\mathrm{A}}}{dt} + \left(\frac{q_1}{V_o} + k\right) C_{\mathrm{A}} = \frac{q_1 C_{\mathrm{A1}}}{V_o} \tag{6.2-18}$$

   whose solution is

$$t = \frac{1}{B} \ln \left\{ \frac{A - BC_{\mathrm{A}o}}{A - BC_{A}} \right\} \tag{6.2-19}$$

where

$$A = \frac{q_1 C_{A1}}{V_o}$$

$$B = \frac{q_1}{V_o} + k$$

Example 6.2-1 Semibatch Operation for Irreversible First-Order Reactions

Acetic anhydride is to be hydrolyzed at 313 K in a semibatch reactor which initially contains 0.01 m^3 of an aqueous anhydride solution with a concentration of 0.05 $kmol/m^3$. After the solution is heated to 313 K, a feed solution of 0.3 $kmol/m^3$ is added at a rate of 3.5×10^{-5} m^3/s. The reaction is

$$r = kC \text{ kmol/m}^3 \cdot \text{s}$$

$$k = 0.006333 \text{ s}^{-1}$$

Find the concentration after 120 s for the following cases:
(a) No product is withdrawn during reaction.
(b) Product is withdrawn at the same rate of the feed.

Solution

(a) No product is withdrawn during reaction. The data are as follows: $k = 0.006333$ s^{-1}, $t = 120$ s, $V_o = 0.01$ m^3 $q_1 = 3.5 \times 10^{-5}$ m^3/s, $C_{Ao} = 0.05$ $kmol/m^3$, $C_{A1} = 0.3$ $kmol/m^3$. Substituting into Eq. (6.2-17)

$$C_A = \frac{e^{-0.006333(120)}}{0.01 + 0.000035(120)} \left\{0.05(0.01) + \left[\frac{0.000035(0.3)}{0.006333}(e^{0.006333(120)} - 1)\right]\right\}$$

$$= 7.86 \times 10^{-2} \text{ kmol/m}^3$$

(b) Product is withdrawn at the same rate of the feed. From the given data, using Eq. (6.2-19)

$$A = \frac{3.5 \times 10^{-5}(0.3)}{0.01} = 1.05 \times 10^{-3}$$

$$B = \frac{3.5 \times 10^{-5}}{0.01} + 0.006333 = 9.83 \times 10^{-3}$$

$$120 = \frac{1}{9.83 \times 10^{-3}} \ln \left[\frac{1.05 \times 10^{-3} - 9.83 \times 10^{-3}(0.05)}{1.05 \times 10^{-3} - 9.83 \times 10^{-3} C_A}\right]$$

or

$$C_A = 0.0894 \text{ kmol/m}^3$$

Second-Order Reactions

$$A + B \xrightarrow{k} D$$

The rate of the reaction is

$$r = kC_A C_B \tag{6.2-20}$$

Eq. (6.2-9) becomes if $C_{B1} = 0$, $C_A = C_{A2}$, and $C_B = C_{B2}$

for $j = A$
$$\frac{dC_A}{dt} + \frac{q_1(C_A - C_{A1})}{V} = -kC_A C_B \tag{6.2-21}$$

for $j = B$
$$\frac{dC_B}{dt} + \frac{q_1 C_B}{V} = -kC_A C_B \tag{6.2-22}$$

Equating Eqs. (6.2-21) and (6.2-22) gives

$$\frac{dC_A}{dt} + \frac{q_1(C_A - C_{A1})}{V} = \frac{dC_B}{dt} + \frac{q_1 C_B}{V} \tag{6.2-23}$$

Since

$$\frac{dn_A}{dt} = \frac{d(VC_A)}{dt} = V\frac{dC_A}{dt} + C_A\frac{dV}{dt} \tag{6.2-24}$$

$$\frac{dn_B}{dt} = \frac{d(VC_B)}{dt} = V\frac{dC_B}{dt} + C_B\frac{dV}{dt} \tag{6.2-25}$$

$$\frac{dV}{dt} = q_1 - q_2 \tag{6.2-7}$$

Eq. (6.2-23) can be simplified to

$$\frac{dn_A}{dt} + C_A q_2 - q_1 C_{A1} = \frac{dn_B}{dt} + C_B q_2 \tag{6.2-26}$$

If no product is withdrawn, $q_2 = 0$ and Eq. (6.2-26) becomes

$$\frac{dn_A}{dt} - q_1 C_{A1} = \frac{dn_B}{dt} \tag{6.2-27}$$

Integrating and rearranging

$$n_A = n_B + M + q_1 C_{A1} t \tag{6.2-28}$$

where

$$M = n_{Ao} - n_{Bo}$$

Substituting Eq. (6.2-28) into Eq. (6.2-22) and using $C_A = n_A/V$.

$$\frac{dC_B}{dt} + \frac{q_1 C_B}{V} = -k\left(\frac{n_B + M + q_1 C_{A1} t}{V}\right) C_B \qquad \textbf{(6.2-29)}$$

Because

$$\frac{dn_B}{dt} = \frac{d(C_B V)}{dt} = C_B \frac{dV}{dt} + V\frac{dC_B}{dt} = C_B (q_1 - q_2) + V\frac{dC_B}{dt}$$

Eq. (6.2-29) can be simplified to (with $q_2 = 0$):

$$\frac{dn_B}{dt} + Pn_B = Qn_B^2 \qquad \textbf{(6.2-30)}$$

where

$$P = \frac{k(M + q_1 C_{A1} t)}{V_o + q_1 t}$$

$$Q = -\frac{k}{V_o + q_1 t}$$

Eq. (6.2-30) is a Bernoulli equation. Dividing Eq. (6.2-30) by n_B^2 and letting $Z = 1/n_B$ gives

$$\frac{dZ}{dt} - PZ = -Q \qquad \textbf{(6.2-31)}$$

Its solution is

$$Z = C \exp(\int P\,dt) + \exp(\int P\,dt)[\int_0^t - Q(\epsilon) \exp(-\int P\,d\epsilon) d\epsilon] \qquad \textbf{(6.2-32)}$$

where C is an integrating constant and $Q(\epsilon)$ is the value of Q which is a function of the dummy variable ϵ. After Z is found, the value of n_B can be calculated as $1/Z$.

n th-Order Reactions

$$n\text{A} \xrightarrow{k} \text{product}$$

The rate of the reaction is

$$r = kC_A^n \tag{6.2-33}$$

Eq. (6.2-9) becomes, if $C_A = C_{A2}$

$$\frac{dC_A}{dt} + \frac{q_1(C_A - C_{A1})}{V_o + (q_1 - q_2)t} = -kC_A^n \tag{6.2-34}$$

which can be solved numerically. In case the feed is inert or the concentration of the limiting reactant in the feed C_{A1} is zero, Eq. (6.2-34) becomes

$$\frac{dC_A}{dt} + \frac{q_1 C_A}{V_o + (q_1 - q_2)t} = -kC_A^n \tag{6.2-35}$$

which is a Bernoulli equation if n is greater than 1. Letting $Z = C_A^{1-n}$, Eq. (6.2-35) reduces to

$$\frac{dZ}{dt} + (1-n)\frac{q_1 Z}{V_o + (q_1 - q_2)t} = -k(1-n) \tag{6.2-36}$$

which is a first-order linear differential equation. Its solution is

$$Z = C\exp(\int Q\,dt) + \exp(\int Q\,dt)\,[\int_0^t R(\epsilon)\exp(-\int Q\,d\epsilon)d\epsilon] \tag{6.2-37}$$

where

$$Q = (n-1)\frac{q_1}{V_o + (q_1 - q_2)t}$$

$$R = k(n-1) \qquad \text{for } n > 1$$

and C is the integrating constant and ϵ is the dummy variable.

Example 6.2-2 Semibatch Operation for *n*th-Order Irreversible Reaction (C1)

Into a stirred tank which originally contains 40.068 g mol triphenyl methyl chloride in dry benzene solution, a stream of 0.054 g mol/cm^3 methyl alcohol is added at a rate of 3.78 liters/min. No product is withdrawn during the reaction. However, pyridine should be added to remove the hydrochloride formed in the reaction so as to prevent the reverse reaction. The reaction which has been reported by Swain (S1) and described by Frost and Pearson (F1) as a third-order reaction, i.e., first-order with respect to the triphenyl methyl chloride and second-order with respect to the methanol is

$$CH_3OH + (C_6H_5)_3CCl \xrightarrow[k]{} (C_6H_5)_3COCH_3 + HCl$$
$$\text{(A)} \qquad \text{(B)} \qquad\qquad \text{(C)} \qquad \text{(D)}$$

The initial volume of mixtures in the tank is 378 liters. The rate of the reaction is

$$r = 0.263\ C_A^2\ C_B$$

where C_A is the concentration of the CH_3OH and C_B is that of the $(C_6H_5)_3CCl$. Determine the concentration of the chloride as a function of time.

Solution

$$CH_3OH + (C_6H_5)CCl \xrightarrow{k} (C_6H_5)_3COCH_3 + HCl$$

or

$$\text{(A)} \ + \ \text{(B)} \xrightarrow{k} \text{(C)} \ + \ \text{(D)}$$

Since no product is withdrawn, $q_2 = 0$, and the mass balances are:

Overall:

$$\frac{dv}{dt} = q_{A1}$$

Component A: Eq. (6.2-6)

$$\frac{dn_A}{dt} = \frac{d(VC_A)}{dt} = q_{A1}C_{A1} - kC_A^2C_BV \tag{B}$$

Component B: Eq. (6.2-6)

$$\frac{dn_B}{dt} = \frac{d(VC_B)}{dt} = -kC_A^2C_BV \tag{C}$$

Integrating Eq. (A) between $t = 0$, $V = 378$ liters, and $t = t$, $V = V$

$$V = 3.78(100 + t) \tag{D}$$

Subtracting Eq. (C) from Eq. (B), rearranging and using $n_A = VC_A$, $n_B = VC_B$

$$\frac{dn_A}{dt} = q_{A1}C_{A1} + \frac{dn_B}{dt} \tag{E}$$

Integrating between limits: $t = 0$, $n_A = n_{Ao} = 0$, $n_B = n_{Bo} = 40.068$ g mol, $Q_{A1} = 3.78$ liters/min, and $C_{A1} = 0.054$ g mol/cm³,

$$n_A = 0.2041t + n_B - 40.0680 \tag{F}$$

Hence the concentration of A is

$$C_A = \frac{n_A}{V} = \frac{0.2041t + n_B - 40.0680}{3.78(100 + t)} \tag{G}$$

Substituting C_A from Eq. (G) into Eq. (C) and using $n_B = VC_B$,

$$\frac{dn_B}{dt} = -0.263\left[\frac{0.2041t + n_B - 40.0680}{3.78(100 + t)}\right]^2 n_B \tag{H}$$

Eq. (H) can be solved numerically by the Runge-Kutta method. However, one must be cautious to calculate the starting value correctly. First, let us approximate the differential by

$$\frac{dn_B}{dt} = \frac{n_{Bf} - n_{Bo}}{t_f - t_o} \tag{I}$$

where n_{Bf} and n_{Bo} are the number of gram moles of component B at the final time t_f and initial time t_o, respectively, during the time increment $t_f - t_o$. For this specific example, let us use 5 min as the time increment. At $t = 0 = t_o$, $n_{Bo} = 40.068$ g mol. Hence Eq. (H) becomes

$$\frac{n_{Bf} - 40.068}{5 - 0} = -0.263\left[\frac{0.2041(5) + n_{Bf} - 40.068}{3.78(100 + 5)}\right]^2 n_{Bf} \tag{J}$$

which is a cubic equation. n_{Bf} can be solved numerically or analytically. However, this procedure is tedious for an nth-degree equation. Hence an approximation of the starting value can be estimated by assuming $n_{Bf} = 40$ in this case, then calculating Eq. (H)

$$\frac{dn_{Bf}}{dt} = -0.263\left[\frac{0.2041(5) + 40 - 40.068}{3.78(100 + 5)}\right]^2 (40) = -0.00006060 \tag{K}$$

$$\frac{dn_{Bo}}{dt} = -0.263\left[\frac{0.2041(0) + 40.068 - 40.068}{3.78(100 + 0)}\right]^2 (40.068) = 0 \tag{L}$$

Hence

$$\left(\frac{dn_B}{dt}\right)_{av1} = \frac{1}{2}\left(\frac{dn_{Bf}}{dt} + \frac{dn_{Bo}}{dt}\right) = \frac{1}{2}(-0.00006060 + 0) = -0.0000303 \tag{M}$$

$$n_{Bf} = n_{Bo} - 0.0000303(5) = 40.068 - 0.0001515 = 40.06784 \tag{N}$$

Repeating the process yields

$$\frac{dn_{Bf}}{dt} = -0.263\left[\frac{0.2041(5) + 40.06784 - 40.068}{3.78(100+5)}\right]^2 (40.06784)$$

$$= -0.0000692594 \qquad \textbf{(O)}$$

$$\left(\frac{dn_B}{dt}\right)_{av2} = \frac{1}{2}(-0.0000692594 + 0) = -0.0000346297 \qquad \textbf{(P)}$$

$$n_{Bf} = 40.068 - 0.0000346297(5) = 40.06783 \qquad \textbf{(Q)}$$

Because the value in Eq. (Q) is very close to the value in Eq. (N), the starting value for the program is $t = 5$ min, $n_B = 40.0678$ g mol and $C_B = 0.1009$ g mol/liter.

The computer program for this problem is given in Fig. 6.2-2 (C1).

6.2-2 Reversible reactions (C1)

First-Order Reactions

$$A \underset{k'}{\overset{k}{\rightleftarrows}} B$$

The rate of reaction is

$$r = kC_A - k'C_B$$

Assuming that no product is withdrawn, then the overall mass balance is

$$\frac{dV}{dt} = -q_1 \qquad \textbf{(6.2-38)}$$

Integrating gives

$$V = V_o + q_1 t \qquad \textbf{(6.2-39)}$$

The component mass balance from Eq. (6.2-6) is

For component A:

$$\frac{d}{dt}(C_A V) - q_1 C_{A1} = -rV \qquad \textbf{(6.2-40)}$$

For component B:

$$\frac{d}{dt}(C_B V) = rV \qquad \textbf{(6.2-41)}$$

```
LNH

00100 PROGRAM SEMIB1(INPUT,OUTPUT)
00110CAK=REACTION RATE CONSTANT,LITER/GMOLE-MIN.
00120CANB=NUMBER OF MOLES OF COMPONENT B,KMOLES
00130CANBIN=STARTING NUMBER OF MOLES OF COMPONENT B,KMOLES
00140CANBO=INITIAL NUMBER OF MOLES OF COMPONENT B,KMOLES
00150CCAF=INLET CONCENTRATION OF COMPONENT A,KMOLES/M**3
00160CCONC=CONCENTRATION OF COMPONENT B,KMOLES/M**3
00170CDELT=TIME INCREMENT,MIN.
00180CQAF=INLET VOLUMETRIC FLOW RATE OF COMPONENT A,LITER/MIN.
00190CN=NUMBER OF TIME INCREMENT
00200CT=TIME OF REACTION,MIN.
00210CV=VOLUME OF REACTION MIXTURE,M**3
00220 DIMENSION ANB(100),T(100),CON(100),V(100),CONC(100)
00230 99 READ*,DELT,ANBIN,AK,QAF,CAF,ANBO,N
00240 ANB(1)=ANBIN
00250 N1=N+1
00260 T(1)=5.0
00270 V(1)=396.9
00280 CONC(1)=ANB(1)/V(1)
00290 DO 5 I=2,N1
00300 K=I-1
00310 T(I)=T(K)+DELT
00320 V(I)=3.78*(100.+T(I))
00330 CON(K)=ANB(K)*((ANB(K)+.20412*T(K)-40.068)/(100.+T(K)))**2
00340 FCONST=-AK/3.78/3.78
00350 D1=DELT*(FCONST*CON(K))
00360 D2=DELT*(FCONST*(CON(K)+0.5*D1))
00370 D3=DELT*(FCONST*(CON(K)+0.5*D2))
00380 D4=DELT*(FCONST*(CON(K)+D3))
00390 ANB(I)=ANB(K)+(1.0/6.0)*(D1+2.*D2+2.*D3+D4)
00400 5 CONC(I)=ANB(I)/V(I)
00410 PRINT 23
00420 23 FORMAT (10X,10HINPUT DATA)
00430 PRINT3,DELT,ANBO,AK,QAF,CAF,ANBIN
00440 3 FORMAT (8X,"TIME INCREMENT = ",E21.4," MIN "/8X,"INITIAL NUMBER
00450+ OF MOLES = ",E19.5," KG-MOLES"/8X,"REACTION RATE CONSTANT = ",E20
00460+.4," LITERS/GMOLES-MIN."/8X,"INLET VOLUMETRIC FLOW RATE OF A = "
00470+,E12.4,"LITER/MIN."/8X,"INLET CONCENTRATION OF A = " ,E18.4,"
00480+KG-MOLE/M**3"/8X,"STARTING NUMBER OF MOLES = " ,E18.6," KG-MOLES")
00490 PRINT 33
```

```
00500 33 FORMAT(10X,11HOUTPUT DATA)
00510 PRINT 14
00520 14 FORMAT(9X,8HINTERVAL,13X,4HTIME,15X,5HMOLES,15X,13HCONCENTRATION)
00530 M=0
00540 TIN=0.
00550 CONCIN=ANBIN/378.
00560 PRINT 15, M,TIN,ANBO,CONCIN
00570 15 FORMAT(11X,I3,11X,F10.4,11X,F10.4,11X,F10.4)
00580 PRINT 15,(I,T(I),ANB(I),CONC(I),I=1,N1)
00590 GO TO 99
00600 END

 READY.

RNH

? 5.,40.0678,0.263,3.78,0.054,40.068,45
        INPUT DATA
      TIME INCREMENT =                    .5000E+01 MIN
      INITIAL NUMBER          OF MOLES =              .40068E+02 KG-MOLES
      REACTION RATE CONSTANT =               .2630E+00 LITERS/GMOLES-MIN.
      INLET VOLUMETRIC FLOW RATE OF A =       .3780E+01LITER/MIN.
      INLET CONCENTRATION OF A =             .5400E-01              KG-MOLE/M**3
      STARTING NUMBER OF MOLES =           .400678E+02 KG-MOLES
        OUTPUT DATA
       INTERVAL             TIME                 MOLES               CONCENTRATION
           0               0.0000               40.0680                 .1060
           1               5.0000               40.0678                 .1010
           2              10.0000               40.0675                 .0964
           3              15.0000               40.0663                 .0922
           4              20.0000               40.0638                 .0883
           5              25.0000               40.0597                 .0848
           6              30.0000               40.0538                 .0815
           7              35.0000               40.0461                 .0785
           8              40.0000               40.0363                 .0757
           9              45.0000               40.0244                 .0730
          10              50.0000               40.0104                 .0706
          11              55.0000               39.9943                 .0683
```

Figure 6.2-2

```
   12        60.0000        39.9761        .0661
   13        65.0000        39.9558        .0641
   14        70.0000        39.9335        .0621
   15        75.0000        39.9091        .0603
   16        80.0000        39.8828        .0586
   17        85.0000        39.8546        .0570
   18        90.0000        39.8246        .0555
   19        95.0000        39.7927        .0540
   20       100.0000        39.7591        .0526
   21       105.0000        39.7237        .0513
   22       110.0000        39.6868        .0500
   23       115.0000        39.6482        .0488
   24       120.0000        39.6081        .0476
   25       125.0000        39.5666        .0465
   26       130.0000        39.5236        .0455
   27       135.0000        39.4792        .0444
   28       140.0000        39.4335        .0435
   29       145.0000        39.3865        .0425
   30       150.0000        39.3383        .0416
   31       155.0000        39.2888        .0408
   32       160.0000        39.2382        .0399
   33       165.0000        39.1865        .0391
   34       170.0000        39.1337        .0383
   35       175.0000        39.0799        .0376
   36       180.0000        39.0251        .0369
   37       185.0000        38.9693        .0362
   38       190.0000        38.9126        .0355
   39       195.0000        38.8550        .0348
   40       200.0000        38.7966        .0342
   41       205.0000        38.7373        .0336
   42       210.0000        38.6772        .0330
   43       215.0000        38.6163        .0324
   44       220.0000        38.5547        .0319
   45       225.0000        38.4924        .0313
   46       230.0000        38.4293        .0308
?
*TERMINATED*

READY.
```

Figure 6.2-2 (continued)

Equating Eqs. (6.2-40) and (6.2-41), integrating with $n_{Bo} = 0$, and remembering that $VC = n$,

$$n_A = n_{Ao} - n_B + q_{A1} C_{A1} t \tag{6.2-42}$$

Eq. (6.2-41) can be written as

$$\frac{dn_B}{dt} = (kC_A - k'C_B)V = kn_A - k'n_B \tag{6.2-43}$$

Substituting Eq. (6.2-42) into Eq. (6.2-43) yields

$$\frac{dn_B}{dt} = kq_1 C_{A1} t - (k + k')n_B + kn_{Ao} \tag{6.2-44}$$

which is a first-order linear differential equation. Its solution is

$$n_B = q_1 C_{A1} k \left[\frac{t}{k + k'} - \frac{1}{(k + k')^2}(1 - e^{-(k+k')t})\right] + \frac{kn_{Ao}}{k + k'}(1 - e^{-(k+k')t}) \tag{6.2-45}$$

Then the concentration of A can be obtained from Eq. (6.2-42) as

$$C_A = \frac{n_A}{V_o + q_1 t} = \frac{\left[n_{Ao} + q_1 C_{A1}\left\{t - k\left[\frac{t}{k + k'} - \frac{1}{(k + k')^2}(1 - e^{-(k+k')t})\right]\right\} - \frac{kn_{Ao}}{k + k'}(1 - e^{-(k+k')t})\right]}{V_o + q_1 t} \tag{6.2-46}$$

Second-Order Reaction

$$A + B \underset{k'}{\overset{k}{\rightleftharpoons}} C + D$$

The rate of reaction is

$$r = kC_A C_B - k'C_C C_D$$

The component mass balances from Eq. (6.2-6) are:

For A $$\frac{dC_A}{dt} - \frac{q_1 C_{A1}}{V} + \frac{q_1 C_A}{V} = -r = -kC_A C_B + k' C_C C_D \quad (6.2\text{-}47)$$

For B $$\frac{dC_B}{dt} - \frac{q_1 C_{B1}}{V} + \frac{q_1 C_B}{V} = -r = -kC_A C_B + k' C_C C_D \quad (6.2\text{-}48)$$

For C $$\frac{dC_C}{dt} - 0 + \frac{q_1 C_C}{V} = r = kC_A C_B - k' C_C C_D \quad (6.2\text{-}49)$$

For D $$\frac{dC_D}{dt} - 0 + \frac{q_1 C_D}{V} = r = kC_A C_B - k' C_C C_D \quad (6.2\text{-}50)$$

Combining Eqs. (6.2-47) and (6.2-48) gives

$$\frac{dC_A}{dt} - \frac{q_1 C_{A1}}{V} + \frac{q_1 C_A}{V} = \frac{dC_B}{dt} - \frac{q_1 C_{B1}}{V} + \frac{q_1 C_B}{V} \quad (6.2\text{-}51)$$

But

$$\frac{dn_j}{dt} = \frac{d(VC_j)}{dt} = V\frac{dC_j}{dt} + C_j \frac{dV}{dt} = V\frac{dC_j}{dt} + C_j(q_1 - q_2) \quad (6.2\text{-}52)$$

hence

$$\frac{dn_A}{dt} - C_A(q_1 - q_2) - q_1 C_{A1} + q_1 C_A = \frac{dn_B}{dt} - C_B(q_1 - q_2) - q_1 C_{B1} + q_1 C_B \quad (6.2\text{-}53)$$

Assuming that no product is withdrawn and no B is in the feed, Eq. (6.2-53) becomes on simplification

$$\frac{dn_A}{dt} - q_1 C_{A1} = \frac{dn_B}{dt} \quad (6.2\text{-}54)$$

Integrating yields

$$n_A = n_{Ao} + n_B - n_{Bo} + q_1 C_{A1} t \quad (6.2\text{-}55)$$

Combining Eq. (6.2-48) with Eq. (6.2-49) or Eq. (6.2-50) gives

$$\frac{dC_B}{dt} + \frac{q_1 C_B}{V} = -\frac{dC_C}{dt} - \frac{q_1 C_C}{V} \quad (6.2\text{-}56)$$

Using Eq. (6.2-52), Eq. (6.2-56) becomes

$$\frac{dn_B}{dt} - C_B(q_1 - q_2) + q_1 C_B = -\frac{dn_C}{dt} + C_C(q_1 - q_2) - q_1 C_C \quad \textbf{(6.2-57)}$$

Simplifying with $q_2 = 0$ and $n_{Co} = 0$ and integrating yields

$$n_C = n_{Bo} - n_B \quad \textbf{(6.2-58)}$$

Similarly,

$$n_D = n_{Bo} - n_B \quad \textbf{(6.2-59)}$$

Eq. (6.2-48) can be rearranged with Eq. (6.2-52) and $q_2 = 0$ as

$$\frac{dn_B}{dt} = -k\frac{n_A}{V} n_B + k' \frac{n_C}{V} n_D \quad \textbf{(6.2-60)}$$

Substituting Eqs. (6.2-55), (6.2-58), and (6.2-59) into Eq. (6.2-60)

$$\frac{dn_B}{dt} = -k\frac{n_B + n_{Ao} - n_{Bo} + q_1 C_{A1} t}{V_o + q_1 t} n_B + k' \frac{(n_{Bo} - n_B)^2}{V_o + q_1 t} \quad \textbf{(6.2-61)}$$

Simplifying results in

$$\frac{dn_B}{dt} + Pn_B^2 + Qn_B + R = 0 \quad \textbf{(6.2-62)}$$

where

$$P = \frac{k - k'}{V_o + q_{1A} t}$$

$$Q = \frac{k(q_{1A} C_{A1} t + M) + 2k' n_{Bo}}{V_o + q_{1A} t}$$

$$R = \frac{-k' n_{Bo}^2}{V_o + q_{1A} t}$$

$$M = n_{Ao} - n_{Bo}$$

which can be solved numerically by the computer program provided. By the same method of approach, a similar equation for a nth-order reaction can be obtained as

$$\frac{dn_j}{dt} + A_n n_j^n + A_{n-1} n_j^{n-1} + \cdots A_o = 0 \tag{6.2-63}$$

or

$$\frac{dn_j}{dt} + \sum_{n=0}^{n=n} A_n n_j^n = 0 \tag{6.2-64}$$

where

n_j = the number of moles of reactive component j

n = the order of the reaction

A_n = the variable coefficient which is a function of the reaction time t, initial volume of the reactant in the reactor V_o, k, k', q_{j1} and C_{jo}, C_{j1}.

Example 6.2-3 Semibatch Operation for Second-Order Reaction by Analytical and Numerical Methods (C1)

Ethyl acetate is to be reacted by adding sodium hydroxide with concentration of C_{A1} = 0.05 g mol/liter continuously to a kettle containing the ethyl acetate. The reactor is initially charged with 0.378 m³ of an aqueous solution containing 10 g/liter ethyl acetate. The sodium hydroxide is added at a rate of 0.0000631 m³/s until stoichiometric amounts are present. By mistake, a technician reported this reaction to be irreversible with a specific rate constant 0.1222 liter/g mol·min. Assuming that the contents of the reactor are well mixed, determine the concentration of the unreacted ethyl acetate as a function of time.

Solution

1. Analytical method

A special case of Eq. (6.2-62) can be solved analytically as follows: First a new variable u is introduced such that

$$n_B = \frac{du}{Pu\,dt} = \frac{u'}{Pu} \tag{A}$$

Differentiating Eq. (A) with respect to t

$$\frac{dn_B}{dt} = \frac{u''}{Pu} - \frac{u'^2}{Pu^2} - \frac{u'}{uP^2}\frac{dP}{dt} \tag{B}$$

Substituting Eq. (B) into Eq. (6.2-62) and simplifying yields

$$u'' + \left(Q - \frac{P'}{P}\right) u' + PRu = 0 \tag{C}$$

which is a second-order differential equation with variable coefficients. Although Eq. (C) can be solved theoretically by the series method, its solution is too tedious and therefore should be obtained more efficiently by the numerical method described above. However, if R in Eq. (6.2-62) is zero, which means that the proposed chemical reaction is irreversible instead of reversible, an analytical solution can be obtained. Let us introduce another variable w

$$w = u' \tag{D}$$

Then

$$u'' = w' \tag{E}$$

and Eq. (C) with $R = 0$ becomes

$$w' + \left(Q - \frac{P'}{P}\right) w = 0 \tag{F}$$

whose solution is

$$w = \exp\left[\int\left(\frac{P'}{P} - Q\right) dt\right] \tag{G}$$

Thus

$$\frac{du}{dt} = u' = w = \exp\left[\int\left(\frac{P'}{P} - Q\right) dt\right] \tag{H}$$

Integrating gives

$$u = \int\left[\exp\int\left(\frac{P'}{P} - Q\right) dt\right] dt + C \tag{I}$$

Substituting Eqs. (H) and (I) into Eq. (A) yields the final result for n_B as

$$n_B = \frac{\exp\left[\int\left(\frac{P'}{P} - Q\right) dt\right]}{P\left[\int \exp\int\left(\frac{P'}{P} - Q\right) dt\, dt + C\right]} \tag{J}$$

The integrating constant C can be evaluated from the initial condition: i.e., $n_B = n_{Bo}$ at $t = 0$.

In this example, the numerical values are:

$$k = 0.1222 \text{ liter/g mol}\cdot\text{min}$$

$$C_{A1} = 0.05 \text{ g mol/liter} = 0.05 \text{ kg}\cdot\text{mol/m}^3$$

$$q_{A1} = 0.0000631(60)(1000) = 3.78 \text{ liters/min}$$

$$C_{Bo} = \frac{10}{88} = 0.1136 \text{ g mol/liter} = 0.1136 \text{ kg}\cdot\text{mol/m}^3$$

$$V_o = 0.378(1000) = 378.0 \text{ liters} = 0.378 \text{ m}^3$$

$$M = C_{Ao}V_o - C_{Bo}V_o = 0 - 0.1136(378.0)$$
$$= -42.94 \text{ g}\cdot\text{mol}$$

$$P = \frac{k}{V_o + q_{A1}t} = \frac{0.1222}{378 + 3.78t} = \frac{0.03233}{100 + t}$$

$$P' = \frac{-0.03233}{(100 + t)^2}$$

$$Q = \frac{k(q_{A1}C_{A1}t + M)}{V_o + q_{A1}t} = \frac{0.1222[3.78(0.05t) - 42.94]}{3.78(100 + t)}$$
$$= \frac{0.00611t - 1.3882}{100 + t}$$

$$Q - \frac{P'}{P} = \frac{0.00611t - 1.3882}{100 + t} + \frac{1}{100 + t} = \frac{0.00611t - 0.3882}{100 + t}$$

$$\int\left(Q - \frac{P'}{P}\right) dt = \int \frac{0.00611t - 0.3882}{100 + t}\, dt$$
$$= 0.00611t - \ln(100 + t) + 0.611$$

$$\exp\left[-\int\left(Q - \frac{P'}{P}\right) dt\right] = \frac{[(100 + t)\, e^{-0.00611t}]}{e^{0.611}}$$

$$\int \exp\left[-\int\left(Q - \frac{P'}{P}\right) dt\right] dt = \int (100 + t)e^{-0.00611t}\, dt$$
$$= \frac{-163.66(263.66 + t)e^{-0.00611t}}{e^{0.611}}$$

Substituting into Eq. (J)

$$n_B = \frac{(100 + t)e^{-0.00611t}}{\dfrac{0.03233}{100 + t}[-163.66(263.66 + t)e^{-0.00611t} + C]}$$

The constant C can be evaluated at $t = 0$, $n_B = 42.94$ g mol as $C = 50{,}355$. The final equation is

$$n_B = \frac{0.189(t + 100)^2}{307.66e^{0.00611t} - (263.66 + t)}$$

As the volume of the reaction mixture at any time is

$$V = V_o + q_{A1}t = 378 + 3.78t = 3.78(100 + t)$$

the concentration of B is then

$$C_B = \frac{n_B}{V} = \frac{0.05(t + 100)}{307.66e^{0.00611t} - (263.66 + t)}$$

2. Numerical Method

The Runge-Kutta method of solving nonlinear ordinary differential equations is employed in the previous computer program. In this example, let us start with Eq. (6.2-62). The previous calculations show that

$$P = \frac{0.03233}{100 + t}$$

$$Q = \frac{0.00611t - 1.3882}{100 + t}$$

$$R = 0 \text{ because } k' = 0$$

Substituting these values into Eq. (6.2-62) yields

$$\frac{dn_B}{dt} + \frac{0.03233}{100 + t} n_B^2 + \frac{0.00611t - 1.3882}{100 + t} n_B = 0$$

First let us approximate the differential by

$$\frac{dn_B}{dt} = \frac{n_{Bf} - n_{Bo}}{t_f - t_o}$$

where $_{Bf}$ and n_{Bo} are the number of gram moles of component B at the final time t_f and the initial time t_o, respectively, during the time increment $t_f - t_o$. For this specific example, let us use 5 min as the time increment. At $t = 0 = t_o$, $n_{Bo} =$ 42.94 g mol. Hence Eq. (6.2-62) becomes

$$\frac{n_{Bf} - 42.94}{5 - 0} + \frac{0.03233}{100 + 5} n_{Bf}^2 + \frac{0.00611(5) - 1.3882}{100 + 5} n_{Bf} = 0$$

which is a quadratic equation. Its solution is $n_{Bf} = 42.88$ g mol.

$$\frac{dn_{Bf}}{dt} = -\left(\frac{0.03233}{105} 42.88^2 + \frac{0.00611(5) - 1.3882}{105} 42.88\right) = -0.01293$$

$$\frac{dn_{Bo}}{dt} = -\left(\frac{0.03233}{100}42.94^2 + \frac{0.00611(0) - 1.3882}{100}42.94\right) = 0$$

Hence

$$\left(\frac{dn_B}{dt}\right)_{av1} = \frac{1}{2}\left(\frac{dn_{Bf}}{dt} + \frac{dn_{Bo}}{dt}\right) = \frac{1}{2}(-0.01293 + 0) = -0.006465$$

and

$$n_{Bf2} = 42.94 - (5)(0.006465) = 42.9077 \text{ g mol}$$

Repeating the process gives

$$\frac{dn_{Bf2}}{dt} = \frac{-0.03233(42.9077)}{105}[42.9077 + 0.189(5) - 42.94] = -0.012075$$

$$\left(\frac{dn_B}{dt}\right)_{av2} = \frac{1}{2}(-0.012075 + 0) = -0.0060375$$

$$n_{Bf3} = 42.94 - (5)(0.0060375) = 42.9098$$

$$\frac{dn_{Bf3}}{dt} = \frac{-0.03233(42.9098)}{105}[42.9098 + 0.189(5) - 42.94] = -0.012086$$

$$\left(\frac{dn_B}{dt}\right)_{av3} = -0.006043$$

$$n_{Bf4} = 42.94 - (5)(0.006043) = 42.9098$$

which checks with n_{Bf3}. We use these values $t = 5$ min, $n_B = 42.9098$ g mol, $C_B = 0.1081$ g mol/liter as the starting values for the previous computer program. The results from the computer printout are given in Fig. 6.2-3. The solution from the analytical method is also included in the following table for comparison.

Time, min	Volume, liters	n_B, g mol Analytical	n_B, g mol Numerical	Concentration C_B Analytical	Concentration C_B Numerical
0	378.0	42.94	42.94	0.1136	0.1136
5	396.9	42.92	42.91	0.1081	0.1081
10	415.8	42.84	42.85	0.1030	0.1030
50	567.0	40.92	41.13	0.0722	0.0725
100	756.0	37.21	37.61	0.0492	0.0497
200	1134.0	29.30	29.95	0.0258	0.0264

```
 81/08/06. 12.09.13.
PROGRAM    SEMIBS

00100 PROGRAM SEMIB2(INPUT,OUTPUT)
00110CAK=REACTION RATE CONSTANT,LITER/GMOLE-MIN.
00120CANB=NUMBER OF MOLES PF COMPONENT B,GMOLES
00130CANBIN=STARTING NUMBER OF MOLES OF COMPONENT B,GMOLES
00140CANBO=INITIAL NUMBER OF MOLES OF COMPONENT B,GMOLES
00150CCAF=INLET CONCENTRATION OF COMPONENT A,GMOLES/LITER
00160CCONC=CONCENTRATION OF COMPONENT B,GMOLES/LITER
00170CDELT=TIME INCREMENT,MIN.
00180CQAF=INLET VOLUMETRIC F;OW RATE PF COMPONENT A,LITER/MIN.
00190CN=NUMBER OF TIME INCREMENT
00200CT=TIME OF REACTION,MIN.
00210CV=VOLUME OF REACTION MIXTURE,LITER
00220 DIMENSION ANB(100),T(100),CON(100),V(100),CONC(100)
00230 99 READ*,DELT,ANBIN,AK,QAF,CAF,ANBO,N
00240 ANB(1)=ANBIN
00250 FCONST=-AK
00260 N1=N+1
00270 T(1)=5.0
00280 V(1)=396.9
00290 CONC(1)=ANB(1)/V(1)
00300 DO 5 I=2,N1
00310 K=I-1
00320 T(I)=T(K)+DELT
00330 V(I)=3.78*(100.+T(I))
00350 CON(K)=ANB(K)*(ANB(K)+.189*T(K)-42.94)/(100.+T(K))
00360 FCONST=-AK/3.78
00400 D1=DELT*(FCONST*CON(K))
00410 D2=DELT*(FCONST*(CON(K)+0.5*D1))
00420 D3=DELT*(FCONST*(CON(K)+0.5*D2))
00430 D4=DELT*(FCONST*(CON(K)+D3))
00440 ANB(I)=ANB(K)+(1.0/6.0)*(D1+2.*D2+2.*D3+D4)
```

Figure 6.2-3

```
00450 5 CONC(I)=ANB(I)/V(I)
00460 PRINT 23
00470 23 FORMAT (10X,10HINPUT DATA)
00480 PRINT3,DELT,ANBO,AK,QAF,CAF,ANBIN
00490 3 FORMAT(15X,"TIME INCREMENT = ",E21.4," MIN "/8X,"INITIAL NUMBER
00500+ OF MOLES = = ",E19.4," G-MOLES"/8X,"REACTION RATE CONSTANT = ",E20
00510+.4," LITERS/GMOLES-MIN."/8X,"INLET VOLUMETRIC FLOW RATE OF A = "
00520+,E12.4,"LITER/MIN."/8X,"INLET CONCENTRATION OF A = " ,E18.4,"
00530+K-MOLE/LITER"/8X,"STARTING NUMBER OF MOLES = " ,E18.4," G-MOLES")
00540 PRINT 33
00550 33 FORMAT(10X,11HOUTPUT DATA)
00560 PRINT 14
00570 14 FORMAT(9X,8HINTERVAL,13X,4HTIME,15X,5HMOLES,9X,13HCONCENTRATION)
00580 M=0
00590 TIN=0.
00600 CONCIN=ANBIN/378.
00610 PRINT 15, M,TIN,ANBO,CONCIN
00620 15 FORMAT(11X,I3,11X,F10.4,11X,F10.4,11X,F10.4)
00630 PRINT 15,(I,T(I),ANB(I),CONC(I),I=1,N1)
00640 GO TO 99
00650 END
READY.
RNH

? 5.,42.9098,0.1222,3.78,0.05,42.94,45
          INPUT DATA
               TIME INCREMENT =              .5000E+01 MIN
        INITIAL NUMBER          OF MOLES = =            .4294E+02 G-MOLES
        REACTION RATE CONSTANT =             .1222E+00 LITERS/GMOLES-MIN.
        INLET VOLUMETRIC FLOW RATE OF A =     .3780E+01LITER/MIN.
        INLET CONCENTRATION OF A =           .5000E-01             K-MOLE/LITER
        STARTING NUMBER OF MOLES =           .4291E+02 G-MOLES
          OUTPUT DATA
         INTERVAL             TIME                MOLES         CONCENTRATION
             0               0.0000              42.9400                 .1135
             1               5.0000              42.9098                 .1081
```

2	10.0000	42.8540	.1031
3	15.0000	42.7491	.0983
4	20.0000	42.6024	.0939
5	25.0000	42.4200	.0898
6	30.0000	42.2070	.0859
7	35.0000	41.9678	.0822
8	40.0000	41.7060	.0788
9	45.0000	41.4247	.0756
10	50.0000	41.1267	.0725
11	55.0000	40.8142	.0697
12	60.0000	40.4892	.0669
13	65.0000	40.1534	.0644
14	70.0000	39.8084	.0619
15	75.0000	39.4555	.0596
16	80.0000	39.0958	.0575
17	85.0000	38.7302	.0554
18	90.0000	38.3598	.0534
19	95.0000	37.9853	.0515
20	100.0000	37.6073	.0497
21	105.0000	37.2265	.0480
22	110.0000	36.8435	.0464
23	115.0000	36.4588	.0449
24	120.0000	36.0727	.0434
25	125.0000	35.6857	.0420
26	130.0000	35.2982	.0406
27	135.0000	34.9104	.0393
28	140.0000	34.5228	.0381
29	145.0000	34.1354	.0369
30	150.0000	33.7486	.0357
31	155.0000	33.3626	.0346
32	160.0000	32.9776	.0336
33	165.0000	32.5937	.0325
34	170.0000	32.2112	.0316
35	175.0000	31.8301	.0306
36	180.0000	31.4506	.0297
37	185.0000	31.0729	.0288
38	190.0000	30.6971	.0280

Figure 6.2-3 (continued)

39	195.0000	30.3232	.0272
40	200.0000	29.9513	.0264
41	205.0000	29.5816	.0257
42	210.0000	29.2141	.0249
43	215.0000	28.8489	.0242
44	220.0000	28.4861	.0235
45	225.0000	28.1257	.0229
46	230.0000	27.7677	.0223

Figure 6.2-3 (continued)

6.2-3 Complex reaction

Reversible Consecutive Reaction

$$\mathrm{A} \underset{k'}{\overset{k}{\rightleftarrows}} \mathrm{B} \xrightarrow{k''} \mathrm{C}$$

The rate of reactions are:

$$r_1 = kC_\mathrm{A} - k'C_\mathrm{B} \qquad \text{for the first reaction}$$

$$r_2 = k''C_\mathrm{B} \qquad \text{for the second reaction}$$

Using Eq. (6.2-6)

$$\frac{dn_j}{dt} + (qC_j)_2 - (qC_j)_1 = \Sigma\alpha_{ij}r_iV \tag{6.2-6}$$

and assuming that no product is withdrawn, i.e., $q_2 = 0$, and that only A enters the reactor which initially contains A, we have

$$\text{For A} \qquad \frac{dn_\mathrm{A}}{dt} - q_1C_{\mathrm{A}1} = -(kC_\mathrm{A} - k'C_\mathrm{B})V \tag{6.2-65}$$

$$\text{For B} \qquad \frac{dn_\mathrm{B}}{dt} = [kC_\mathrm{A} - (k' + k'')C_\mathrm{B}]V \tag{6.2-66}$$

$$\text{For C} \qquad \frac{dn_\mathrm{C}}{dt} = k''C_\mathrm{B}V \tag{6.2-67}$$

Converting to moles by $n_j = C_jV$

$$\text{For A} \qquad \frac{dn_\mathrm{A}}{dt} + kn_\mathrm{A} = k'n_\mathrm{B} + q_1C_{\mathrm{A}1} \tag{6.2-68}$$

$$\text{For B} \qquad \frac{dn_\mathrm{B}}{dt} + (k' + k'')n_\mathrm{B} = kn_\mathrm{A} \tag{6.2-69}$$

$$\text{For C} \qquad \frac{dn_\mathrm{C}}{dt} - k''n_\mathrm{B} = 0 \tag{6.2-70}$$

Equations (6.2-68) to (6.2-70) are a system of simultaneous differential equations which can be solved by the matrix method. However, for this simple specific case, an analytical solution can be obtained by (1) Differentiating Eq. (6.2-69) with respect to t

$$\frac{d^2n_B}{dt^2} + (k' + k'')\frac{dn_B}{dt} = k\frac{dn_A}{dt} \tag{6.2-71}$$

(2) Substituting Eq. (6.2-68) into Eq. (6.2-71) and simplifying

$$\frac{d^2n_B}{dt^2} + P\frac{dn_B}{dt} + Qn_B = R \tag{6.2-72}$$

where

$$P = k + k' + k''$$

$$Q = kk''$$

$$R = q_1C_{A1}k$$

Eq. (6.2-72) is a second-order ordinary differential equation whose solution is

$$n_B = Ae^{m_1t} + Be^{m_2t} + \frac{kq_1C_{A1}}{kk''} \tag{6.2-73}$$

where m_1 and m_2 are the roots of the auxiliary equation

$$m^2 + Pm + Q = 0 \tag{6.2-74}$$

After n_B is solved, substituting into Eq. (6.2-69) yields the solution for n_A. Then the integrating constants A and B can be determined from the initial conditions that $t = 0$, $C_A = C_{Ao}$, $C_B = 0$.

Example 6.2-4 Semibatch Operation for Reversible Consecutive Reaction

It is desired to produce a substance B from a compound A in a semibatch reactor into which a solution of A of concentration of 1 kmol/m³ enters at a rate of 0.05 m³/s. The volume of the mixture is 10 m³. The initial concentration of A in the reactor is 0.1 kmol/m³. The rate constants for the following reactions

$$A \underset{k'}{\overset{k}{\rightleftarrows}} B \xrightarrow{k''} C$$

are $k = 0.35\ s^{-1}$, $k' = 0.02\ s^{-1}$, $K'' = 0.13\ s^{-1}$. Determine the time to obtain maximum production of B.

Solution

1. By differential equation

Using Eq. (6.2-73) with $k = 0.35\ s^{-1}$, $k' = 0.02\ s^{-1}$, $k'' = 0.13\ s^{-1}$, $q_1 = 0.05$ m^3/s, $C_{A1} = 1\ kmol/m^3$, $C_{Ao} = 0.1\ kmol/m^3$, $V_o = 10\ m^3$ we calculate

$$P = k + k' + k'' = 0.35 + 0.02 + 0.13 = 0.50$$
$$Q = kk'' = 0.35(0.13) = 0.0455$$
$$R = q_1 C_{A1} k = 0.05(1)(0.35) = 0.0175$$

The roots m_1 and m_2 in Eq. (6.2-73) are obtained from Eq. (6.2-74) below

$$m^2 + Pm + Q = m^2 + 0.5m + 0.0455 = 0$$

Solving

$$m_1 = -0.38 \qquad m_2 = -0.12$$

Substituting into Eq. (6.2-73) yields

$$n_B = Ae^{-0.38t} + Be^{-0.12t} + 0.385 \tag{A}$$

Differentiating Eq. (A) with respect to t, substituting into Eq. (6.2-69) and simplifying gives

$$n_A = -0.6571Ae^{-0.38t} + 0.0857Be^{-0.12t} + 0.165 \tag{B}$$

With the initial condition that $C_{Ao} = 0.1$ or $n_{Ao} = 0.1(10) = 1$, and $n_{Bo} = 0$ at $t = 0$, the integrating constants can be evaluated from Eqs. (A) and (B). Then the equations for A, B, and C are:

$$n_A = 0.7678e^{-0.38t} + 0.06715e^{-0.12t} + 0.165 \tag{C}$$
$$n_B = -1.1685e^{-0.38t} + 0.7835e^{-0.12t} + 0.385 \tag{D}$$
$$n_C = 0.3997e^{-0.38t} - 0.8488e^{-0.12t} + 0.05t + 0.449 \tag{E}$$

where Eq. (E) is obtained by integrating Eq. (6.2-70) with $n_{Co} = 0$ at $t = 0$. Differentiating Eq. (D) with respect to t, equating to zero and solving for t yields the time for the production of maximum of B

$$t_{max} = 5.97\ s$$

2. By matrices

Sometimes it is more convenient to use matrices to solve systems of differential equations than to use the classical method of differential equations. Besides, the computer can solve matrix differential equations. Let us use the same example for illustration. Substituting the numerical values into Eqs. (6.2-68) and (6.2-69) yields

$$\frac{dn_A}{dt} + 0.35n_A - 0.02n_B = 0.05 \tag{F}$$

$$\frac{dn_B}{dt} + 0.15n_B - 0.35n_A = 0 \tag{G}$$

which can be arranged in matrix form as

$$\frac{d}{dt}[n] - [A][n] = [b] \tag{H}$$

where

$$[n] = \begin{bmatrix} n_A \\ n_B \end{bmatrix}$$

$$[A] = \begin{bmatrix} -0.35 & +0.02 \\ +0.35 & -0.15 \end{bmatrix}$$

$$[b] = \begin{bmatrix} 0.05 \\ 0 \end{bmatrix}$$

From Appendix 6, it is shown that the solution of Eq. (H) is

$$[n]_t = e^{[A]t}([n]_o + [A]^{-1}[b]) - [A]^{-1}[b] \tag{I}$$

where $e^{[A]t}$ can be evaluated by the Sylvester theorem which is also described in Appendix 6. Following the method in Appendix 6, we can calculate

$$[A]^{-1} = \frac{\begin{bmatrix} -0.15 & -0.35 \\ -0.02 & -0.35 \end{bmatrix}^T}{0.35(0.15) - 0.35(0.02)} = \frac{\begin{bmatrix} -0.15 & -0.02 \\ -0.35 & -0.35 \end{bmatrix}}{0.0455} \tag{J}$$

$$|\lambda I - A| = \begin{vmatrix} \lambda + 0.35 & -0.02 \\ -0.35 & \lambda + 0.15 \end{vmatrix} = (\lambda + 0.35)(\lambda + 0.15) - 0.007 \tag{K}$$

$$= \lambda^2 + 0.5\lambda + 0.0455 = 0$$

Solving

$$\lambda_1 = -0.38 \qquad \text{and} \qquad \lambda_2 = -0.12$$

$$\text{adj}[\lambda I - A] = \begin{bmatrix} \lambda + 0.15 & 0.35 \\ 0.02 & \lambda + 0.35 \end{bmatrix}^T = \begin{bmatrix} \lambda + 0.15 & 0.02 \\ 0.35 & \lambda + 0.35 \end{bmatrix}$$

$$\text{adj}(-0.38) = \begin{bmatrix} -0.23 & 0.02 \\ 0.35 & -0.03 \end{bmatrix} \qquad \text{adj}(-0.12) = \begin{bmatrix} 0.03 & 0.02 \\ 0.35 & 0.23 \end{bmatrix}$$

$$\det(\lambda) = \lambda^2 + 0.5\lambda + 0.0455$$

$$\det'(\lambda) = 2\lambda + 0.5$$

$$\det'(-0.38) = 2(-0.38) + 0.5 = -0.26 \qquad \det'(-0.12) = 2(-0.12) + 0.5 = 0.26$$

$$e^{[A]t} = \frac{e^{-0.38t}\,\text{adj}(-0.38)}{\det'(-0.38)} + \frac{e^{-0.12t}\,\text{adj}(-0.12)}{\det'(-0.12)}$$

$$= \frac{e^{-0.38t}}{-0.26}\begin{bmatrix} 0.23 & 0.02 \\ 0.35 & -0.03 \end{bmatrix} + \frac{e^{-0.12t}}{0.26}\begin{bmatrix} 0.03 & 0.02 \\ 0.35 & 0.23 \end{bmatrix} \tag{L}$$

$$[A]^{-1}[b] = \frac{\begin{bmatrix} -0.15 & -0.02 \\ -0.35 & -0.35 \end{bmatrix}}{0.0455}\begin{bmatrix} 0.05 \\ 0 \end{bmatrix} = \begin{bmatrix} -0.1648 \\ -0.3846 \end{bmatrix} \tag{M}$$

$$[n]_o + [A]^{-1}[b] = \begin{bmatrix} 1 \\ 0 \end{bmatrix} + \begin{bmatrix} -0.1648 \\ -0.3846 \end{bmatrix} = \begin{bmatrix} 0.8352 \\ -0.3846 \end{bmatrix} \tag{N}$$

Substituting Eqs. (L), (M), and (N) into Eq. (I) and simplifying gives

$$[n] = \begin{bmatrix} 0.06715e^{-0.12t} + 0.7678e^{-0.38t} + 0.165 \\ 0.7835e^{-0.12t} - 1.1685e^{-0.38t} + 0.385 \end{bmatrix} \tag{O}$$

which are the same as Eqs. (C) and (D). After n_A and n_B are found from Eq. (O), n_C can be calculated as before by Eq. (6.2-70) or conveniently by adding Eqs. (6.2-68), (6.2-69), and (6.2-70) and integrating with $n_{Ao} = 1$, $n_{Bo} = n_{Co} = 0$ at $t = 0$ to give

$$n_C = 1 + q_1 C_{A1} t - n_A - n_B \tag{P}$$

Consequently all the values of n_A, n_B, and n_C can be plotted versus t in Fig. 6.2-4 in which the maximum of n_B occurs at approximately t = 6s.

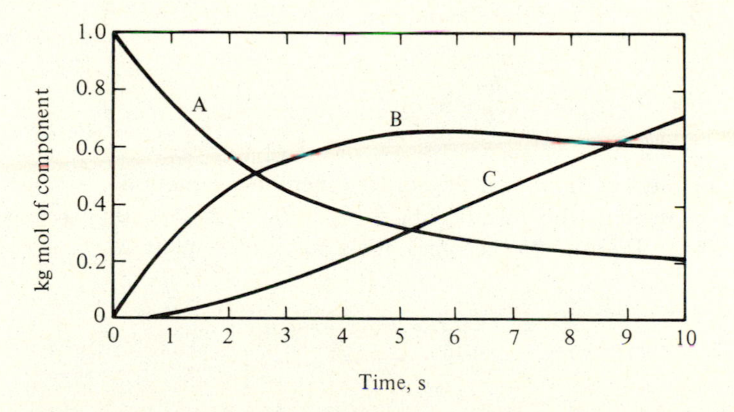

Figure 6.2-4 Composition vs. time in a semibatch reactor for

$$A \underset{k'}{\overset{k}{\rightleftharpoons}} B \xrightarrow{k''} C$$

Parallel Consecutive Reactions Let us consider the following reactions:

$$\mathrm{B} + \mathrm{C} \xrightarrow{k_1} \mathrm{M} + \mathrm{H} \qquad r_1 = k_1 BC$$

$$\mathrm{M} + \mathrm{C} \xrightarrow{k_2} \mathrm{D} + \mathrm{H} \qquad r_2 = k_2 MC$$

$$\mathrm{D} + \mathrm{C} \xrightarrow{k_3} \mathrm{T} + \mathrm{H} \qquad r_3 = k_3 DC$$

where B, M, D, and C are the concentrations of components B, M, D, and C, respectively. Using Eq. (6.2-6)

$$\frac{dn_j}{dt} + (qC_j)_2 - (qC_j)_1 = \Sigma \alpha_{ij} r_i V \tag{6.2-6}$$

and assuming that a stream of pure C enters a reactor initially charged with pure B at a rate of F kmol per unit time, the equations for each component become

$$\frac{dn_\mathrm{C}}{dt} = F - \frac{k_1}{V} n_\mathrm{B} n_\mathrm{C} - \frac{k_2}{V} n_\mathrm{M} n_\mathrm{C} - \frac{k_3}{V} n_\mathrm{D} n_\mathrm{C} \tag{6.2-75}$$

$$\frac{dn_\mathrm{B}}{dt} = -\frac{k_1}{V} n_\mathrm{B} n_\mathrm{C} \tag{6.2-76}$$

$$\frac{dn_\mathrm{M}}{dt} = +\frac{k_1}{V} n_\mathrm{B} n_\mathrm{C} - \frac{k_2}{V} n_\mathrm{M} n_\mathrm{C} \tag{6.2-77}$$

$$\frac{dn_\mathrm{D}}{dt} = +\frac{k_2}{V} n_\mathrm{M} n_\mathrm{C} - \frac{k_3}{V} n_\mathrm{D} n_\mathrm{C} \tag{6.2-78}$$

$$\frac{dn_\mathrm{T}}{dt} = +\frac{k_3}{V} n_\mathrm{D} n_\mathrm{C} \tag{6.2-79}$$

which is a system of nonlinear first-order differential equations. Because initially only B is present in the reactor, $n_\mathrm{B} = n_{\mathrm{B}o}$, $n_\mathrm{C} = n_{\mathrm{C}o} = 0$, $n_\mathrm{M} = n_{\mathrm{M}o} = 0$, $n_\mathrm{D} = n_{\mathrm{D}o} = 0$, Eq. (6.2-75) becomes, for a small increment Δt,

$$\frac{dn_\mathrm{C}}{dt} \cong F - \frac{k_1}{V} n_{\mathrm{B}o} n_\mathrm{C} \tag{6.2-80}$$

Integrating gives

$$n_\mathrm{C} \cong (1 - e^{-k_1 n_{\mathrm{B}o} t/V}) \left(\frac{VF}{k_1 n_{\mathrm{B}o}} \right) \tag{6.2-81}$$

After C is found from Eq. (6.2-81), the values of B, M, D, and T at the end of the time increment can be calculated from Eqs. (6.2-76) to (6.2-79) which are the starting values of B, M, D, T for the next increment of time *t*.

Example 6.2-5 Semibatch Operation for Parallel Consecutive Reactions

Dry chlorine gas is fed at a rate of 1.4 lb mole/h per pound mole of initial charge of benzene to a stirred tank reactor which contains an initial charge of pure benzene N_o (assuming $N_o = 1$ in this example). The reactor is operated at a temperature of 55°C and a pressure of 2 atm. The unit is equipped with a reflux condenser which returns all vaporized benzene and chlorbenzenes to the reactor while allowing the generated hydrogen chloride gas and excess chlorine gas to leave the system. The rates of the following reactions were first reported by McMullin (M1) and later used by Peters and Timmerhaus (P1)

$$C_6H_6 + Cl_2 \xrightarrow{k_1} C_6H_5Cl + HCl \qquad k_1 = 510\ (\text{lb mol/ft}^3)^{-1}\text{h}^{-1}$$

$$C_6H_5Cl + Cl_2 \xrightarrow{k_2} C_6H_4Cl_2 + HCl \qquad k_2 = 64\ (\text{lb mol/ft}^3)^{-1}\text{h}^{-1}$$

$$C_6H_4Cl_2 + Cl_2 \xrightarrow{k_3} C_6H_3Cl_3 + HCl \qquad k_3 = 2.1\ (\text{lb mol/ft}^3)^{-1}\text{h}^{-1}$$

```
 81/08/03. 15.14.51.
PROGRAM    CHLO

00100 PROGRAM CHLO(INPUT,OUTPUT)
00110CDT=TIME INCRFEMENT,HR
00120CT=TIME,HR
00130CC=CONCENTRATION OF CHLORINE,LBMOLE/FT**3
00140CB=CONCENTRATION OF BENZENE,LBMOLE/FT**3
00150CM=CONCENTRATION OF MONOCLOROBENZENE,LBMOLE/FT**3
00160CD=CONCENTRATION OF DICHLOROBENZENE,LBMOLE/FT**3
00170CTR=CONCENTRATION OF TRICHLOROBENZENE,LBMOLE/FT**3
00180 REAL M
00181 PRINT 11
00182 11 FORMAT(3X,'T',6X,'C',9X,'B',9X,'M',9X,'D',9X,'TR')
00190 DT=0.01
00200 T=0.
00210 B=1.
00220 M=0.
00221 K=1
00230 D=0.
00240 C=0.
00250 TR=0.
00260 I=1
00270 5 FB=349.315*B+43.8356*M+1.4383*D
00280 C=(1.4-(1.4-FB*C)/EXP(FB*DT))/FB
00290 7 B=B/EXP(349.315*C*DT)
00300 M=(B**0.125-B)*8./7.
```

```
00310 D=0.0049467*(29.*B-239.*B**0.125+210.*B**0.004167)
00320 TR=1.-B-M-D
00330 IF (I.GT.200) GO TO 99
00332 J=10*K
00333 IF (I.NE.J) GO TO 4
00334 T=FLOAT(I)*DT
00340 PRINT 14,T,C,B,M,D,TR
00341 K=K+1
00350 4 IF(C.LT.0.12) GO TO 6
00360 I=I+1
00361 C=0.12
00382 GO TO 7
00383 6 I=I+1
00386 GO TO 5
00387 14 FORMAT(1X,F3.1,5(1X,E9.4))
00400 99 STOP
00410 END
READY.
RNH
```

T	C	B	M	D	TR
.1	.4492E-02	.8627E+00	.1360E+00	.1290E-02	.3246E-06
.2	.5177E-02	.7281E+00	.2663E+00	.5596E-02	.2671E-05
.3	.6081E-02	.5977E+00	.3886E+00	.1373E-01	.1052E-04
.4	.7311E-02	.4725E+00	.5006E+00	.2684E-01	.3024E-04
.5	.9048E-02	.3547E+00	.5986E+00	.4662E-01	.7399E-04
.6	.1160E-01	.2469E+00	.6773E+00	.7557E-01	.1660E-03
.7	.1548E-01	.1536E+00	.7287E+00	.1174E+00	.3567E-03
.8	.2140E-01	.8036E-01	.7421E+00	.1768E+00	.7538E-03
.9	.2990E-01	.3256E-01	.7077E+00	.2582E+00	.1575E-02
1.0	.4053E-01	.9382E-02	.6268E+00	.3606E+00	.3185E-02
1.1	.5241E-01	.1819E-02	.5173E+00	.4748E+00	.6055E-02
1.2	.6636E-01	.2251E-03	.3997E+00	.5893E+00	.1071E-01
1.3	.8514E-01	.1575E-04	.2868E+00	.6953E+00	.1787E-01
1.4	.1126E+00	.4903E-06	.1859E+00	.7855E+00	.2863E-01
1.5	.1200E+00	.7449E-08	.1102E+00	.8469E+00	.4293E-01
1.6	.1200E+00	.1126E-09	.6523E-01	.8768E+00	.5801E-01
1.7	.1200E+00	.1703E-11	.3863E-01	.8879E+00	.7345E-01
1.8	.1200E+00	.2574E-13	.2287E-01	.8882E+00	.8897E-01
1.9	.1200E+00	.3892E-15	.1354E-01	.8820E+00	.1044E+00
2.0	.1200E+00	.5885E-17	.8021E-02	.8722E+00	.1198E+00

```
SRU        1.250 UNTS.
```

The volume of liquid in the reactor remains constant at 1.46 ft^3/lb mol of initial benzene charge. The solubility limit of the chlorine gas in the solution is 0.12 lb mol chlorine per pound mole of initial charge of benzene. Plot the pound moles of the components versus time and thus determine the time to obtain maximum concentration of the monochlorobenzene.

Solution

(Assume $\Delta t = 0.01$ h.) Substituting the numerical values into Eq. (6.2-81) gives

$$C = (1 - e^{-510(1)(0.01)/1.46})\left[\frac{1.46(1.4)}{510/1}\right] = 0.003886 \text{ lb mol/ft}^3$$

Using Eq. (6.2-76)

$$\frac{dn_B}{dt} = -\frac{510}{1.46}(0.003886)n_B$$

and integrating from $n_B = 1$ at $t = 0$ to $n_B = n_B$ at $t = 0.01$ gives

$$n_B = 0.9865 \text{ lb mol}$$

The relationship of the composition of monochlorobenzene n_M with the composition of benzene n_B can be determined by dividing Eq. (6.2-77) by Eq. (6.2-76) to eliminate the variable t, and then integrating to yield

$$n_M = \frac{(n_B^{0.125} - n_B)(8)}{7}$$

$$= \frac{(0.9865^{0.125} - 0.9865)(8)}{7} = 0.01347 \text{ lb mol}$$

The relationship of the composition of dichlorobenzene n_D with the composition of benzene n_B can be determined by dividing Eq. (6.2-78) by Eq. (6.2-76), using the result from Eq. (6.2-77) and integrating to give

$$n_D = \frac{240}{7(29)(239)}(29n_B - 239n_B^{0.125} + 210n_B^{1/240})$$

$$= 0.0049467[29(0.9865) - 239(0.9865)^{1/8} + 210(0.9865)^{1/240}]$$

$$= 0.00001145 \text{ lb mol}$$

The relationship of the composition of trichlorobenzene n_T with the compositions of other components is

$$n_B + n_M + n_D + n_T = 1$$

which is obtained by adding Eqs. (6.2-76) to (6.2-79) and integrating. Thus

$$n_T = 1 - 0.9865 - 0.01347 - 0.00001145$$

$$= 0.0000186 \text{ lb mol}$$

For the next time increment 0.01 to 0.02 h, repeat the process with the new values of n_C, n_B, n_M, n_D, and n_T. This procedure is carried on for 2 h or 200 time increments. A digital computer is used to produce the results, which check very well with the results of Peters and Timmerhaus (P1) by analog computer or MIMIC program. The output of the computer program is:

	Composition, lb mol				
Time, h	Benzene	Monochloro-benzene	Dichloro-benzene	Trichloro-benzene	Chlorine
0	1.0000	0.	0.	0.	0.
0.1	8.6269E-1	1.3602E-1	1.2904E-3	3.3640E-7	4.4917E-3
0.2	7.2814E-1	2.6626E-1	5.5964E-3	2.6890E-6	5.1775E-3
0.3	5.9767E-1	3.8859E-1	1.3729E-2	1.0529E-5	6.0810E-3
0.4	4.7255E-1	5.0058E-1	2.6843E-2	3.0247E-5	7.3109E-3
0.5	3.5467E-1	5.9863E-1	4.6624E-2	7.3996E-5	9.0481E-3
0.6	2.4692E-1	6.7734E-1	7.5573E-2	1.6600E-4	1.1600E-2
0.7	1.5356E-1	7.2873E-1	1.1736E-1	3.5625E-4	1.5479E-2
0.8	8.0364E-2	7.4208E-1	1.7680E-1	7.5384E-4	2.1400E-2
0.9	3.2561E-2	7.0765E-1	2.5821E-1	1.5746E-3	2.9904E-2
1.0	9.3816E-3	6.2685E-1	3.6059E-1	3.1853E-3	4.0534E-2
1.1	1.8190E-3	5.1728E-1	4.7484E-1	6.0548E-3	5.2406E-2
1.2	2.2510E-4	3.9972E-1	5.8934E-1	1.0711E-2	6.6361E-2
1.3	1.5747E-5	2.8682E-1	6.9529E-1	1.7867E-2	8.5136E-2
1.4	4.9027E-7	1.8591E-1	7.8547E-1	2.8626E-2	1.1256E-1
1.5	7.4126E-9	1.1009E-1	8.4690E-1	4.3013E-2	1.2000E-1
1.6	1.1207E-10	6.5189E-2	8.7680E-1	5.8011E-2	1.2000E-1
1.7	1.6944E-12	3.8603E-2	8.8790E-1	7.3497E-2	1.2000E-1
1.8	2.5619E-14	2.2859E-2	8.8820E-1	8.8941E-2	1.2000E-1
1.9	3.8734E-16	1.3536E-2	8.8200E-1	1.0446E-1	1.2000E-1
2.0	5.8564E-18	8.0157E-3	8.7220E-1	1.1978E-1	1.2000E-1

From Figure 6.2-5, it is seen that the maximum compositions of monochlorobenzene and dichlorobenzene occur at about 0.8 h (0.78 h by Peters and Timmerhaus) and about 1.8 h (1.75 h by Peters and Timmerhaus), respectively.

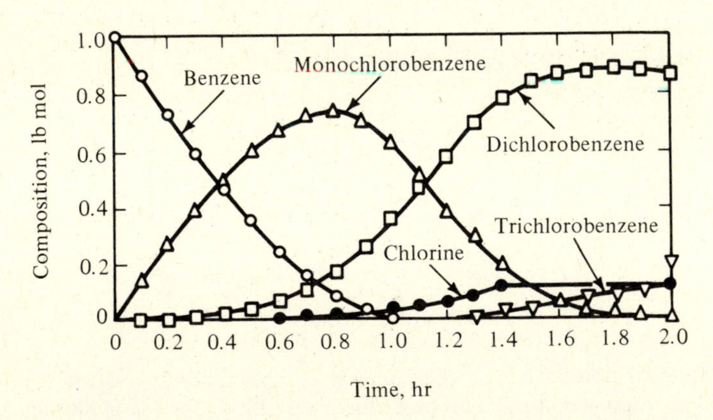

Figure 6.2-5 Concentration profile of chlorination of benzene in a semibatch reactor

6.3 Nonisothermal Condition

If the reaction is carried out at a temperature which is not constant, the generalized energy equation of Eq. (4.2-25) in Chapter 4 should be used

$$\frac{\partial\Sigma(C_jh_j)}{\partial t}+\frac{\partial(u\Sigma C_jh_j)}{\partial x}=k\frac{\partial^2 T}{\partial x^2}+G-Ua(T-T^*) \tag{4.2-25}$$

Neglecting the diffusion and the generation terms, multiplying by dV, and integrating yields

$$\int\frac{\partial}{\partial t}(\Sigma C_jh_j)dV+\int\frac{\partial}{\partial x}(u\Sigma C_jh_j)dV=-Ua(T-T^*)V \tag{6.3-1}$$

By divergence theorem, Eq. (6.3-1) becomes

$$\frac{\partial}{\partial t}(\Sigma C_jh_j)V+(uS\Sigma C_jh_j)_2-(uS\Sigma C_jh_j)_1=-Ua(T-T^*)V \tag{6.3-2}$$

where S is the cross-sectional area of the reactor and u is the linear velocity of the fluid flowing through the reactor. Expanding and using the volumetric flow rate $q=uS$, we get

$$V\Sigma C_j\frac{\partial h_j}{\partial t}+V\Sigma h_j\frac{\partial C_j}{\partial t}+\Sigma C_jh_j\frac{\partial V}{\partial t}+(q\Sigma C_jh_j)_2-(q\Sigma C_jh_j)_1=-Ua(T-T^*)V \tag{6.3-3}$$

But

$$\frac{\partial h_j}{\partial t}=\frac{\partial T\partial h_j}{\partial t\partial T}=C_{pj}\frac{\partial T}{\partial t} \tag{6.3-4}$$

and

$$\Sigma C_jC_{pj}=\mathfrak{z}_p \tag{6.3-5}$$

and the mass balance equation

$$V\frac{dC_j}{dt}=\Sigma\alpha_{ij}r_iV+q_1(C_{j1}-C_{j2})$$

and the total flow balance

$$\frac{dV}{dt}=q_1-q_2$$

Eq. (6.3-3) becomes

$$V\mathfrak{z}_p\frac{dT}{dt}+\Sigma h_j[\Sigma\alpha_{ij}r_iV+q_1(C_{j1}-C_{j2})]+\Sigma C_jh_j(q_1-q_2)$$
$$+(q\Sigma C_jh_j)_2-(q\Sigma C_jh_j)_1=-Ua(T-T^*)V \quad \textbf{(6.3-6)}$$

Simplifying and using the definition of the heat of reaction $\Delta H = \Sigma\alpha_j h_j$,

$$V\mathfrak{z}_p\frac{dT}{dt}+\Sigma\Delta H_ir_iV+q_1\Sigma C_{j1}(h_j-h_{j1})=-Ua(T-T^*)V \quad \textbf{(6.3-7)}$$

Simplifying by the definition of enthalpy and $\mathfrak{z}_p = \Sigma C_jC_{pj}$ results in

$$V\mathfrak{z}_p\frac{dT}{dt}+\Sigma\Delta H_ir_iV+q_1\mathfrak{z}_{p1}(T_2-T_1)=-Ua(T-T^*)V \quad \textbf{(6.3-8)}$$

Dividing by $q_1\mathfrak{z}_{p1}$ and using $\theta = V/q_1$ gives

$$\theta\frac{dT}{dt}+\frac{\Sigma\Delta H_ir_i}{\mathfrak{z}_{p1}}\theta+T_2-T_1=-\frac{Ua(T_2-T^*)\theta}{\mathfrak{z}_{p1}} \quad \textbf{(6.3-9)}$$

where θ is defined as the residence time.

Example 6.3-1 Nonisothermal Semibatch Operations

Initially a 1500-liter reactor contains 900 liters formalin solution at 25°C (42 percent by weight formaldehyde). To produce hexamethylenetetramine (HMT), an aqueous ammonia solution (25 weight percent NH_3) at 25°C is added gradually to the reactor at a rate of 8 liters/min while the solution in the reactor is raised to 50°C in order to start the reaction. The reactor is allowed to operate not over 100°C by passing cooling water through the internal coils. Then the rate of reaction is much faster than the rate of heat transfer with the surroundings. Calculate the length of the internal cooling coils (2.5 cm O.D.). The rate of the reverse reaction is negligible. Data: Heat of reaction = −530 cal/g of HMT

Specific heat of reaction mixture C_p = 1.0 cal/g·°C.

Specific heat of 25 weight percent NH_3 solution = 1.0 cal/g·°C.

Density of NH_3 solution = 0.91 g/cm^3

Density of formalin (42 percent) at 25°C = 1.1 g/cm^3

Overall heat transfer coefficient = 1.1535×10^{-2} cal/s·cm^2·°C

Solution

$$4NH_3 + 6HCHO \rightarrow N_4(CH_2)_6 + 6H_2O$$

For complete conversion, the total amount of NH_3 required is

$$(NH_3)_t = \frac{900{,}000(1.1)(0.42)}{30}\frac{4}{6}(17) = 157{,}080 \text{ g}$$

From the ammonia feed rate of 8 liters/min, the total time of reaction is

$$t_t = \frac{157{,}080}{8{,}000(0.91)(0.25)} = 86.31 \text{ min}$$

In order to calculate the cooling surface required, the assumption of steady state should be made. Hence from the generalized energy balance equation, Eq. (6.3-9) becomes

$$\Sigma \Delta H_i r_i V + q_{13p}(T_2 - T_1) = -Ua(T - T^*)V \quad \textbf{(A)}$$

The heat of reaction is

$$\Delta H = -530 \text{ cal/g HMT} = -\frac{530(140)}{4(17)} = -1{,}091 \text{ cal/g } NH_3$$

Since the reaction is so fast that the reaction rate can be assumed to be equal to the rate of the addition of NH_3. Thus

$$rV = F_o w_o \quad \textbf{(B)}$$

where w_o is the weight percent of NH_3 in the feed, or 0.25 in this case. F_o is the mass feed rate in lb/h or

$$F_o = 8{,}000(60)(0.91) = 436{,}800 \text{ g/h}$$

Then

$$rV = F_o w_o = 436{,}800(0.25) = 109{,}200 \text{ g } NH_3 \text{ per hour}$$

$$q_{13p}(T_2 - T_1) = q_1 \Sigma C_j C_{pj}(T_2 - T_1) = F_o \Sigma C_{pj}(T_2 - T_1) = 436{,}800(1)(100 - 25)$$
$$= 32{,}760{,}000 \text{ cal/h}$$

Substituting into Eq. (A) gives

$$-(1{,}091)(109{,}200) + 32{,}760{,}000 = -1.1535 \times 10^{-2}(3{,}600)(100 - 25)aV$$

or

$$aV=\frac{P}{A}V=\frac{PAL}{A}=\pi DL=2.77\times 10^4$$

$$L=\frac{2.77\times 10^4}{2.5\pi}=3{,}526.87 \text{ cm}$$

In order to calculate the time required to raise the temperature from 50°C to 100°C, it is assumed that the flow of cooling water should be shut off, i.e., $Ua(T-T^*)=0$. Equation (6.3-8) can be modified to another convenient form

$$m_t C_v \frac{dT_2}{dt}+\Sigma\Delta H_i r_i V+C_p F_o(T_2-T_1)=0$$

where C_v and C_p are assumed to be 1 since the solution is very dilute. m_t is the total mass of the mixture at any time which can be calculated by

$$\frac{dm}{dt}=8{,}000(60)(0.91)=436{,}800$$

Integrating

$$m_t=m_o+436{,}800t$$

where

$$m_o=\text{initial mass of the solution}$$
$$=900{,}000(0.11)(10)=9.9\times 10^5 \text{ g}$$

Substituting with $T_1=25°\text{C}$ gives

$$(9.9\times 10^5+436{,}800t)\frac{dT_2}{dt}-(1{,}091)(109{,}200)+436{,}800(T_2-25)=0$$

Separating variables yields

$$\frac{dT_2}{1.1914\times 10^8-4.368\times 10^5(T_2-25)}=\frac{dt}{9.9\times 10^5+4.368\times 10^5 t}$$

Integrating between $t=0$, $T_2=50$ and $t=t$, $T_2=100$ results in $t=0.5729$ h or 34 min.

6.4 *Optimization*

The determination of maximum production of B in a reversible consecutive reaction and of monochlorobenzene in a parallel consecutive reaction have been described in Examples 6.2-4 and 6.2-5, respectively. The other methods described for batch reactors in Chapter 5 and CSTR in Chapter 7 are applicable to the semibatch as well.

Problems

1. Initially a reactor is charged with 378 liters of an aqueous solution of 10 g ethyl acetate per liter. The ester in the reactor is then to be saponified by continuously adding 0.1 *N* sodium hydroxide solution at a rate of 3.78 liter/min until stoichiometric amounts are present. At 20°C, the reaction is relatively fast and irreversible with a rate constant of 92 liters/g mol·min. Assuming complete mixing, determine the concentration of unreacted ethyl acetate as a function of time. When will the maximum concentration occur?

2. At 173°C, the rate constant for the reaction $X \longrightarrow Y$ is $k = 0.8\ h^{-1}$. Into a continuous stirred-tank reactor which originally contains 3.5 kg Y, a stream of isomer X enters at the rate which varies with time as follows:

Time t, h	0–2	2–5	5–6	6–7	7–10	10–11	11–12	12–13	13–14	14–15	15–16	16+
F_A, kg/h	80.0	100.0	125	148	180	148	125	100	80	45	23	0

Assuming constant temperature at 173°C, determine the total amount of X and Y in the reactor as function of time.

3. Initially a tank is charged with 180 kg pure ethyl alcohol (density = 0.789 g/cm³). An aqueous solution of acetic acid (density = 0.958 g/cm³) which contains 42.6 percent by weight acid is then added at a rate of 1.8 kg/min for 120 min. Assuming that the density is constant and equal to that of water, compute the conversion of acetic acid to ester as a function of time for the complete range of operation. At the constant operating temperature of 100°C, the rate constants for the following reaction are

$$CH_3COOH + C_2H_5OH \underset{k'}{\overset{k}{\rightleftharpoons}} CH_3COOC_2H_5 + H_2O$$

$$k = 4.76 \times 10^{-4} \text{ liter/g mol}\cdot\text{min}$$
$$k' = 1.63 \times 10^{-4} \text{ liter/g mol}\cdot\text{min}$$

4. Into a stirred tank which originally contains 10 liters aqueous acetic anhydride solution at a concentration of 0.00005 g mol/cm³ is added a stream of 0.0003 g mol/cm³ of anhydride at a rate of 2 liters/min. No product is withdrawn. The rate constant for the hydrolysis of the acetic anhydride k is 0.38 min⁻¹. Determine the concentration of the anhydride solution as a function of time.

5. Initially a tank contains V_o of a fluid A with a concentration of C_{Ao}. The fluid A decomposes by a second-order reaction. With ideal stirring, the reaction mixture flows out at a rate of q. What is the concentration of A at time t?

6. Into an empty holding tank, a stream containing component A is pumped at a volumetric rate of q. A decomposes by a first-order reaction. After the tank is full, the outlet valve is opened. Then both flow rates are equal. Find the concentration of A (a) for the filling period (b) after the valve is opened.

7. Products C, D, and E are produced from A and B by chemical reaction on semibatch basis. Initially a reactor is charged with 25 mol A and then B is added at a rate of 0.8 mol/min. If the following reactions are all second order

$$A + B \xrightarrow{k_1} C \qquad k_1 = 2.0 \times 10^{-4}\ \text{m}^3/\text{mol} \cdot \text{min}$$

$$C + B \xrightarrow{k_2} D \qquad k_2 = 0.5 \times 10^{-4}\ \text{m}^3/\text{mol} \cdot \text{min}$$

$$D + B \xrightarrow{k_3} E \qquad k_3 = 0.2 \times 10^{-3}\ \text{m}^3/\text{mol} \cdot \text{min}$$

Plot the moles of the products versus time assuming $V = 1\ \text{m}^3$.

8. Ethylene oxide is gradually added at a rate of 1 g mol/min to a reactor of 1 m³ originally containing 1,322.8313 g mol ethylene oxide, 14.7521 g mol water, 10.8344 g mol/ethylene glycol, 29.3171 g mol diethylene glycol, and 0.6534 g mol triethylene glycol. Show the product distribution for the following reactions every 10 min.

$$H_2O + \underset{\text{Ethylene oxide}}{C_2H_4O} \xrightarrow{25°C} \underset{\text{Ethylene glycol}}{C_2H_6O_2} \qquad k_1 = 7.37 \times 10^{-7}\ \text{liters/g mol} \cdot \text{min}$$

$$C_2H_6O_2 + C_2H_4O \xrightarrow{25°C} \underset{\text{Diethylene glycol}}{C_4H_{10}O_3} \qquad k_2 = 14.74 \times 10^{-7}\ \text{liters/g mol} \cdot \text{min}$$

$$C_4H_{10}O_3 + C_2H_4O \xrightarrow{25°C} \underset{\text{Triethylene glycol}}{C_6H_{14}O_4} \qquad k_3 = 14.74 \times 10^{-7}\ \text{liters/g mol} \cdot \text{min}$$

9. Initially a reactor is charged with 15 m³ of reactant X. The solubility of reactant Y is limited to 0.03 m³ Y per m³ X, so it is pumped into the reactor just fast enough to keep the solution saturated. Within this range of concentration of interest, the reaction is considered to be pseudo first order with respect to Y (k = 20 h^{-1}). The molal density of each reactant is 0.5 kmol/m³. After 2 h, at what rate is Y being pumped in?

References

(C1). Chen, N.H. *AIChE Module Instruction,* Series E: Kinetics, Vol. 2: *Reactors & Rate Data, E2.2 Semibatch Reactors II,* New York: American Institute of Chemical Engineers, 1981.

(F1). Frost, A.A., and Pearson, R.G. *Kinetics and Mechanism,* New York: John Wiley & Sons, Inc., 1953.

(M1). Macmullin, R.B. *Chem. Eng. Progr.,* 44:183 (1948).

(P1). Peters, M.S., and Timmerhaus, K.D. *Plant Design and Economics for Chemical Engineers,* 3rd ed. New York: McGraw-Hill Book Company, 1980.

(S1). Swain, C.G. *J. Am. Chem. Soc.,* 70:1119 (1948).

CHAPTER SEVEN

Ideal Continuous Stirred-Tank Reactors (CSTR)

7.1 *Introduction*

Continuous stirred-tank reactors are widely used in the chemical process industry. Although single reactors may be used, it is usually preferable to employ a battery of such reactors connected in series. The CSTR are simply tanks into which reactants flow and from which the product stream is removed on a continuous basis. In each tank, a stirrer is provided to give uniform compositions and temperature. Figure 7.1-1 is a typical example of a cascade of CSTR. In this chapter, the design of a single tank to carry out a chemical reaction on a continuous basis is first discussed, after which the cascade of continuous stirred tanks for these chemical reactions will be discussed. Then the non-steady-state behavior of these tanks will be described, followed by a discussion of the stability of the reactions and the optimization of these reactors.

7.2 *Single Continuous Stirred-Tank Reactor*

From Chapter 4, the general mass balance is Eq. (4.2-17)

$$\frac{\partial C_j}{\partial t} + \frac{\partial (uC_j)}{\partial x} = D\frac{\partial^2 C_j}{\partial x^2} + \Sigma\, \alpha_{ij} r_i - k_m a(C_j - C_j^*) \qquad \textbf{(4.2-17)}$$

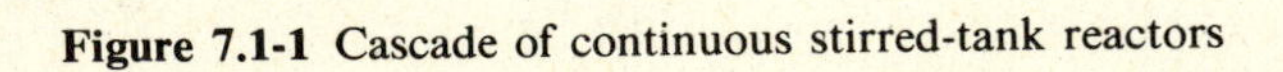

Figure 7.1-1 Cascade of continuous stirred-tank reactors

Assuming that the process is steady, i.e., $\partial C_j/\partial t = 0$, the diffusion is negligible, $D = 0$, there is not interphase transfer, $k_m = 0$, Eq. (4.2-17) becomes

$$\frac{\partial(uC_j)}{\partial x} = \Sigma\alpha_{ij}r_i \tag{7.2-1}$$

Multiplying by dV and integrating gives

$$\int \frac{\partial(uC_j)}{\partial x} dV = \Sigma\alpha_{ij}r_i V \tag{7.2-2}$$

By divergence theorem and using $q = uS$, we get

$$(qC_j)_2 - (qC_j)_1 = \Sigma\alpha_{ij}r_i V \tag{7.2-3}$$

or

$$\dot{m}_2 - \dot{m}_1 = \Sigma\alpha_{ij}r_i V \tag{7.2-4}$$

The overall mass balance is

$$\frac{d(\rho V)}{dt} = (\rho q)_1 - (\rho q)_2 \tag{7.2-5}$$

Assuming that the density is constant, Eq. (7.2-5) becomes

$$\frac{dV}{dt} = q_1 - q_2 \tag{7.2-6}$$

Under steady-state condition, $dV/dt = 0$, and therefore

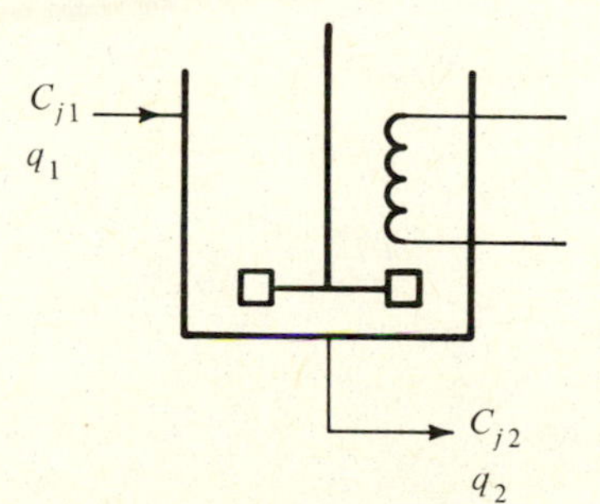

Figure 7.2-1 Single CSTR

$$q_1 = q_2 \tag{7.2-7}$$

With this identity, Eq. (7.2-3) becomes

$$C_{j2} - C_{j1} = \frac{\Sigma \alpha_{ij} r_i V}{q} = \Sigma \alpha_{ij} r_i \theta \tag{7.2-8}$$

where $\theta = V/q$ is defined as the residence time. Eq. (7.2-8) is the design equation for CSTR.

7.3 *Irreversible Reaction*

7.3-1 *First-order reaction*

$$\text{A} \longrightarrow \text{product}$$

The rate of reaction for a first-order reaction is

$$r = kC_A$$

Substituting into Eq. (7.2-8) gives

$$C_{A2} - C_{A1} = -kC_{A2}\theta \tag{7.3-1}$$

Rearranging results in

$$\frac{C_{A2}}{C_{A1}} = \frac{1}{1 + k\theta} \tag{7.3-2}$$

But

$$C_{A2} = C_{A1}(1 - f) \tag{7.3-3}$$

hence

$$f = \frac{k\theta}{1 + k\theta} \tag{7.3-4}$$

7.3-2 *Second-order reaction*

Type I:

$$2\text{A} \longrightarrow \text{product}$$

The reaction rate is

$$r = kC_A^2$$

Substituting into Eq. (7.2-8) yields

$$C_{A2} - C_{A1} = -kC_A^2\theta \tag{7.3-5}$$

Rearranging gives a quadratic equation

$$kC_{A2}^2\theta + C_{A2} - C_{A1} = 0 \tag{7.3-6}$$

whose solution is

$$C_{A2} = \frac{1}{2k\theta}[-1 \pm (1 + 4k\theta C_{A1})^{1/2}] \tag{7.3-7}$$

Type II:

$$A + B \rightarrow \text{product}$$

The reaction rate is

$$r = kC_A C_B$$

The mass balance yields

$$C_{B2} = C_{A2} - C_{A1} + C_{B1} \tag{7.3-8}$$

Substituting into Eq. (7.2-8) yields

$$C_{A2} - C_{A1} = -kC_{A2}(C_{A2} - C_{A1} + C_{B1})\theta \tag{7.3-9}$$

which is also a quadratic equation whose solution can be easily obtained.

7.4 Reversible Reaction

7.4-1 First-order reaction

$$A \underset{k'}{\overset{k}{\rightleftarrows}} B$$

The reaction rate is

$$r = kC_A - k'C_B$$

Substituting into Eq. (7.2-8) gives

$$C_{A2} - C_{A1} = -(kC_{A2} - k'C_{B2})\theta \tag{7.4-1}$$

The stoichiometric equations for the reaction are:

$$C_{A2} = C_{A1} - \lambda$$

$$C_{B2} = C_{B1} + \lambda$$

Eliminating λ yields

$$C_{B2} = C_{A1} + C_{B1} - C_{A2} \tag{7.4-2}$$

Substituting into Eq. (7.4-1) gives

$$C_{A2} - C_{A1} = -[kC_{A2} - k'(C_{A1} + C_{B1} - C_{A2})]\theta \tag{7.4-3}$$

Solving for C_{A2}

$$C_{A2} = \frac{1}{1 + k + k'}[(1 + k')C_{A1} + k'C_{B1}] \tag{7.4-4}$$

7.4-2 Second-order reaction

$$A + B \underset{k'}{\overset{k}{\rightleftharpoons}} C + D$$

The reaction rate is

$$r = kC_AC_B - k'C_CC_D$$

Equation (7.2-8) becomes

$$C_{A2} - C_{A1} = -(kC_{A2}C_{B2} - k'C_{C2}C_{D2})\theta \tag{7.4-5}$$

The stoichiometric equations are

$$C_{A2} = C_{A1} - \lambda$$

$$C_{B2} = C_{B1} - \lambda$$

$$C_{C2} = C_{C1} + \lambda$$

$$C_{D2} = C_{D1} + \lambda$$

Eliminating λ yields

$$C_{B2} = C_{A2} - C_{A1} + C_{B1} \tag{7.4-6}$$

$$C_{C2} = -C_{A2} + C_{A1} + C_{C1} \tag{7.4-7}$$

$$C_{D2} = -C_{A2} + C_{A1} + C_{D1} \tag{7.4-8}$$

Substituting into Eq. (7.4-5) with $C_{C1} = C_{D1} = 0$ gives

$$C_{A1} = C_{A2} + \theta[kC_{A2}(C_{A2} + C_{B1} - C_{A1}) - k'(C_{A1} - C_{A2})^2] \tag{7.4-9}$$

which is also a quadratic equation.

7.5 *Parallel Reaction*

7.5-1 *Type I: Side Reaction*

$$\begin{array}{ccc} A & \xrightarrow{k_1} & B \\ {\scriptstyle k_2}\downarrow & & \\ C & & \end{array}$$

The reaction rate is

$$r = k_1C_A + k_2C_A = (k_1 + k_2)C_A$$

Substituting into Eq. (7.2-8) gives

$$\text{For A} \qquad C_{A2} = C_{A1} - (k_1 + k_2)C_{A2}\theta \tag{7.5-1}$$

$$\text{For B} \qquad C_{B2} = C_{B1} + k_1C_{A2}\theta \tag{7.5-2}$$

$$\text{For C} \qquad C_{C2} = C_{C1} + k_2C_{A2}\theta \tag{7.5-3}$$

Solving results in

$$C_{A2} = \frac{C_{A1}}{1 + \theta(k_1 + k_2)} \tag{7.5-4}$$

$$C_{B2} = \frac{C_{B1} + \theta[k_1(C_{A1} + C_{B1}) + k_2C_{B1}]}{1 + \theta(k_1 + k_2)} \tag{7.5-5}$$

$$C_{C2} = \frac{C_{C1} + \theta[k_2(C_{A1} + C_{C1}) + k_1C_{C1}]}{1 - \theta(k_1 + k_2)} \tag{7.5-6}$$

7.5-2 *Type II*

$$B + C \xrightarrow{k_1} M + H \qquad \lambda_1$$

$$M + C \xrightarrow{k_2} D + H \qquad \lambda_2$$

$$D + C \xrightarrow{k_3} T + H \qquad \lambda_3$$

Using

$$C_j = C_{jo} + \Sigma \alpha_{ij} r_i \theta \qquad \text{and} \qquad C_j = C_{jo} + \Sigma \alpha_{ij} \lambda_i$$

we get

For B	$C_{B2} = C_{B1} - k_1 C_{C2} C_{B2} \theta$	$C_{B2} = C_{B1} - \lambda_1$
For M	$C_{M2} = 0 \quad + k_1 C_{C2} C_{B2} \theta - k_2 C_{M2} C_{C2} \theta$	$C_{M2} = 0 \quad + \lambda_1 - \lambda_2$
For D	$C_{D2} = 0 \quad + k_2 C_{C2} C_{M2} \theta - k_3 C_{D2} C_{C2} \theta$	$C_{D2} = 0 \quad + \lambda_2 - \lambda_3$
For T	$C_{T2} = 0 \quad + k_3 C_{C2} C_{D2} \theta$	$C_{T2} = \quad \lambda_3$
For C		$C_{C2} = C_{C1} - \lambda_1 - \lambda_2 - \lambda_3$

Solving these equations, we obtain

$$C_{B2} = \frac{C_{B1}}{1 + k_1 C_{C2} \theta} \qquad \lambda_2 = \lambda_3 + C_{D2} = C_{T2} + C_{D2}$$

$$C_{M2} = \frac{k_1 C_{C2} C_{B2} \theta}{1 + k_2 C_{C2} \theta} \qquad \text{Hence} \qquad \lambda_1 = \lambda_2 + C_{M2} = C_{T2} + C_{D2} + C_{M2}$$

$$C_{D2} = \frac{k_2 C_{C2} C_{M2} \theta}{1 + k_3 C_{C2} \theta} \qquad C_{C2} = C_{C1} - (C_{T2} + C_{D2} + C_{M2}) - (C_{T2} + C_{D2}) - C_{T2}$$

$$= C_{C1} - C_{M2} - 2C_{D2} - 3C_{T2}$$

If the initial concentration of C (C_{C1}) is related to the initial concentration of B (C_{B1}) by ϵ, i.e.,

$$C_{C1} = \epsilon C_{B1}$$

then

$$\epsilon C_{B1} = C_{C2} + C_{M2} + 2C_{D2} + 3C_{T2}$$

Therefore

$$\epsilon C_{B1} = C_{C2} + \frac{k_1 C_{C2} C_{B2} \theta}{1 + k_2 C_{C2} \theta} + 2\frac{k_2 C_{C2} C_{M2} \theta}{1 + k_3 C_{C2} \theta} + 3k_3 C_{C2} C_{D2} \theta$$

$$= C_{C2} + \frac{k_1 C_{C2} C_{B1} \theta}{(1 + k_1 C_{C2} \theta)(1 + k_2 C_{C2} \theta)}$$

$$+ \frac{2k_1 k_2 C_{C2}^2 C_{B1} \theta^2}{(1 + k_1 C_{C2} \theta)(1 + k_2 C_{C2} \theta)(1 + k_3 C_{C2} \theta)} + 3k_3 C_{C2} C_{D2} \theta \quad \textbf{(7.5-7)}$$

which is a quadratic equation. Its solution can be obtained analytically or numerically. Another approach is to use a trial and error method.

Example 7.5-1 Parallel Reaction CSTR

Calculate the concentrations of C_6H_6, C_6H_5Cl, $C_6H_4Cl_2$, $C_6H_3Cl_3$, and Cl_2 for the following reactions if $k_1/k_2 = 8$, $k_2/k_3 = 30$, $C_{B1} = 11.2$ g mol/liter ϵ (the chlorine-to-benzene feed ratio) = 1.4, and $k_2\theta = 1$.

$$C_6H_6 + Cl_2 \xrightarrow{k_1} C_6H_5Cl + HCl$$

$$C_6H_5Cl + Cl_2 \xrightarrow{k_2} C_6H_4Cl_2 + HCl$$

$$C_6H_4Cl_2 + Cl_2 \xrightarrow{k_3} C_6H_3Cl_3 + HCl$$

Solution

Assume

$$C_{C2} = C_{Cl_2,2} = 0.926$$

Since

$$k_1\theta / k_2\theta = 8, \qquad k_2\theta / k_3\theta = 30$$

Then

$$C_{B2} = C_{C_6H_6,2} = \frac{C_{B1}}{1 + k_1\theta C_{C2}} = \frac{11.2}{1 + 8(0.926)} = 1.3321$$

$$C_{M2} = C_{C_6H_5Cl,2} = \frac{k_1\theta C_{C2} C_{B2}}{1 + k_2\theta C_{C2}} = \frac{8(0.926)(1.3321)}{1 + 0.926} = 5.1235$$

$$C_{D2} = C_{C_6H_4Cl_2,2} = \frac{k_2\theta C_{C2} C_{M2}}{1 + k_3\theta C_{C2}} = \frac{1(0.926)(5.1235)}{1 + (1/30)(0.926)} = 4.6023$$

$$C_{T2} = C_{C_6H_3Cl_3,2} = k_3\theta C_{C2} C_{D2} = (1/30)0.926(4.6023) = 0.1421$$

Substituting into

$$\epsilon C_{B1} = C_{C2} + C_{M2} + 2C_{D2} + 3C_{T2}$$

$$(1.4)(11.2) = 0.926 + 5.1235 + 2(4.6023) + 3(0.1421)$$

$$15.68 \quad = \quad 15.68 \qquad \text{check}$$

7.6 *Series Reaction*

$$A \xrightarrow{k_1} B \xrightarrow{k_2} C$$

For A $\quad C_{A2} = C_{A1} - k_1 C_{A2}\theta$ **(7.6-1)**

For B $\quad C_{B2} = 0 + k_1 C_{A2}\theta - k_2 C_{B2}\theta$ **(7.6-2)**

For C $\quad C_{C2} = 0 + k_2 C_{B2}\theta$ **(7.6-3)**

Solving gives

$$C_{A2} = \frac{C_{A1}}{1 + k_1\theta} \tag{7.6-4}$$

$$C_{B2} = \frac{k_1\theta C_{A1}}{(1 + k_1\theta)(1 + k_2\theta)} \tag{7.6-5}$$

$$C_{C2} = \frac{k_1 k_2 \theta^2 C_{A1}}{(1 + k_1\theta)(1 + k_2\theta)} \tag{7.6-6}$$

7.7 *Multiple-Tank Cascade*

A major shortcoming of a single stirred tank is that all of the reaction takes place at the low final reactant concentration and hence an unduly large reactor hold-up is required. If a number of smaller well-stirred reactors are arranged in series, only the last one will have a reaction rate governed by the final reactant concentration and all of the others will have higher rates. Hence for a given duty, the total reactor hold-up will be less than for a single tank. In order to estimate the volume of the reactors required for a chemical reaction, let us make total and mass-flow balances for the cascade in Fig. 7.7-1.

7.7-1 *Total flow balance*

Overall:

$$q_0 - q_N = \frac{dV}{dt} \tag{7.7-1}$$

$$\dot{m}_0 - \dot{m}_N = \frac{dm}{dt} \tag{7.7-2}$$

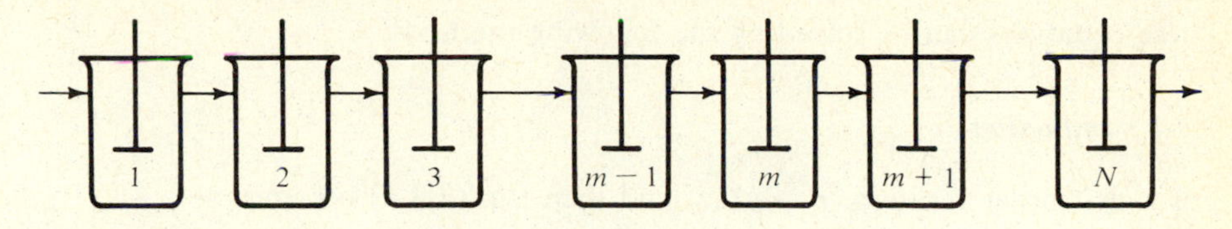

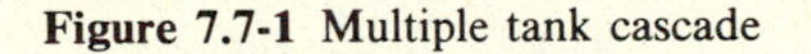
Figure 7.7-1 Multiple tank cascade

Around m-tank

$$q_{m-1} - q_m = \frac{dV_m}{dt} \quad \textbf{(7.7-3)}$$

$$\dot{m}_{m-1} - \dot{m}_m = \frac{dm_m}{dt} \quad \textbf{(7.7-4)}$$

7.7-2 Component flow balance

Overall:

$$\theta \frac{dC_j}{dt} + C_{jN} - C_{j0} = \Sigma\alpha_{ij} r_i \theta \quad \textbf{(7.7-5)}$$

Around m-tank

$$\theta \frac{dC_{jm}}{dt} + C_{jm} - C_{j(m-1)} = \Sigma\alpha_{ij} r_{im} \theta_m \quad \textbf{(7.7-6)}$$

7.8 Analytical Methods

For simplicity, using a single reaction under steady-state conditions reduces Eqs. (7.7-5) and (7.7-6) to

$$C_{jN} - C_{j0} = \alpha_j r\theta \quad \textbf{(7.8-1)}$$

and

$$C_{jm} - C_{j(m-1)} = \alpha_j r\theta_m \quad \textbf{(7.8-2)}$$

If j is the limiting component k, these equations become

$$C_{kN} - C_{k0} = -r\theta \quad \textbf{(7.8-3)}$$

and

$$C_{km} - C_{k(m-1)} = -r_m \theta_m \quad \textbf{(7.8-4)}$$

These equations can be solved by the following methods.

7.8-1 Induction

For a first-order reaction, $r = kC_{km}$ and then Eq. (7.8-4) becomes

$$C_{km} - C_{k(m-1)} = -kC_{km}\theta_m \tag{7.8-5}$$

Rearranging gives

$$(1 + k\theta_m)C_{km} = C_{k(m-1)} \tag{7.8-6}$$

$$\text{For } m = 1 \qquad C_{k1} = \frac{C_{k0}}{(1 + k\theta_1)} \tag{7.8-7}$$

$$m = 2 \qquad C_{k2} = \frac{C_{k1}}{(1 + k\theta_2)} = \frac{1}{(1 + k\theta_2)} \frac{C_{k0}}{(1 + k\theta_1)} \tag{7.8-8}$$

$$= m \qquad \frac{C_{km}}{C_{k0}} = \frac{1}{(1 + k\theta_m)} \cdots \frac{1}{(1 + k\theta_1)} \tag{7.8-9}$$

If $\theta_1 = \theta_2 = \ldots \theta_m$,

$$\frac{C_{km}}{C_{k0}} = \frac{1}{\Pi(1 + k\theta_m)} \qquad m = 1,2, \ldots m \tag{7.8-10}$$

For a second-order reaction, $r_m = kC_{km}^2$, Eq. (7.8-4) becomes

$$C_{km} + kC_{km}^2\theta_m = C_{k(m-1)} \tag{7.8-11}$$

which is a quadratic equation to give

$$C_{km} = \frac{-1 \pm (1 + 4k\theta_m C_{m-1})^{1/2}}{2k\theta_m} \tag{7.8-12}$$

By induction

$$\text{For } m = 1 \qquad C_{k1} = \frac{-1 \pm (1 + 4k\theta_1 C_{k0})^{1/2}}{2k\theta_1} \tag{7.8-13}$$

$$\text{For } m = 2 \qquad C_{k2} = \frac{-1 \pm (1 + 4k\theta_2 C_{k1})^{1/2}}{2k\theta_2} \tag{7.8-14}$$

Because C_{k1} can be obtained in terms of C_{k0} in Eq. (7.8-13), substituting C_{k1} from Eq. (7.8-13) into Eq. (7.8-14) gives C_{k2} in terms of C_{k0}. Repeat the process until $m = m$.

For higher-order reactions, Eq. (7.8-4) is better solved by the graphical methods which will be discussed later.

7.8-2 Method of finite difference

Let

$$C_{km} = AC_k^m \tag{7.8-11}$$

where A is a constant to be determined from the boundary condition, m is the number of reactors in the cascade. Substituting Eq. (7.8-11) into Eq. (7.8-6) gives

$$(1 + k\theta_m)AC_k^m = AC_k^{m-1} \tag{7.8-12}$$

or

$$C_k = \frac{1}{1 + k\theta_m} \tag{7.8-13}$$

Thus Eq. (7.8-11) becomes

$$C_{km} = A\left(\frac{1}{1 + k\theta}\right)^m \tag{7.8-14}$$

When $m = 0$,

$$C_{km} = C_{k0} = A$$

Therefore

$$C_{km} = C_{k0}\left(\frac{1}{1 + k\theta}\right)^m \tag{7.8-15}$$

For second-order reactions, Eq. (7.8-11) cannot be solved by the method of finite difference because it is a nonlinear equation due to the presence of the term C_{km}^2. For second- and higher-order reactions, the graphical method discussed below will be employed.

7.8-3 Method of generating functions

Eq. (7.8-6) can be written in another equivalent form

$$(1 + k\theta_{m+1})C_{k(m+1)} = C_{km} \tag{7.8-16}$$

Assuming the residence time in each reactor is the same, the subscript $m + 1$ in the θ can be dropped off. Multiplying Eq. (7.8-16) by S_m and summing from 0 to m yields

$$\sum_{0}^{\infty} (1 + k\theta)C_{k(m+1)}S_m = \Sigma C_{km}S_m \tag{7.8-17}$$

From the properties of generating functions, we have*

$$\frac{(1 + k\theta)}{s} (\mathbf{C}_k - C_{k0}) = \mathbf{C}_k \tag{7.8-18}$$

Rearranging and simplifying gives

$$\mathbf{C}_k = \frac{C_{k0}}{1 - \dfrac{s}{1 + k\theta}} \tag{7.8-19}$$

Inverting yields the result

$$C_{km} = C_{k0}\left(\frac{1}{1 + k\theta}\right)^m \tag{7.8-20}$$

For second- and higher-order reactions, the following graphical methods will be employed.

7.9 Graphical Methods

There are two kinds of graphical methods depending on the use of the rate of reaction r or the rate constant k.

7.9-1 Using the reaction rate r

If the reaction rate is plotted against the concentration, a curve is generated (see Fig. 7.9-1). The number of reactors required can be found by drawing a line through the point C_{k0} on the abscissa to meet the curve at C_{k1} with a slope $-1/\theta_1$. Repeat the process until the final concentration is reached. For example, the rate equation for a first-order reaction is

$$r_m = -\frac{1}{\theta_m} (C_{km} - C_{k(m-1)}) \tag{7.9-1}$$

---

* See N. H. Chen, *New Mathematics for Chemical Engineers* (P.O. Box 385, Lowell, Mass.: Hoover Book Co., 1977).

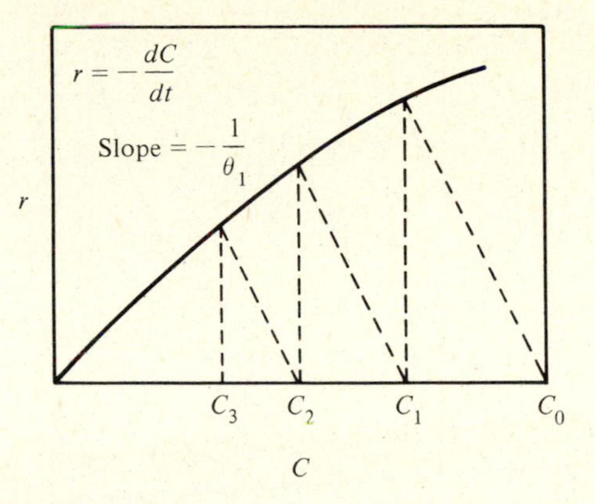

Figure 7.9-1 Number of reactors by reaction rate

$$r_1 = -\frac{1}{\theta_1}(C_{k1} - C_{k0}) \tag{7.9-2}$$

$$= -\frac{1}{\theta_1}(C_1 - C_0) \tag{7.9-3}$$

7.9-2 Using rate constant **k**

If the component mass balance equation is plotted in the graph with C_{m-1} as the ordinate and the C_m as the abscissa (Fig. 7.9-2), the number of reactors required can be found by drawing triangles until the final concentration is reached. In general, the mass balance equation can be represented by:

$$C_{km} - C_{k(m-1)} = -kf(C_{km})\theta \tag{7.9-4}$$

or

$$C_{k(m-1)} = C_{km} + kf(C_{km})\theta \tag{7.9-5}$$

Example 7.9-1 Number of Reactors by Reaction Rate

A stream of A at concentration 24.03 g mol/liter flows at a rate of 2831.7 liter/h to a continuous stirred-tank battery in which the following reaction is carried out

$$2A \rightleftarrows C + D$$

The forward rate constant is 0.4162 m^3/kmol · h; and the equilibrium constant K is 16.0. (a) Determine the volume of one reactor for 80 percent of the equilibrium

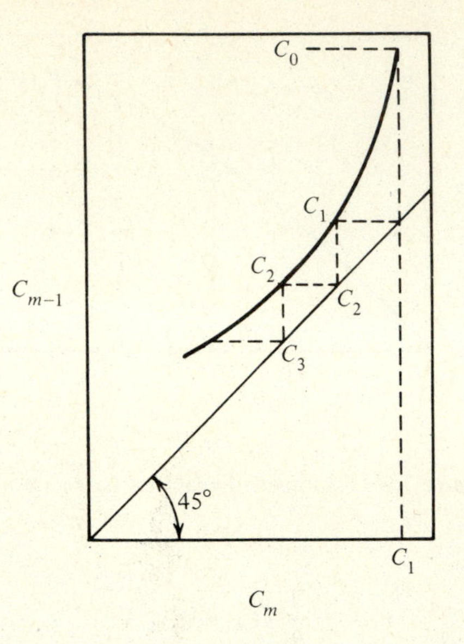

Figure 7.9-2 Number of reactors by rate constant

conversion. (b) If the reactors are limited to one-tenth the capacity in (a), how many reactors in series are required?

Solution

Let x_e be the equilibrium amount of C or D formed

$$K_e = \frac{x_e^2}{(24.03 - 2x_e)^2} = 16$$

Solving for x_e

$$x_e = 10.68 \text{ g mol/liter}$$

Then

$$x_2 = 0.8(10.68) = 8.54 \text{ g mol/liter}$$
$$C_{A2} = 24.03 - 2(8.54) = 6.94 \text{ g mol/liter}$$

The reaction rate is

$$r = -\frac{dC_A}{dt} = k\left(C_A^2 - \frac{C_C C_D}{K_e}\right)$$

The stoichiometric equations are

$$C_A = 24.03 - 2\lambda$$

$$C_C = \quad + \lambda$$

$$C_D = \quad + \lambda$$

Hence

$$C_A + 2C_C = 24.03$$

$$C_C = C_D = 12.02 - 0.5C_A$$

The rate equation becomes

$$r = -\frac{dC_A}{dt} = k\left(C_A^2 - \frac{(24.03 - C_A)^2}{4(16)}\right)$$

$$= k\left(C_A - \frac{(24.03 - C_A)}{8}\right)\left(C_A + \frac{(24.03 - C_A)}{8}\right)$$

For each value of C_A, the value of r, or $-dC_A/dt$ can be computed. The curve is thus generated as shown in the figure.

(a) $C_{A2} - C_{A1} = -r\theta$

$$6.942 - 24.03 = -0.4162\left[1.125(6.942) - \frac{24.03}{8}\right]\left[0.875(6.942) + \frac{24.03}{8}\right]\frac{V}{2831.7}$$

$$V = 2664.79 \text{ liters}$$

(b) One-tenth of 2665 is 266.5

$$\text{Slope} = \frac{-1}{\theta} = \frac{-1}{V/q} = \frac{-1}{266.5/2831.7} = -10.63$$

A line through $C_0 = 24.03$ with this slope can be drawn to meet the curve generated above. Repeat the process until $C_{A2} = 6.94$ is reached. From Fig. 7.9-3, the number of stages required is 3.8. Hence four stages are used.

Example 7.9-2 Number of Reactors by Rate Constant

In a series of continuous stirred-tank reactors, 2.2 kg/h ethanol is reacted with 1.8 kg/h acetic acid. Each reactor has a capacity of 0.01 m^3 and the reaction is carried out at 100°C, the reaction rate constant for the esterification is 4.76×10^{-4} liters/g mol · min and for the hydrolysis of the ester is 1.63×10^{-4} liters/g mol · min. The density of the mixture is about 864 kg/m^3. Find the number of reactors required to reduce the acid concentration to 60 percent conversion.

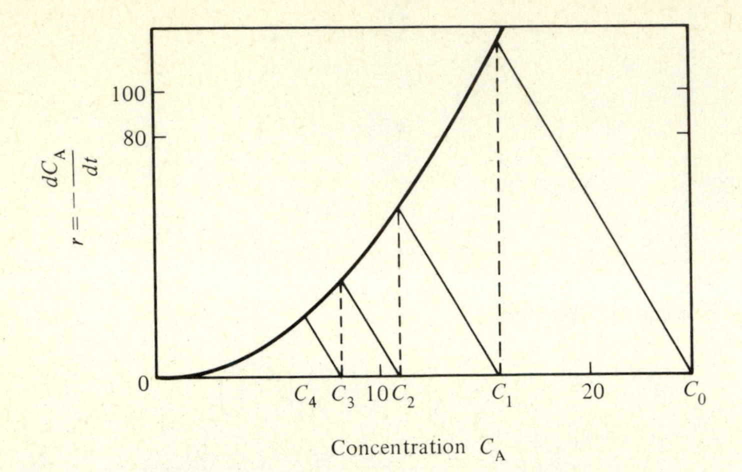

Figure 7.9-3 Solution to Example 7.9-1

Solution

$$C_2H_5OH + CH_3COOH \rightleftharpoons CH_3COOC_2H_5 + H_2O$$
$$\text{(B)} \qquad \text{(A)} \qquad \text{(C)} \qquad \text{(D)}$$

$$C_{A(m-1)} - C_{Am} = (k_1 C_A C_B - k_2 C_C C_D)_m \theta \qquad \textbf{(A)}$$

The stoichiometric equations are

$$C_A = C_{Ao} - \lambda$$
$$C_B = C_{Bo} - \lambda$$
$$C_C = C_{Co} + \lambda$$
$$C_D = C_{Do} + \lambda$$

Eliminating λ gives

$$C_B = C_A + C_{Bo} - C_{Ao}$$
$$C_C = C_D = -C_A + C_{Ao}$$

Letting $c = C_{Bo} - C_{Ao}$ and substituting into Eq. (A) gives

$$C_{A(m-1)} = C_{Am} + [k_1 C_{Am}(C_{Am} + c) - k_2(C_{Ao} - C_{Am})^2]\theta_m \qquad \textbf{(B)}$$

From the problem, the numerical values are:

$$C_{Ao} = \frac{1.8(864)}{4.0(60)} = 6.48 \text{ g mol/liter}$$

$$C_{Bo} = \frac{2.2(864)}{4.0(46)} = 10.33 \text{ g mol/liter}$$

$$c = 10.33 - 6.48 = 3.85 \text{ g mol/liter}$$

$$\theta = \frac{0.01(864)(60)}{4.0} = 129.6 \text{ min}$$

Substituting into Eq. (B) and simplifying gives

$$C_{A(m-1)} = C_{Am}^2(0.040564) + 1.51128 C_{Am} - 0.887047 \tag{C}$$

From Eq. (C), the following table can be prepared:

C_{Am}	6.5	6.0	5.0	4.0	3.0	2.0
$C_{A(m-1)}$	10.65	9.64	7.62	5.81	4.01	2.30

The values of $C_{A(m-1)}$ and C_{Am} are plotted in Fig. 7.9-4. The acid concentration in the final effluent = 0.4(6.48) = 2.59 g mol/liter. From the Fig. 7.9-4, the result is about three reactors.

Example 7.9-3 Analytical Solution

A stream of ethyl alcohol (C_{Ao} = 6.70 g mol/liter) and acetic acid (C_{Bo} = 10.20 g mol/liter) enters the first of two tanks in series. The rate constants of the esterification for these two tanks are $k_1 = 4.76 \times 10^{-4}$ and $k_2 = 1.63 \times 10^{-4}$ liter/g mol · min. The residence times are $\theta_1 = 53.0 \text{ min}^{-1}$ and $\theta_2 = 27.0 \text{ min}^{-1}$. Calculate the concentrations in these two tanks.

Solution

Using Eq. (B) in Example (7.9-2) for the first tank

$$6.70 = C_{A1}\left\{1 + 53.0\left[4.76 \times 10^{-4}(C_{A1} + 3.5) - \frac{1.63 \times 10^{-4}(6.7 - C_{A1})^2}{C_{A1}}\right]\right\}$$

Solving gives

$$C_{A1} = 5.47 \text{ g mol/liter}$$

For the second tank

$$5.47 = C_{A2} + [4.76 \times 10^{-4} C_{A2}(C_{A2} + 3.5) - 1.63 \times 10^{-4}(6.7 - C_{A2})^2]27.0$$

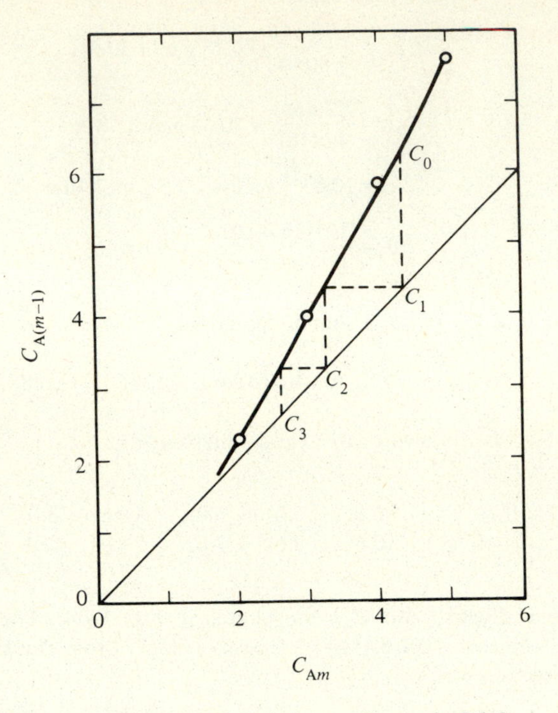

Figure 7.9-4 Solution to Example 7.9-2

Solving gives

$$C_{A2} = 5.07 \text{ g mol/liter}$$

7.10 Non-Steady-State Operations

For non-steady-state operations, the accumulation term in Eq. (4.2-17) is significant. If the diffusion and interphase transfer are negligible, Eq. (4.2-17) becomes

$$\frac{\partial C_j}{\partial t} + \frac{\partial (uC_j)}{\partial x} = \Sigma \alpha_{ij} r_i \tag{7.10-1}$$

Multiplying by dV and integrating gives

$$\int \frac{\partial C_j}{\partial t}\, dV + \int \frac{\partial (uC_j)}{\partial x}\, dV = \Sigma \alpha_{ij} r_i V \tag{7.10-2}$$

By divergence theorem and using $q = uS$, we obtain

$$V\frac{\partial C_j}{\partial t} + (qC_j)_2 - (qC_j)_1 = \Sigma \alpha_{ij} r_i V \qquad (7.10\text{-}3)$$

For reactor m, dividing Eq. (7.10-3) and using $\theta = V/q$ gives

$$\theta_m \frac{dC_{jm}}{dt} + C_{jm} - C_{j(m-1)} = \Sigma \alpha_{ij} r_{im} \theta_m \qquad (7.10\text{-}4)$$

or

$$\frac{dC_{jm}}{dt} + \left(\frac{C_{jm}}{\theta_m} - \Sigma \alpha_{ij} r_{im}\right) = \frac{C_{j(m-1)}}{\theta_m} \qquad (7.10\text{-}5)$$

7.10-1 First-order reactions

For first-order chemical reactions, the rate is

$$r = kC_A$$

Substituting into Eq. (7.10-5) gives without the subscript j or A

$$\frac{dC_m}{dt} + \left(\frac{1}{\theta_m} + k\right) C_m = \frac{C_{m-1}}{\theta_m} \qquad (7.10\text{-}6)$$

There are many methods to solve Eq. (7.10-6) which are described as follows:

Differential Method Defining $Q_m = 1/\theta_m + k$, Eq. (7.10-6) becomes

$$\frac{dC_m}{dt} + Q_m C_m = \frac{C_{m-1}}{\theta_m} \qquad (7.10\text{-}7)$$

For the first reactor

$$\frac{dC_1}{dt} + Q_1 C_1 = \frac{C_0}{\theta_1} \qquad (7.10\text{-}8)$$

which is a first-order linear differential equation. Its solution is

$$C_1 = \frac{1}{Q_1}\left[\frac{C_0}{\theta_1}(1 - e^{-Q_1 t})\right] + C_{10} e^{-Q_1 t} \qquad (7.10\text{-}9)$$

For the second reactor,

$$\frac{dC_2}{dt} + Q_2 C_2 = \frac{C_1}{\theta_2} \qquad (7.10\text{-}10)$$

Substituting C_1 from Eq. (7.10-9) into Eq. (7.10-10) gives also a first-order linear differential equation which can also be solved.

$$C_2 = \frac{C_0}{\theta_1\theta_2 Q_1 Q_2} + \frac{A_1 e^{-Q_1 t}}{\theta_2(Q_2 - Q_1)} + A_2 e^{-Q_2 t} \qquad \textbf{(7.10-11)}$$

where

$$A_1 = C_{1o} - \frac{C_0}{\theta_1 Q_1}$$

$$A_2 = C_{2o} - \frac{C_0}{\theta_1\theta_2 Q_1 Q_2} - \frac{A_1}{\theta_2(Q_2 - Q_1)}$$

For the third reactor,

$$\frac{dC_3}{dt} + Q_3 C_3 = \frac{C_2}{\theta_3} \qquad \textbf{(7.10-12)}$$

Substituting C_2 from Eq. (7.10-11) into Eq. (7.10-12) and solving gives

$$C_3 = \frac{C_0}{\theta_1\theta_2\theta_3 Q_1 Q_2 Q_3} + \frac{A_1 e^{-Q_1 t}}{\theta_2\theta_3(Q_2 - Q_1)(Q_3 - Q_1)} + \frac{A_2 e^{-Q_2 t}}{\theta_3(Q_3 - Q_2)} + A_3 e^{-Q_3 t} \qquad \textbf{(7.10-13)}$$

where

$$A_3 = C_{3o} - \frac{C_0}{\theta_1\theta_2\theta_3 Q_1 Q_2 Q_3} - \frac{A_1}{\theta_2\theta_3(Q_2 - Q_1)(Q_3 - Q_1)} - \frac{A_2}{\theta_3(Q_3 - Q_2)}$$

Example 7.10-1 Non-Steady-State Operation by the Differential Method

Acetic anhydride is hydrolyzed at 40°C in a two-CSTR system, the first of which has 10 liters of an aqueous solution containing 0.5×10^{-4} g mol anhydride per cubic centimeter. The second has 8 liters solution of 0.625×10^{-4} g mol anhydride per cubic centimeter. The vessel is heated to 40°C and at that time a feed solution containing 3×10^{-4} g mol anhydride per cubic centimeter is added at the rate of 2 liters/min. The product is withdrawn at the same rate. The rate constant is $k = 0.38\ \text{min}^{-1}$. Find the effluent concentration of the anhydride from the first and second reactors after 2 min.

Solution

Using Eq. (7.10-9) for the first reactor

$$\theta_1 = \frac{V}{q} = \frac{10}{2} = 5 \text{ min}$$

$$Q_1 = \frac{1}{5} + 0.38 = 0.58 \text{ min}^{-1}$$

$$C_0 = 3 \times 10^{-4} \text{ g mol/cm}^3$$

$$C_{1o} = 0.5 \times 10^{-4} \text{ g mol/cm}^3$$

Thus

$$C_1 = \frac{1}{0.58}\left[\frac{3 \times 10^{-4}}{5}(1 - e^{-0.58t})\right] + 0.5 \times 10^{-4} e^{-0.58t}$$

$$= 1.0345 \times 10^{-4} - 5.3448 \times 10^{-5} e^{-0.58t}$$

At $t = 2$ min,

$$C_1 = 1.0345 \times 10^{-4} - 5.3448 \times 10^{-5} e^{-0.58(2)} = 8.67 \times 10^{-5} \text{ g mol/cm}^3$$

Using Eq. (7.10-11) for the second reactor

$$\theta_2 = \frac{8}{2} = 4 \text{ min}$$

$$Q_2 = \frac{1}{4} + 0.38 = 0.63 \text{ min}^{-1}$$

$$A_1 = 0.5 \times 10^{-4} - \frac{3 \times 10^{-4}}{(5)(0.58)} = -5.3448 \times 10^{-5}$$

$$A_2 = 0.625 \times 10^{-4} - \frac{3 \times 10^{-4}}{5(4)(0.58)(0.63)} - \frac{-5.3448 \times 10^{-5}}{4(0.63 - 0.58)} = 3.2 \times 10^{-5}$$

Thus

$$C_2 = \frac{3 \times 10^{-4}}{5(4)(0.58)(0.63)} + \frac{-5.3448 \times 10^{-5} e^{-0.58t}}{4(0.63 - 0.58)} + 3.2 \times 10^{-5} e^{-0.63t}$$

$$= 4.1051 \times 10^{-5} - 2.6724 \times 10^{-4} e^{-0.58t} + 2.8869 \times 10^{-4} e^{-0.63t}$$

At $t = 2$ min,

$$C_2 = 3.9163 \times 10^{-5} \text{ g mol/cm}^3$$

Sequential Method Taking the Laplace transform of Eq. (7.10-6) with $C_m(0) = C_{mo}$ gives

$$[s\mathbf{C}_m - C_m(0)] + \left(\frac{1}{\theta_m} + k\right)\mathbf{C}_m = \frac{\mathbf{C}_{m-1}}{\theta_m}$$

or

$$(s + Q_m)\mathbf{C}_m = \frac{\mathbf{C}_{m-1}}{\theta_m} + C_{mo} \tag{7.10-14}$$

where

$$Q_m = \frac{1}{\theta_m} + k$$

For $m = 1$

$$\mathbf{C}_1 = \frac{\mathbf{C}_0}{\theta_1(s + Q_1)} + \frac{C_{1o}}{s + Q_1} = \frac{C_0}{\theta_1 s(s + Q_1)} + \frac{C_{1o}}{s + Q_1} \tag{7.10-15}$$

For $m = 2$

$$\mathbf{C}_2 = \frac{\mathbf{C}_1}{\theta_2(s + Q_2)} + \frac{C_{2o}}{s + Q_2}$$

$$= \frac{C_{2o}}{s + Q_2} + \frac{C_{1o}}{\theta_2(s + Q_2)(s + Q_1)} + \frac{C_0}{\theta_1\theta_2 s(s + Q_1)(s + Q_2)} \tag{7.10-16}$$

Likewise for $m = m$,

$$\mathbf{C}_m = \frac{C_{mo}}{s + Q_m} + \frac{C_{(m-1)o}}{\theta_m(s + Q_m)(s + Q_{m-1})} + \cdots + \frac{C_{2o}}{\prod_3^m \theta_i \prod_2^m (s + Q_i)} + \frac{C_{1o}}{\prod_2^m \theta_i \prod_1^m (s + Q_i)} + \frac{C_0}{s \prod_1^m \theta_i(s + Q_i)} \tag{7.10-17}$$

If $\theta_1 = \theta_2 = \ldots \theta_m$ and $Q_1 = Q_2 = \ldots Q_m$, all the subscripts can be dropped off,

$$\mathbf{C}_m = \frac{C_{mo}}{s + Q} + \frac{C_{(m-1)o}}{\theta(s + Q)^2} + \cdots + \frac{C_{1o}}{\theta^{m-1}(s + Q)^m} + \frac{C_0}{s\theta^m(s + Q)^m} \tag{7.10-18}$$

Resolving the term in the second bracket into partial fractions and then inverting yields

$$C_m = e^{-Qt}\left[C_{mo} + \frac{t}{\theta}C_{(m-1)o} + \cdots \left(\frac{t}{\theta}\right)^{m-1}\frac{C_{1o}}{(m-1)!}\right] + \frac{C_0}{(1 + k\theta)}m\left[1 - e^{-Qt}\left(\frac{(Qt)^{m-1}}{(m-1)!} + \cdots + Qt + 1\right)\right] \tag{7.10-19}$$

Example 7.10-2 Non-Steady-State Operation by the Sequential Method

A stream of acetic anhydride ($C_{Ao} = 0.14$ g mol/liter) at a rate of 700 cm³/min enters a battery of three continuous stirred tanks in series, each of which initially contains the anhydride at a concentration of 0.22 g mol/liter. Each tank has a capacity of 2,100 cm³. The rate constant is 0.38 min⁻¹ at 40°C. Assuming isothermal conditions, calculate the concentration in the third tank as a function of time.

Solution

$$C_{3o} = C_{2o} = C_{1o} = 0.22 \text{ g mol/liter}$$

$$C_0 = 0.14 \text{ g mol/liter}$$

$$\theta = 3 \text{ min}$$

$$Q = \frac{1}{3} + 0.38 = 0.713$$

$$1 + k\theta = 1 + 0.38(3) = 2.14$$

Substituting into Eq. (7.10-19)

$$C_3 = e^{(-0.713t)}\left[0.22 + \frac{t}{3}(0.22) + \left(\frac{t}{3}\right)^2\left(\frac{0.22}{2!}\right)\right] + \frac{0.14}{2.14^3}\left\{1 - e^{-0.713t}\left[\left(\frac{0.713t}{2!}\right)^2 + 0.713t + 1\right]\right\}$$

$$= \frac{0.1^4}{2.14^3} + e^{-0.713t}\left[0.22\left(1 + \frac{t}{3} + \frac{t^2}{18}\right) - \frac{0.14}{2.14^3}(1 + 0.713t + 0.127t^2)\right]$$

$$= 0.01428 + (0.2057 + 0.0631t + 0.0104t^2)e^{-0.713t}$$

Laplace Transform Taking the Laplace transform of Eq. (7.10-7) with $C_m(0) = C_{mo}$ gives

$$(s + Q_m)\mathbf{C}_m = \frac{\mathbf{C}_{m-1}}{\theta_m} + C_{mo} \qquad \textbf{(7.10-14)}$$

where $Q_m = \frac{1}{\theta_m} + k$. This equation can be rearranged as

$$\theta_m(s + Q_m)\mathbf{C}_m - \mathbf{C}_{m-1} = C_{mo}\theta_m \qquad \textbf{(7.10-20)}$$

which is a first-order finite difference equation. Its solution is

$$\mathbf{C}_m = \frac{C_0}{s}\left[\frac{1}{\theta(s+Q)}\right]^m + \frac{C_{mo}\theta}{\theta(s+Q)-1} \tag{7.10-21}$$

provided $\theta_1 = \theta_2 = \cdots \theta_m$ and $Q_1 = Q_2 = \cdots Q_m$. Expanding and inverting gives

$$C_m = \frac{C_0}{(\theta Q)^m} - \frac{C_0}{(\theta Q)^m}\left[1 + Qt + \frac{(Qt)^2}{2!} + \cdots \frac{(Qt)^{m-1}}{(m-1)!}\right]e^{-Qt}$$
$$+ C_m(0)\left[1 + \frac{t}{\theta} + \cdots \frac{1}{(m-1)}\left(\frac{t}{\theta}\right)^{m-1}\right]e^{-Qt} \tag{7.10-22}$$

which is the same as Eq. (7.10-19).

Generating Function Equation (7.10-6) can be expressed as

$$\frac{dC_m}{dt} + QC_m = \frac{C_{m-1}}{\theta} \tag{7.10-23}$$

subject to

$$C(0, t) = C_0$$
$$C(m, 0) = C^o$$

Taking the generating function (see N.H. Chen: *New Mathematics for Chemical Engineers,* 1977, p. 389) gives

$$\frac{d}{dt}\left(\frac{\mathbf{C}_m - C_0}{s}\right) + Q\left(\frac{\mathbf{C}_m - C_0}{s}\right) = \frac{\mathbf{C}_m}{\theta} \tag{7.10-24}$$

and

$$\mathbf{C}(m, 0) = \frac{C^o}{1-s}$$

Differentiating

$$\frac{d\mathbf{C}_m}{dt} + \left(Q - \frac{s}{\theta}\right)\mathbf{C}_m = QC_0 \tag{7.10-25}$$

yields a first-order linear differential equation and the solution is

$$\mathbf{C}_m = \frac{C^o e^{st/\theta}}{(1-s)e^{Qt}} + \frac{QC_0}{Q - s/\theta}\left(1 - \frac{e^{st/\theta}}{e^{Qt}}\right)$$

$$= \sum_{m=0}^{\infty} \left\{ \sum_{j=0}^{m} \frac{C^o(t/\theta)^j}{j!e^{Qt}} + C_0\left(\frac{1}{(\theta Q)^m}\right) - \frac{C_0}{e^{Qt}} \sum_{j=0}^{m} \frac{1}{(\theta Q)^{m-j}} \left(\frac{t}{\theta}\right)^j \frac{1}{j!} \right\} s^m \tag{7.10-26}$$

Inverting gives

$$C_m = \sum_{j=0}^{m} \frac{C^o}{e^{Qt}j!}\left(\frac{t}{\theta}\right)^j + C_o\left(\frac{1}{\theta Q}\right)^m - \frac{C_0}{e^{Qt}} \sum_{j=0} \left[\frac{1}{(\theta Q)^{m-j}}\left(\frac{t}{\theta}\right)^j \frac{1}{j!}\right]$$

$$= \frac{C_0}{(\theta Q)^m} - \frac{C_0}{(\theta Q)^m}\left(1 + Qt + \frac{(Qt)^2}{2!} + \cdots\right)e^{-Qt} + C^o\left(1 + \frac{t}{\theta} + \frac{1}{2}\left(\frac{t}{\theta}\right)^2 + \frac{1}{3!}\left(\frac{t}{\theta}\right)^3 + \cdots\right)e^{-Qt} \tag{7.10-27}$$

Matrices Let us consider the following reaction to be carried out in a series of N stirred-tank reactors

$$\mathrm{A} \underset{k_1'}{\overset{k1}{\rightleftharpoons}} \mathrm{B} \underset{k_2'}{\overset{k2}{\rightleftharpoons}} \mathrm{C}$$

Each reactor has a volume of V m³. Initially, the tank contains pure solvent and a catalyst. At start, q m³/h of reactant A of concentration C_0 kmol/m³ is fed to the first tank. Estimate the time required for the concentration of A leaving the Nth tank to be C_N kmol/m³.

Data:

$$C^o_{A1} = 1 \qquad C^o_{B1} = 0 \qquad C^o_{C1} = 0$$

$$C^o_{A,2,3,\ldots N} = 0 \qquad C^o_{B,2,3,\ldots N} = 0 \qquad C^o_{C,2,3,\ldots N} = 0$$

A component balance on any tank m gives the following three equations:

$$\frac{dC_{Am}}{dt} = \frac{C_{A(m-1)}}{\theta} - \left(\frac{1}{\theta} + k_1\right)C_{Am} + k_1' C_{Bm} \tag{7.10-28}$$

$$\frac{dC_{Bm}}{dt} = \frac{C_{B(m-1)}}{\theta} - \left(\frac{1}{\theta} + k_2 + k_1'\right)C_{Bm} + k_1 C_{Am} + k_2' C_{Cm} \tag{7.10-29}$$

$$\frac{dC_{Cm}}{dt} = \frac{C_{C(m-1)}}{\theta} - \left(\frac{1}{\theta} + k_2'\right)C_{Cm} + k_2 C_{Bm} \tag{7.10-30}$$

These equations can be written in matrix form as

$$\frac{d[C_m]}{dt} = \frac{1}{\theta}[C_{m-1}] - [D][C_m] \qquad \textbf{(7.10-31)}$$

where

$$[D] = \begin{bmatrix} \left(\frac{1}{\theta} + k_1\right) & -k'_1 & 0 \\ -k_1 & \left(\frac{1}{\theta} + k_2 + k'_1\right) & -k'_2 \\ 0 & -k_2 & \left(\frac{1}{\theta} + k'_2\right) \end{bmatrix}$$

The initial conditions are

$$[C^o_m] = \begin{bmatrix} C^o_1 \\ C^o_2 \\ C^o_3 \end{bmatrix} = \begin{bmatrix} 1 \\ 0 \\ 0 \end{bmatrix}$$

and the boundary conditions are $[C^t_0]$. Equation (7.10-31) is equivalent to

$$\frac{d[C_{m+1}]}{dt} = \frac{1}{\theta}[C_m] - [D][C_{m+1}] \qquad \textbf{(7.10-32)}$$

Taking generating functions of Eq. (7.10-32) and the boundary conditions yields

$$\frac{d}{dt}\left[\frac{C_n(s, t) - C_0}{s}\right] = -[D]\left[\frac{C_n(s, t) - C_0}{s}\right] + \frac{1}{\theta}[C_n(s, t)] \qquad \textbf{(7.10-33)}$$

Also, the boundary conditions become

$$[C^o_s] = \frac{[C^o_n]}{1 - s}$$

Equation (7.10-33) is a first-order differential equation which can be solved to give

$$C_n e^{([D]-s/\theta)t} = \frac{[C^o_m]}{1 - s} + \int_0^t e^{([D]-s/\theta)w}[D][C_0]\, dw \qquad \textbf{(7.10-34)}$$

Simplification of Eq. (7.10-34) yields

$$C_n = \frac{[C_m^o]}{1-s}\frac{e^{st/\theta}}{e^{[D]t}} + \frac{[D][C_0]}{[D](s/\theta)}\left(1 - \frac{e^{st/\theta}}{e^{[D]t}}\right) \tag{7.10-35}$$

Taking the inverse gives

$$[C_m] = \sum_{j=0}^{m} \frac{[C_m^o]}{e^{[D]t}}\frac{1}{j!}\left(\frac{t}{\theta}\right)^j + \frac{[C_0]}{(\theta[D])^m} - \frac{[C_0]}{e^{[D]t}}\sum_{j=0}^{m}\left[\frac{1}{(\theta[D])^{n-j}}\left(\frac{t}{\theta}\right)^j\frac{1}{j!}\right] \tag{7.10-36}$$

Expanding and rearranging yields

$$[C_m] = \frac{[C_0]}{(\theta[D])^m} - \frac{[C_0]}{(\theta[D])^m}\left(1 + [D]t + \ldots\frac{[D]^3t^3}{3!} + \ldots\right)e^{-[D]t}$$
$$+ [C^o]\left(1 + \frac{t}{\theta} + \ldots\frac{1}{3!}\left(\frac{t}{\theta}\right)^3 + \ldots\right)e^{-[D]t} \tag{7.10-37}$$

7.10-2 Second- and higher-order reactions

For a second-order reaction, Eq. (7.10-4) becomes

$$\theta_m\frac{dC_{jm}}{dt} + C_{jm} - C_{j(m-1)} = -kC_{jm}^2\theta_m \tag{7.10-38}$$

which is a nonlinear Riccati differential difference equation. For an nth-order reaction, Eq. (7.10-4) becomes

$$\theta_m\frac{dC_{jm}}{dt} + C_{jm} - C_{j(m-1)} = -kC_{jm}^n\theta_m \tag{7.10-39}$$

Both Eqs. (7.10-38) and (7.10-39) are best solved by the computer program in Chapter 6, PROGRAM SEMIBS.

7.11 Stability of CSTR

Since temperature is an important factor in operating a chemical reactor, it is desirable to establish a stable operating temperature. In this section, we shall discuss the method of achieving this goal. This can be done by mass and energy balances.

7.11-1 Mass balance of CSTR

The mass balance for component j is Eq. (7.2-8)

$$C_{j2} - C_{j1} = \Sigma \alpha_{ij} r_i \theta \qquad \textbf{(7.2-8)}$$

Using A as the limiting component and assuming a first-order reaction, we get

$$C_{A2} - C_{A1} = -kC_{A2}\theta \qquad \textbf{(7.11-1)}$$

But

$$k = k_o e^{-E/RT_2} \qquad \textbf{(7.11-2)}$$

then Eq. (7.11-1) becomes

$$qC_{A2} - qC_{A1} = -k_o e^{-E/RT_2} C_{A2} V \qquad \textbf{(7.11-3)}$$

or

$$C_{A1} - C_{A2} - \theta k_o e^{-E/RT_2} C_{A2} = 0 \qquad \textbf{(7.11-4)}$$

7.11-2 Energy balance of a CSTR

Under steady-state conditions, Eq. (6.3-8) in Chapter 6 becomes

$$\frac{\Sigma \Delta H_i r_i \theta}{\mathfrak{z}_p} + T_2 - T_1 + \frac{Ua(T_2 - T^*)\theta}{\mathfrak{z}_p} = 0 \qquad \textbf{(7.11-5)}$$

But

$$\mathfrak{z}_p = \Sigma C_j C_{pj} = \mathbf{C}_p \rho$$

where $\mathbf{C}_p$ is the mean heat capacity of the mixture at constant pressure and ρ is the density of the mixture. Then Eq. (7.11-5) can be rearranged to

$$-\frac{\Sigma \Delta H_i r_i V}{\rho \mathbf{C}_p q} + (T_1 - T_2) - \frac{Ua(T_2 - T^*)V}{\rho \mathbf{C}_p q} = 0 \qquad \textbf{(7.11-6)}$$

Letting $J = -\Delta H/\rho \mathbf{C}_p$ and $K = Ua/\rho \mathbf{C}_p$ and assuming first-order reaction, Eq. (7.11-6) becomes

$$Jk_o e^{-E/RT_2} C_{A2}\theta + (T_1 - T_2) - K\theta(T_2 - T^*) = 0 \qquad \textbf{(7.11-7)}$$

Multiplying Eq. (7.11-4) by J and adding to Eq. (7.11-7) gives

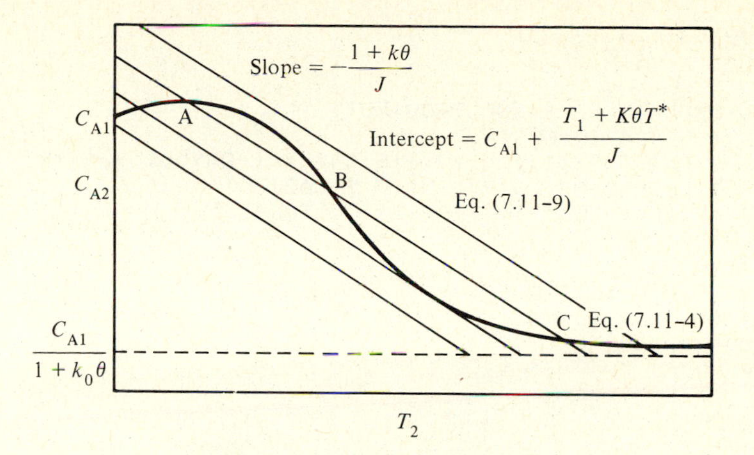

Figure 7.11-1 Stability of CSTR

$$J(C_{A1} - C_{A2}) + T_1 - T_2 - K\theta(T_2 - T^*) = 0 \qquad \textbf{(7.11-8)}$$

or

$$C_{A2} = C_{A1} - \frac{(1 + K\theta)T_2}{J} + \frac{(T_1 + K\theta T^*)}{J} \qquad \textbf{(7.11-9)}$$

Equations (7.11-4) and (7.11-9) are plotted in Fig. 7.11-1 in which C_{A2} is the ordinate and T_2 is the abscissa. The curve represents Eq. (7.11-4). The straight lines are obtained for Eq. (7.11-9). It is seen that the curve intersects the straight line at three points: *A, B,* and *C.* Point *B* is unstable while points *A* and *C* are stable. If the starting temperature is below point *A,* the reaction will gradually reach point *A* as the operating temperature. Similarly, if the starting temperature is above *C,* the reaction will gradually reach point *C.* If the starting temperature is between point *A* and *B,* it will approach *A* gradually. If the temperature starts between *B* and *C,* it will approach *C* as the final operating temperature.

Example 7.11-1 Stability of CSTR

Find the operating temperatures for the following data:

$q = 4.00$ gal/h	$C_{A1} = 0.9$ g/ml	$V = 100$ gal
$k_o = 2.61 \times 10^{14}$ h^{-1}	$E = 28950.59$ cal/g mol	$R = 1.987$ cal/g mol · K
$\rho = 0.9$ g/ml	$C_p = 0.5$ cal/g · °C	$T_1 = 20°C = 293$ K
$\Delta H = -83$ cal/g	$U = 0$	

```
 81/08/03. 15.17.43.
PROGRAM   CSTR

00100 PROGRAM CSTR(INPUT,OUTPUT)
00110 1 READ*,T2,EPSI
00120 2 CA2=0.9/(1.+25.*2.61E+14*EXP(-28950.59/1.987/T2))
00130 T2C=293.+83./0.9/0.5*(0.9-CA2)
00140 IF(ABS(T2C-T2).LT.EPSI) GO TO 4
00150 T2=T2C
00170 GO TO 2
00180 4 PRINT*,T2,CA2
00181 GO TO 1
00190 STOP
00200 END

READY.
RNH


 - * TRIVIAL PROGRAM UNIT IGNORED
? 250.,1.E-4
 293.0002745001 .8999985116772
? 300.,1.E-4
 293.0002745409 .8999985116772
? 410.,1.E-4
 457.266099297 .009400495446835
? 450.,1.E-4
 457.2661057494 .009400491264348
? 457.2661057494 1.E-4
 457.2661057494 .009400491264346
? 500.,1.E-4
 457.2661721576 .009400448218342
? 600.,1.E-4
 457.2661745876 .009400446643241
? 409.8705875,1.E-4
 409.8705875 .2663638496863
? 409.87,1.E-4
 293.0002745372 .8999985116772
```

Figure 7.11-2

Solution

Substituting the numerical values into Eqs. (7.11-4) and (7.11-9) gives

$$C_{A2}[1 + 25(2.61) \times 10^{14} e^{-28{,}950.59/1.987/T_2}] - 0.9 = 0 \tag{A}$$

$$T_2 - 293 + (0.9 - C_{A2})(-83/0.9/0.5) = 0 \tag{B}$$

Solving Eqs. (A) and (B) by either the trial-and-error method (PROGRAM CSTR) or by the Newton-Raphson method (PROGRAM CSTRN) yields the following points:

1. $C_{A2} = 0.9000 \qquad T_2 = 293.00 \text{ K}$
2. $C_{A2} = 0.2664 \qquad T_2 = 409.87 \text{ K}$
3. $C_{A2} = 0.0094 \qquad T_2 = 457.27 \text{ K}$

In this case, points 1 and 3 are stable and point 2 is unstable. If the reaction starts at a temperature below 409.87, the operating temperature will approach 293 K at which the conversion is very low. If the starting temperature is between 409.87 and 457.27, the operating temperature will approach 457.27 at which the conversion is very high. Since Eqs. (A) and (B) are nonlinear, they are better solved by a computer. The programs are described below. The first program CSTR (Fig. 7.11-2) uses the trial-and-error method and the second program CSTRN (Fig. 7.11-3) uses the Newton-Raphson method. It is interesting to note that the point 2 can be achieved by the first program only with the exact starting value (otherwise, the result will turn to point 1 or point 3), whereas point 2 can be obtained by the second program with a nearby starting value.

7.12 Optimization

The optimization of the operating variables for the performance of a CSTR is just as important as the design of the reactor. Although there is only one type of reactor, we have many kinds of reactions. To describe various cases would be lengthy and cumbersome. Hence, for each type of optimization, only one kind of reaction is selected for illustration. By the same reasoning, the method can be applied to other reactions. Furthermore, the generalization of the optimization of each type, although more useful, would make the description unclear. Therefore, the case approach is used here. By keeping in mind that the method can be applied to other reactions, we may derive the equations as needed. In this section, we are going to discuss the following: minimum volume, maximum yield, optimum temperature, minimum total cost, and maximum profit.

7.12-1 Minimum volume

Isothermal Case Consider a cascade of three stirred-tank reactors in series in which a first-order irreversible reaction is carried out isothermally. Calculate the minimum reaction volume (see Fig. 7.12-1). For an irreversible nth-order reaction

$$A \xrightarrow{k} \text{product}$$

the reaction rate is

$$r = kC_A^n \qquad \textbf{(7.12-1)}$$

```
00100 PROGRAM CSTRN(INPUT,OUTPUT)
00110CT1=FEED TEMPERATURE,DEGK
00120CCA1=FEED CONCENTRATION OF A,GM/CM**3
00130CDELH=HEAT OF REACTION,CAL/GM
00140CRHO=SOLUTION DENSITY,GM/CM**3
00150CCP=SOLUTION HEAT CAPACITY,CAL/GM-DEGC
00160CTHETA=RESIDENCE TIME,HR
00170CAK=FREQUENCY FACTOR,HR**(-L)
00180CE=ACTIVATION ENERGY,CAL/GMOLE
00190CR=GAS CONSTANT,CAL/GMOLE-DEGK
00200CT20=ASSUMED EFFLUENT TEMPERATURE,DEGK
00210 READ*,T1,CA1,DELH,RHO,CP,THETA,AK,E,R
00220 11 READ*,C,T
00230 I=0
00240 PRINT 93
00250 93 FORMAT(6X,1HI,9X,1HC,16X,1HT,16X,1HF,16X,1HG)
00260 1 IF(I-30) 5,5,33
00270 5 I=I+1
00280 F=C*(1.+THETA*AK*EXP(-E/R/T))-CA1
00290 G=T-T1+(CA1-C)*DELH/RHO/CP
00300 FC=1.+THETA*AK*EXP(-E/R/T)
00310 FT=THETA*AK*C*EXP(-E/R/T)*(E/R/T**2)
00320 GC=-DELH/RHO/CP
00330 GT=1.
00340 A=FC*GT-FT*GC
00350 DC=(-F*GT+G*FT)/A
00360 DT=(-G*FC+F*GC)/A
00370 PRINT 91,I,C,T,F,G
00380 C=C+DC
00390 T=T+DT
00400 IF(ABS(DC)-1.E-3) 2,2,1
00410 2 IF(ABS(DT)-1.E-3) 3,3,1
00420 91 FORMAT(5X,I2,4E17.8)
00430 3 GO TO 11
00440 33 STOP
00450 END
READY.
RNH
```

```
? 293.,0.9,-83.,0.9,0.5,25.,2.61E+14,28950.59,1.987
? 0.95,300.
       I        C                 T                 F                 G
       1    .95000000E+00    .30000000E+03    .50005013E-01    .16222222E+02
       2    .90000093E+00    .29299983E+03    .24199717E-05   -.10942358E-11
? 0.2625,410.57
       I        C                 T                 F                 G
       1    .26250000E+00    .41057000E+03    .25927114E-01   -.13333333E-01
       2    .26628648E+00    .40988494E+03    .52744603E-03   -.45474735E-12
       3    .26636447E+00    .40987055E+03    .23396632E-06   -.13642421E-11
? 0.0033,458.
       I        C                 T                 F                 G
       1    .33000000E-02    .45800000E+03   -.56768191E+00   -.39133333E+00
       2    .90908004E-02    .45732325E+03   -.26213672E-01   -.18189894E-11
       3    .93991714E-02    .45726638E+03   -.10965151E-03   -.90949470E-12
```

Figure 7.11-3

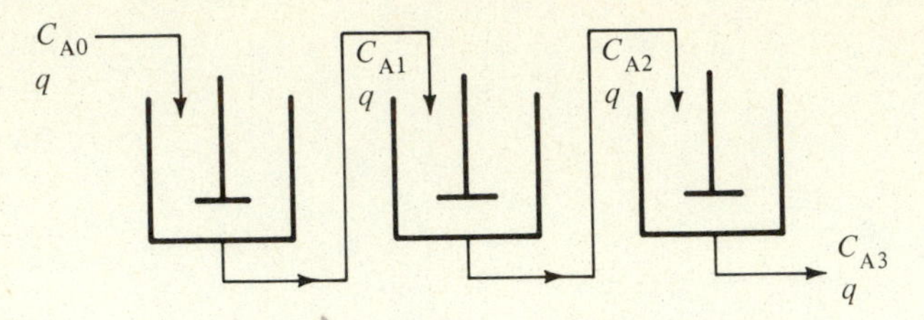

Figure 7.12-1 Minimum reaction volume

and Eq. (7.8-4) becomes

$$C_{Am} - C_{A(m-1)} = -kC_A^n\theta_m \tag{7.12-2}$$

Then the design equation for each reactor is

$$V_1 = \frac{q}{k}\left(\frac{C_{A0} - C_{A1}}{C_{A1}^n}\right) \tag{7.12-3}$$

$$V_2 = \frac{q}{k}\left(\frac{C_{A1} - C_{A2}}{C_{A2}^n}\right) \tag{7.12-4}$$

$$V_3 = \frac{q}{k}\left(\frac{C_{A2} - C_{A3}}{C_{A3}^n}\right) \tag{7.12-5}$$

The total volume of the cascade is

$$V = V_1 + V_2 + V_3 = \frac{q}{k}\left[\frac{C_{A0} - C_{A1}}{C_{A1}^n} + \frac{C_{A1} - C_{A2}}{C_{A2}^n} + \frac{C_{A2} - C_{A3}}{C_{A3}^n}\right] \tag{7.12-6}$$

$$= \frac{q}{k}\left[\frac{C_{A0}}{C_{A1}^n} - \frac{1}{C_{A1}^{n-1}} + \frac{C_{A1}}{C_{A2}^n} - \frac{1}{C_{A2}^{n-1}} + \frac{C_{A2}}{C_{A3}^n} - \frac{1}{C_{A3}^{n-1}}\right] \tag{7.12-7}$$

Differentiating this V with respect to C_{A1} and C_{A2} and equating to zero yields

$$\frac{\partial V}{\partial C_{A1}} = \frac{q}{k}\left(-\frac{nC_{A0}}{C_{A1}^{n+1}} - \frac{(1-n)}{C_{A1}^n} + \frac{1}{C_{A2}^n}\right] = 0 \tag{7.12-8}$$

$$\frac{\partial V}{\partial C_{A2}} = \frac{q}{k}\left(-\frac{nC_{A1}}{C_{A2}^{n+1}} - \frac{(1-n)}{C_{A2}^n} + \frac{1}{C_{A3}^n}\right] = 0 \tag{7.12-9}$$

With this system of two finite difference equations, C_{A1} and C_{A2} can be solved for different order of reaction n by noting that q, k, C_{A0}, and C_{A3} are given.

However, the solution is too complicated. An induction method is employed as follows: For $n = 1$, that is, for first-order reaction, Eqs. (7.12-8) and (7.12-9) become

$$-\frac{C_{A0}}{C_{A1}^2}+\frac{1}{C_{A2}}=0 \qquad \text{or} \qquad C_{A1}^2 = C_{A0}C_{A2} \tag{7.12-10}$$

$$-\frac{C_{A1}}{C_{A2}^2}+\frac{1}{C_{A3}}=0 \qquad \text{or} \qquad C_{A2}^2 = C_{A1}C_{A3} \tag{7.12-11}$$

Rearranging Eqs. (7.12-10) and (7.12-11) yields

$$\frac{C_{A1}}{C_{A0}}=\frac{C_{A2}}{C_{A1}}=\frac{C_{A3}}{C_{A2}} \tag{7.12-12}$$

Substituting back into Eqs. (7.12-3) to (7.12-5) indicates that

$$V_1 = V_2 = V_3 \tag{7.12-13}$$

at the optimum. This method of solution can be extended to m reactors.

$$V_1 = V_2 = V_3 = \cdots V_m \tag{7.12-14}$$

Hence, in order to minimize the total volume of the reactors required to carry out a first-order irreversible reaction isothermally in a series of continuous stirred-tank reactors, each volume of the reactors should be of equal size.

For second-order reactions, equal tanks do not give the optimum condition. In this case, substituting $n = 2$ into Eqs. (7.12-8) and (7.12-9) gives

$$-\frac{2C_{A0}}{C_{A1}^3}+\frac{1}{C_{A1}^2}+\frac{1}{C_{A2}^2}=0 \tag{7.12-15}$$

$$-\frac{2C_{A1}}{C_{A2}^3}+\frac{1}{C_{A2}^2}+\frac{1}{C_{A3}^2}=0 \tag{7.12-16}$$

Solving C_{A1} from Eq. (7.12-16) and substituting into Eq. (7.12-15) yields

$$2C_{A0}=\frac{1}{2}\left(C_{A2}+\frac{C_{A2}^3}{C_{A3}^2}\right)+\frac{1}{2}\left[C_{A2}+\left(\frac{C_{A2}^3}{C_{A3}^2}\right)\right]^3\Big/C_{A2}^2 \tag{7.12-17}$$

in which the only unknown is C_{A2}, which can be solved numerically by the Newton-Raphson method. Then C_{A1} can be calculated from Eq. (7.12-15). Finally all the volumes required for the reaction can be computed from Eqs. (7.12-3) to (7.12-5). For other higher-order reactions the method is the same.

Example 7.12-1 Minimizing Total Volume for Second-Order Reaction in CSTR

Determine the minimum total volume for the reaction (W1)

$$\underset{(A)}{\text{Benzoquinone}} + \underset{(B)}{\text{cyclopentadiene}} \longrightarrow \underset{(C)}{\text{adduct}}$$

to be carried out in two CSTR at 25°C if a feed containing an equimolal concentration of reactants $C_{A0} = C_{B0} = 0.07$ kmol/m³ is employed, $k = 9.92$ m³/kmol · ks, the liquid feed rate is 0.30 m³/ks and 90 percent is converted.

Solution

The concentration of benzoquinone at the outlet $C_{A2} = 0.07(1 - 0.9) = 0.007$ kmol/m³. Hence the total volume can be formulated as

$$V = \frac{0.30}{9.92}\left(\frac{0.07 - C_{A1}}{C_{A1}^2} + \frac{C_{A1} - 0.07}{0.007^2}\right) \tag{A}$$

Differentiating V with respect to C_{A1} and equating to zero

$$\frac{dV}{dC_{A1}} = -\frac{2(0.07)}{C_{A1}^3} + \frac{1}{C_{A1}^2} + \frac{1}{0.007^2} = 0$$

Solving C_{A1} by trial and error gives $C_{A1} = 0.01814$ kmol/m³. Substituting back into Eq. (A) yields

$$V = \frac{0.3}{9.92}\left(\frac{0.07 - 0.01814}{0.01814^2} + \frac{0.01814 - 0.007}{0.007^2}\right)$$

$$= 4.766 + 6.875 = 11.642 \text{ m}^3$$

It is seen that $V_1 = 4.766$ and $V_2 = 6.875$ m³. Both are not equal.

Nonisothermal Case In this case, the rate constant in each reactor is different. Denote k_1 to be the rate constant in reactor 1, k_2 in reactor 2, etc. Then the total volume required to carry out an nth-order irreversible reaction for three reactors in series is

$$V = V_1 + V_2 + V_3 = q\left(\frac{C_{A0} - C_{A1}}{k_1 C_{A1}^n} + \frac{C_{A1} - C_{A2}}{k_2 C_{A2}^n} + \frac{C_{A2} - C_{A3}}{k_3 C_{A3}^n}\right) \tag{7.12-18}$$

Set the first derivatives of V with respect to C_{A1} and C_{A2} to zero. Solving for C_{A1} and C_{A2} and rearranging as before gives

$$\frac{k_1 C_{A1}}{C_{A0}} = \frac{k_2 C_{A2}}{C_{A1}} = \frac{k_3 C_{A3}}{C_{A2}} \tag{7.12-19}$$

This equation can be extended to m reactors as

$$\frac{k_1 C_{A1}}{C_{A0}} = \frac{k_2 C_{A2}}{C_{A1}} = \cdots = \frac{k_m C_{Am}}{C_{A(m-1)}} \tag{7.12-20}$$

Again the individual volume in each reactor and the total volume of all reactors can be calculated from Eq. (7.12-18). It is seen that at optimum the individual volumes are not equal.

7.12-2 Maximum yield

It is always desirable to obtain maximum yield. Every chemical engineer should adjust the operating variables to achieve this goal. In a continuous stirred-tank reactor, the governing equation for every chemical species j in the system is

$$C_{jm} - C_{j(m-1)} = \sum_{j=1}^{s} \alpha_{ij} r_{im} \theta_m \tag{7.12-21}$$

This indicates that a system of finite difference equations occurs. If the reaction is first-order, this system of equations can be solved most of the time by various mathematical techniques, such as the finite difference method, Laplace transform, generating functions, or matrices, to express the concentrations of the j species as functions of initial reactant concentrations, rate constants, residence times, and the number of reactors. Then by classical differential calculus, the maximum value of the product concentration can be obtained by differentiating the function with respect to the variables, equating to zero, solving for the variables, and finally substituting back into the function. If the reaction is of a higher order, the resulting equations would be very complicated, and numerical techniques should be employed.

Example 7.12-2 Maximum Yield in CSTR

Pure liquid X ($C_{X0} = 1$ kg/m^3) flows continuously at a rate of 4,000 kg/min into the first of a battery of two equal-sized stirred-tank reactors in series. Both tanks are maintained at the same temperature at which the following reactions take place

$$X \xrightarrow{k_1} Y \xrightarrow{k_2} Z$$

where $k_1 = 6\ \text{min}^{-1}$ and $k_2 = 3\ \text{min}^{-1}$. The fluid density is 1,060 kg/m³. Determine the size of the vessels that give the maximum yield of Y.

Solution

For component X, Eq. (7.12-21) becomes

$$C_{Xm} - C_{X(m-1)} = -k_1 C_{Xm}\theta_m \tag{A}$$

For component Y, Eq. (7.12-21) becomes

$$C_{Ym} - C_{Y(m-1)} = k_1 C_{Xm}\theta_m - k_2 C_{Ym}\theta_m \tag{B}$$

Now we have a system of finite difference equations. Our objective is to solve these two equations simultaneously for an expression of the concentration of the product, C_Y as a function of the initial reactant concentration C_{X0}, the rate constants k_1 and k_2, and the number of reactors m. Let us illustrate this procedure by the following four methods:

1. Method of Finite Difference

Eqs. (A) and (B) can be rearranged to

$$C_{X(m-1)} - C_{Xm} = k_1 C_{Xm}\theta \tag{C}$$

$$C_{Y(m-1)} - C_{Ym} = k_2 C_{Ym}\theta - k_1 C_{Xm}\theta \tag{D}$$

Here $\theta_1 = \theta_2 = \theta$ is assumed.

Letting $k_1\theta = \gamma$ and solving Eq. (C) yields

$$[(1+\gamma)E - 1]C_{X(m-1)} = 0 \tag{E}$$

or

$$C_{Xm} = K_1\left(\frac{1}{1-\gamma}\right)^m \tag{F}$$

in which K_1 is a constant to be determined from the boundary conditions. Letting $k_2\theta = \delta$, substituting C_X from Eq. (E) into Eq. (D), and rearranging,

$$(1+\delta)C_{Ym} - C_{Y(m-1)} - \gamma K_1\left(\frac{1}{1+\gamma}\right)^m = 0 \tag{G}$$

The complementary solution of Eq. (G) is

$$C_{Ym} = K_2\left(\frac{1}{1+\delta}\right)^m \tag{H}$$

in which K_2 is another constant to be determined. The particular solution of Eq. (G) can be determined by the method of operators and is

$$C_{Y(m-1)} = \frac{1}{(1+\delta)E - 1}\gamma K_1 \left(\frac{1}{1+\gamma}\right)^m \tag{I}$$

$$= \frac{\gamma K_1 \left(\frac{1}{1+\gamma}\right)^m}{(1+\delta)\left(\frac{1}{1+\gamma}\right) - 1} = \frac{\gamma K_1 \left(\frac{1}{1+\gamma}\right)^{m-1}}{\delta - \gamma} \tag{J}$$

Thus, the complete solution for C_Y is

$$C_{Ym} = K_2 \left(\frac{1}{1+\delta}\right)^m + \frac{\gamma K_1 \left(\frac{1}{1+\gamma}\right)^m}{\delta - \gamma} \tag{K}$$

The constants K_1 and K_2 are determined from the boundary conditions:

At $m = 0$, $\quad C_X = C_{X0}$

At $m = 0$, $\quad C_Y = 0$

Hence

$$K_1 = C_{X0} \qquad K_2 = \frac{\gamma}{\gamma - \delta} C_{X0}$$

and the concentration of Y becomes

$$C_{Ym} = \frac{\gamma}{\gamma - \delta} C_{X0} \left(\frac{1}{1+\delta}\right)^m + \frac{\gamma C_{X0} \left(\frac{1}{1+\gamma}\right)^m}{\delta - \gamma} \tag{L}$$

$$= \frac{\gamma}{\delta - \gamma} C_{X0} \left[\left(\frac{1}{1+\gamma}\right)^m - \left(\frac{1}{1+\delta}\right)^m\right] \tag{M}$$

2. Method of Matrices

If the system of the finite equations has more than two equations, it is easier to use matrices. In this example, Eqs. (C) and (D) can be rearranged to

$$C_{Xm} = \rho_1 C_{X(m-1)} \tag{N}$$

$$C_{Ym} = \rho_2 C_{Y(m-1)} + \gamma \rho_1 \rho_2 C_{X(m-1)} \tag{O}$$

where

$$\gamma = k_1\theta \qquad \delta = k_2\theta \qquad \delta_1 = (1+\gamma)^{-1} \qquad \rho_2 = (1+\delta)^{-1}$$

In matrix form, Eqs. (N) and (O) become

$$[C_m] = [A][C_{m-1}] = \begin{bmatrix} \rho_1 & 0 \\ \gamma\rho_1\rho_2 & \rho_2 \end{bmatrix} [C_{m-1}] \tag{P}$$

The characteristic equation is

$$|\lambda I - A| = \begin{vmatrix} \lambda - \rho_1 & 0 \\ -\gamma\rho_1\rho_2 & \lambda - \rho_2 \end{vmatrix} = 0$$

Thus

$$\lambda = \rho_1 \text{ and } \rho_2$$

$$\det(\lambda) = (\lambda - \rho_1)(\lambda - \rho_2)$$

$$\det{}'(\lambda) = (\lambda - \rho_1) + (\lambda - \rho_2)$$

$$\text{adj}[\lambda I - A] = \begin{bmatrix} \lambda - \rho_2 & \gamma\rho_1\rho_2 \\ 0 & \lambda - \rho_1 \end{bmatrix}^T = \begin{bmatrix} \lambda - \rho_2 & 0 \\ \gamma\rho_1\rho_2 & \lambda - \rho_1 \end{bmatrix}$$

$$\text{adj}(\rho_1) = \begin{bmatrix} \rho_1 - \rho_2 & 0 \\ \gamma\rho_1\rho_2 & 0 \end{bmatrix} \qquad \text{adj}(\rho_2) = \begin{bmatrix} 0 & 0 \\ \gamma\rho_1\rho_2 & \rho_2 - \rho_1 \end{bmatrix}$$

$$[C_m] = \left\{ \frac{\rho_1^m}{\rho_1 - \rho_2} \begin{bmatrix} \rho_1 - \rho_2 & 0 \\ \gamma\rho_1\rho_2 & 0 \end{bmatrix} + \frac{\rho_2^m}{\rho_2 - \rho_1} \begin{bmatrix} 0 & 0 \\ \gamma\rho_1\rho_2 & \rho_2 - \rho_1 \end{bmatrix} \right\} \begin{bmatrix} C_{X0} \\ 0 \end{bmatrix}$$

$$\begin{bmatrix} \rho_1^m & 0 \\ \dfrac{\gamma\rho_1\rho_2(\rho_1^m - \rho_2^m)}{\rho_1 - \rho_2} & \rho_2^m \end{bmatrix} \begin{bmatrix} C_{X0} \\ 0 \end{bmatrix} = \begin{bmatrix} C_{X0}\rho_1^m \\ \dfrac{\gamma\rho_1\rho_2 C_{X0}(\rho_1^m - \rho_2^m)}{\rho_1 - \rho_2} \end{bmatrix} \tag{Q}$$

Hence

$$C_{Ym} = \frac{\gamma\rho_1\rho_2 C_{X0}}{\rho_1 - \rho_2}(\rho_1^m - \rho_2^m) = \frac{\gamma}{\delta - \gamma} C_{X0} \left[\left(\frac{1}{1+\gamma} \right)^m - \left(\frac{1}{1+\delta} \right)^m \right] \tag{M}$$

3. Method of Generating Functions

Analogous to the Laplace transform in handling differential equations, generating functions are easier for finite difference equations. In this example, Eqs. (N) and (O) can be rewritten as

$$C_{X(m+1)} = \rho_1 C_{Xm} \tag{R}$$

$$C_{Y(m+1)} = \rho_2 C_{Ym} + \gamma\rho_1\rho_2 C_{Xm} \tag{S}$$

with $C_{X0} = C_{X0}$ and $C_{Y0} = 0$. Taking generating functions, we get

$$\frac{X(s) - X(0)}{s} = \rho_1 X(s) \tag{T}$$

$$\frac{Y(s) - 0}{s} = \rho_2 Y(s) + \rho_1\rho_2 X(s) \tag{U}$$

Solving

$$X(s) = \frac{X(0)}{1 + s\rho_1}; \qquad Y(s) = \frac{s\rho_1\rho_2 X(0)}{(1 - s\rho_1)(1 - s\rho_2)} \tag{V}$$

Inverting

$$C_{Xm} = C_{X0}\, \rho_1^m \tag{W}$$

$$C_{Ym} = C_{X0}\, \rho_1\rho_2 \left(\frac{\rho_2^m - \rho_1^m}{\rho_2 - \rho_1}\right) = C_{X0}\gamma \left(\frac{\rho_2^m - \rho_1^m}{\gamma - \delta}\right) \tag{M}$$

From these three methods, it is seen that

$$\begin{aligned} C_{Ym} &= \frac{\gamma}{\delta - \gamma}\, C_{X0}(\rho_1^m - \rho_2^m) \\ &= \frac{k_1}{k_2 - k_1}\, C_{X0}\left[\left(\frac{1}{1 + k_1\theta}\right)^m - \left(\frac{1}{1 + k_2\theta}\right)^m\right] \end{aligned} \tag{X}$$

In order to maximize C_{Ym}, differentiating Eq. (X) with respect to θ and equating to zero

$$\begin{aligned} \frac{dC_{Ym}}{d\theta} &= -\frac{mk_1C_{X0}}{k_2 - k_1}\left[\frac{k_1}{(1 + k_1\theta)^{m+1}} - \frac{k_2}{(1 + k_2\theta)^{m+1}}\right] \\ &= 0 \end{aligned} \tag{Y}$$

or

$$\frac{k_1}{k_2} = \left(\frac{1 + 6\theta}{1 + 3\theta}\right)^3 = \frac{6}{3}$$

Hence $\theta = 0.117$ min. The volume of the reactors is then

$$V = \frac{0.117(4{,}000)}{1{,}060} = 0.4415 \text{ m}^3$$

The concentration of Y at the exit is

$$\begin{aligned} C_{Y2} &= \frac{6}{3 - 6}(1)\left\{\left[\frac{1}{1 + 6(0.117)}\right]^2 - \left[\frac{1}{1 + 3(0.117)}\right]^2\right\} \\ &= 0.4054 \text{ kg/m}^3 \end{aligned}$$

4. Method of Maximum Principle

The problem is to maximize C_{Y2} subject to

1. $C_{X(m-1)} - C_{Xm} = k_1 C_{Xm}\theta$

2. $C_{Y(m-1)} - C_{Ym} = (k_2 C_{Ym} - k_1 C_{Xm})\theta$

The Hamiltonian function is

$$H = \lambda_{Xm} C_{Xm} + \lambda_{Ym} C_{Ym} \tag{AA}$$

$$= \lambda_{Xm} \frac{C_{X(m-1)}}{1 + k_1\theta} + \frac{\lambda_{Ym} C_{Y(m-1)}}{1 + k_2\theta} + \frac{\lambda_{Ym} k_1 \theta C_{X(m-1)}}{(1 + k_1\theta)(1 + k_2\theta)} \tag{BB}$$

Applying the optimal conditions

$$\frac{\partial H}{\partial \theta} = \frac{\lambda_{Xm} C_{X(m-1)}(-k_1)}{(1 + k_1\theta)^2} + \frac{\lambda_{Ym} C_{Y(m-1)}(-k_2)}{(1 + k_2\theta)^2} + \frac{\lambda_{Ym} k_1 \theta C_{X(m-1)}(-k_1)}{(1 + k_2\theta)(1 + k_1\theta)^2}$$

$$+ \frac{\lambda_{Ym} C_{X(m-1)} k_1 \theta (-k_2)}{(1 + k_1\theta)(1 + k_2\theta)^2} + \frac{\lambda_{Ym} C_{X(m-1)} k_1}{(1 + k_1\theta)(1 + k_2\theta)} \tag{CC}$$

$$\frac{\partial H}{\partial C_{X(m-1)}} = \lambda_{X(m-1)} = \frac{\lambda_{Xm}}{1 + k_1\theta} + \frac{\lambda_{Ym} k_1 \theta}{(1 + k_1\theta)(1 + k_2\theta)} \qquad \lambda_X(2) = 0 \tag{DD}$$

$$\frac{\partial H}{\partial C_{Y(m-1)}} = \lambda_{Y(m-1)} = \frac{\lambda_{Ym}}{1 + k_2\theta} \qquad \lambda_Y(2) = 1 \tag{EE}$$

$$\frac{\partial H}{\partial \lambda_{Xm}} = C_{Xm} = \frac{C_{X(m-1)}}{1 + k_1\theta} \qquad C_X(0) = 1 \tag{FF}$$

$$\frac{\partial H}{\partial \lambda_{Ym}} = C_{Ym} = \frac{C_{Y(m-1)}}{1 + k_2\theta} + \frac{k_1 \theta C_{X(m-1)}}{(1 + k_1\theta)(1 + k_2\theta)} \qquad C_Y(0) = 0 \tag{GG}$$

Now solve these equations numerically with $k_1 = 6$, $k_2 = 3$, arbitrarily starting with $\theta = 0.152$. From (FF),

$$C_{X1} = \frac{1}{1 + 6(0.152)} = 0.5230 \qquad C_{X2} = \frac{C_{X1}}{1 + 6(0.152)} = \frac{0.523}{1.912} = 0.2735$$

From (GG)

$$C_{Y1} = \frac{0}{1 + 3(0.152)} + \frac{6(0.152)(1)}{[1 + 6(0.152)][1 + 3(0.152)]} = 0.3276$$

$$C_{Y2} = \frac{0.3276}{1.456} + \frac{6(0.152)(0.523)}{1.456(1.912)} = 0.3963$$

From (EE)

$$\lambda_{Y1} = \frac{1}{1.456} = 0.6868 \qquad \lambda_{Y0} = \frac{0.6868}{1.456} = 0.4717$$

From (DD)

$$\lambda_{X1} = \frac{0}{1.912} + \frac{6(0.152)}{1.456(1.912)} = 0.3276 \qquad \lambda_{X0} = \frac{0.3276}{1.456} + \frac{0.6868(6)(0.152)}{1.456(1.912)} = 0.3963$$

Substituting these values into (CC)

for $m = 1$
$$\frac{\partial H}{\partial \theta} = -0.227$$

for $m = 2$
$$\frac{\partial H}{\partial \theta} = -0.227$$

The values of θ for the next iterations can be estimated by assuming $K = 0.154185$. Then for $m = 1$

$$\theta = 0.152 - 0.154185(0.227) = 0.117 \text{ min}$$

for $m = 2$

$$\theta = 0.152 - 0.154185(0.227) = 0.117 \text{ min}$$

The whole process is repeated as follows. From (FF)

$$C_{X1} = \frac{1}{1 + 6(0.117)} = 0.5875 \qquad C_{X2} = \frac{0.5875}{1.702} = 0.3452$$

From (GG)

$$C_{Y1} = \frac{0}{1 + 3(0.117)} + \frac{6(0.117)(1)}{1.351(1.702)} = 0.3053 \qquad C_{Y2} = \frac{0.3053}{1.351} + \frac{6(.117)(0.5875)}{1.351(1.702)} = 0.4054$$

From (EE)

$$\lambda_{Y1} = \frac{1}{1.351} = 0.7401 \qquad \lambda_{Y0} = \frac{0.7401}{1.351} = 0.5479$$

From (DD)

$$\lambda_{X1} = \frac{6(0.117)}{1.351(1.702)} = 0.3053 \qquad \lambda_{X0} = \frac{0.3053}{1.702} + \frac{0.7401(6)(0.117)}{1.351(1.702)} = 0.4054$$

Substituting these values into (CC)

for $m = 1$
$$\frac{\partial H}{\partial \theta} = +0.0006$$

for $m = 2$
$$\frac{\partial H}{\partial \theta} = +0.0007$$

Since these partial derivatives are very close to the theoretical value of zero, no further iteration is necessary. The volume of the reactors is then

$$V = \frac{0.117(4{,}000)}{1{,}060} = 0.4415 \text{ m}^3$$

and the concentration of the product C_Y at the exit has been calculated as 0.4054.

7.12-3 Optimal temperature

It is important to find out the optimal operating temperature to minimize the residence time.

Single CSTR Let us consider a simple case. Determine the minimum reactor volume that will be required to obtain a fractional conversion f of A in the following first-order reversible reaction if the feed is pure A and if the input volumetric flow rate is q_0.

$$\text{A} \underset{k_2}{\overset{k_1}{\rightleftarrows}} \text{B}$$

The rate equation is

$$r = k_1 C_A - k_2 C_B \tag{7.12-22}$$

The design equation is

$$C_A - C_{A0} = -r\theta \tag{7.12-23}$$

Convert the component concentrations by

$$C_A = A_{A0}(1 - f) \tag{7.12-24}$$

$$C_B = C_{A0} f \tag{7.12-25}$$

where C_{A0} is the inlet concentration of A and f is the fractional conversion of A. Thus, combining Eqs. (7.12-22) to (7.12-25) yields

$$\theta = \frac{f}{k_1 - (k_1 + k_2) f} \tag{7.12-26}$$

Differentiating θ with respect to temperature T and equating to zero gives

$$\frac{dk_1}{dT} = \left(\frac{dk_1}{dT} + \frac{dk_2}{dT}\right)f \tag{7.12-27}$$

By the Arrhenius equation

$$k_i = k_{io}e^{-E_i/RT} \tag{7.12-28}$$

Equation (7.12-27) can be simplified to

$$\frac{k_1}{k_2} = \frac{E_2}{E_1}\left(\frac{f}{1-f}\right) \tag{7.12-29}$$

Then the optimal operating temperature can be calculated as

$$T^* = \frac{E_2 - E_1}{R \ln \dfrac{k_{2o}E_2}{k_{1o}E_1}\left(\dfrac{f}{1-f}\right)} \tag{7.12-30}$$

Substituting into Eq. (7.12-26) with the aid of Eq. (7.12-28) results in

$$V^* = \frac{q_o f}{k_{1o}(1-f)\beta^{-E_1/(E_2-E_1)} - k_{2o}f\beta^{-E_2/(E_2-E_1)}} \tag{7.12-31}$$

where

$$\beta = \frac{E_2 k_{2o}}{E_1 k_{1o}}\left(\frac{f}{1-f}\right)$$

CSTR Cascade Similarly, let us consider a specific case. It is desired to find the optimal temperatures and holding times to increase the product B from $C_{B4} = 0$ to C_{B1} g mol/ml in a three-tank reactor sequence (see Fig. 7.12-2) for

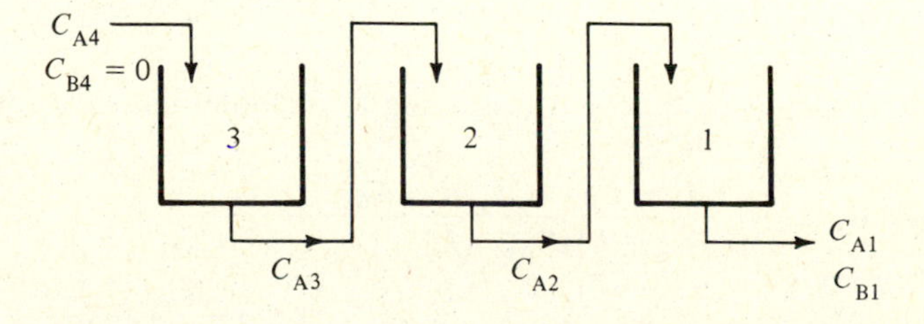

Figure 7.12-2 Optimal temperature for CSTR cascade

a first-order reversible reaction $A \underset{k_2}{\overset{k_1}{\rightleftharpoons}} B$. We will solve this problem by the finite difference method and by the method of maximum principle.

Finite difference method. The rate of the reaction is

$$r_m = k_1^m C_{Am} - k_2^m C_{Bm} \tag{7.12-32}$$

in which the superscript m denotes the number of the reactor. The design equation is

$$C_{Am} - C_{A(m+1)} = -r_m \theta_m \tag{7.12-33}$$

Therefore

$$A_{Am} - C_{A(m+1)} = (k_2^m C_{Bm} - k_1^m C_{Am})\theta_m \tag{7.12-34}$$

But

$$-\frac{dC_{Am}}{dt} = \frac{dC_{Bm}}{dt} \tag{7.12-35}$$

or

$$-\int_{C_{A(m+1)}}^{C_{Am}} dC_{Am} = \int_{C_{B(m+1)}}^{C_{Bm}} dC_{Bm} \tag{7.12-36}$$

Integrating

$$C_{A(m+1)} - C_{Am} = C_{Bm} - C_{B(m+1)}$$

or

$$C_{Am} + C_{Bm} = C_{A(m+1)} + C_{B(m+1)} \tag{7.12-37}$$

Eliminating C_{Bm} by Eq. (7.12-37) in Eq. (7.12-34) gives

$$\theta_m = \frac{C_{Am} - C_{A(m+1)}}{k_2^m(C_{A(m+1)} + C_{B(m+1)}) - (k_1^m + k_2^m)C_{Am}} \tag{7.12-38}$$

Hence the total residence time θ is (assuming a unit inlet concentration of A)

$$\theta = \theta_1 + \theta_2 + \theta_3 = \frac{C_{A1} - C_{A2}}{k_2^1 - (k_1^1 + k_2^1)C_{A1}} + \frac{A_{A2} - C_{A3}}{k_2^2 - (k_1^2 + k_2^2)C_{A2}} + \frac{C_{A3} - 1}{k_2^3 - (k_1^3 + k_2^3)C_{A3}} \tag{7.12-39}$$

because of Eq. (7.12-37). Then differentiating Eq. (7.12-39) with respect to C_{A3}, C_{A2}, T_1, T_2, and T_3, and equating to zero, simplifying yields

$$\frac{1}{k_2^3-(k_1^3+k_2^3)C_{A3}}+\frac{(C_{A3}-1)[-(k_1^3+k_2^3)]}{[k_2^3-(k_1^3+k_2^3)C_{A3}]^2}-\frac{1}{k_2^2-(k_1^2+k_2^2)C_{A2}}=0 \quad \textbf{(7.12-40)}$$

$$\frac{1}{k_2^2-(k_1^2+k_2^2)C_{A2}}+\frac{(C_{A2}-C_{A3})[-(k_1^2+k_2^2)]}{[k_1^2-(k_1^2+k_2^2)C_{A2}]^2}-\frac{1}{k_2^1-(k_1^1+k_2^1)C_{A1}}=0 \quad \textbf{(7.12-41)}$$

$$\frac{dk_2^1}{dT_1}-\left(\frac{dk_1^1}{dT_1}+\frac{dk_2^1}{dT_1}\right)C_{A1}=0 \quad \textbf{(7.12-42)}$$

$$\frac{dk_2^2}{dT_2}-\left(\frac{dk_1^2}{dT_2}+\frac{dk_2^2}{dT_2}\right)C_{A2}=0 \quad \textbf{(7.12-43)}$$

$$\frac{dk_2^3}{dT_3}-\left(\frac{dk_1^3}{dT_3}+\frac{dk_2^3}{dT_3}\right)C_{A3}=0 \quad \textbf{(7.12-44)}$$

with the Arrhenius equation, these five unknowns C_{A2}, C_{A3}, T_1, T_2, and T_3 can be solved from Eqs. (7.12-40) to (7.12-44). However, the solution is so complicated that the next method, solution by maximum principle, should be used.

Method of maximum principle. The complete detail of the method of maximum principle is described in Appendix 9.

Example 7.12-3 Maximum Principle for CSTR Cascade

We want to find the optimal temperatures and holding times to increase the concentration of the product C_B from 0 g mol/ml to 0.8 g mol/ml in a three-tank reactor sequence for a first-order reversible reaction $A \underset{k_2}{\overset{k_1}{\rightleftharpoons}} B$ with the following data:

$$E_1 = 9.2 \text{ kcal/g mol} \qquad E_2 = 12.5 \text{ kcal/g mol} \qquad k_{1o} = 2.5119 \times 10^5 \text{ min}^{-1}$$
$$k_{2o} = 1.9953 \times 10^7 \text{ min}^{-1} \qquad C_{A4} = 1 \text{ g mol/ml}$$

Solution

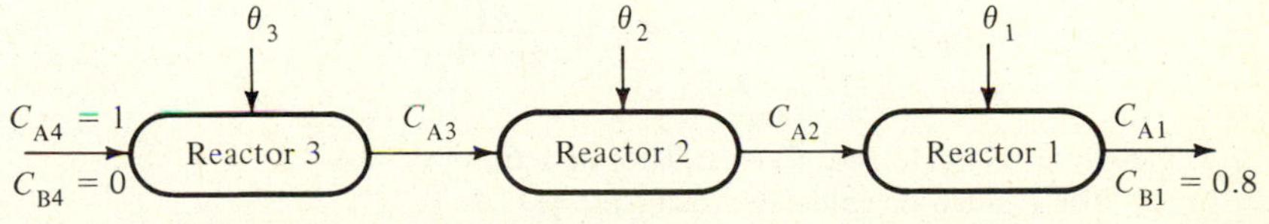

The problem is to minimize $\Sigma\theta_m$ subject to $C_{m+1} = C_m - [k_1 - (k_1 + k_2)C_m]\theta_m$. Hence the Hamiltonian function is

$$H = \theta_m + \lambda_{m+1} C_{m+1} = \theta_m + \lambda_{m+1}\{C_m - [k_1 - (k_1 + k_2)C_m]\theta_m\} \qquad \textbf{(A)}$$

Applying the optimal conditions

$$\frac{\partial H}{\partial \theta_m} = 1 - \lambda_{m+1}[k_1 - (k_1 + k_2)C_m] = 0 \qquad \textbf{(B)}$$

$$\frac{\partial H}{\partial T} = -\lambda_{m+1}\theta_m[k_1' - (k_1' + k_2')C_m] = 0 \qquad \textbf{(C)}$$

$$\frac{\partial H}{\partial \lambda_{m+1}} = C_{m+1} = C_m - [k_1 - (k_1 + k_2)C_m]\theta_m \qquad \textbf{(D)}$$

$$\frac{\partial H}{\partial C_m} = \lambda_m = \lambda_{m+1}[1 + (k_1 + k_2)\theta_m] \qquad \textbf{(E)}$$

Assume $T_1 = 274.1$ K, $T_2 = 300$ K, and $T_3 = 335$ K. The values of k_1, k_2, k_1', k_2' can be calculated from

$$k_1 = 2.5119 \times 10^5 e^{-9{,}200/1.98/T}$$

$$k_2 = 1.9953 \times 10^7 e^{-12{,}500/1.987/T}$$

$$k_1' = \frac{9{,}200 k_1}{1.987/T^2}$$

$$k_2' = \frac{12{,}500 k_2}{1.987/T^2}$$

	Reactor Temperature °K	k_1 min^{-1}	k_2 min^{-1}	k_1' min^{-1}	k_2' min^{-1}
1	274.1	1.1585×10^{-2}	2.1502×10^{-3}	7.1395×10^{-4}	1.8000×10^{-4}
2	300.0	4.98×10^{-2}	1.5596×10^{-2}	2.5621×10^{-3}	1.0901×10^{-3}
3	335.0	2.4977×10^{-1}	1.3947×10^{-1}	1.0304×10^{-2}	7.8181×10^{-3}

From Eq. (C), it is seen that

$$C_m = \frac{k_1'}{(k_1' + k_2')}$$

Therefore $C_1 = 0.7986$, $C_2 = 0.7015$, and $C_3 = 0.5686$.

The θ's can be calculated from Eq. (D) as

$$\theta_m = \frac{(C_m - C_{m+1})}{[k_1 - (k_1 + k_2)C_m]}$$

for $m = 1$ $$\theta_1 = \frac{0.7986 - 0.7015}{0.01158 - (0.01158 + 0.00215)0.7986} = 157.80 \text{ min}$$

for $m = 2$ $$\theta_2 = \frac{0.7015 - 0.5686}{0.0498 - (0.0498 + 0.01559)0.7015} = 33.868 \text{ min}$$

for $m = 3$ $$\theta_3 = \frac{0.5686}{0.2497 - (0.2497 + 0.1395)0.5686} = 19.9895 \text{ min}$$

$$\text{Total } \theta = \theta_1 + \theta_2 + \theta_3 = 211.6575 \text{ min}$$

The λ's can be calculated from Eq. (B) and Eq. (E):

From Eq. (B)

$$\lambda_4 = 1/[0.2497 - (0.2497 + 0.1395)0.5686] = 35.1618$$

From Eq. (E)

$$\lambda_3 = 35.1618[1 + (0.2497 + 0.1395)19.9895] = 308.7178$$

$$\lambda_2 = 308.7178[1 + (0.0498 + .01559)33.868] = 991.9948$$

$$\lambda_1 = 991.9948[1 + (.01158 + .00215)157.8] = 3{,}141.2450$$

The $\partial H/\partial\theta$'s can be calculated from Eq. (B).

$$\frac{\partial H}{\partial \theta_1} = 1 - 991.9948[0.01158 - (0.01158 + 0.00215)0.7986] = 0.3897$$

$$\frac{\partial H}{\partial \theta_2} = 1 - 308.7178[0.0498 - (0.0498 + 0.01559)0.70152] = -0.2125$$

$$\frac{\partial H}{\partial \theta_3} = 1 - 35.1618[0.2497 - (0.2497 + 0.1395)0.5685] = 0$$

Assume $K = 136$, which is arbitrarily chosen If K is small, more iteration will be used For illustration purposes, this value is taken. Then

$$\theta_{m,\text{new}} = \theta_{m,\text{old}} - K\frac{\partial H}{\partial \theta_m}$$

For $m = 1$ $$\theta_{1,\text{new}} = 157.8 - 136(0.3897) = 104 \text{ min}$$

From Eq. (D),

$$C_2 = 0.7986 - [0.01158 - (0.01158 + 0.00215)0.7986]104 = 0.7346 \text{ g mol/ml}$$

Then T_2 can be obtained by using trial and error on Eq. (C). Assume $T_2 = 291.5$; then $k_1' = 1.7303 \times 10^{-3}\ \text{min}^{-1}$ and $k_2' = 6.2645 \times 10^{-4}\ \text{min}^{-1}$.

$$C_2 = \frac{1.7303 \times 10^{-3}}{[1.7303 \times 10^{-3} + 6.2645 \times 10^{-4}]} = 0.7346 \quad \text{check}$$

For $m = 2$ $\quad \theta_2 = 33.868 + 136(0.2125) = 62.77$ min

$$C_3 = 0.7346 - [0.03175 - (0.03175 + 0.00846)0.7346]62.77$$

$$= 0.5957 \text{ g mol/ml}$$

Assume $T_3 = 328.0$; then $k'_1 = 0.0080006$, $k'_2 = 0.0054615$

$$C_3 = \frac{0.0080006}{(0.0080006 + 0.0054615)} = 0.5957 \quad \text{check}$$

For $m = 3$ $\quad 0 = 0.5957 - [0.1859 - (0.1859 + 0.0934)0.5957]\theta_3$

or

$$\theta_3 = 30.51 \text{ min}$$

Total $\theta = \theta_1 + \theta_2 + \theta_3 = 197.28$ min which shows improvement over the previous value of 211.6575 min.

Repeat the calculation as follows:

$$\lambda_4 = 1/[0.1859 - (0.1859 + 0.0934)0.5957] = 51.2269$$

$$\lambda_3 = 51.2269[1 + (0.1859 + 0.0934)30.51] = 487.75$$

$$\lambda_2 = 487.75[1 + (0.03175 + 0.00846)62.77] = 1718.84$$

$$\lambda_1 = 1{,}718.84[1 + (0.01158 + 0.00215)104] = 4173.19$$

$$\frac{\partial H}{\partial \theta_1} = 1 - 1{,}718.84[0.01158 - (0.01158 + 0.00215)0.7986] = -0.05747$$

$$\frac{\partial H}{\partial \theta_2} = 1 - 487.75[0.03175 - (0.03175 + 0.00846)0.7346] = -0.07877$$

The partial derivatives are quite small so that no further iteration seems necessary. The answer of the problem is:

$T_1 = 274.1$ K	$T_2 = 291.5$ K	$T_3 = 328.0$ K
$C_1 = 0.7986$ g mol/ml	$C_2 = 0.7346$ g mol/ml	$C_3 = 0.5957$ g mol/ml
$\theta_1 = 104$ min	$\theta_2 = 62.77$ min	$\theta_3 = 30.51$ min
	Total $\theta = 197.28$ min	

7.12-4 Minimum total cost

The total cost in operating a chemical reactor consists of (1) cost of reactor volume (2) cost of raw material or reactants.

$$\text{Total cost } C_T = \text{cost of reactor volume} + \text{cost of reactants} \qquad \textbf{(7.12-45)}$$

If the unit cost of the reactor volume, including installation, auxiliary equipment, instrumentation, overhead, labor, depreciation, interest, etc. is C_v in dollars per hour per cubic meter, then the cost of the reactor required for the operation is

$$\text{Cost of reactor volume per hour} = C_v V \tag{7.12-46}$$

where V is the reactor volume in cubic meters. If the unit cost of the reactant A is C_m in dollars per kilogram mole of A, then

$$\text{Cost of reactant A} = C_m q C_{A0} \tag{7.12-47}$$

where q is the volumetric flow rate of the reactants, in cubic meters per hour, C_{A0} is the inlet concentration of the reactant A in kilogram moles per cubic meter. Thus

$$C_T = C_v V + C_m q C_{A0} \tag{7.12-48}$$

Since the reactor volume relates to other variables by the design and the reaction rate equation, and the reactants have some relationship with the product, we can minimize the total cost. The following two examples are illustrations.

Example 7.12-4 Minimum Total Cost

It is required to determine the optimal reactor volume, feed rate, and conversion for the following reaction to be carried out in a CSTR.

$$X \rightarrow Y \qquad r = (0.1\ \text{min}^{-1})C_X$$

The production of Y is 100 kmol/min with an inlet concentration of X, $C_{X0} = 1$ kmol/m³ at which the cost of the reactant, C_m is \$1.00 per kilogram mole of X. Cost of reactor including installation, auxiliary equipment, instrumentation, overhead, labor, depreciation, interest, etc. C_v is \$0.1 per minute per cubic meter. If unreacted X is discarded, what is the unit cost of Y for these conditions?

Solution

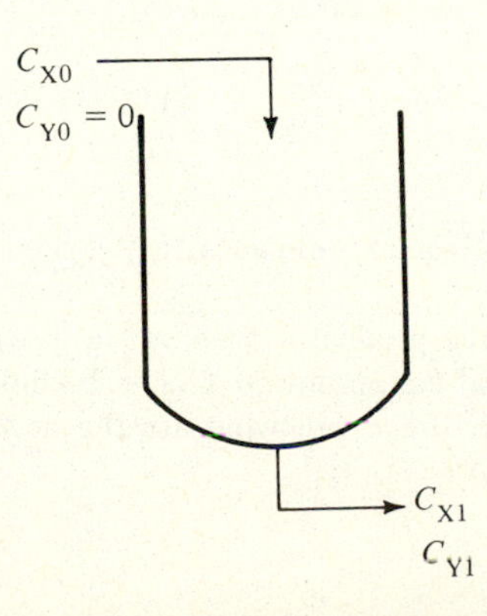

$$C_T = VC_v + C_m q C_{X0} \tag{A}$$

Since

$$C_{X1} - C_{X0} = -r\theta$$

or

$$C_{X1} - C_{X0} = -kC_{X1}\frac{V}{q} \tag{B}$$

$$V = \frac{q(C_{X0} - C_{X1})}{kC_{X1}} \tag{C}$$

Hence

$$C_T = C_v q \frac{(C_{X0} - C_{X1})}{kC_{X1}} + C_m q C_{X0} \tag{D}$$

But

$$C_{X1} = C_{X0}(1 - f) \quad \text{and} \quad C_Y = C_{X0} f \quad \text{and} \quad C_{Yq} = 100 \text{ kmol/min}$$
$$= C_{X0} - C_Y$$

Putting the numerical values and simplifying gives

$$C_T = \frac{100}{(1 - C_Y)} + \frac{100}{C_Y} \tag{E}$$

Differentiating with respect to C_Y, equating to zero and solving for C_Y

$$C_{Y,\text{opt}} = 0.5 \text{ kmol/m}^3$$

Then

$$C_{T,\text{opt}} = \$400.00$$

$$V_{\text{opt}} = 2000 \text{ m}^3$$

$$q_{\text{opt}} = 200 \text{ m}^3/\text{min}$$

$$F_{X0,\text{opt}} = q_{\text{opt}} C_{X0} = 200(1) = 200 \text{ kmol/min}$$

$$\text{Cost of Y} = \frac{C_T}{F_Y} = \frac{\$400.00}{100} = \$4.00 \text{ per kilogram mole of Y}$$

Example 7.12-5 Best Reactor Volume with Recycle

If all unreacted X of the product stream in the previous example is recycled and brought up to the initial concentration $C_{X0} = 1$ kmol/m³ at a cost of $C_r = \$0.15$ per kilogram mole of X processed, what are the new operating conditions and the unit cost of producing Y?

Solution

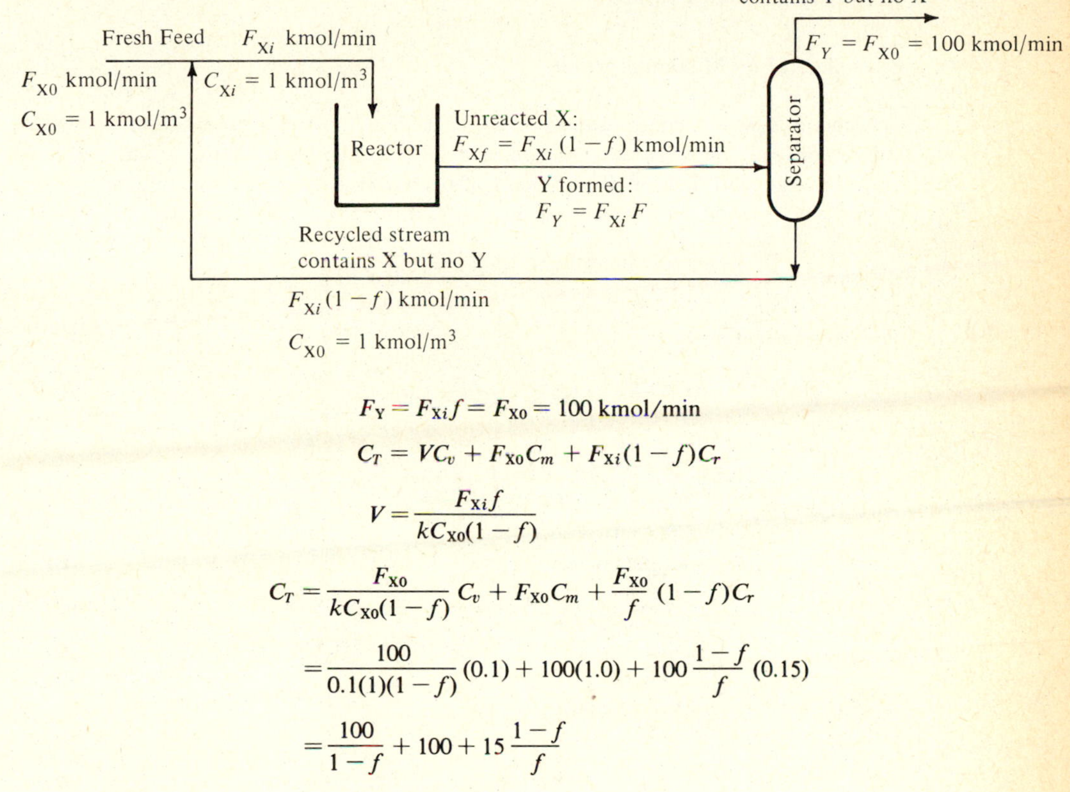

$$F_Y = F_{Xi}f = F_{X0} = 100 \text{ kmol/min}$$

$$C_T = VC_v + F_{X0}C_m + F_{Xi}(1-f)C_r$$

$$V = \frac{F_{Xi}f}{kC_{X0}(1-f)}$$

$$C_T = \frac{F_{X0}}{kC_{X0}(1-f)}C_v + F_{X0}C_m + \frac{F_{X0}}{f}(1-f)C_r$$

$$= \frac{100}{0.1(1)(1-f)}(0.1) + 100(1.0) + 100\frac{1-f}{f}(0.15)$$

$$= \frac{100}{1-f} + 100 + 15\frac{1-f}{f}$$

Differentiating with respect to f, equating to zero and solving

$$f_{\text{opt}} = 0.279$$

Then

$$V_{\text{opt}} = 1{,}387 \text{ m}^3 \qquad F_{Xi,\text{opt}} = 358.4 \text{ kmol per minute}$$

$$\text{Recycle rate} = F_{Xi} - F_{X0} = 258.4 \text{ kmol/min}$$

$$\text{Cost of product } \frac{C_T}{F_Y} = \$2.77 \text{ per kilogram mole of B}$$

7.12-5 Maximum profit

In general, the profit of operating a chemical reaction is the difference between the sales of the product and the cost of operation, assuming all of the product is sold.

$$\text{Profit} = \text{sales of product} - \text{cost of operation} \qquad \textbf{(7.12-49)}$$

Then the profit can be maximized.

Example 7.12-6 Maximum Profit

Compound X with a concentration of $C_{Xo} = 2.5$ kmol/m³ at a flow rate of 12 m³/min is converted to Y by the reaction in a CSTR. The value of the product Y is \$1.50 per kilogram mole of Y. The cost of operation is \$2.50 per minute per cubic meter. The reaction rate constant is 30 min⁻¹. Find the maximum profit.

Solution

The design equation is

$$12(2.5 - C_X) = 30 C_X V$$

or

$$C_X = \frac{2.5}{1 + 30V/12} \qquad \textbf{(A)}$$

$$\text{Sales income per minute} = \$1.50(12)(2.5 - C_X) \qquad \textbf{(B)}$$

$$\text{Cost of operation per minute} = \$2.50V \qquad \textbf{(C)}$$

$$\text{Profit per minute, } P = \$1.50(12)(2.5 - C_X) - \$2.50V \qquad \textbf{(D)}$$

Substituting C_X from Eq. (A) into Eq. (D)

$$P = 112.5V\left(\frac{1}{1 + 30V/12}\right) - 2.5V \qquad \textbf{(E)}$$

Differentiating with respect to V, equating to zero and solving for V

$$V_{opt} = 2.283 \text{ m}^3$$

Substituting into Eq. (A) gives

$$C_{X,opt} = 0.3727 \text{ kmol/m}^3$$

Substituting into Eq. (D) gives

$$P_{opt} = \$32.58 \text{ per minute}$$

The yield is

$$\text{Yield} = \frac{12(2.5 - 0.3728)}{12(2.5)} = 0.8509 \quad \text{or} \quad 85.09\%$$

Example 7.12-7 Maximum Profit with Recycle

In Example 7.12-6, the effluent from the reactor blows to a separator in which unreacted X is separated as a solid, half of which is cycled back to the reactor. Using the same values of the parameters as in Example 7.12-6, calculate the yield to give maximum profit.

Solution

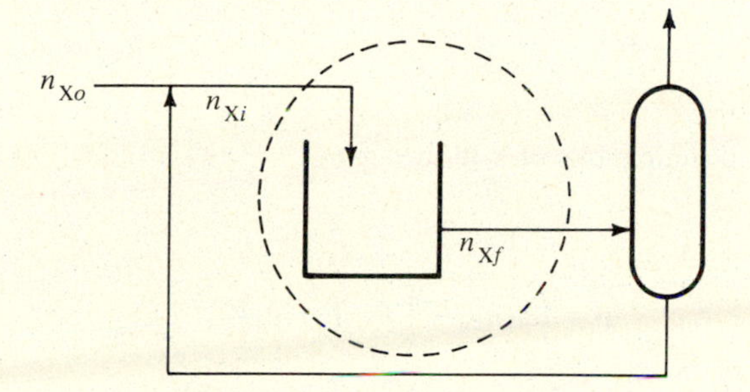

Mass balance around the envelop

$$q(C_{X_i} - C_{X_f}) = rV = kC_{X_f}V \tag{A}$$

Because

$$C_{Xi} = \frac{n_{Xi}}{V} \qquad C_{Xf} = \frac{n_{Xf}}{V}$$

then

$$\frac{n_{Xi} - n_{Xf}}{V} = \frac{kn_{Xf}}{q} \tag{B}$$

or

$$n_{Xi} - n_{Xf} = \frac{kn_{Xf}V}{q} \tag{C}$$

But

$$n_{Xi} = n_{Xo} + \frac{1}{2}n_{Xf} \tag{D}$$

Therefore

$$n_{Xo} + \frac{1}{2}n_{Xf} - n_{Xf} = \frac{kn_{Xf}V}{q} \tag{E}$$

or

$$n_{Xo} - \frac{1}{2}n_{Xf} = \frac{kn_{Xf}V}{q} \tag{F}$$

Substituting the numerical values

$$2.5(12) - \frac{1}{2} n_{Xf} = \frac{30}{1^2} n_{Xf} V \tag{G}$$

or

$$30 = (1 + 5V)(0.5) n_{Xf} \tag{H}$$

$$\text{kilogram moles of X reacted} = n_{Xi} - n_{Xf}$$

$$= n_{X0} + \frac{1}{2} n_{Xf} - n_{Xf} = n_{X0} - \frac{1}{2} n_{Xf} \tag{I}$$

$$\text{kilogram moles of Y formed} = n_{X0} - \frac{1}{2} n_{Xf}$$

Thus

$$P = 1.5\left(n_{X0} - \frac{1}{2} n_{Xf}\right) - 2.5V$$

$$= 1.5\left[30 - \frac{1}{2}\left(\frac{30}{0.5(1+5V)}\right)\right] - 2.5V \tag{J}$$

$$= 45 - \frac{45}{1+5V} - 2.5V$$

Differentiating P with respect to V, equating to zero, and solving for V

$$V_{opt} = 1.6974 \text{ m}^3$$

$$n_{Xf} = \frac{30}{0.5[1 + 5(1.6974)]} = 6.3245 \text{ kmol}$$

$$\text{kilogram moles of Y produced} = 30 - \frac{1}{2}(6.3245) = 26.8377 \text{ kmol}$$

$$\text{Yield} = \frac{26.8377}{30} = 0.8945 \quad \text{or} \quad 89.45\%$$

Example 7.12-8 Maximum Profit for Reversible Reaction and Recycle

100 kg/m³ X in aqueous solution enters at a rate of 35 m³/h to a CSTR of capacity 80 m³. Species A undergoes a first-order reversible reaction with $k_{A-B} = 0.5 \text{ h}^{-1}$ and $K = 5$. The effluent passes to a separator in which the product Y is recovered. A fraction δ of the unreacted X is recycled to the reactor as a solution containing 100 kg/m³ X; the remaining X and water are rejected. The value of Y is $0.05 per kilogram. The operating costs are $0.50 per cubic meter of solution entering the separator. Calculate δ to maximize the operational profit. What fraction of the X fed to the plant is converted?

Solution

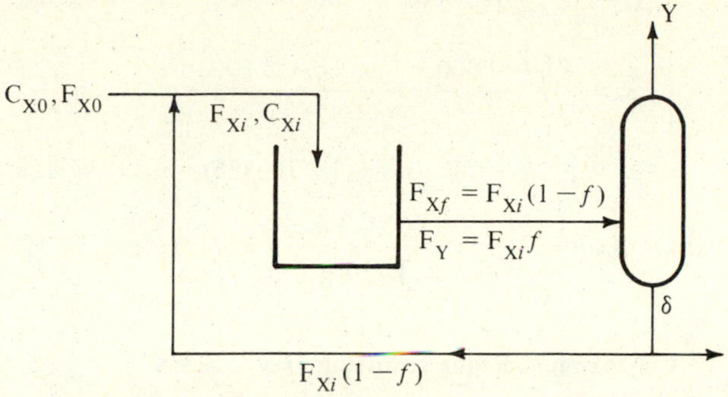

$$P = 0.05F_Y - \left(35 + \frac{\delta F_{X_i}(1-f)}{100}\right)0.05 = 0.05F_{X_i}f - \left(35 + \frac{\delta F_{X_i}(1-f)}{100}\right)0.50 \quad \textbf{(A)}$$

$$X \rightleftarrows Y$$

$$\begin{aligned} r &= k_1C_X - k_2C_Y \\ &= k_1C_{X0}(1-f) - k_2(C_{X0}M + C_{X0}f) \end{aligned} \quad \textbf{(B)}$$

where

$$M = \frac{C_{Y0}}{C_{X0}} = 0$$

Since

$$F_{X_i}f = rV$$

thus

$$F_{X_i}f = C_{X0}[k_1(1-f) - k_2(M+f)]V \quad \textbf{(C)}$$

$$= 8{,}000[0.5(1-f) - 0.1f] = 8{,}000(0.5 - 0.6f) \quad \textbf{(D)}$$

But

$$3{,}500 + \delta F_{Xi}(1-f) = F_{Xi}$$

therefore

$$\begin{aligned} P &= 400(0.5 - 0.6f) - 0.50\left[35 + \frac{F_{X_i} - 3{,}500}{100}\right] \\ &= 40(0.5 - 0.6f)\left(-\frac{1}{f} + 10\right) \end{aligned} \quad \textbf{(E)}$$

Differentiating with respect to f, equating to zero, and solving for f

$$f_{opt} = 0.2886$$

$$F_{X_i} = \frac{1}{0.2886}[8{,}000(0.5 - 0.6(0.2886)] = 9{,}060 \text{ kg/h}$$

$$\delta = \frac{F_{X_i} - 3{,}500}{F_{X_i}(1-f)} = \frac{9{,}060 - 3{,}500}{9{,}060(1 - 0.2886)} = 0.8626$$

$$F_{Xf} = F_{Xi}(1-f) = 9{,}060(1 - 0.2886) = 6{,}445 \text{ kg/h}$$

$$\text{Conversion} = \frac{F_{Xi}f}{F_{Xo}} = \frac{9{,}060(0.2886)}{35(100)} = 0.7471 \quad \text{or} \quad 74.71\%$$

Example 7.12-9 Maximum Profit Involving Heat Losses

X and Y each at a concentration of 1 lb mol/ft³ enter a CSTR at a rate of 55 ft³/min. The product of the heat capacity and the density of the mixture is 45 Btu/ft³·°F. The feed temperature is 450/K and the operating temperature is 500/K. The rate constants in pound moles per cubic foot per minute of the reaction.

$$X + Y \underset{k'}{\overset{k}{\rightleftharpoons}} Z + W$$

are

$$k = 1.5e^{10-5{,}000/T}$$

$$k' = 2.2e^{20-10{,}000/T}$$

where T is K. The values of 1 lb mol X, Y, Z, and W are \$1.5, \$1.5, \$15.00, and \$3.00, respectively. Cost of reactor of V ft³ is $2.5V$ in dollars per hour. Cost of removing Q Btu/min is $0.05Q$ cent/min. Find the value of the fractional conversion and thus the volume of the reactor that maximizes the profit.

Solution

The rate of the reaction is

$$r = kC_XC_Y - k'C_ZC_W \tag{A}$$

The initial concentrations of the components are: $C_{Xo} = C_{Yo} = 1$; $C_{Zo} = C_{Wo} = 0$ Hence the rate of the reaction becomes

$$r = k(1-f)^2 - k'f^2 \tag{B}$$

where f is the fractional conversion of X and the rate constants k and k' at the operating temperature 500 K can be calculated from the given equation as $k = 1.5$ and $k' = 2.2$. Since the reaction is carried out in a CSTR, Eq. (7.4-5) gives

$$C_{X2} - C_{X1} = -r\theta \tag{C}$$

But $C_{X2} = C_{X1}(1 - f)$ and $\theta = V/q$, where V is the volume of the reactor and q is the volumetric flow rate of the mixture, Eq. (C) becomes

$$rV = qf \tag{D}$$

Then the cost of the reactor is

$$C_{\text{reactor}} = 2.5V = 2.5\frac{qf}{r} = 2.5\frac{qf}{1.5(1-f)^2 - 2.2f^2} \tag{E}$$

The heat removal can be calculated from Eq. (7.11-5)

$$Q = Ua(T^* - T_2)V = \Delta HrV + \rho C_p q(T_2 - T_1) \tag{F}$$

Using

$$\Delta H = E - E'$$

Eq. (F) becomes

$$\begin{aligned} Q &= 1.987(10{,}000 - 5{,}000)(1.8)rV + 45(55)(500 - 450)(1.8) \\ &= 17{,}883qf + 222{,}750 \text{ Btu/min} \end{aligned} \tag{G}$$

and the cost of cooling is then

$$\begin{aligned} C_{\text{cooling}} &= 0.05Q = \$0.0005\ Q \text{ dollars/min} \\ &= \$0.0005(17{,}883qf + 222{,}750)(60) \\ &= 6{,}682\ (1 + 4.45f) \text{ dollars/h} \end{aligned} \tag{H}$$

The value of the products and the unreacted reactants is

$$\begin{aligned} \text{Sales} = 55(1.5C_X + 1.5C_Y + 15C_Z + 3C_W) &= 55[1.5(1-f) + 1.5(1-f) + 15f + 3f] \\ &= 55(3 + 15f) \text{ dollars/min} \end{aligned} \tag{I}$$

The cost of the reactants is

$$C_{\text{reactants}} = 55(1.5C_{X0} + 1.5C_{Y0}) = 55(1.5 + 1.5) = 165 \text{ dollars/min} \tag{J}$$

Hence the net profit becomes

$$\begin{aligned} P &= \text{Sales} - \text{cost of reactants} - \text{cost of reactors} - \text{cost of cooling} \\ &= [55(3 + 15f) - 165]60 - 2.5\frac{55f}{1.5(1-f)^2 - 2.2f^2} - 6{,}682(1 + 4.45f) \qquad \text{(K)} \\ &= 19{,}752f - \frac{137.5f}{1.5 - 3f - 0.7f^2} - 6{,}682 \end{aligned}$$

Differentiating P with respect to f and equating to zero

$$\frac{dP}{df} = 19{,}752 - \frac{137.5}{1.5 - 3f - 0.7f^2} - \frac{137.5(3 + 1.4f)f}{(1.5 - 3f - 0.7f^2)^2} = 0 \tag{L}$$

Solving for f^* gives

$$f^* = 0.4225$$

Then the optimum volume is

$$\begin{aligned} V^* &= \frac{qf^*}{r^*} = \frac{55f^*}{1.5 - 3f^* - 0.7f^{*2}} \\ &= \frac{55(0.4225)}{1.5 - 3(0.4225) - 0.7(0.4225)^2} \\ &= 216.07 \text{ ft}^3 \end{aligned} \tag{M}$$

The optimum profit is

$$\begin{aligned} P^* &= 19{,}752(0.4225) - \frac{137.5(0.4225)}{1.5 - 3(0.4225) - 0.7(0.4225)^2} - 6{,}682 \\ &= \$1{,}123 \text{ per hour} \end{aligned}$$

Problems

1. 450 cm^3/min acetic anhydride flows into the first tank (V = 1 liter) of three CSTR in series. The rate constant of this first-order irreversible reaction at 25°C is 0.158 min^{-1}. If the second and third tanks have volume 2.2 and 1.75 liters, respectively, determine the fraction hydrolyzed in the effluent from the third tank by graphical method.

2. Suppose in Problem 1 that the first tank is operated at 15°C (k = 0.567 min^{-1}), the second at 40°C (k = 0.38 min^{-1}), and the third at 25°C (k = 0.158 min^{-1}). Determine the fraction hydrolyzed in the effluent from the third tank.

3. For the reaction

$$N_2 + O_2 \underset{k'}{\overset{k}{\rightleftharpoons}} 2NO$$

k = 1.11 × 10^4 liter/g mol·s and K = 0.01 at 2,700 K (E1). Calculate the CSTR volume to achieve 70 percent of the equilibrium conversion at 2,700/K for the following conditions: (a) volumetric flow rate = 25 liter/s, (b) pressure of the reactor = 12 atm, (c) feed composition: 80 percent N_2, 12 mol percent O_2, and 8 mol percent inerts.

4. 600 cm^3/min acetic anhydride flows to the first of three continuous stirred tanks in series each of which has a capacity of 2,000 cm^3. The rate equation is $r = 0.158C$ where C is the concentration of the anhydride in gram moles per cubic centimeter. Assuming isothermal conditions, calculate the degree of hydrolysis in each tank and the total fraction hydrolyzed at the exit of each tank.

5. For the first-order irreversible reaction

$$1,1\text{-}C_2H_4Cl_2 \rightarrow C_2H_3Cl + HCl$$

$k = 0.0414$ min^{-1} at 705 K (F1). If pure 1,1-dichloroethane is fed to a reactor CSTR at 705/K and 9 atm at a rate of 7 liters/h, what volume of the reactor would be required for 80 percent conversion of the 1,1-dichloroethane?

6. Substance X is to be converted into Y in solution by a first-order irreversible reaction ($k = 0.2$ h^{-1}). Calculate the residence times for 97 percent conversion of X for (a) three tanks in series of equal volume and (b) a large tank followed by a second one of 10 percent capacity of the large one.

7. A liquid feed X (70 mol percent X and 30 mol percent inerts) enters a CSTR at 27°C, at a rate of 50 liter/min, in which the following elementary reaction takes place

$$X \rightarrow Y + Z$$

The heat of reaction is −300 cal/g mol X at 273 K. The specific heats of X, Y, Z and inerts are 4, 2.4, 2.4, and 5 cal/g mol·°C, respectively; $k = 0.12$ h^{-1} at 25°C. $E = 25{,}000$ cal/gmol. Determine the volume of the reactor required for 90 percent conversion.

8. The following first-order reaction is carried out in a continuous stirred reactor

$$X \xrightarrow{k_1} Y \underset{k_2'}{\overset{k_2}{\rightleftarrows}} Z$$

Show that the outlet concentration of Y is

$$C_Y = \frac{k_1\theta(1 + k_2'\theta C_{X\phi})}{(1 + k_1\theta)(1 + k_2\theta + k_2'\theta)}$$

where $C_{X\phi}$ is the inlet concentration of X, $\theta = V/q$, V is the volume of the reactor and q is the inlet volumetric flow rate of X. Assuming that the irreversible reaction is exothermic and the reversible one is endothermic, derive the energy balance for the system.

9. For the reaction

$$X \xrightarrow{k_1} Y \xrightarrow{k_2} Z$$

$k_1 = 0.16\ \text{min}^{-1}$, $k_2 = 0.06\ \text{min}^{-1}$. A stream of pure A is fed to a continuously stirred tank at a rate of 140 liters/min. Which of the following reactors would give the highest production rate of Y?
(a) A single stirred tank of volume of 280 liters.
(b) Two stirred tanks in series, each with a volume of 140 liters.
(c) Two stirred tanks in parallel, each of 140 liters volume and with feed stream split equally between them.

10. A stream of styrene ($C_{X0} = 0.8$ kmol/m³) and butadiene ($C_{Y0} = 3.6$ kmol/m³) enters a series of CSTR, each of volume 27 m³, at a rate of 40,000 kg/h. Determine the number of reactors required for polymerization of 80 percent of the limiting reactant from the following data: molecular weight of styrene and butadiene = 104 and 54, respectively. Density of reaction mixture = 0.87 g/cm³. Rate of reaction $r = 10^{-5} C_X C_Y$ liter/g mol·s, assuming $V = 1$ liter.

11. A pure liquid reactant at 25°C (density = 1.25 g/cm³, heat capacity at constant pressure = 0.95 cal/g·°C) enters a CSTR at a rate of 250 cm³/s. The heat of reaction is constant at −47,000 cal/g mol. Calculate the heat removal from the reactor in order to maintain an isothermal operation at 163°C for 95 percent conversion. The molecular weight of the reactant is 250.

12. The first-order exothermic reaction

$$X \rightarrow Y$$

is carried out in a stirred reactor under the following conditions: Mean residence time = 1.02 min. Feed concentration = 1.0 g mol/liter. Feed temperature = 352 K. Rate constant $k = e^{(25.2-10{,}000/T}\ \text{min}^{-1}$. Coolant temperature = 352 K. $UA/\rho C_p = 1.02\ \text{min}^{-1}$ and $\Delta H/\rho r C_p = -205$ K·liter/g mol. Determine the stable and unstable temperatures.

13. The first-order ($k = 10\ \text{ks}^{-1}$) liquid reaction $X \rightarrow Y + Z$ is carried out in a CSTR with a volume of 0.25 m³. A feed of concentration of X (1 kmol/m³) flows to the reactor under steady-state conditions at a rate of 1.8 m³/ks. Suddenly the concentration is changed to 2.2 kmol/m³. Assuming isothermal conditions, (a) calculate the effluent composition C_X, C_Y, and C_Z prior to the sudden change, (b) repeat (a) after the change, (c) derive an equation for the time variation of the effluent concentration of component X, (d) calculate the effluent composition 50 s after the change in the inlet composition.

14. A stream of 500 kg/h of pure liquid X is fed to the first of two equal-sized stirred tanks in series in which the following reaction takes place

$$\begin{array}{ccccc} X & \xrightarrow{k_1} & Y & \xrightarrow{k_2} & Z \\ & & \downarrow{\scriptstyle k_3} & & \\ & & W & & \end{array}$$

where $k_1 = k_2 = k_3 = 0.15 \text{ min}^{-1}$. Determine the volume of the reactor required to give maximum yield of product Y.

15. 100 kmol X per hour with a concentration of 0.02 kmol/m³ reacts with Y to produce Z and W

$$X + Y \rightarrow Z + W \qquad k = 550 \text{ m}^3/\text{kmol}\cdot\text{h}$$

The amount of Z required is 100 kmol/h. No recycle of the unreacted reactants is necessary. Calculate the optimum CSTR reactor size and feed composition. Data: Y costs $130 per kilogram mole; operating cost is $0.005 per hour per cubic meter.

16. A plant produces 50 kmol Y per hour by hydrolysis in a CSTR of a feed stream containing 0.06 kmol/m³ of reactant X in the following first order reaction

$$X \rightarrow 2Y$$

The system effluent stream from the reactor goes to a countercurrent extraction column in which Y is quantitatively extracted. 2.5 percent of the incoming X passes through the system unreacted. Fixed and operating costs for this process are $12.00 per hour; reactant cost is $1.2 per kilogram mole and Y can be sold at $0.88 per kilogram mole.

(a) What are the present profits per hour?

(b) In order to maximize the profits, what are the optimum feed rate of X, optimum conversion of X, and optimum production of Y?

17. 800,000 kg Z is produced per year from the following elementary aqueous phase reaction

$$X + Y \rightarrow 2Z \qquad k = 0.002 \text{ liter/g mol}\cdot\text{s}$$

(a) Using a CSTR, find the optimum reactor size and feed composition to the reactor.

(b) Determine the cost of production of Y.

Data: Cost of reactants: X = $8.0 per kilogram Y = $0.02 per kilogram
Molecular weights: X = 40 Y = 80 Z = 60
Solubility of Y = 9 g/liter
Operating cost = $15 per year per gallon
Hours of operation: 7,200 h/year

References

(E1). Ermenc, E.O. *Chem. Engr.*, 77, 193 (1970).

(F1). Fogler, H.S. *The Elements of Chemical Kinetics and Reactor Calculations.* Englewood Cliffs, N.J.: Prentice-Hall, Inc., 1974.

CHAPTER EIGHT

Ideal Tubular Reactors

8.1 Introduction

In Chapter 7 the ideal continuous stirred-tank reactor (CSTR) was described, in which complete mixing within the tank is assumed. The composition and temperature of the reaction mixture are uniform everywhere in the tank. The CSTR is good for liquid reactions but is not convenient for gas-phase reactions. In industry, many gaseous reactions are carried out in a tubular reactor in which no mixing in the direction of flow and complete mixing in the radial direction are assumed. Hence concentration and temperature, except for isothermal operation, and consequently reaction rate, will vary with the length of the reactor. Since this kind of flow acts like a plug or piston in the turbulent region, the reactor is termed a *piston-flow* or *plug-flow reactor* (PFR). If the flow is in the laminar region, it is called a *laminar-flow reactor* (LFR). For both types of flow, let us define the space-time τ

$$\tau = \frac{z}{u} \tag{8.1-1}$$

where z is the distance through which the component j of a reaction mixture travels in the reactor with a linear velocity u. This space-time τ is also equal to time t in the component rate equation Eq. (3.3-3). Hence

$$r = \frac{dC_j}{\alpha_j dt} = \frac{dC_j}{\alpha_j d\tau} = \frac{u}{\alpha_j}\frac{dC_j}{dz} \tag{8.1-2}$$

But

$$F_j = qC_j \tag{8.1-3}$$

where F_j is the molal flow rate of the j component, and q is the volumetric flow rate. Then

$$\frac{dF_j}{dz} = q\frac{dC_j}{dz} = uA\frac{dC_j}{dz}$$

or

$$u\frac{dC_j}{dz}=\frac{1}{A}\frac{dF_j}{dz}=\frac{dF_j}{dV} \tag{8.1-4}$$

Substituting into Eq. (8.1-2) gives

$$r=\frac{1}{\alpha_j}\frac{dF_j}{dV} \tag{8.1-5}$$

In this chapter, we are going to discuss both the plug- and laminar-flow reactors as well as recycle reactors. Optimization will also be described.

8.2 Isothermal Condition

The design equation for an ideal plug-flow reactor operated isothermally can also be derived from the generalized equation Eq. (4.2-17)

$$\frac{\partial C_j}{\partial t}+\frac{\partial(uC_j)}{\partial x}=D\frac{\partial^2 C_j}{\partial x^2}+\Sigma\alpha_{ij}r_i-k_m a(C_j-C_j^*) \tag{4.2-17}$$

At steady state, $\partial C_j/\partial t=0$. Neglecting diffusion and interphase transfer gives

$$\frac{\partial(uC_j)}{\partial x}=\Sigma\alpha_{ij}r_i \tag{8.2-1}$$

Multiplying by the cross-sectional area A yields with $q=uA$

$$\frac{d(C_jq)}{dx}=\Sigma\alpha_{ij}r_iA$$

or

$$\frac{dF_j}{dx}=\Sigma\alpha_{ij}r_iA \tag{8.2-2}$$

where $F_j=C_jq$ is the flow rate of j component. Rearranging gives

$$dF_j=d(C_jq)=\Sigma\alpha_{ij}r_iA\,dx=\Sigma\alpha_{ij}r_i\,dV \tag{8.2-3}$$

Integrating yields

$$q_2C_{j2}-q_1C_{j1}=\Sigma\alpha_{ij}r_iV$$

or

$$F_{j2} - F_{j1} = \Sigma \alpha_{ij} r_i V \tag{8.2-4}$$

But

$$q_2 C_{j2} = q_1 C_{j1}(1 - f)$$

or

$$F_{j2} = F_{j1}(1 - f) \tag{8.2-5}$$

Then Eq. (8.2-4) becomes

$$-C_{j1} q_1 f = \Sigma \alpha_{ij} r_i V$$

or

$$-F_{j1} f = \Sigma \alpha_{ij} r_i V \tag{8.2-6}$$

In differential form

$$-C_{j1} q_1 \, df = \Sigma \alpha_{ij} r_i \, dV$$

or

$$-F_{j1} \, df = \Sigma \alpha_{ij} r_i \, dV \tag{8.2-7}$$

For a single reaction

$$-C_{A1} q_1 \, df = -r \, dV$$

or

$$-F_{A1} \, df = -r \, dV \tag{8.2-8}$$

where C_{A1} is the inlet concentration of the limiting component A. Integrating Eq. (8.2-8) yields the final design equation

$$\frac{V}{q_1} = C_{A1} \int_0^f \frac{df}{r} \tag{8.2-9}$$

where V/q_1 is defined as space-time. The reciprocal of space-time, i.e., q_1/V is defined as space-velocity.

Example 8.2-1 Isothermal Tubular Reactor

The reaction rate for the decomposition of ethanol

$$C_2H_5OH \rightarrow C_2H_4 + H_2O$$

at 423 K and 2 atm in a plug-flow reactor is (A1)

$$r = 0.52C^2 \text{ g mol/liter}\cdot\text{s}$$

where C is the ethanol concentration in gram moles per liter. Find the space-time and mean residence time for 50 percent conversion of the pure feed.

Solution

Since the volume is changing, the variable-volume method in Chapter 3 should be used. From Eq. (3.8-4)

$$V = V_o(1 + \epsilon_k f_k) \qquad \textbf{(3.8-4)}$$

where

$$\epsilon_k = \frac{\alpha N_{ko}}{\alpha_k N_o} = \frac{(2-1)(1)}{(1)(1)} = 1$$

$$V_o = \frac{N_o RT}{P} = \frac{RT}{P}$$

thus

$$V = (1 + f_k)\frac{RT}{P}$$

But

$$N_j = N_{jo} + \frac{\alpha_j N_{ko} f_k}{\alpha_k} = 1 - (1)(1)f_k/1$$

hence

$$C_j = \frac{N_j}{V} = \frac{(1 - f_k)}{(1 + f)RT/P}$$

Substituting into Eq. (8.2-9) gives

$$\frac{V}{q_1} = C_{A1}\int_0^f \frac{df}{0.52C_A^2}$$

$$= \frac{P}{RT}\int_0^f \frac{df}{0.52\{1 - f/[(1+f)RT/P]\}^2}$$

Integrating for $f = 0.5$ at $T = 423$ K and $P = 2$ atm gives $V/q_1 = 57.7$ s. If $q_1 = 10$ liter/s, $V = 57.7(10) = 577$ liters.

For the mean residence time, let us choose F g mol ethanol entering the reactor per hour as the basis of calculation. Then

$$\text{Mean residence time} = \frac{\text{distance travel}}{\text{velocity}}$$

or

$$\text{Mean residence time } t = \frac{\text{distance travel } x}{\text{volumetric flow rate of mixture } q/\text{cross-sectional area } A_c}$$

In differential form, this becomes

$$dt = \frac{A_c\,dx}{q} = \frac{dV}{q}$$

where q is the volumetric flow rate of the mixture. The number of moles of mixture at any time per hour is the sum of the moles of unreacted C_2H_5OH and the moles of C_2H_4 and H_2O formed. That is, the number of moles of mixture per hour $= F(1 - f) + Ff + Ff = F(1 + f)$. Therefore q, the volumetric flow rate $= RTF(1 + f)/P$. Then

$$dt = \frac{dV}{RTF(1+f)/P}$$

But

$$dV = \frac{F\,df}{r}$$

Thus

$$dt = \frac{F\,df}{r[RTF(1+f)/P]} = \frac{F\,df}{0.52\{1 - f/[RT/P(1+f)]\}^2\,[RTF(1+f)/P]}$$

Integrating for $f = 0.5$ at $T = 423$ K and 2 atm gives $t = 43.6$ s.

Example 8.2-2 Tubular Reactor by Numerical Method

Fluid A of concentration 16 g/liter flows through an insulated reactor tube 1 m long at a linear velocity of 15 m/h. The first-order irreversible reaction ($k = 30$ h^{-1}) A $\longrightarrow$ B takes place in the tube. Calculate the concentration of A as a function of length of the tube by the analytical method, neglecting the diffusion effect. Check the first two points (10 segments of 10 cm length) by the Runge-Kutta numerical method.

Solution

1. Analytical method:

Since

$$C_A = C_{A1}(1 - f)$$

differentiating gives

$$dC_A = -C_{A1}\,df$$

Substituting into Eq. (8.2-9) yields

$$\frac{V}{q_1} = -\int_{C_{A1}}^{C_A} \frac{dC_A}{r}$$

But

$$r = kC_A = 30C_A \qquad \text{and} \qquad q_1 = 15A_c$$

The design equation becomes

$$-\frac{30}{15}dx = \frac{dC_A}{C_A}$$

where x is the length of the reactor. Integrating gives

$$C_A = 16e^{-2x}$$

At $x = 0.1$ m $\quad C_A = 16e^{-0.2} = 13.10$ g/liter

At $x = 0.2$ m $\quad C_A = 16e^{-0.4} = 10.72$ g/liter

2. Numerical method:

Runge-Kutta fourth-order method is used in the computer program in Fig. 8.2-1.

8.3 Nonisothermal Condition

The generalized energy equation is Eq. (4.3-8) as

$$\frac{\partial \Sigma(C_j h_j)}{\partial t} = -\frac{\partial(u \Sigma C_j h_j)}{\partial x} + k\frac{\partial^2 T}{\partial x^2} + G - Ua(T - T^*) \tag{4.3-8}$$

By neglecting diffusion and generation under steady-state conditions, we get

$$\frac{\partial}{\partial x}(u \Sigma C_j h_j) = -Ua(T - T^*) \tag{8.3-1}$$

Multiplying both sides by A, and recalling that $a = P/A$, $q = uA$ gives

$$\frac{\partial}{\partial x}(q \Sigma C_j h_j) = UP(T - T^*) \tag{8.3-2}$$

Expanding yields

$$q\, \Sigma C_j \frac{\partial h_j}{\partial x} + q\, \Sigma h_j \frac{\partial C_j}{\partial x} = -UP(T - T^*) \tag{8.3-3}$$

```
 81/08/03. 15.29.03.
PROGRAM   RUNKA4

00100 PROGRAM RUNKA4(INPUT,OUTPUT)
00110 DIMENSION AC(20),AZ(20)
00120CDAZ=INCREMENT OF X,METERS
00130CACON=RATE CONSTANT,HR**(-1)
00140CVEL=LINEAR VELOCITY,M/HR
00150CACIN=INITIAL CONCENTRATION OF A,GM/LITER
00160CN=TOTAL NUMBER OF INCREMENTS
00170 READ*,DAZ,ACON,VEL,ACIN,N
00180 AC(1)=ACIN
00190 FUN=-ACON/VEL
00200 N1=N+1
00210 AZ(1)=0.
00220 DO 55 I=2,N1
00230 J=I-1
00240 AZ(I)=AZ(J)+DAZ
00250 DK1=DAZ*(FUN*AC(J))
00260 DK2=DAZ*(FUN*(AC(J)+.5*DK1))
00270 DK3=DAZ*(FUN*(AC(J)+.5*DK2))
00280 DK4=DAZ*(FUN*(AC(J)+DK3))
00290 55 AC(I)=AC(J)+(1./6.)*(DK1+2.*DK2+2.*DK3+DK4)
00300 PRINT 20, VEL,ACON,ACIN,AZ(N1),DAZ
00310 20 FORMAT(5F14.4)
00320 M=0
00330 PRINT 21,M,(AZ(I),AC(I),I,I=1,N),AZ(N1),AC(N1)
00340 21 FORMAT(/,11X,I3,11X,F10.4,11X,F10.4)
00350 STOP
00360 END
READY.
RNH

?  .1,30.,15.,16.,10
      15.0000        30.0000        16.0000         1.0000          .1000

            0              0.0000             16.0000

            1               .1000             13.0997

            2               .2000             10.7252

            3               .3000              8.7811

            4               .4000              7.1894

            5               .5000              5.8862

            6               .6000              4.8192

            7               .7000              3.9456

            8               .8000              3.2304

            9               .9000              2.6449

           10              1.0000              2.1654

SRU      0.991 UNTS.
```

Figure 8.2-1

From mass balance equation, we obtain

$$q\frac{dC_j}{dx} = \Sigma\alpha_{ij}r_iA \tag{8.3-4}$$

Hence

$$q\Sigma C_j\frac{\partial h_j \partial T}{\partial T\,\partial x} + \Sigma h_j\ \Sigma\alpha_{ij}r_iA = -UP(T - T^*) \tag{8.3-5}$$

Using $\partial h_j/\partial T = C_{pj}$ and $\Sigma h_j\alpha_j = \Delta H$, and $\Sigma C_jC_{pj} = \mathfrak{z}_p = \rho C_p$ where ρ and C_p are the density and heat capacity at constant pressure of the reaction mixture, respectively, we get

$$q\mathfrak{z}_p\frac{dT}{dx} + \Sigma\Delta H_ir_iA = -UP(T - T^*) \tag{8.3-6}$$

Recalling $A\,dx = dV$ and $P\,dx = dA_s$ where A_s is the surface area, we obtain

$$q\mathfrak{z}_p\,dT + \Sigma\Delta H_ir_i\,dV = -U\,dA_s\,(T - T^*) \tag{8.3-7}$$

which is the design equation for a nonisothermal plug-flow reactor.

Example 8.3-1 Nonisothermal Tubular Reactor

10 kmol/h of an ideal gas mixture at 900 K and 5 atm flows through a tube 10 cm long. The data for the reaction $A + B \rightarrow D$ are

1. Feed composition (mole percent) A = 40, B = 40, inert = 20
2. Specific heat, cal/g mol·°C, reactants 6, product 10, inerts 5
3. Heat of reaction ΔH = 12,500 cal/g mole A at 300 K
4. Rate constants

T, K	777	806	834	861	889
$10^{-6}k$, cal/g mol·h	0.00487	0.00949	0.0187	0.0365	0.0699

Find reactor volume as function of gram moles of A unconverted under adiabatic conditions.

Solution

Basis of calculation: 1 g mol of feed mixture per hour. Under adiabatic conditions, Eq. (8.3-7) becomes

$$q\rho C_p \, dT + \Delta H r \, dV = 0 \tag{A}$$

Let F = gram moles of mixture per hour and X = gram moles of A converted per gram mole of mixture

Then

$$r \, dV = F \, dX$$

and Eq. (A) becomes

$$q\rho C_p \, dT = -\, \Delta H(T) F \, dX \tag{B}$$

where $\Delta H(T)$ is the heat of reaction at any temperature T. Since 1 g mol of feed mixture per hour is chosen as the basis of calculation and is equal to $q\rho$, Eq. (B) becomes

$$[(0.4 - X)6 + (0.4 - X)6 + 10X + 0.2(5)]dT = -\, \Delta H(T) dX \tag{C}$$

But

$$\begin{aligned} \Delta H(T) &= \Delta H(300 \text{ K}) + \Delta C_p(T - 300) \\ &= 12{,}500 + [1(10) - 1(6) - 1(6)](T - 300) \\ &= 11{,}900 - 2T \end{aligned} \tag{D}$$

Substituting into Eq. (C) and simplifying yields

$$(5.8 - 2X) dT = -(11{,}900 - 2T) dX \tag{E}$$

Separating variables and integrating from T = 900 K, X = 0 to $T = T$, $X = X$ gives

$$T = \frac{5{,}220 - 11{,}900X}{5.8 - 2X} \tag{F}$$

The mass balance equation is

$$r \, dV = F \, dX$$

or

$$V = F \int_0^X \frac{dX}{r} \tag{G}$$

with

$$r = kC_A C_B$$

$$\begin{aligned} \text{Total 1 g mol of mixture} &= n_A + n_B + n_D + n_I \\ &= (0.4 - X) + (0.4 - X) + X + 0.2 \\ &= 1 - X \end{aligned}$$

$$V = \frac{NRT}{P} = \frac{(1-X)(0.08206T)}{P}$$

Hence

$$r = k\left[\frac{0.4-X}{(1-X)(0.08206T/5)}\right]^2 = \frac{k}{(0.0164T)^2}\left[\frac{0.4-X}{1-X}\right]^2 \tag{H}$$

The integral in Eq. (G) can be evaluated by Simpson's method as follows. For each value of X, T can be calculated by Eq. (F). Then k can be evaluated from the given table by interpolation and r is calculated by Eq. (H). Finally the integral in Eq. (G) can be evaluated by Simpson's method. Then the volume of the reactor required can be calculated for the feed F. The calculation is shown below:

X	T [Eq. (F)]	$10^{-3}k$, liter/g mol·h (table)	$1/r$ [1/Eq. (H)]
0	900.00	0.087	15.65
0.01	882.53	0.060	22.49
0.02	864.93	0.040	33.45
0.03	847.21	0.026	51.03
0.04	829.37	0.0167	78.78

Hence

$$\int_0^{0.04} \frac{dX}{r} = \frac{0.01}{3}[1(15.65) + 4(22.49 + 51.03) + 2(33.45) + 1(78.78)] = 1.518$$

$$V = F\int_0^{0.04} \frac{dX}{r} = 10{,}000(1.518) = 15{,}180 \text{ liters}$$

Example 8.3-2 Nonisothermal Tubular Reactor by the Numerical Method

A fluid (density = 1,120 g/liter, C_p = 1.0 cal/g·°C) flowing through a tube at a linear velocity of 3 m/s and at a concentration of 112 g/liter and a temperature of 30°C undergoes a first-order irreversible reaction, the heat of which is −70 cal/g. The k_0 and E in the Arrhenius equation are 3.00 s^{-1} and 1,200 cal/g, respectively. Determine the temperature and concentration profiles in the reactor for a length of 1.5 m by the Runge-Kutta method, assuming constant properties of the fluid and no axial diffusion.

Solution

From Eq. (8.2-2), the first-order reaction and the Arrhenius equation, the mass balance equation becomes

```
00100 PROGRAM RUTA4(INPUT,OUTPUT)
00110CDENS=FLUID DENSITY,GM/LITER
00120CVEL=FLUID VELOCITY,M/SEC
00130CHOR=HEAT OF REACTION,CAL/GM
00140CE=ACTIVATION ENERGY,CAL/GM
00150CR=GAS CONSTANT,CAL/GM-DEGK
00160CSH=SPECIFIC HEAT,CAL/GM-DEGC
00170CH=INCREMENT
00180CT1=INITIAL TEMPERATURE,DEGK
00190CC1=INITIAL CONCENTRATION,GM/LITER
00200CX1=INITIAL LENGTH OF TUBE,METER
00210CA=FREQUENCY FACTOR,SEC**(-1)
00220CXMAX=MAXIMUM LENGTH OF TUBE,METER
00230 DIMENSION T(125),C(125),X(125)
00240 READ*,DENS,VEL,HOR,E,R,SH,H,T1,C1,X1,A,XMAX
00250 PRINT 4
00260 4 FORMAT(//,4X," X(M) ",5X, " T(DEGREE C) ", 3X, " C(GM/LITER)")
00270 CALL RK(M,T1,C1,X1,DENS,VEL,HOR,E,R,SH,H,A,XMAX,T,C,X)
00271 STOP
00272 END
00273 SUBROUTINE RK(M,T1,C1,X1,DENS,VEL,HOR,E,R,SH,H,A,XMAX,TR,C,X)
00280 DIMENSION T(125),C(125),X(125)
00290 T(1)=T1
00300 C(1)=C1
00310 X(1)=X1
00320 XXH=XMAX/H
00330 ITA=XXH
00340 VELROS=VEL*DENS*SH
00350 DELHA=-HOR*A
00360 DO 33 I=1,ITA
00370 DELAC=DELHA*C(I)
00380 EX=EXP(E/(R*T(I)))
00390 XO=DELAC/VELROS/EX
00400 YO=-A*C(I)/EXP(E/(R*T(I)))/VEL
00410 X1=-HOR*A*(C(I)+YO/2.)/EXP(E/(R*(T(I)+XO/2.)))/VELROS
00420 Y1=-A*(C(I)+YO/2.)/EXP(E/(R*(T(I)+XO/2.)))/VEL
00430 X2=-HOR*A*(C(I)+Y1/2.)/EXP(E/(R*(T(I)+X1/2.)))/VELROS
00440 Y2=-A*(C(I)+Y1/2.)/EXP(E/(R*(T(I)+X1/2.)))/VEL
00450 X3=-HOR*A*(C(I)+Y2)/EXP(E/(R*(T(I)+X2)))/VELROS
00460 Y3=-A*(C(I)+Y2)/EXP(E/(R*(T(I)+X2)))/VEL
00470 T(I+1)=T(I)+(1./6.)*(XO+2.*X1+2.*X2+X3)
00480 C(I+1)=C(I)+(1./6.)*(YO+2.*Y1+2.*Y2+Y3)
00490 X(I+1)=X(I)+H
00500 IF(X(I+1)-XMAX) 33,33,10
00510 33 CONTINUE
00520 10 M=ITA+1
00530 DO 8 I=1,M
00540 T(I)=T(I)-273.
00550 PRINT 5,X(I),T(I),C(I)
00560 5 FORMAT(F9.3,6X,F10.3,5X,F10.3/)
00570 8 CONTINUE
00580 RETURN
00590 END
READY.
RNH

? 1120.,3.,-70.,1200.,1.987,1.,.1,303.,112.,0.,3.,2.0
```

Figure 8.3-1

X(M)	T(DEGREE C)	C(GM/LITER)
0.000	30.000	112.000
.100	30.894	97.692
.200	31.678	85.149
.300	32.365	74.167
.400	32.965	64.564
.500	33.489	56.176
.600	33.946	48.857
.700	34.345	42.475
.800	34.693	36.914
.900	34.996	32.072
1.000	35.259	27.858
1.100	35.488	24.192
1.200	35.687	21.005
1.300	35.860	18.235
1.400	36.011	15.827
1.500	36.141	13.736
1.600	36.255	11.920
1.700	36.354	10.343
1.800	36.439	8.974
1.900	36.513	7.785

Figure 8.3-1 ***(continued)***

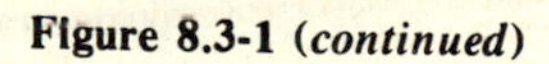

$$\frac{dC_A}{dx} = -\frac{k_o C_A e^{-E/RT}}{u} \tag{A}$$

From Eq. (8.3-6), the energy balance equation under adiabatic conditions is

$$\frac{dT}{dx} = -\frac{\Delta H k_o C_A e^{-E/RT}}{u \rho C_p} \tag{B}$$

Equations (A) and (B) can be solved by the fourth-order Runge-Kutta method given in the computer program in Fig. 8.3-1.

8.4 Pressure-Drop Consideration

Neglecting the diffusion and the gravity, the generalized momentum equation (4.4-7) under steady-state conditions becomes

$$\rho u_x \frac{\partial u}{\partial x} = -\frac{\partial P}{\partial x} - \frac{2f_f u^2 \rho}{D} \tag{8.4-1}$$

The change of velocity with the distance x is usually small. Hence Eq. (8.4-1) can be further simplified to

$$\frac{dP}{dx} + \frac{2f_f u^2 \rho}{D} = 0$$

or

$$\frac{dP}{dx} = -\frac{2f_f u^2 \rho}{D} \tag{8.4-2}$$

Now the density of the mixture ρ is a function of conversion. Then the volume of reactor required for conversion is also affected by the pressure drop. The incorporation of the pressure drop into the reactor design can be seen from the following examples.

Example 8.4-1 Isothermal Tubular Reactor with a Pressure Drop

3,991.47 kg/h of a mixture (50 mol percent A and 50 mol percent inerts) at 333.33 K and 5 atm entering a tube of 6.09 cm ID undergoes the following gas-phase reaction:

$$A \rightarrow 2B$$

of which the rate constant is 2,000 h^{-1}. Assuming that molecular weights of A and inerts are 40 and 20, respectively, and viscosity of the mixture is 0.0206 cP, determine the relation between conversion, pressure drop, and volume of the reactor ($f_D = 0.046/Re^{0.2}$ is assumed).

Solution

The calculations are based on SI units. Equation (8.4-2) can be written in another form as

$$\frac{dP}{dx} = -\frac{f_D \rho u^2}{2D} \tag{A}$$

where f_D is the Darcy friction factor. Assuming

$$f_D = \frac{0.046}{Re^{0.2}} = 0.046\left(\frac{\pi\mu D}{4w}\right)^{0.2} \tag{B}$$

substituting into Eq. (A), using $u = 4w/(\pi D^2\rho)$, and simplifying gives

$$dP + 0.036\frac{\mu^{0.2}w^{1.8}}{D^{4.8}\rho}\,dx = 0 \tag{C}$$

For the chemical reaction

$$A \rightarrow 2B$$

the law of conservation of mass indicates that

$$NM = N_o M_o \tag{D}$$

where N and M are the total number of moles and the average molecular weight of the gas mixture, respectively. The subscript o refers to the initial conditions. However, the final moles relate to the initial moles by

$$N = N_o + \alpha\beta \tag{E}$$

where

$$\alpha = \Sigma\alpha_j = 2 - 1 = 1$$

Hence the average molecular weight of the gas mixture at any position in the reactor is from Eq. (D)

$$M = \frac{N_o M_o}{N_o + \beta} \tag{F}$$

The density of the mixture is then

$$\rho = \frac{PM}{RT} = \frac{PN_o M_o}{RT(N_o + \beta)} \tag{G}$$

Let X = moles of A converted per mole of mixture. It is seen that

$$\beta = N_{Ao} f = X$$

and

$$\frac{w\,dX}{M_o} = r\,dV = r\frac{\pi D^2}{4}\,dx \tag{H}$$

Eliminating dx in Eq. (C) by Eq. (H) gives

$$dP + \frac{0.036w^{1.8}\mu^{0.2}RT(N_o + X)}{D^{4.8}PN_oM_o}\left(\frac{4w\,dX}{\pi D^2M_or}\right) = 0 \qquad \text{(I)}$$

Simplifying yields

$$dP + \Omega\frac{(N_o + X)}{PN_o}\frac{dX}{r} = 0 \qquad \text{(J)}$$

where

$$\Omega = \frac{0.046w^{2.8}\mu^{0.2}RT}{D^{6.8}M_o^2}$$

Substituting the numerical values from this problem, we get

$$\Omega = \frac{0.046(3991.47/3600)^{2.8}(2.0668 \times 10^{-5})^{0.2}(8,314.34)333.33}{0.0609^{6.8}\ 30^2} = 4.0212 \times 10^9$$

Equation (J) becomes

$$dP + 4.0212 \times 10^9\frac{N_o + X}{PN_o}\frac{dX}{r} = 0 \qquad \text{(K)}$$

where P is expressed in kg/m·s². Now $N_o = 1$ and

$$r = kC_A = k\frac{N_A}{V} = k\frac{N_{Ao} - X}{N_o + X}\left(\frac{P}{RT}\right) = k\left(\frac{0.5 - X}{1 + X}\right)\frac{P}{RT} \qquad \text{(L)}$$

Substituting into Eq. (K) and integrating yields

$$\int_{P_0}^{P} P^2\,dP + \frac{4.0212 \times 10^9(8,314.34)(333.33)}{0.5555}\int_0^X\frac{(1 + X)^2}{0.5 - X}\,dX = 0 \qquad \text{(M)}$$

Simplifying gives

$$P^3 = 1.3014 \times 10^{17} - 6.018 \times 10^{16}\int_0^X\frac{(1 + X)^2}{0.5 - X}\,dX \qquad \text{(N)}$$

Substituting Eq. (L) into Eq. (H), we obtain

$$V = \frac{wRT}{M_ok}\int_0^X\frac{1 + X}{P(0.5 - X)}\,dX = \frac{1.1087(8,314.34)(333.33)}{30(0.5555)}\int_0^X\frac{1 + X}{P(0.5 - X)}\,dX$$

$$= 184,378.77\int_0^X\frac{1 + X}{P(0.5 - X)}\,dX \qquad \text{(O)}$$

Eqs. (N) and (O) can be solved numerically as follows: For each value of X, calculate $(1 + X)^2/(0.5 - X)$ and then evaluate P by Eq. (N). Once P is found, V can be calculated by integrating Eq. (O)

X	$\frac{(1+X)^2}{0.5-X}$	$\int_0^X \frac{(1+X)^2}{0.5-X}dX$	P [Eq. (N)]	$\int_0^X \frac{1+X}{P(0.5-X)}$	V [Eq. (O)]
0	2.00	0	5.0609×10^5	0	0
0.05	2.45	0.11	4.9737×10^5	2.1608×10^{-7}	0.0398
0.10	3.02	0.25	4.8579×10^5	4.7488×10^{-7}	0.0876
0.15	3.87	0.42	4.7095×10^5	7.9082×10^{-7}	0.1458
0.20	4.80	0.64	4.5023×10^5	1.1873×10^{-6}	0.2189
0.25	6.22	0.89	4.2411×10^5	1.7041×10^{-6}	0.3142
0.30	8.45	1.26	3.7822×10^5	2.4285×10^{-6}	0.4478
0.35	12.10	1.77	2.8656×10^5	3.6433×10^{-6}	0.6717

Example 8.4-2 Pressure-Drop Consideration—Computer Solution

The decomposition of ethane to ethylene and hydrogen between 650 and 930°C is a first-order irreversible reaction

$$C_2H_6 \rightarrow C_2H_4 + H_2$$

$k = 5.764 \times 10^{16} e^{-41,310/T}\ s^{-1}$ where T is in K; 800 kg/h pure ethane at 650°C flows into a 15-cm tube. Heat is supplied to the tube at a rate of 15,773 W/m². The initial pressure is 2 atm. The Fanning friction factor is assumed to be 0.0055. Determine the length of the tube required for 75 percent conversion of the ethane. Data: (T is in K).

	ΔH_f at 298 K, cal/g mol	C_p, cal/g mol·K
$C_2H_6(g)$	−20,236	$3.75 + 35.7 \times 10^{-3}T - 10.12 \times 10^{-6}T^2$
$C_2H_4(g)$	12,496	$5.25 + 24.2 \times 10^{-3}T - 6.88 \times 10^{-6}T^2$
$H_2(g)$	0	$7.00 - 0.385 \times 10^{-3}T + 0.60 \times 10^{-6}T^2$

Rate constant: $k = 5.764 \times 10^{16} e^{-41,310/T}\ s^{-1}$

Molecular weight: $C_2H_6 = 30$, $C_2H_4 = 28$, $H_2 = 2$

Gas constant: $R = 8{,}314.34\ kg \cdot m^2/s^2 \cdot kmol \cdot K$

Solution

For this problem, we have to set up the following three differential equations which are then solved by the improved Euler method.

1. Mass Balance: From Eq. (8.2-9), we have

$$r\,dV = q_1 C_{A1}\,df = F\,df \tag{A}$$

where F is the flow rate of the feed in kilogram moles per hour. Let us use 1 kmol C_2H_6 as the basis of calculation. Then the stoichiometric equations for the reaction are

$$N_{C_2H_6} = 1 - \beta \tag{B}$$

$$N_{C_2H_4} = \beta \tag{C}$$

$$N_{H_2} = \beta \tag{D}$$

Total number of moles of the gas mixture $= 1 - \beta + \beta + \beta = 1 + \beta$. From the ideal gas law, the volume of the gas mixture $= (1 + \beta)RT/P$. Thus for a first-order reaction, the rate of the reaction is

$$r = kC_{C_2H_6} = k\frac{(1-\beta)P}{(1+\beta)RT} \tag{E}$$

```
 81/08/03.  15.36.50.
PROGRAM    IMEULR

00100 PROGRAM IMEULR(INPUT,OUTPUT)
00110CTC=INLET TEMPERATURE,DEGC
00120CW=MOLECULAR WEIGHT OF ETHANE
00130CFR=FLOW RATE,KG/HR
00140CDI=DIAMETER OF TUBE.CM
00150CQ=HEAT FLOW RATE,W/M**2
00160CCONV=FRACTIONAL CONVERSION
00170CDL=INCREMENT OF TUBE LENGTH,CM
00180CAL=LENGTH OF TUBE,CM
00190CTK=TEMPERATURE,DEGK
00200CF=FRACTIONAL   CONVERSION
00210CP=PRESSURE,N/M**2
00220 READ*,TC,W,FR,DI,Q,CONV,DL
00230 AQ=Q*W/FR*3.14159*DI*3600./4.1868E5
00231 FFATOR=2.07504E20*3.14159*DI*DI*W/(4.*10000.*8314.34*FR)
00240 AL=0.
00250 F=0.
00260 P=2.*1.01325E+5
00270 TK=TC+273.
00280 PRINT 201,AL,TK,F,P
00290 2 DH=(32732.+8.5*(TK-298.)-5.942E-3*(TK*TK-298.*298.)+1.28E-6*(
00300+    TK*TK*TK-298.*298.*298.))
00310 CP=(1.0-F)*(3.75+35.7E-3*TK-10.12E-6*TK*TK)+F*(12.25+23.815E-3*TK-
00320+ 6.28E-6*TK*TK)
00330 FACTOR=FFATOR*P
00340 RHO=(W/(1.+F))*P/(8314.34*TK)
00350 U=800./(.785*DI*DI)*10000./RHO
```

Figure 8.4-1

```
00360 DFDL=FACTOR*EXP(-41310./TK)*(1.-F)/(TK*(1.+F))
00370 DTDL=(AQ-DH*DFDL)/CP
00380 DPDL=-0.022/2./(DI/100.)*RHO*U*U/3600.**2
00390 TKP=TK+DTDL*DL
00400 PP=P+DPDL*DL
00410 FP=F+DFDL*DL
00420 DHP=(32732.+8.50*(TKP-298.)-5.942E-3*(TKP*TKP-298.*298.)+1.28E
00430+-6*(TKP*TKP*TKP-298.*298.*298.))
00440 CPP=(1.0-FP)*(3.75+35.7E-3*TKP-10.12E-6*TKP*TKP)+FP*(12.25+23.815E
00450+-3*TKP-6.28E-6*TKP*TKP)
00460 RHOP=(W/(1.+FP))*PP/(8314.34*TKP)
00470 UP=800./(.785*DI*DI)*10000./RHOP
00480 DFDLP=FACTOR*EXP(-41310./TKP)*(1.-FP)/(TKP*(1.+FP))
00490 DTDLP=(AQ-DHP*DFDLP)/CPP
00500 DPDLP=-0.022/2./(DI/100.)*RHOP*UP*UP/3600.**2
00510 TK=TK+0.5*DL*(DTDL+DTDLP)
00520 P=P+0.5*DL*(DPDL+DPDLP)
00530 F=F+0.5*DL*(DFDL+DFDLP)
00540 AL=AL+DL
00550 PRINT 201,AL,TK,F,P
00551 IF(F-CONV) 2,4,4
00552 4 TC=TK-273.
00553 PRINT 201,AL,TK,F,P
00560 201 FORMAT(1H ,4E18.4)
00570 99 STOP
00580 END
READY.
RNH


? 650.,30.,800.,15.,15773.,.75,7.5
       0.                .9230E+03          0.                .2027E+06
       .7500E+01         .9757E+03          .8539E-02         .2025E+06
       .1500E+02         .9870E+03          .5140E-01         .2024E+06
       .2250E+02         .9907E+03          .1007E+00         .2023E+06
       .3000E+02         .9941E+03          .1502E+00         .2022E+06
       .3750E+02         .9974E+03          .1998E+00         .2020E+06
       .4500E+02         .1001E+04          .2492E+00         .2019E+06
       .5250E+02         .1004E+04          .2986E+00         .2017E+06
       .6000E+02         .1008E+04          .3478E+00         .2016E+06
       .6750E+02         .1011E+04          .3970E+00         .2014E+06
       .7500E+02         .1015E+04          .4459E+00         .2012E+06
       .8250E+02         .1019E+04          .4946E+00         .2010E+06
       .9000E+02         .1024E+04          .5431E+00         .2008E+06
       .9750E+02         .1028E+04          .5914E+00         .2006E+06
       .1050E+03         .1033E+04          .6392E+00         .2004E+06
       .1125E+03         .1039E+04          .6867E+00         .2002E+06
       .1200E+03         .1045E+04          .7335E+00         .2000E+06
       .1275E+03         .1052E+04          .7796E+00         .1998E+06
       .1275E+03         .1052E+04          .7796E+00         .1998E+06

SRU       1.719 UNTS.
```

Figure 8.4-1 (*continued*)

Since

$$dV = A_c\,dL \tag{F}$$

where A_c is the cross-sectional area of the tube and L is the length. Therefore Eq. (A) becomes

$$\frac{df}{dL} = \frac{3{,}600k(1-f)PA_c}{F(1+f)RT} \tag{G}$$

because $f = \beta$ in this case.

2. Energy Balance: From Eq. (8.3-7) we have

$$q\rho\mathbf{C}_p\,dT + \Delta Hr\,dV = -U(T - T^*)dA_s \tag{H}$$

where $\mathbf{C}_p$ is the average heat capacity at constant pressure of the gas mixture

$$\mathbf{C}_p = (1-f)C_{p,C_2H_6} + fC_{p,\,C_2H_4} + fC_{p,\,H_2} \tag{I}$$

Since

$$dA_s = \pi D\,dL \tag{J}$$

and

$$q\rho = F \tag{K}$$

Equation (H) becomes

$$F[(1-f)C_{p,\,C_2H_6} + f(C_{p,\,C_2H_4} + C_{p,\,H_2})]dT + \Delta HrA_c\,dL + U(T - T^*)\pi D\,dL = 0 \tag{L}$$

But

$$r\,dV = rA_c\,dL = F\,df$$

and we have

$$F[(1-f)C_{p,\,C_2H_6} + f(C_{p,\,C_2H_4} + C_{p,\,H_2})]\frac{dT}{dL} + \Delta HF\frac{df}{dL} + U(T - T^*)\pi D = 0 \tag{M}$$

3. Momentum Balance: From Eq. (8.4-2), we have

$$\frac{dP}{dL} = -\frac{f_D u^2 \rho}{2D} \tag{N}$$

Putting all the numerical values from the problem into Eqs. (G), (M), and (N), we have the computer program in Fig. 8.4-1.

8.5 *Analytical Solution of Adiabatic Irreversible Reactions*

If the system in question is at constant density, the analytical solutions, as well as Problems 5-14 and 5-15 by the exponential integrals discussed in Chapter 5, are equally well applied to adiabatic plug-flow reactors because the reaction time t in a batch reactor is equal to the space-time τ in a plug-flow reactor.

If the system in question is at constant pressure with variable volume, the use of exponential integral can also be employed. Following the same approach as in Chapter 5, the following equations can be developed.

Zero-order reaction

$$X_1 \xrightarrow{k} X_2$$

The rate of the reaction is

$$r = k$$

The final design equation is

$$\frac{V}{F} = \frac{C_p}{k(-\Delta H)}(Te^{E/RT} - T_o e^{E/RT_o}) - \frac{EC_p}{Rk(-\Delta H)}\left[Ei\left(\frac{E}{RT}\right) - Ei\left(\frac{E}{RT_o}\right)\right]$$

First-order reaction

$$X_1 \xrightarrow{k_1} \alpha X_2$$

The rate of the reaction is

$$r = k_1 P\left(\frac{N_1}{N_T}\right)$$

The final design equation is

$$\frac{V}{F} = \frac{-\alpha C_p}{A(-\Delta H)P}(Te^{E/RT} - T_o e^{E/RT_o}) + \left[N_o + \alpha N_{1o} + \frac{\alpha E C_p}{R(-\Delta H)}\right](AP)^{-1}\left[Ei\left(\frac{E}{RT}\right) - Ei\left(\frac{E}{RT_o}\right)\right] - \frac{(N_o + \alpha N_{1o})e^{E/RJ}}{AP}[Ei(Z) - Ei(Z_o)] \quad \textbf{(8.5-2)}$$

where

$$\boldsymbol{\alpha} = \alpha - 1$$

$$J = T_o + \frac{(-\Delta H)N_{1o}}{\mathbf{C}_p}$$

$$Z = \frac{E}{R}\left(\frac{1}{T} - \frac{1}{J}\right)$$

$$Z_o = \frac{E}{R}\left(\frac{1}{T_o} - \frac{1}{J}\right)$$

$\mathbf{C}_p$ is the average heat capacity of the mixture in Btu/lb · °F

F is the feed rate in lb/h

$-\Delta H$ is the heat of reaction in Btu/mole of X_1 reacted

A is the frequency factor in moles X_1/atm · ft^3 · s

N_{1o} is the initial moles of X_1 per pound of feed

N_o is the total moles of feed per pound of feed

N_T is the total moles of mixture per pound of feed

P is the pressure in atm

8.6 Laminar-Flow Reactors

For laminar flow in a tube (Fig. 8.6-1), the velocity distribution is

$$u_r = u_o\left(1 - \frac{r^2}{R^2}\right) \tag{8.6-1}$$

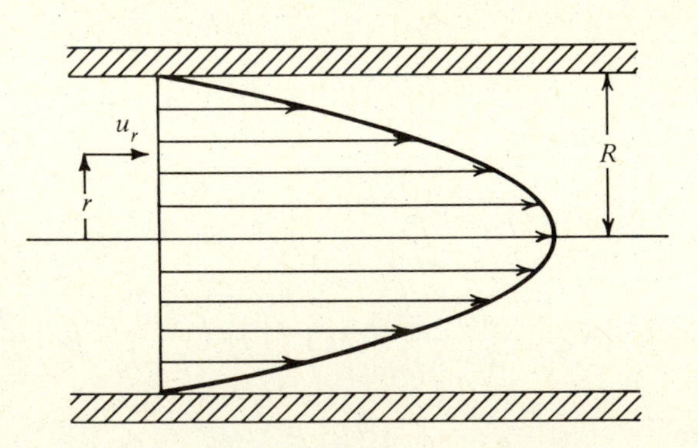

Figure 8.6-1 Laminar flow in a tube

where u_r and u_o are the velocities at radius r and center, respectively. R is the radius of the tube. Thus the average or mean velocity is

$$u_m = u_{av} = \frac{\int_0^R u_r 2\pi r\,dr}{\int_0^R 2\pi r\,dr}$$

$$= \frac{2\pi \int_0^R u_o\left[1-\left(\frac{r}{R}\right)^2\right] r\,dr}{2\pi \int_0^R r\,dr} = \frac{u_o}{2} \tag{8.6-2}$$

The volumetric flow rate q is

$$q = \pi R^2 u_m = \frac{\pi R^2 u_o}{2} \tag{8.6-3}$$

The residence time for fluid at radius r for length L is

$$t = \frac{L}{u_r} = \frac{L}{u_o[1-(r/R)^2]} \tag{8.6-4}$$

Differentiating and simplifying gives

$$r\,dr = \frac{LR^2\,dt}{2u_o t^2} \tag{8.6-5}$$

The flow through an annular element of radius r is

$$2\pi u_r r\,dr \qquad \text{or} \qquad \frac{2\pi L r\,dr}{t} \tag{8.6-6}$$

Amount of material converted $= f 2\pi L r\,dr/t$ where f is fractional conversion. The mean fractional conversion is

$$f_m = \frac{\int_0^R f 2\pi L r\,dr/t}{\int_0^R 2\pi u_r r\,dr} \tag{8.6-7}$$

or

$$f_m \frac{R^2 u_o}{2} = \int_0^R f \frac{2\pi L r}{t} \, dr \tag{8.6-8}$$

Eliminating $r\,dr$ by Eq. (8.6-5) gives

$$f_m \frac{R^2 u_o}{2} = \int_{t_o}^{t_\infty} f \frac{2\pi L}{t} \frac{LR^2}{2u_o} \frac{dt}{t^2} \tag{8.6-9}$$

At center $r = 0$; therefore

$$t = \frac{L}{u_o} = t_o$$

At wall $r = R$; therefore

$$t = \infty = t_\infty$$

Further simplification of Eq. (8.6-9) gives

$$f_m \frac{\pi R^2 u_o}{2} = \frac{\pi L^2 R^2}{u_o} \int_{t_o}^{t_\infty} f \frac{dt}{t^3} \tag{8.6-10}$$

8.6-1 *First-order reaction*

$$u \frac{dC_A}{dx} = -r = -kC_A \tag{8.6-11}$$

Substituting Eq. (8.6-1), separating variables, and integrating gives

$$\int_{C_{Ao}}^{C_A} \frac{dC_A}{C_A} = -\int_0^L \frac{k}{u_o[1-(r/R)^2]} \, dx \tag{8.6-12}$$

or

$$C_A = C_{Ao} \exp\left\{-\frac{kL}{u_o[1-(r/R)^2]}\right\} \tag{8.6-13}$$

But

$$C_A = C_{Ao}(1-f)$$

Then

$$f = 1 - \exp\left\{-\frac{kL}{u_o[1-(r/R)^2]}\right\} = 1 - e^{-kt} \tag{8.6-14}$$

Substituting into Eq. (8.6-10) gives

$$f_m \frac{\pi R^2 u_o}{2} = \frac{\pi L^2 R^2}{u_o} \int_{t_o}^{t_\infty} \left(1 - e^{-kt} \frac{dt}{t^3}\right) \quad \textbf{(8.6-15)}$$

Simplifying results in

$$f_m = 2t_o^2 \int_{t_o}^{t_\infty} \left(\frac{1}{t^3} - \frac{e^{-kt}}{t^3}\right) dt \quad \textbf{(8.6-16)}$$

The last integral should be evaluated by exponential integral.

8.6-1 Second-order reaction

$$u \frac{dC_A}{dx} = -kC_A^2 \quad \textbf{(8.6-17)}$$

Substituting Eq. (8.6-1) into Eq. (8.6-17) yields

$$u_o\left[1 - \left(\frac{r}{R}\right)^2\right] \frac{dC_A}{dx} = -kC_A^2 \quad \textbf{(8.6-18)}$$

Separating variables and integrating, we obtain

$$\frac{1}{C_{Ao}} - \frac{1}{C_A} = -\frac{kL}{u_o\left[1 - \left(\frac{r}{R}\right)^2\right]} = -kt \quad \textbf{(8.6-19)}$$

or

$$\frac{(C_A - C_{Ao})}{C_{Ao}C_A} = -kt \quad \textbf{(8.6-20)}$$

Since

$$C_A = C_{Ao}(1 - f) \quad \textbf{(8.6-21)}$$

or

$$C_A - C_{Ao} = -C_{Ao}f$$

Equation (8.6-20) becomes

$$-\frac{f}{C_{Ao}(1 - f)} = -kt$$

or

$$f = \frac{ktC_{Ao}}{1 + ktC_{Ao}} \tag{8.6-22}$$

Substituting into Eq. (8.6-10) yields

$$f_m \frac{\pi R^2 u_o}{2} = \frac{\pi L^2 R^2}{u_o} \int_{t_o}^{t_\infty} \frac{ktC_{Ao}}{1 + ktC_{Ao}} \frac{dt}{t^3} \tag{8.6-23}$$

Expanding the integral gives

$$f_m = kC_{Ao}t_m\left(1 + \frac{kC_{Ao}t_m}{2} \ln \frac{kC_{Ao}t_m}{2 + kC_{Ao}t_m}\right) \tag{8.6-24}$$

Example 8.6-1 PFR and LFR

A mixture of 5.00 mol butanol per mole of acetic acid with 0.032 percent H_2SO_4 by weight (as catalyst) flows to a tubular reactor (10 cm ID) at a rate of 454 kg/h, which is operated isothermally at 100°C. Estimate the length of the reactor required for 45 percent conversion under plug-flow and laminar-flow conditions for the following data:

1. $r = kC_A^2$ where $k = 17.4$ ml/g mol acetic acid per minute (L^2)
   $r =$ g mol acid/ml·min
2. Specific gravity of feed = 0.75, which is assumed to apply throughout the conversion
3. Viscosity of mixture = 0.54 cP

Solution

1. Plug flow:

$$C_{A1} = \frac{0.75}{5.00(74) + (1)(60)} = 0.001744 \text{ g mol/cm}^3$$

Using Eq. (8.2-9)

$$\frac{V}{q_1} = C_{A1} \int_0^{0.45} \frac{df}{r} = 0.001744 \int_0^{0.45} \frac{df}{17.4(0.001744)^2(1-f)^2}$$

$$= 26.96 \text{ min}$$

Hence

$$V = \frac{454{,}000(26.96)}{60(0.75)} = 271{,}996.4 \text{ cm}^3$$

and

$$\frac{\pi(10)^2}{4} L = 271{,}996.4 \quad \text{or} \quad L = 3{,}463 \text{ cm}$$

2. Laminar flow:

$$\text{Re} = \frac{DG}{\mu} = 10\left[\frac{454{,}000}{3{,}600(\pi 100/4)}\right]\frac{1}{0.54 \times 10^{-2}} = 2{,}973$$

The flow is in the transition between laminar and turbulent. For laminar flow, the mean conversion is obtained from Eq. (8.6-24)

$$kC_{A0}t_m = 17.4(0.001744)(26.96) = 0.8181$$

Substituting into Eq. (8.6-42) gives

$$f_m = 0.8181\left[1 + \frac{0.8181}{2}\ln\left(\frac{0.8181}{2 + 0.8181}\right)\right] = 0.4042$$

That means the above reactor is unable to reach the required conversion of 0.45. To obtain the residence time for 0.45 conversion, Eq. (8.6-42) must be solved by trial and error. Assuming $t_m = 33$ min, $kC_{A0}t_m = 1.0014$

$$f_m = 1.0014\left[1 + \frac{1.0014}{2}\ln\left(\frac{1.0014}{2 + 1.0014}\right)\right] = 0.451 \qquad \text{check}$$

Then

$$\frac{V}{q_1} = 33$$

or

$$V = 33\left(\frac{454{,}000}{60(0.75)}\right) = 332{,}933 \text{ cm}^3$$

Then

$$L = 4239 \text{ cm}$$

8.7 Recycle Reactors

In order to increase the yield, it is sometimes advantageous to have some part of the effluent stream return to the inlet stream. Figure 8.7-1 illustrates that the stream of the fresh feed combines with a stream of the recycle effluent which comes from a part of the effluent stream, the major part of which becomes the net product. Due to the addition of this recycle, the number of moles of the

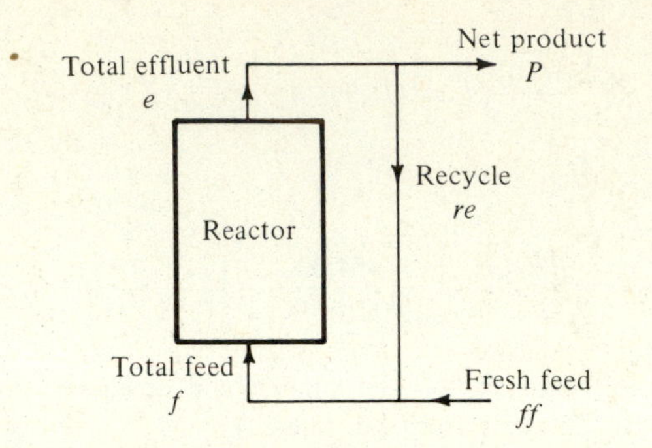

Figure 8.7-1 Recycle reactor

components and the volume of the feed stream entering the reactor can be calculated. From the figure, it is seen that

$$\begin{pmatrix}\text{No. of moles}\\ \text{of A in } f \text{ stream}\end{pmatrix} = \begin{pmatrix}\text{No. of moles}\\ \text{of A recycled}\end{pmatrix} + \begin{pmatrix}\text{No. of moles of}\\ \text{A in fresh feed}\end{pmatrix}$$

$$N_{Af} = N_{Are} + N_{Aff} \tag{8.7-1}$$

Similarly for the volumetric flow rates

$$q_f = q_{re} + q_{ff} \tag{8.7-2}$$

Since $N = Cq$, we have

$$C_{Af}q_f = C_{Are}q_{re} + C_{Aff}q_{ff} \tag{8.7-3}$$

Then

$$C_{Af} = C_{Are}\left(\frac{q_{re}}{q_{re} + q_{ff}}\right) + C_{Aff}\left(\frac{q_{ff}}{q_{re} + q_{ff}}\right) \tag{8.7-4}$$

Letting

$$R = \frac{q_{re}}{q_{ff}}$$

and since

$$C_{Are} = C_{Ae} = C_{AP}$$

we get

$$C_{Af} = \frac{1}{1 + R}(RC_{AP} + C_{Aff}) \tag{8.7-5}$$

Using this C_{Af} for C_{A1} in Eq. (8.2-9), we can evaluate the space time V/q_1 where q_1 is q_f in this case. These two streams can be applied to the corresponding values in the CSTR.

Example 8.7-1 Recycle Reactor

n-Pentane isomerizes to isobutane in a first-order reaction with rate constant $k = 0.00364\ \text{min}^{-1}$. 95 mol *n*-pentane per cubic meter can produce isobutane in such a way that 55 mol/m³ *n*-pentane remains in the product stream. Find the space-time in the reactor for the following cases: (1) once-through process and (2) recycle process with $R = 5$.

Solution

1. Ideal once-through plug-flow reactor

Equation (8.2-9) can be arranged to

$$\frac{V}{q_f} = \frac{1}{k} \ln \frac{C_{Af}}{C_{AP}} = \frac{1}{0.00364} \ln \frac{95}{55} = 150.15 \text{ min} \tag{A}$$

2. Recycled plug-flow reactor with $R = 5$

Using Eq. (8.7-5)

$$C_{Af} = \frac{1}{1+5}[5(55) + 95] = 61.6666 \text{ kmol/m}^3$$

Substituting into Eq. (A) gives

$$\frac{V}{q_f} = \frac{1}{0.00364} \ln \frac{61.6666}{55} = 31.43 \text{ min}$$

8.8 *Optimization*

As in the optimization of the operating conditions in a CSTR, case studies are used here. The same is true that other reactions may be applied to each case.

8.8-1 *Minimum volume*

The design equation (8.2-9) indicates that the minimization of the volume of the reactor is equivalent to the maximization of the reaction rate for a fixed conversion.

We know that for a given inlet gas composition and given operating pressure, the reaction rate is a function of temperature and conversion

$$r = \phi(f, T) \tag{8.8-1}$$

For a reversible reaction, the rate of reaction can be expressed as

$$r = k\Pi C_j^{\alpha j} - k'\Pi C_j^{\alpha_j'} \tag{3.2-6}$$

By means of the Arrhenius equation, Eq. (3.2-6) becomes

$$r = k_o e^{-E_o/RT}\Pi C_j^{\alpha j} - k_o' e^{-E_o'/RT}\Pi C_j^{\alpha_j'} \tag{8.8-2}$$

Differentiating Eq. (8.8-2) with respect to T at constant concentration, i.e., at fixed conversion gives

$$\frac{\partial r}{\partial T} = \frac{E_o}{RT^2} k\Pi C_j^{\alpha j} - \frac{E_o'}{RT^2} k'\Pi C_j^{\alpha_j'} = \frac{1}{RT^2}(E_o r_f - E_o' r_b) \tag{8.8-3}$$

Thus if the reaction is irreversible ($r_b = 0$) or endothermic ($E_o > E_o'$), $\partial r/\partial T > 0$ and r increases continuously with temperature. However, if the reaction is exothermic ($E_o' > E_o$), the derivative will vanish and the optimum temperature T^* can be obtained as

$$T^* = -\frac{\Delta H}{R}\left(\ln \frac{E_o' k_o'}{E_o k_o} \frac{\Pi C_j^{\alpha_j'}}{\Pi C_j^{\alpha j}}\right)^{-1} \tag{8.8-4}$$

where $-\Delta H = E_o' - E_o$. After T^* is calculated, r^* can be evaluated from Eq. (8.8-2) and finally the minimum volume can be determined from Eq. (8.2-9).

Example 8.8-1 Minimum Tubular Volume for Second-Order Reversible Reaction

Determine the optimum operating temperature and then the minimum volume of a tubular reactor for a conversion of 70 percent at the outlet. An equimolar amount of A and B at a pressure of 1 atm undergoes the following reaction (B1)

$$A + B \underset{k'}{\overset{k}{\rightleftharpoons}} C + D \qquad k = 4.35 \times 10^{13} e^{-25{,}000/RT}$$
$$k' = 7.42 \times 10^{14} e^{-30{,}000.RT}$$

in the tubular reactor at which the maximum permissible temperature is 870 K.

Solution

By ideal gas law, the initial volume of the mixture is

$$V_o = \frac{N_o RT}{P} = \frac{2RT}{P} \tag{A}$$

Because there is no volume change

$$V = V_o \text{ since } \epsilon = 0 \tag{B}$$

By Eq. (1.4-30) and Eq. (1.4-41)

$$N_j = N_{jo} + \frac{\alpha_j f_k N_{ko}}{\alpha_k} \tag{C}$$

Choosing A as the limiting component gives

$$N_A = 1 - f_A \tag{D}$$

$$N_B = 1 - f_A \tag{E}$$

$$N_C = 0 + f_A \tag{F}$$

$$N_D = 0 + F_A \tag{G}$$

Then the concentration of the components are

$$C_A = C_B = \frac{(1 - f_A)P}{2RT} \tag{H}$$

$$C_C = C_D = \frac{f_A P}{2RT} \tag{I}$$

The reaction rate is

$$r = kC_A C_B - k' C_C C_D \tag{J}$$

$$= k_o e^{-E_o/RT} C_A C_B - k'_o e^{-E'_o/RT} C_C C_D \tag{K}$$

Substituting Eqs. (H) and (I) into Eq. (K) yields

$$r = \left(\frac{P}{2RT}\right)^2 [k_o e^{-E_o/RT}(1 - f_A)^2 - k'_o e^{-E'_o/RT} f_A^2] \tag{L}$$

Differentiating with respect to T and equating to zero gives

$$k_o e^{-E_o/RT^*}(1 - f_A)^2(E_o - 2RT^*) - k'_o e^{-E'_o/RT^*} f_A^2(E'_o - 2RT^*) = 0 \tag{M}$$

where T^* is the optimum temperature. Rearranging results in

$$\frac{k_o e^{-E_o/RT^*}(E_o - 2RT^*)}{k'_o e^{-E'_o/RT^*}(E'_o - 2RT^*)} = \frac{f_A^2}{(1 - f_A)^2} \tag{N}$$

With the numerical values, Eqs. (N) and (L) become

$$\frac{4.35 \times 10^{13} e^{-25,000/1.986\,T^*}[25,000 - 2(1.986)T^*]}{7.42 \times 10^{14} e^{-30,000/1.986\,T^*}[30,000 - 2(1.986)T^*]} = \left(\frac{f_A}{1 - F_A}\right)^2 \tag{O}$$

or

$$5.8625 \times 10^{-2} e^{2,517/T^*} \left[\frac{6,291 - T^*}{7,553 - T^*}\right] = 5.4444 \tag{P}$$

and

$$r^* = \frac{1.615 \times 10^{15}}{T^{*2}} [e^{-12,588/T^*}(1 - f_A)^2 - 17.057 e^{-15,105/T^*} f_A^2] \tag{Q}$$

Thus the optimum volume is

$$V^* = qC_{A1} \int_0^{f_A} \frac{1}{r^*}\, df_A = F_{A1} \int_0^{f_A} \frac{1}{r^*}\, df_A \tag{R}$$

where F_{A1} is the molar flow rate of component A at the inlet of the reactor. Hence Eqs. (P), (Q), and (R) can be solved. The optimum temperatures for the range of f from 0 to 0.48 calculated from Eq. (N) exceed the allowable temperature of 870 K. Hence the temperature used for the determination of V/F_{A1} is 870 K for this range. Then analytical solution can be developed by integrating Eq. (R) by the method of partial fractions. The result is

$$\frac{V}{F_{A1}} = \frac{1}{2AB} \ln \frac{A(1 - f) + Bf}{A(1 - f) - BF} \tag{S}$$

where

$$A^2 = \frac{1.615 \times 10^{15}}{T^2} e^{-12,588/T}$$

$$B^2 = \frac{1.615 \times 10^{15}(17.057)}{T^2} e^{-15,105/T}$$

The values of V/F_{A1} for this range are listed in the following table. For the range of $f = 0.48$ to 0.70, the optimum temperatures for each f are first evaluated by Eq. (P), then the optimum rate of reaction is calculated, and the final value of the minimum V/F_{A1} is determined by the Simpson's rule. All these values are also listed in the same table. Lastly, it is interesting to compare the result from above to that achieved by isothermal condition. It is seen that an optimum temperature can be determined to minimize the value of V/F_{A1} in Eq. (S). All the calculations are listed in the other table.

f_A	T^*	r^*	$1/r^*$	V/F_{A1}
0.0	(870)			
0.1	(870)			1.005×10^{-4}
0.2	(870)			2.299×10^{-4}
0.3	(870)			4.111×10^{-4}
0.4	(870)			7.154×10^{-4}
0.48	(870)			1.351×10^{-3}
0.49	849	39.82	0.0251	
0.50	825	27.04	0.0369	
0.51	805	18.34	0.0540	
0.52	785	12.40	0.0806	
0.53	765	8.36	0.1196	
0.54	748	5.628	0.1776	
0.55	730	3.7717	0.2651	
0.56	713	2.5188	0.3970	
0.57	698	1.6772	0.5960	
0.58	683	1.1116	0.8990	
0.59	668	0.7333	1.3637	
0.60	653	0.4812	2.0781	
0.61	640	0.3143	3.1812	
0.62	626	0.2040	4.9008	
0.63	613	0.1316	7.5951	
0.64	600	0.0843	11.8546	
0.65	588	0.0537	18.6226	
0.66	576	0.0339	29.4943	
0.67	565	0.0212	47.0700	
0.68	554	0.01318	75.8200	
0.69	543	0.008106	123.3500	
0.70	532	0.004930	202.8200	

Simpson's rule
$V/F_{A1} = 4.36$

For isothermal reaction, the values of V/F_{A1} from Eq. (S) are:

T	532	543	550	554
V/F_{A1}	12.78	9.57	8.72	9.54

The result is $V/F_{A1} = 8.72$ at 550 K.

8.8-2 Maximum yield

Similar to the cases for batch and continuous stirred-tank reactors, the operating conditions to obtain maximum yield of certain product in a tubular reactor can be developed in a similar manner.

Example 8.8-2 Maximum Yield in a Tubular Reactor (L1)

A gaseous mixture containing 0.53 g mol/liter A and 0.43 g mol/liter B flows to a tubular reactor at a velocity such that the total holding time is 8 min. The following first-order reaction occurs in the reactor

$$A \xrightarrow{k_1} B \xrightarrow{k_2} C$$

where

$$k_1 = 0.535 \times 10^{11} e^{-18{,}000/RT} \text{ min}^{-1} \text{ and } T \text{ in K}$$

$$k_2 = 0.461 \times 10^{18} e^{-30{,}000/RT} \text{ min}^{-1} \text{ and } T \text{ in K}$$

Maximize the concentration of B by finding the optimal temperature distribution in the reactor.

Solution

The problem is best solved by the method of maximum principle. Maximize C_B subject to

$$\frac{dC_A}{dt} = -k_1 C_A \qquad C_A(0) = 0.53 \tag{A}$$

$$\frac{dC_B}{dt} = k_1 C_A - k_2 C_B \qquad C_B(0) = 0.43 \tag{B}$$

By the method of maximum principle, the Hamiltonian function is

$$H = \lambda_1 \frac{dC_A}{dt} + \lambda_2 \frac{dC_B}{dt} = -\lambda_1 k_1 C_A + \lambda_2(k_1 C_A - k_2 C_B)$$

Then

$$\frac{\partial H}{\partial C_A} = -\frac{d\lambda_1}{dt} = -k_1\lambda_1 + k_2\lambda_2 \qquad \frac{d\lambda_1}{dt} = k_1\lambda_1 - k_1\lambda_2 \qquad \lambda_1(T) = 0 \tag{C}$$

$$\frac{\partial H}{\partial C_B} = -\frac{d\lambda_2}{dt} = -k_2\lambda_2 \qquad \frac{d\lambda_2}{dt} = k_2\lambda_2 \qquad \lambda_2(T) = 1 \tag{D}$$

$$\frac{\partial H}{\partial T} = -\lambda_1 C_A \frac{dk_1}{dT} + \lambda_2 C_A \frac{dk_1}{dT} - \lambda_2 C_B \frac{dk_2}{dT} = 0$$

Solving

$$T(t) = \frac{E_1 - E_2}{R \ln\left\{\dfrac{k_{10}E_1}{k_{20}E_2}[C_A(\lambda_2 - \lambda_1)/(C_B\lambda_2)]\right\}} \tag{E}$$

Now we have five equations (A), (B), (C), (D), and (E) to solve for five unknowns C_A, C_B, λ_1, λ_2, and T. The temperature profile at various values of t can be calculated from these five equations. For example, find the optimal temperature at $t = 6.4$ min = distance from inlet divided by linear velocity. Assume $T = 339$ K, $k_1 = 0.1579\ \text{min}^{-1}$ and $k_2 = 0.02799\ \text{min}^{-1}$ from the Arrhenius equation.

From Eq. (A)

$$\frac{dC_A}{dt} = -k_1 C_A \qquad C_A = 0.53e^{-0.1579t} = 0.53e^{-0.1579(6.4)} = 0.1929\ \text{g}\cdot\text{mol/liter}$$

From Eq. (B)

$$\frac{dC_B}{dt} = k_1 C_A - k_2 C_B \qquad C_B = 0.6635\ \text{g}\cdot\text{mol/liter}$$

From Eq. (D)

$$\frac{d\lambda_2}{dt} = 0.02799\lambda_2 \qquad \lambda_2 = e^{-0.02799(6.4)} = 0.9562$$

From Eq. (C)

$$\frac{d\lambda_1}{dt} = 0.1579(\lambda_1 - e^{-0.02799t}) \qquad \lambda_1 = 0.0242$$

Substituting into Eq. (E) gives

$$T = \frac{18{,}000 - 30{,}000}{1.987 \ln \{[0.535 \times 10^{11} 18{,}000/0.461 \times 10^{18}(30{,}000)] [0.1929(0.9562 - 0.0242)/0.6635(0.9562)]\}} = 340.42 \quad \text{against 339 assumed}$$

8.8-3 Minimum total cost

The total cost in operating a tubular reactor consists mainly of two items: the cost of the reactor, including installation, auxiliary equipment, instrumentation, overhead, labor, depreciation, and interest, and the cost of the reactants. Hence if the unit cost of the reactor on an hourly basis is C_v in dollars per cubic meter per hour and C_m is the unit cost of the reactants in dollars per kilogram mole of X, then the total cost is

$$C_T = C_v V + C_m q C_{X0} \tag{8.8-5}$$

Since the reactor volume relates to other variables by the design and the reaction rate equation, and the reactants have some relationship with the product, we can minimize the total cost.

Example 8.8-3 Minimize Total Cost Without Recycle

100 kmol of X with a concentration of 0.02 kmol/m³ reacts with Y to produce Z and W per hour

$$X + Y \rightarrow Z + W \qquad k = 450 \text{ m}^3/\text{kmol} \cdot \text{h}$$

The amount of Z required is 90 kmol/h. No recycle of the unreacted reactant is necessary. Calculate the optimum tubular reactor size and as well as the feed composition. The data are: Y costs $1.50 per kilogram mole, cost of X is negligible, operating cost of the reactor is $0.05 per hour per cubic meter.

Solution

The design equation is

$$\frac{V}{q} = C_{X0} \int_0^f \frac{df}{r} \tag{A}$$

The reaction rate is

$$r = kC_X C_Y \tag{B}$$

But

$$C_X = C_{X0}(1 - f)$$

and

$$C_B = C_{Y0} - C_{X0} f$$

Substituting into Eq. (B) gives

$$r = kC_{X0}^2(1 - f)(M - f) \tag{C}$$

where

$$M = C_{Y0}/C_{X0}$$

With Eq. (C), Eq. (A) becomes

$$\frac{V}{q} = \frac{1}{kC_{X0}} \int_0^f \frac{df}{(1 - f)(M - f)} \tag{D}$$

Integrating by partial fractions yields

$$\frac{V}{F_{X0}} = \frac{1}{kC_{X0}^2(M - 1)} \ln \frac{M - f}{M(1 - f)} \tag{E}$$

where

$$F_{X0} = qC_{X0}$$

Substituting Eq. (E) into Eq. (8.8-5) gives

$$C_T = C_v \frac{F_{X0}}{kC_{X0}^2(M-1)} \ln \frac{M-f}{M(1-f)} + C_m F_{Y0} \tag{F}$$

Inserting the numerical values from the example results in

$$C_T = 0.05 \frac{100}{450(0.0004)(M-1)} \ln \frac{M-0.90}{M(1-0.90)} + 1.50(100M)$$

$$= \frac{27.78}{(M-1)} \ln \frac{M0.9}{0.15M} + 150M \tag{G}$$

Differentiating Eq. (G) with respect to M, equating to zero, and solving for M, or calculating C_T graphically, we get

$$M^* = \frac{C_{Y0}^*}{C_{X0}^*} = 2.179$$

Then the optimal volume can be determined from Eq. (E) as

$$V^* = 833 \text{ m}^3$$

Example 8.8-4 Minimum Total Cost with Recycle

A mixture of X and Y in equal moles reacts in a tubular reactor to produce 800 g mol Z per hour with $C_{Y0} = 0.1$ g mol/liter. Unreacted A leaves the top of the separator with the product Z but unreacted Y is recycled to the reactor. Its recovery is ignored. The reaction is

$$X + Y \rightarrow 2Z \qquad k = 0.002 \text{ liter/g mol} \cdot \text{s}$$

The cost of X is \$0.75 per gram mole, of Y is \$0.002 per gram mole, of producing Z is \$230 per hour, and of operating the reactor is \$80 × 10^{-4} per liter per hour. Find the optimal ratio of C_{Y0}/C_{X0} and the optimum fractional conversion.

Solution

The total cost of this operation is

$$C_T = C_v V + \text{cost of producing Z} + (\text{gmol X wasted per h})C_X$$

The reactor volume required as determined in Example 8.8-3 is

$$V = \frac{F_{X0}}{kC_{X0}^2(M-1)} \ln \frac{M-f}{M(1-f)} \tag{A}$$

$$= \frac{(F_{X0}f)M^2}{fkC_{Y0}^2(M-1)} \ln \frac{M-f}{M(1-f)} \tag{B}$$

Then the total cost is

$$C_T = \left[\frac{F_{X0}f(M)(M)}{fkc_{Y0}C_{Y0}(M-1)} \ln \frac{M-f}{M(1-f)}\right](80)(10)^{-4}$$
$$+ 230 + \left[F_{X0}f\left(\frac{1-f}{f}\right)\right](0.75) \tag{C}$$

But

$$fF_{X0} = 400 \text{ g mol/h} \qquad k = 0.002 \text{ liter/g mol}\cdot\text{s} = 7.2 \text{ liters/g mole}\cdot\text{h}$$

$$C_{X0} = 0.1 \text{ g mol/liter}$$

Substituting and simplifying

$$C_T = \frac{44.44M^2}{f(M-1)} \ln \frac{M-f}{M(1-f)} + 230 + 300\left(\frac{1-f}{f}\right) \tag{D}$$

By trial and error it is found that the optimum occurs at

$$M^* = 1.13 \qquad f^* = 0.74 \qquad \text{and} \qquad C_T^* = 502.48$$

8.8-4 Maximum profit

As in Section 7.12-5, the profit in operating a reactor to carry out a certain kind of reaction can be obtained as

$$\text{Profit} = \text{sales of product} - \text{cost of operation} \tag{8.8-6}$$

Then the profit can be maximized.

Example 8.8-5 Maximum Profit

Compound A with a concentration of C_{Xo} = 2.5 kmol/m³ at a flow rate of 12 m³/min is converted to Y in a tubular reactor. The value of the product Y is \$1.5 per kilogram mole of Y. The cost of operation is \$2.50 per minute per cubic meter. The reaction rate constant is 30 min⁻¹. Find the maximum profit.

Solution

The design equation is

$$\frac{V}{q} = -\int_{C_{X0}}^{C_X} \frac{dC_X}{r} \tag{A}$$

Since

$$r = kC_X$$

Eq. (A) after integrating and simplifying becomes

$$C_X = C_{X0}e^{-kV/q} \tag{B}$$

The profit per minute

$$P = \$1.5(12)(2.5 - C_X) - \$2.5V \tag{C}$$

Substituting Eq. (B) into Eq. (C) gives

$$P = 18(2.5 - C_{X0}e^{-kV/q}) - 2.5V \tag{D}$$

Putting the numerical values into Eq. (D) yields

$$P = 45(1 - e^{-2.5V}) - 2.5V \tag{E}$$

For maximum P, the derivative of P with respect to V should be zero.

$$\frac{\partial P}{\partial V} = -45e^{-2.5V}(-2.5) - 2.5 = 0 \tag{F}$$

Solving for V gives

$$V^* = 1.5226 \text{ m}^3$$

Substituting into Eq. (B) yields

$$C_X^* = 0.0555 \text{ kmol/m}^3$$

Substituting into Eq. (E) yields

$$P^* = \$40.19$$

The percent yield is

$$\% \text{ yield} = \frac{12(2.5 - 0.0555)}{12(2.5)} = 97.78\%$$

Example 8.8-6 Maximum Profit with Recycle

Solve Example 7.12-7 by using a tubular reactor to replace the CSTR.

Solution

The design equation is

$$\frac{V}{q} = -\int_{C_{Xi}}^{C_{Xf}} \frac{dC_X}{r} \tag{A}$$

But

$$r = kC_X$$

Equation (A) after integrating and simplifying gives

$$\frac{C_{Xf}}{C_{Xi}} = e^{-kV/q} \tag{B}$$

or

$$\frac{n_{Af}}{n_{Xi}} = e^{-kV/q}$$

From Example 7.12-7

$$n_{Xi} = n_{X0} + \frac{1}{2} n_{Xf}$$

Therefore

$$n_{Xf} = \left(30 + \frac{1}{2} n_{Xf}\right) e^{-30V/12}$$

or

$$n_{Xf} = \frac{30e^{-2.5V}}{1 - 0.5e^{-2.5V}} \tag{C}$$

$$\text{kmoles X reacted} = n_{Xi} - n_{Xf} = n_{X0} + \frac{1}{2} n_{Xf} - n_{Xf} = n_{X0} - \frac{1}{2} n_{Xf} \tag{D}$$

$$\text{kmoles Y formed} = n_{X0} - \frac{1}{2} n_{Xf}$$

$$P = 1.5\left(n_{X0} - \frac{1}{2} n_{Xf}\right) - 2.5V$$

$$= 1.5\left[30 - \frac{1}{2}\left(\frac{30e^{-2.5V}}{1 - 0.5e^{-2.5V}}\right)\right] - 2.5V \tag{E}$$

$$= 45 - 2.5V - \frac{6.25}{e^{2.5V} - 0.5} \tag{F}$$

Differentiating P with respect to V, equating to zero, and solving for V gives

$$V^* = 1.2626 \text{ m}^3$$

Substituting into Eq. (C) yields

$$n^*_{Xf} = 1.305 \text{ kmol}$$

The yield is then

$$\text{Yield} = \frac{30 - \frac{1}{2}(1.305)}{30} = 97.82\%$$

Problems

1. For the following first order, irreversible andothermic reaction

$$4PH_3 \rightarrow P_4 + 6H_2$$

the rate constant is

$$\log k = -\frac{18963}{T} + 2 \log T + 12.13$$

where k is in s^{-1} and T in K; 25 kg/h of phosphine enters a tubular reactor operating at 1 atm and 945 K. The highest temperature allowable in the reactor is 680°C, at which the phosphine is a vapor. Determine the size of the reactor for 10 percent conversion (a) isothermally at 680°C, (b) adiabatically with an inlet temperature of 680°C.

Data: Heat of reaction = 5,665 cal/g mol phosphine

Molal heat capacities: T in K.

P_4 : $C_p = 5.9 + 0.0096T$

PH_3: $C_p = 6.7 + 0.0063T$

H_2 : $C_p = 6.424 + 0.1039 \times 10^{-2}T - 0.007804 \times 10^{-5}T^2$

2. 0.5 kmol steam per kilogram mole of butadiene is polymerized reversibly to dimer in a noncatalytic tubular reactor operating at 640°C and 1 atm. The forward rate constant is

$$\log k = -\frac{5{,}470}{T} + 8.063$$

where k is in gmole/liter · h · atm². The equilibrium constant at 640°C is 1.27. Determine the length of 10-cm ID tube for 40 percent conversion of the butadiene with a total feed rate of 9 kmole/h.

3. For Problem 9 in Chapter 7, add the following reactor: A plug-flow (ideal tubular-flow) reactor with a volume of 280 liters.

4. The pyrolysis of gaseous acetone at 520°C and 1 atm is carried out in a tubular-flow reactor of 85 cm long and 4 cm ID.

$$CH_3COCH_3 \rightarrow CH_2{=}C{=}O + CH_4$$

The rate constant is

$$\ln k = 34.34 - \frac{68000}{RT}$$

where k is in s^{-1} and T in K. Determine the flow rate in g/h for a conversion of 35 percent.

5. The reaction rate for the decomposition of acetaldehyde

$$CH_3CHO \rightarrow CH_4 + CO$$

is (C1)

$$r = 0.00033C^2 e^{31.71-(25,080/T)} \text{ g mol/cm}^3 \cdot \text{s}$$

and the heat of reaction at 518°C is

$$\Delta H = -4.55 \text{ kcal/g mol}$$

1 kmol/h pure CH_3CHO at 780 K enters a tubular reactor which is operated at 1 atm. Determine the volume of the reactor for 50 percent conversion under adiabatic conditions. Pressure drop is neglected.

Data: Heat capacities C_p with T in K

CH_3CHO: $C_p = 4.19 + 3.164 \times 10^{-2}T - 0.515 \times 10^{-5}T^2 - 3.8 \times 10^{-9}T^3$

CH_4 : $C_p = 4.75 + 1.2 \times 10^{-2}T + 0.303 \times 10^{-5}T^2 - 2.63 \times 10^{-9}T^3$

CO : $C_p = 6.726 + 0.04 \times 10^{-2}T + 0.1283 \times 10^{-5}T^2 - 0.5307 \times 10^{-9}T^3$

6. It has been shown in Example 8.6-1 that 4,239 cm of 10-cm diameter tube would be suitable for the conversion under laminar-flow conditions of a throughput of 10,089 cm³/min containing 0.001744 g mol/cm³ of reactant. For the given velocity constant of 17.4 cm³/g mole · min, evaluate the point fractional conversions (a) on the tube axis at 2,119 cm and 4,239 cm, (b) at the reactor exit at radii 1.25, 2.5, 3.75 cm (c) at the coordinate L = 2,119 cm, r = 2.5 cm.

7. 100 kmol B per hour is produced in a tubular reactor into which an aqueous feed of A at a concentration of 1 kmol/m³ enters. A first-order reaction A → B takes place ($k = 2\ h^{-1}$). Unused A is discarded. Find the volume of the reactor, fractional conversion, initial molar feed rate of the reactant, and unit cost of producing B at optimum condition. The data are

Cost of reactant stream = \$0.40 per kmole A

Cost of reactor including installation etc = \$0.20/m³ · h

8. Aqueous feed at a concentration of 1 kmol/m³ valued at \$1.00 per kilogram mole of A enters at a rate of 1000 m³/h to a tubular reactor in which a first-order reaction takes place

$$A \longrightarrow B \qquad k = 2\ h^{-1}$$

The product B can be sold at a price of \$1.60 per kilogram mole of B. The reactor including installation, etc., costs \$0.20 per cubic meter per hour. Unused A is discarded. Find the volume of the reactor, fractional conversion, molar flow rate of product B, and hourly profit for obtaining maximum profit.

9. Aqueous A at C_{Ao} = 1 kmol/m³ (cost of A is \$1.00 per kilogram mole) enters a tubular reactor of 1 m³ to produce B (worth \$4.00 per kilogram mole) by the following reaction

$$A \longrightarrow B \qquad r = 0.2C_A^2$$

Determine the most profitable operating conditions if:
(a) Unreacted A is discarded.
(b) Unreacted A is separated from the reactor's product stream, concentrated to 1 kmol/m³ and then reused at a cost of \$0.375 per kilogram mole recycled.
(c) Same as (b) but the cost is negligible.

10. 3 kmol/m³ X enters a tubular reactor in which the following reaction

$$X \rightleftarrows 2Y$$

takes place. Find the temperature at which the reaction rate is a maximum for 50% conversion. The data are

Equilibrium constant $K = 100$ at 300 K

Heat of reaction $= -2.5 \times 10^7$ J at 300 K

The forward rate constant k in s^{-1} is $\ln k = 20.7233 - 10000/T$

References

(A1). Aris, R. *Elementary Chemical Reactor Analysis.* Englewood Cliffs, N.J.: Prentice-Hall, Inc., 1969.

(B1). Bilous, O., and Amundson, N.R. *Chem. Eng. Sci.,* 5:81,115 (1956).

(C1). Cooper, A.R., and Jeffreys, G.V. *Chemical Kinetics and Reactor Design.* Englewood Cliffs, N.J.: Prentice-Hall, Inc., 1973.

(L1). Lee, E.S. *A.I.Ch.E. J.,* 10:3, 309 (1964).

(L2). Leyes, C.E., and Othmer, D.F., *Ind. Eng. Chem.,* 37:968 (1945).

CHAPTER NINE

Heterogeneous Reactors

9.1 *Introduction*

In the previous chapters, we were concerned about the design of the reactors for a single phase. However, many industrial processes involve reactants in more than one phase, which are often called *heterogeneous-phase processes.* These reactions are complicated by the fact that before substances in different phases can react, they must move to at least the interface. Hence, in addition to affecting the chemical reaction, physical transport processes between phases also affect the overall rate of the heterogeneous reaction. In this chapter, the types of heterogeneous reactions are classified, and the types of reactors to carry out such reactions are described. Since most such reactions involve catalysts, the phenomenon of the interaction of a gas with a catalyst is then investigated. Next, the mechanism of each phenomenon is developed. Finally, the design of different types of the heterogeneous reactors is illustrated with examples.

9.2 *Types of Heterogeneous Reactions*

In industry, there are many processes involving heterogeneous reactions. These can be classified as follows:

1. Gas-solid catalytic reactions: Some important industrial processes employing solid catalysts are the manufacture of gasoline by cracking hydrocarbons in the presence of alumina-silica, the production of ethylbenzene by the same catalyst, the hydrogenation by promoted iron oxide to produce styrene, the manufacture of methanol by hydrogenation in the presence of zinc chromate, the production of phthalic anhydride by oxidation with vanadium pentoxide, and the manufacture of sulfuric acid by oxidation with platinum or vanadium pentoxide.
2. Gas-solid noncatalytic reactions: Some industrial uncatalyzed reactions are: the combustion of coal, the manufacture of hydrogen by action of steam on iron, and the reaction of chlorine or uranium oxide to recover volatile uranium chloride.
3. Liquid-solid reactions: Some important industrial processes are the leaching of ura-

nium ores with sulfuric acid, the manufacture of acetylene by the action of water on calcium carbide and the hydration of lime.

4. Liquid-liquid reactions: Examples include the manufacture of caustic soda by reaction of sodium amalgam and water, and the formation of soaps by action of aqueous alkaline on fats or fatty acids.

5. Solid-solid reactions: Examples are the manufacture of cement and the production of calcium carbide by action of lime and carbon.

6. Gas-liquid reactions: Some important processes are the manufacture of nitric acid by absorption of nitric oxide in water, the hydrogenation of vegetable oils with gaseous hydrogen, the production of sodium thiosulfate by action of sulfur dioxide on aqueous sodium carbonate and sodium sulfide.

9.3 *Types of Heterogeneous Reactors*

In industry, there are two major types of heterogeneous reactors:

1. Solid catalysts remain in fixed position: Among this type, we have
   (a) Fixed bed: Sometimes this is called a *packed-bed reactor.* It consists of many cylindrical tubes filled with catalyst pellets. Reactants flow from the bottom of the reactor through the catalyst bed and are converted into products which flow out the top of the reactor.
   (b) Trickle bed: This is a fixed bed of catalyst pellets with both gas and liquid flow. The liquid flows downward over a bed of catalyst pellets while gas simultaneously passes either upward or downward through the bed. The liquid wets the pellets and flows in thin layers from pellet to pellet and the gas flows through the remaining voids.
   (c) Moving bed: In this type, a fluid phase passes upward through a packed bed

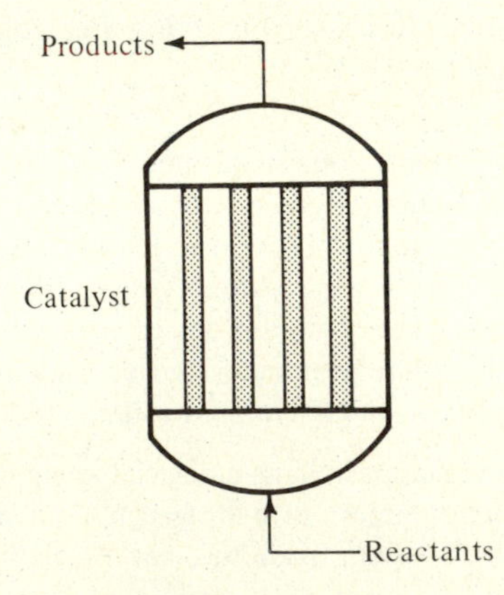

Figure 9.3-1 Fixed bed

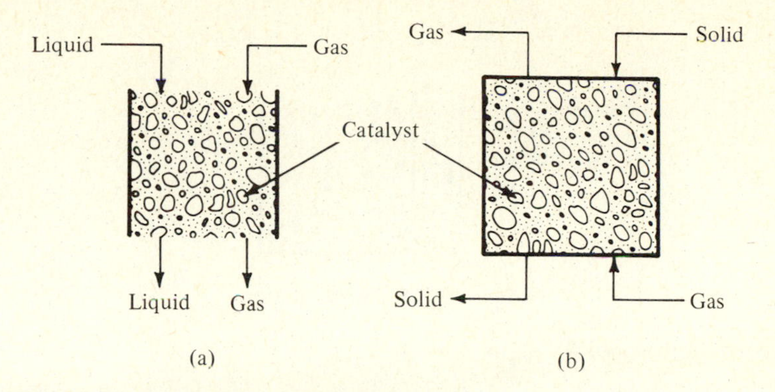

Figure 9.3-2 (a) Trickle bed; (b) moving bed

of catalyst pellets. Solid is fed to the top of bed, moves downward under the influence of gravity in a manner approximating plug flow, and is removed from the bottom. The catalyst pellets are then continuously transferred to the top of the reactor in external equipment by pneumatic or mechanical means. During the period when the catalyst pellets are not in the reactor proper, they may be regenerated or reconditioned in an auxiliary facility.

2. Particles suspended in a fluid: There are two types:
   (a) Fluidized bed: Usually this consists of a reactor and a generator. The reactants enter the bottom of the reactor, which is filled with solid catalyst. During the reaction, products leave the top of the reactor and the spent catalyst leaves the bottom of the reactor and travels to a generator in which the impurities in the spent catalyst are burned off as combustible gas by introducing fresh air to the bottom of the generator. The combustible gas flows from the top of the generator and the reactivated catalyst flows by gravity to the reactor for further use.

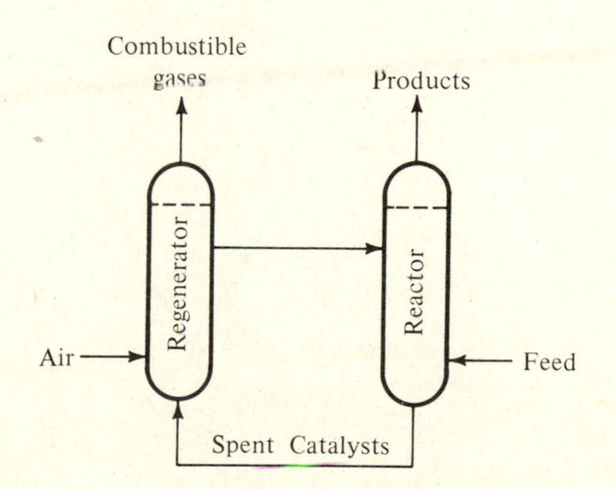

Figure 9.3-3 Fluidized bed

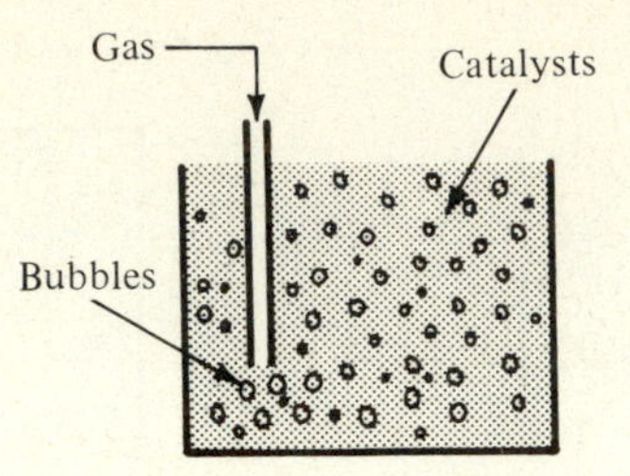

Figure 9.3-4 Slurry reactor

(b) Slurry reactor: Gas is bubbled into and dissolved in the liquid in which catalyst particles are suspended.

9.4 The Phenomenon of Catalytic Reactions

There are seven steps involved in a gas reacting on the surface of a catalyst particle.

1. The molecule of X_1 must flow through a boundary layer from the bulk stream to the exterior surface of the particle. If the concentration of X_1 in the bulk phase is denoted by x, the concentration at the surface would be x_s. The rate of mass transfer is then $k_{cx}(x - x_s)$ in moles per unit time per unit external surface area where k_{cx} is the mass transfer coefficient.
2. Internal diffusion of x within the porous pellet takes place. Let r be position within the catalyst particle, the concentration of X within the particle is then $x(r)$, a function of the position r, and obeys the partial differential equation for diffusion, that is, $x(r) = x_s$ when r is the position on the exterior surface. x is expressed in moles per cubic meter.

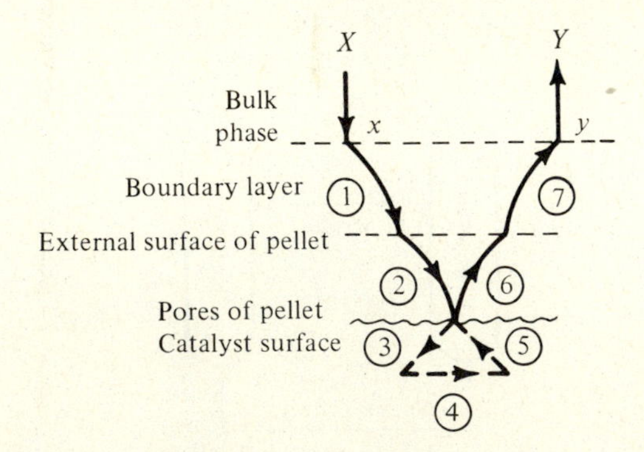

Figure 9.4-1 Catalytic reaction

3. When the gas molecule X reaches the surface, it is adsorbed on the surface. The concentration is denoted by $x(r)$ and the rate of adsorption r_{ad}. $x(r)$ is expressed in moles per unit surface area and r_{ad} in moles per unit time per unit surface area.

4. After the gas molecule X is absorbed, reaction occurs on the catalyst surface to produce Y.

5. Desorption of the product Y from the catalyst into the porous interior of the pellet.

6. Diffusion of the product within the pellet to the external surface.

7. Transfer of product from the external surface back to the main stream of the reaction mixture at a rate $k_{cy}(y_s - y)$.

9.5 *Adsorption and Desorption*

When a gas enters a bed packed with solid catalyst, adsorption occurs in which the gas molecules fill the void space within the solid. Part of the molecules will be attached to the surface of the catalyst. The process is called *adsorption.* The catalyst is named the *adsorbent* and the gas *absorbate.* There are two kinds of adsorptions: (a) Physical adsorption in which the interaction between the adsorbed molecules and the solid is weak and (b) Chemical adsorption in which the interaction between the gas and the surface is strong.

All physical adsorption and most chemical adsorptions are exothermic. In this chemisorption, there are two kinds of isotherms:

Langmuir adsorption isotherm

1. Single species: Langmuir assumed that the usual mass-action laws could describe the individual steps. Let σ be an adsorption site. The reaction is

$$X + \sigma = X\sigma \tag{9.5-1}$$

where $X\sigma$ represents adsorbed X. The rates are:

$$r_a = k_a C_X C_\sigma \qquad k_a = A_a e^{-E_a/RT} \tag{9.5-2}$$

$$r_d = k_d C_{X\sigma} \qquad k_d = A_d e^{-E_d/RT} \tag{9.5-3}$$

where C_X and $C_{X\sigma}$ are surface concentrations in kilogram moles per kilogram of catalyst. The total sites are either vacant or contain adsorbed X:

$$C_t = C_\sigma + C_{X\sigma} \tag{9.5-4}$$

At equilibrium, $r_a = r_d$. Therefore

$$k_a C_X C_\sigma = k_d C_{X\sigma} \tag{9.5-5}$$

Thus the amount adsorbed is

$$C_{X\sigma} = \frac{k_a C_X C_\sigma}{k_d} = \frac{C_t K_X C_X}{1 + K_X C_X} \tag{9.5-6}$$

where $K_X = k_a/k_d$ is the adsorption equilibrium constant. Equation (9.5-6) may also be expressed as

$$\begin{aligned} \theta &= \text{fractional coverage} \\ &= \frac{C_{X\sigma}}{C_t} = \frac{K_X C_X}{(1 + K_X C_X)} \end{aligned} \tag{9.5-7}$$

2. Two species:

$$X + \sigma = X\sigma \tag{9.5-8}$$

$$Y + \sigma = Y\sigma \tag{9.5-9}$$

$$\frac{dC_{X\sigma}}{dt} = k_{aX} C_X C_\sigma - k_{dX} C_{X\sigma} \tag{9.5-10}$$

$$\frac{dC_{Y\sigma}}{dt} = k_{aY} C_Y C_\sigma - k_{dY} C_{Y\sigma} \tag{9.5-11}$$

$$C_t = C_\sigma + C_{X\sigma} + C_{Y\sigma} \tag{9.5-12}$$

At equilibrium:

$$C_{X\sigma} = K_X C_X C_\sigma \tag{9.5-13}$$

$$C_{Y\sigma} = K_Y C_Y C_\sigma \tag{9.5-14}$$

$$C_t = C_\sigma + K_X C_X C_\sigma + K_Y C_Y C_\sigma \tag{9.5-15}$$

and

$$C_\sigma = \frac{C_t}{1 + K_X C_X + K_Y C_Y} \tag{9.5-16}$$

Thus the adsorbed amounts are

$$C_{i\sigma} = \frac{C_i K_i C_t}{1 + K_X C_X + K_Y C_Y} \tag{9.5-17}$$

3. Dissociated molecules on adsorption:

$$X_2 + 2\sigma = 2X\sigma \tag{9.5-18}$$

At equilibrium

$$C_{X\sigma}^2 = K_X C_{X_2} C_\sigma^2 \tag{9.5-19}$$

Then

$$C_t = C_\sigma + \sqrt{K_X C_{X_2}}\, C_\sigma \tag{9.5-20}$$

and

$$C_{X\sigma} = \frac{C_t \sqrt{K_X C_{X_2}}}{1 + \sqrt{K_X C_{X_2}}} \tag{9.5-21}$$

"BET" (Brunauer-Emmett-Teller) adsorption isotherm

This is for a nonuniform surface.

$$\frac{x}{v(1-x)} = \frac{1}{v_m c} + \left(\frac{c-1}{v_m c}\right) x$$

where x = the normalized pressure (P/P_o)
P_o = the saturation pressure
v = the volume of the gas actually absorbed
v_m = the volume which would be adsorbed in a monolayer
c = a constant exponentially related to the heats of adsorption of the first layer and the heat of liquefaction

9.6 Intrinsic Rate Equation

Steps 3, 4, and 5 in the section "The Phenomenon of Catalytic Reactions" can be grouped together to obtain an intrinsic rate equation. For simplicity, let us illustrate a simple reversible reaction

$$X \rightleftarrows Y \tag{9.6-1}$$

For the adsorption, it is assumed that

$$X + \sigma = X\sigma \tag{9.6-2}$$

where σ represents a vacant site. Then the rate of adsorption is

$$r_a = k_X\left(C_X C_\sigma - \frac{C_{X\sigma}}{K_X}\right) \tag{9.6-3}$$

where k_X = the chemisorption rate constant
C_σ = the concentration of vacant site
$C_{X\sigma}$ = the concentration of chemisorbed X
K_X = the adsorption equilibrium constant

For the surface reaction, the mechanism is

$$X\sigma = Y\sigma \tag{9.6-4}$$

Assuming that both reactions are first-order, the net rate of the surface reaction of $X\sigma$ is

$$r_s = k_s\left(C_{X\sigma} - \frac{C_{Y\sigma}}{K_s}\right) \tag{9.6-5}$$

where k_s = the surface reaction rate constant

K_s = the surface reaction equilibrium constant

For the desorption, the mechanism is

$$Y\sigma = Y + \sigma \tag{9.6-6}$$

and the desorption rate is

$$r_d = k'_Y\left(C_{Y\sigma} - \frac{C_Y \cdot C_\sigma}{K_d}\right)$$

or

$$r_d = k_Y\left(\frac{C_{Y\sigma}}{K_Y} - C_Y C_\sigma\right) \tag{9.6-7}$$

where k_Y = the desorption rate constant

K_Y = the reciprocal of the desorption equilibrium constant = $1/K_d$

The overall equilibrium constant for these three steps is

$$K = \frac{K_X K_s}{K_Y} \tag{9.6-8}$$

The total number of sites C_t is the sum of the vacant and the occupied sites

$$C_t = C_\sigma + C_{X\sigma} + C_{Y\sigma} \tag{9.6-9}$$

The rates of change of the various species are

$$\frac{dC_X}{dt} = -r_a \tag{9.6-10}$$

$$\frac{dC_{X\sigma}}{dt} = r_a - r_s \tag{9.6-11}$$

$$\frac{dC_{Y\sigma}}{dt} = r_s - r_d \tag{9.6-12}$$

$$\frac{dC_Y}{dt} = r_d \tag{9.6-13}$$

Thus, a steady-state approximation of the middle two equations indicates·

$$r_a = r_s = r_d = r_X \tag{9.6-14}$$

Eliminating C_σ, $C_{X\sigma}$, and $C_{Y\sigma}$ from Eqs. (9.6-3), (9.6-5), and (9.6-7) in terms of the fluid-phase concentrations C_X and C_Y yields

$$r_X = \frac{C_t(C_X - C_Y/K)}{\left(\frac{1}{K_X k_s} + \frac{1}{k_X} + \frac{1}{Kk_Y}\right) + \left(\frac{1}{K_X k_s} + \frac{1+K_s}{Kk_Y}\right)K_X C_X + \left(\frac{1}{K_X k_s} + \frac{1+K_s}{K_s k_X}\right)K_Y C_Y} \tag{9.6-15}$$

Surface reaction rate controlling: k_X, $k_Y \gg k_s$. Then

$$r_X = \frac{K_X k_s C_t(C_X - C_Y/K)}{1 + K_X C_X + K_Y C_Y} \tag{9.6-16}$$

Adsorption of X controlling: k_Y, $k_s \gg k_X$. Then

$$r_X = \frac{k_X C_t(C_X - C_Y/K)}{1 + (1 + 1/K_s)K_Y C_Y} = \frac{k_X C_t(C_X - C_Y/K)}{1 + (K_X/K)C_Y + K_Y C_Y} \tag{9.6-17}$$

Desorption of Y controlling: k_X, $k_s \gg k_Y$. Then

$$r_X = \frac{Kk_Y C_t(C_X - C_Y/K)}{1 + K_X C_X + K_X C_X k_s} \tag{9.6-18}$$

9.7 *Solid Catalysts*

A catalyst is a substance that accelerates or slows the rate of a reaction without itself being consumed. It also directs the reaction toward a particular product, eliminates unwanted side reactions, and initiates a new reaction. Industrial catalysts are seldom pure substances. Often the active constituent is mixed with other substances of varying catalytic activity to improve its effectiveness, to increase the available surface, to stabilize against crystal growth and sintering, to create favorable orientation of surface molecules, and to improve mechanical strength. The substance added to the active constituent, or additive, which possesses no catalytic ability of its own but enhances the activity of a catalyst, is called a *promoter.* Substances present in small amounts which reduce the activity are called *poisons.* During reaction, some substances are added to the system to maintain the activity of the catalyst by decreasing poisoning. These are called *accelerators.* Those additives that can reduce the activity are called *inhibitors.* To function effectively, the catalyst frequently must be deposited on other substances of adequate mechanical strength but with no catalytic effects of their own. These substances are called *carriers,* and they can increase the effective surface of the catalyst, improve the stability, reduce sensitivity to poisons, and prevent local overheating which might cause sintering.

A good industrial solid catalyst should possess the following characteristics: long life, low cost, ease of regeneration, capability of being manufactured and reproduced in large batches, resistance to poisons, strong physical strength and resistance to thermal shock, intrinsic activity and selectivity, stable structure, appropriate pore size distribution, and proper fluid-flow characteristics.

Carriers are used as a framework on which the catalyst is deposited. They may be highly porous so that the effective surface of the catalyst can be increased by spreading out the catalyst in a thin film. They may improve stability by keeping crystals of the catalyst far enough apart to prevent their fusing together. Carriers can prevent local overheating.

Catalysts may be classified as (1) metals (conductors) which are commonly used in hydrogenation, including ammonia synthesis and Fisher-Tropsch reactions; in dehydrogenation; in hydrogenolysis, etc. (2) Metallic oxides and sulfides (semiconductors) which are usually employed in oxidation (including contact SO_2 processes), reduction, dehydrogenation, cyclization (hydrogenation). (3) Salts and acid-site catalysts (usually insulators) which are applied in polymerization, isomerization, cracking, dehydration, alkylation, hydrogen transfer, halogenation, dehalogenation, etc.

The usual steps in the manufacture of a catalyst are (1) preparation of the active constituent, (2) depositing it on a carrier, (3) activation. The first two steps can be carried out by the following methods:

1. Precipitation—This involves soaking the carrier with a solution and then adding a chemical agent to precipitate the desirable constituents on the surface of the carrier.
2. Impregnation—Active constituent is usually deposited by soaking or spraying the carrier with the solution of the catalyst.
3. Thermal decomposition—Metallic nitrates, carbonates, hydrates, formates, oxalates,

and acetates decompose upon moderate heating to yield volatile products and metallic oxides or metals which are the active constituents of a catalyst and a porous material.

4. Leaching—For example, catalyst Ni can be obtained as a highly porous metallic catalyst from Ni-Al alloy by treating with alkali to remove the aluminum.
5. Thermal fusion—Components, including promoters, are melted in an electric furnace.

The last step, activation of the surface, is necessary for producing a good catalyst. This operation will remove adsorbed or deposited foreign material and will change the physical or chemical nature in some way so as to increase the activity. The procedure, in general, involves (1) moderate heating to drive off the adsorbed poisons, (2) calcination, (3) treatment with chemicals, and (4) reduction with hydrogen or treatment with oxygen, hydrogen sulfide, carbon monoxide, or chlorinated hydrocarbons.

Prior to describing the design of catalytic reactor, it is essential to know the physical properties of the bed.

1. Specific surface S: It is defined as the external surface area per unit volume.
   (a) Spherical particle: If the diameter of the spherical particle is d_p, the specific surface can be calculated by

$$S = \frac{\pi d_p^2}{\pi d_p^3/6} = \frac{6}{d_p} \tag{9.7-1}$$

   (b) Cylindrical particle: If the diameter and the weight of the cylinder are d_c and h_c, respectively, then

$$S = \frac{\pi d_c h_c + 2\pi d_c^2/4}{\pi d_c^2 h_c/4} = \frac{(4 + 2d_c/h_c)}{d_c} \tag{9.7-2}$$

2. Surface area per unit mass of particle S_g or A_m
   (a) Spherical particle: $S_g = S/\rho_p = 6/(d_p \rho_p)$ **(9.7-3)**
   (b) Cylindrical particle: $S_g = S/\rho_p = (4 + 2d_c/h_c)/(d_c \rho_p)$ **(9.7-4)**
3. Densities:
   (a) Particle density ρ_p: mass per unit volume of a particle.
   (b) Solid density ρ_s: mass per unit volume of solid phase (skeletal density)
   (c) Bulk density ρ_b: mass of catalyst per unit volume of bed.
4. Void fraction or porosity of the particle ϵ_p

$$\epsilon = \frac{\text{void or pore volume of particle}}{\text{total volume of particle}} = \frac{m_p V_g}{m_p V_g + m_p(1/\rho_s)} \tag{9.7-5}$$

$$= \frac{V_g \rho_s}{1 + V_g \rho_s}$$

where V_g = the void volume per gram of particles

m_p = the mass of the particles

or

$$\epsilon_p = \frac{\text{void volume}}{\text{total volume}} = \frac{V_g}{1/\rho_p} = \rho_p V_g \tag{9.7-6}$$

5. Average pore radius **r**: If n_p is the number of pores per particle, and **L** is the average pore length, then

$$\text{Void volume per particle} = m_p V_g = n_p(\pi \mathbf{r}^2 \mathbf{L}) \tag{9.7-7}$$

$$\text{Surface area per particle} = m_p S_g = n_p(2\pi \mathbf{r} \mathbf{L}) \tag{9.7-8}$$

Dividing gives

$$\mathbf{r} = \frac{2V_g}{S_g} \tag{9.7-9}$$

where V_g = the void volume per gram of particles

m_p = the mass of the particles

6. Pore length **L**: If V_p is the gross geometric volume of particle, then

$$\epsilon_p = \frac{m_p V_g}{V_p} = \frac{n_p \pi \mathbf{r}^2 \mathbf{L}}{\mathbf{L} S_x} \tag{9.7-10}$$

where S_x is the geometric surface area. Substituting Eq. (9.7-7) into Eq. (9.7-10) gives

$$\frac{n_p(\pi \mathbf{r}^2 \mathbf{L})}{V_p} = \frac{n_p \pi \mathbf{r}^2}{S_x} \tag{9.7-11}$$

or

$$\mathbf{L} = \frac{V_p}{S_x} \tag{9.7-12}$$

For a spherical catalyst

$$\mathbf{L} = \frac{(4/3)\pi R^3}{4\pi R^2} = \frac{R}{3} \tag{9.7-13}$$

Solving Eq. (9.7-10) for n_p and using Eq. (9.7-9) to eliminate **r** yields

$$n_p = \frac{\epsilon_p S_x S_g^2}{4\pi V_g^2} \tag{9.7-14}$$

7. Superficial velocity V_∞: If V_i is the interstitial velocity, then

$$V_\infty = \epsilon V_i \tag{9.7-15}$$

V_i may be called the fluid velocity through the bed.

Catalysts become poisoned when they meet impurities deleterious to their activity. The reactant participating in the desired reaction must now be moved to the unpoisoned part of the surface before any further reaction can take place. The net effect is to decrease the reaction surface and to increase the average distance over which the reactants must diffuse prior to reaction at the surface. There are two types of poisoning.

1. Homogeneous poisoning in which the active surface is evenly poisoned. The ratio of activity F of the poisoned sphere to the unpoisoned portion will be equal to the ratio of the reactant fluxes

$$F = \frac{\left(\dfrac{dC}{dr}\right)'_{r=r_o}}{\left(\dfrac{dC}{dr}\right)_{r=r_o}} \tag{9.7-16}$$

where prime is for the poisoned portion. It will be seen in Eq. (9.9-10) that the concentration of reactant is a function of the distance the reactant has traveled through the sphere is

$$C_A = C_\infty \frac{\sinh(\lambda r)}{\sinh(\lambda r_o)} \tag{9.9-10}$$

where $\lambda = (k/D_e)^{1/2}$. If the sphere were poisoned, the activity would be $k(1-\omega)$ rather than k where k is the first-order constant and ω is the fraction of the active surface of the porous sphere-shaped catalyst pellets being poisoned. Then Eq. (9.9-10) becomes

$$C_A = C_\infty \frac{\sinh(\lambda' r)}{\sinh(\lambda' r_o)} \tag{9.7-17}$$

where $\lambda' = [k(1-\omega)/D_e]^{1/2}$. Evaluating the respective fluxes at $r = r_o$ gives

$$F = \frac{\sqrt{(1-\omega)}\coth[\phi\sqrt{(1-\omega)}]}{\coth(\phi)} \qquad \text{where } \phi = r_o(k/D_e)^{1/2} \tag{9.7-18}$$

2. Selective poisoning in which the exterior surface of the catalyst is first poisoned and then progressively moved towards the center of the catalyst. Assuming that the reaction is diffusion limited, the flux of reactant past the boundary between poisoned and unpoisoned surfaces in the steady state is equal to the chemical reaction rate.

$$\text{Flux of reactant at the boundary between poisoned and unpoisoned portion} = D_e \frac{C_\infty - C_{ro}}{\omega r_o} \tag{9.7-19}$$

$$\text{Reaction rate in unpoisoned portion } (1-\omega)r_o = D_e\left(\frac{dC}{dr}\right)_{r=(1-\omega)r_o} \tag{9.7-20}$$

Since the concentration profile in the unpoisoned portion is

$$C = C_{ro}\frac{\sinh(\lambda r)}{\sinh[\lambda(1-\omega)r_o]} \tag{9.7-21}$$

Therefore

$$D_e\left(\frac{dC}{dr}\right)_{r=(1-\omega)r_o} = \frac{D_e}{r_o} C_{ro}\phi \coth[\phi(1-\omega)] \tag{9.7-22}$$

where

$$\phi = \lambda r_o = r_o(k/D_e)^{1/2}$$

In the steady-state condition, equating Eqs. (9.7-19) and (9.7-22) and solving for C_{ro} gives

$$C_{ro} = \frac{C_\infty}{1 + \phi\omega \coth[\phi(1-\omega)]} \tag{9.7-23}$$

Thus the reaction rate in the partially poisoned sphere is

$$\frac{D_e(C_\infty - C_{ro})}{r_o} = \frac{C_\infty D_e}{r_o}\,\frac{\phi \coth[\phi(1-\omega)]}{1 + \phi\omega \coth[\phi(1-\omega)]} \tag{9.7-24}$$

and the reaction rate in the unpoisoned sphere is

$$\left(\frac{C_\infty D_e}{r_o}\right)\phi \coth \phi \tag{9.7-25}$$

Hence

$$F = \frac{\coth[\phi(1-\omega)]}{1 + \phi\omega \coth[\phi(1-\omega)]}\left(\frac{1}{\coth \phi}\right) \tag{9.7-26}$$

9.8 *Internal Diffusion*

When the mean free path of a molecule is small compared with the diameter of the pore, *ordinary or bulk diffusion* occurs. The rate of mass transfer in the pore can be predicted by Fick's law

$$N_A = -D_{AB}\frac{dC_A}{dz} \tag{9.8-1}$$

where N_A is the molar flux of A in moles per unit pore cross section per unit time, and z is the diffusion length along the pore. The diffusivity D_{AB} is the ordinary fluid molecular diffusivity which can be estimated by the equation of Chen (C1)

$$D_{12} = \frac{0.43(T/100)^{1.81}[(1/M_1) + (1/M_2)]^{0.5}}{P(T_{c_1}T_{c_2}/10{,}000)^{0.1405}[(V_{c_1}/100)^{0.4} + (V_{c_2}/100)^{0.4}]^2} \tag{9.8-2}$$

where D_{12} = expressed in square centimeters per second
P = the pressure of the system in atmospheres
T_{c_1}, T_{c_2} = the critical temperature in Kelvin of components 1 and 2
V_{c_1}, V_{c_2} = the critical volumes in cubic centimeters per gram of components 1 and 2
M_1, M_2 = the molecular weights of components 1 and 2
T = the temperature of the system

When the pore size gets so small that its dimensions are less than the mean path of the fluid, the so-called Knudsen diffusion prevails. From kinetic theory of gases, the Knudsen diffusivity can be defined as

$$D_K = \frac{2}{3}\mathbf{r}\sqrt{\frac{8RT}{M}} = 9.7 \times 10^{-3}\mathbf{r}\sqrt{\frac{T}{M}} \tag{9.8-3}$$

where D_K = expressed in square centimeters per second
$\mathbf{r}$ = the average pore radius in centimeters
T = the absolute temperature in Kelvin
M = the molecular weight

From kinetic theory, it is seen that

$$\frac{\lambda}{2r} = \frac{D_{AB}}{D_K} \tag{9.8-4}$$

where λ is the mean free path. There will be some range of pressure or molecular concentrations over which the transition from ordinary diffusion to Knudsen diffusion can take place. Within this region, both diffusions can be combined as

$$D_c = \frac{1}{1/D_K + (1 - \alpha Y_A)/D_{AB}} \tag{9.8-5}$$

where Y_A is the mole fraction of component A in the gas phase and

$$\alpha = 1 + \frac{N_B}{N_A} \quad \textbf{(9.8-6)}$$

where N_A and N_B are the molar fluxes of A and B, respectively. For equimolar counterdiffusion, $N_B = -N_A$ and then Eq. (9.8-5) reduces to

$$D_c = \frac{1}{1/D_K + 1/D_{AB}} \quad \textbf{(9.8-7)}$$

The third mechanism is *surface diffusion,* in which a gas is absorbed on the inner surface of a porous solid. This solid usually has pores of nonuniform cross section which pursue a very tortuous path through the particle and which may interact with many other pores. Hence the flux predicted by the first two mechanisms should be multiplied by a geometric factor which takes into account the tortuosity and the fact that the flow is affected by that fraction of the total pellet volume which is solid or simply by the porosity. Thus an effective diffusivity can be defined as

$$D_{eff} = \frac{D_c \epsilon_p}{\tau} \quad \textbf{(9.8-8)}$$

where τ is the tortuosity factor.

Example 9.8-1 Calculation of Effective Diffusivity

In cracking cumene by means of silica-alumina catalyst at 1 atm and 510°C, the following data were used (H1): $S_g = 342\ m^2/g$, $\tau = 3$, $\epsilon_p = 0.51$, equivalent particle diameter = 0.43, density of the particle $\rho_p = 1.14\ g/cm^3$, $D_{AB} = 0.15\ cm^2/s$. Calculate the effective diffusivity D_{eff}.

Solution

$$V_g = \frac{\epsilon_p}{\rho_p} = \frac{0.51}{1.14} = 0.447\ cm^3/g$$

$$\mathbf{r} = \frac{2V_g}{S_g} = \frac{2(0.447)}{342 \times 10^4} = 2.61 \times 10^{-7}\ cm = 2.61\ mm = 26.1\ \text{Å}$$

$$D_K = 9.7 \times 10^3\ \mathbf{r} \sqrt{\frac{T}{M}} = 9.7 \times 10^3 (2.61)(10^{-7}) \sqrt{\frac{783}{120.10}}$$

$$= 6.46 \times 10^{-3}\ cm^2/s$$

$$D_c = \frac{1}{1/D_K + 1/D_{AB}} = \frac{1}{1/0.00646 + 1/0.15} = 0.00619 \text{ cm}^2/\text{s}$$

$$D_{\text{eff}} = \frac{D_c \epsilon_p}{\tau} = \frac{(0.00619)(0.51)}{3} = 0.00105 \text{ cm}^2/\text{s}$$

9.9 *Isothermal Reactions in Porous Catalysts*

When molecules diffuse through pores of catalysts, reaction also occurs. The effect can be accounted for by the use of an effectiveness factor η, which is defined as the ratio of the average rate of the reaction in the pellet to that which would be obtained if the total catalyst were exposed to the fluid at exterior surface conditions

$$\eta = \frac{\text{Average rate for the entire catalyst}}{\text{Rate at exterior surface conditions}} \tag{9.9-1}$$

Assuming isothermal conditions and first-order irreversible reactions, this factor can be derived as follows. Suppose molecules of A which decompose to B diffuse through surface area S_x in one dimension only (see Fig. 9.9-1). A component mass balance within the increment volume under steady-state conditions is

$$\begin{pmatrix}\text{Molar flow rate}\\ \text{of A in}\end{pmatrix} - \begin{pmatrix}\text{Molar flow rate}\\ \text{of A out}\end{pmatrix} - \begin{pmatrix}\text{Rate of disappearance of}\\ \text{A by chemical reaction}\end{pmatrix} = 0$$

or

$$-\left(S_x D_{\text{eff}} \frac{dC_A}{d\omega}\right)_\omega - \left(-S_x D_{\text{eff}} \frac{dC_A}{d\omega}\right)_{\omega+\Delta\omega} - r_v S_x \, \Delta\omega = 0 \tag{9.9-2}$$

where S_x is the surface area of the catalyst through which the molecules diffuse, ω is the one-dimensional direction, and r_v is the rate of the chemical reaction

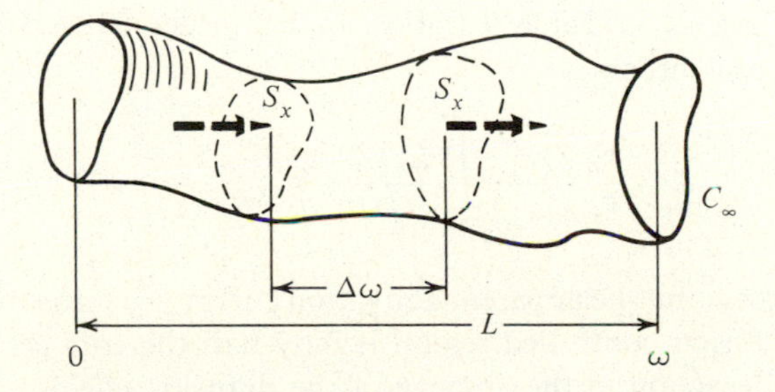

Figure 9.9-1 Mass balance in a porous catalyst

based on the total volume of the catalyst. Dividing Eq. (9.9-2) through by $S_x \Delta\omega$ and taking limit as $\Delta\omega$ approaches zero gives

$$\frac{1}{S_x}\frac{d}{d\omega}\left(S_x D_{\text{eff}} \frac{dC_{\text{A}}}{d\omega}\right) - r_v = 0 \tag{9.9-3}$$

The boundary conditions are:

$$\text{At } \omega = 0,\ dC_{\text{A}}/d\omega = 0;\ \text{at } \omega = L,\ C_{\text{A}} = C_\infty \tag{9.9-4}$$

By the definition of effectiveness factor in Eq. (9.9-1), we obtain

$$\eta = \frac{\dfrac{1}{V}\displaystyle\int_0^V r_v\, dV}{r_v(C_\infty)} \tag{9.9-5}$$

where $r_v(C_\infty)$ is the rate of the chemical reaction evaluated at the exterior surface concentration C_∞. Integrating Eq. (9.9-3) gives

$$\int_0^L S_x r_v\, d\omega = \int_0^L S_x D_{\text{eff}} \frac{dC_{\text{A}}}{d\omega}\, d\omega = S_x D_{\text{eff}} \frac{dC_{\text{A}}(L)}{d\omega} \tag{9.9-6}$$

Substituting into Eq. (9.9-5) gives

$$\eta = \frac{D_{\text{eff}}(C_\infty)\dfrac{dC_{\text{A}}(L)}{d\omega}}{L r_v(C_\infty)} \tag{9.9-7}$$

Thus Eqs. (9.9-3) to (9.9-7) constitute the generalized equations for the calculation of the effectiveness factor in a porous catalyst under isothermal condition. For a second- or nth-order reaction, the numerical method is preferred. For a first-order reaction, the solution for rectangular slab, cylindrical, and spherical pellets can be summarized as in Table 9.1. For an nth-order irreversible reaction, the generalized modulus becomes

$$\phi = \frac{V_p}{S_x}\left(\frac{(n+1)}{2}\frac{k_v C_\infty^{n-1}}{D_{\text{eff}}}\right)^{1/2} \qquad n > -1 \tag{9.9-11}$$

It can also be shown that the apparent activation energy E_D measured when reaction occurs in the diffusion controlled region is only half the true activation energy E for the chemical reaction in the absence of the diffusion effects.

$$E_D = \frac{E}{2} \tag{9.9-12}$$

Table 9.1 Effectiveness factor for slab, cylinder, and sphere (solution of Eqs. (9.9-3) to (9.9-5) for first-order irreversible reaction)

	ω	S_x	V_p	$\phi = V_p\lambda/S_x$	C_A/C_∞	η	
Slab	x	A_c	A_cL	λL	$\cosh \lambda x/\cosh \lambda L$	$(\tanh \phi)/\phi$	**(9.9-8)**
Cylinder	r	$2\pi r_o L$	$\pi r_o^2 L$	$\lambda r_o/2$	$I_o(\lambda r)/I_o(\lambda r_o)$	$I_1(2\phi)/[\phi I_o(2\phi)]$	**(9.9-9)**
Sphere	r	$4\pi r_o^2$	$4\pi r_o^3/3$	$\lambda r_o/3$	$\sinh \lambda r/\sinh \lambda r_o$	$[\coth (3\phi) - 1/(3\phi)]/\phi$	**(9.9-10)**

where V_p = the volume of the pellet
$\lambda = (k/D_{eff})^{1/2}$
I_o, I_1 = Bessel functions
k = first-order rate constant
ϕ = Thiele modulus
A_c = cross sectional area of slab
r_o = radius of cylinder or sphere

Evidently the apparent order of reaction n_D measured with the diffusion effects is related to the true order n as:

$$n_D = \frac{(n+1)}{2} \tag{9.9-13}$$

Example 9.9-1 Effectiveness Factor

Gaseous A at 700 K and 1 atm partial pressure decomposes under steady-state conditions at a rate of 0.25 kmol/s by a first-order irreversible reaction in an isothermal tubular reactor (volume = 100 m^3), packed with spherical catalysts of 2.5 mm. The voidage is 0.5 and the effective diffusion coefficient of the reactant within the catalyst is 1.0×10^{-6} m^2/s. Determine the effectiveness factor of the catalyst.

Solution

For first-order reaction, the Thiele modulus is

$$\begin{aligned}\phi &= \frac{V}{S_x}\left(\frac{n+1}{2}\frac{k_v C_\infty}{D_{\text{eff}}}\right)^{1/2} \\ &= \frac{4\pi r_o^3/3}{4\pi r_o^2}\left(\frac{k_v C_\infty}{D_{\text{eff}}}\right)^{1/2} = \frac{r_o}{3}\left(\frac{k_v}{D_{\text{eff}}}\right)^{1/2}\end{aligned} \tag{A}$$

Since the observed reaction rate $\dot{N}$ in k mol/s is

$$\dot{N} = k_v C_\infty \eta V(1-\epsilon) \tag{B}$$

Eliminating k_v gives

$$\dot{N} = D\left(\frac{3\phi}{r_o}\right)^2 C_\infty \eta V(1-\epsilon) \tag{C}$$

or

$$\begin{aligned}\phi^2\eta &= \frac{\dot{N} r_o^2}{9V(1-\epsilon)D_{\text{eff}}C_\infty} \\ &= \frac{1}{9}\frac{0.25(1000)(2.5)^2(1/1000)^2}{100(1-0.5)(1.0)(10^{-6})[1/82.056(700)](100)^3} \\ &= 0.1994416\end{aligned} \tag{D}$$

Solving with

$$\eta = \frac{\coth(3\phi) - 1/3\phi}{\phi} \tag{E}$$

by trial and error gives

$$\eta = 0.887$$

Example 9.9-2 Volume of Packing and Size of Catalyst

A new reactant for which k_v is twice as large as in the previous example will be used to carry out the same reaction to the same extent as in the previous example. Because of the different pore characteristics, the effective diffusion coefficient is reduced to 7×10^{-7} m²/s. What volume of packing for the same catalyst of radius 2.5 mm will be required? What size would the particles have to be to have the same effectiveness factor as in the previous example?

Solution

(A)

$$\phi_1 = \frac{r_o}{3}\left(\frac{k_v}{D_{\text{eff}}}\right)^{1/2} = \frac{2.5}{3}\left(\frac{k_{v1}}{1.2 \times 10^{-6}}\right)^{1/2} \tag{A}$$

$$\phi_2 = \frac{2.5}{3}\left(\frac{2k_{v1}}{7 \times 10^{-7}}\right)^{1/2} \tag{B}$$

Hence

$$\frac{\phi_2}{\phi_1} = 1.8516402 \tag{C}$$

Then

$$2k_{v1}V_2(1-\epsilon)C\eta_2 = \mathrm{k}_{v1}(100)(1-\epsilon)\mathrm{C}\eta_1 \tag{D}$$

or

$$V_2 = 50\,\frac{\eta_1}{\eta_2} \tag{E}$$

$$= 50\,\frac{[(\coth 3\phi_1 - 1/3\phi_1)/\phi_1]}{[(\coth 3\phi_2 - 1/3\phi_2)/\phi_2]} \tag{F}$$

From Example 9.9-1

$$\phi_1 = 0.4741503 \qquad 3\phi_1 = 1.4224511$$

Thus

$$\phi_2 = 1.8516402\phi_1 = 0.8779557 \qquad 3\phi_2 = 2.6338673$$

Substituting into Eq. (F) gives

$$V_2 = 61.74 \text{ m}^3$$

(B) $\eta = 0.887$

$$\phi_1 = 0.4741503$$

$$\frac{\phi_2}{\phi_1} = \frac{r_2}{r_1}\left(2\,\frac{D_{\text{eff}\,1}}{D_{\text{eff}\,2}}\right)^{1/2} = \frac{r_2}{2.5}\left[\frac{2(1.2 \times 10^{-6})}{7 \times 10^{-7}}\right]^{1/2} \tag{G}$$

or

$$\phi_2 = 0.3511822 r_2 \tag{H}$$

Solving with

$$0.887 = \frac{[\coth(3\phi_2) - 1/3\phi_2]}{\phi_2} \tag{I}$$

gives

$$r_2 = 1.35 \text{ mm}$$

9.10 Nonisothermal Reaction in Porous Catalysts

Analogous to Eq. (9.9-3) for the mass balance, an equation for the energy balance can be developed as

$$-\frac{1}{S_x}\frac{d}{d\omega}\left(S_x k_{\text{eff}}\frac{dT}{d\omega}\right) = (-\Delta H) r_v \tag{9.10-1}$$

where k_{eff} = the effective thermal conductivity = $k_s^{1-\epsilon} k_f^{\epsilon}$ **(9.10-2)**

k_s = the thermal conductivity of the solid

k_f = the thermal conductivity of the bulk fluid

ϵ = the porosity of the pellet

Δh = the heat of reaction

Dividing Eq. (9.10-1) by $(-\Delta H)$ and subtracting from Eq. (9.9-3) yields

$$\frac{d}{d\omega}\left(S_x D_{\text{eff}}\frac{dC_A}{d\omega} + \frac{k_{\text{eff}}}{-\Delta H}\,S_x\frac{dT}{d\omega}\right) = 0 \tag{9.10-3}$$

Further integration for constant D_{eff} and k_{eff} gives

$$D_{\text{eff}} C_A + \frac{k_{\text{eff}}}{-\Delta H}\,T = \text{constant} \tag{9.10-4}$$

or

$$T - T_\infty = \frac{D_{eff}(-\Delta H)}{k_{eff}}(C_{A\infty} - C_A) \tag{9.10-5}$$

Hence either T or C_A in Eq. (9.9-3) and Eq. (9.10-1) can be eliminated by Eq. (9.10-5) and then can be solved preferably by the numerical method. The solution will consist of the following parameters:

1. A Thiele-type modulus

$$\phi = r_o \sqrt{\frac{k_1 S_x}{D_{eff}}} \tag{9.10-6}$$

2. An Arrhenius number

$$\gamma = \frac{E}{RT_\infty} \tag{9.10-7}$$

where E is the intrinsic activation energy and R is the gas constant.

3. An energy generation function

$$\beta = \frac{(-\Delta H) D_{eff} C_\infty}{k_{eff} T_\infty} \tag{9.10-8}$$

The results from the numerical solution are usually presented in a figure in which the effectiveness factor is plotted against the Thiele modulus using the β as the parameter for each value of γ.

9.11 External Mass and Heat Transfer

For equimolar counterdiffusion of a component A in the gas phase, the rate of mass transfer of A from the bulk fluid to the interface is

$$N_A = k_G(p_A - p_{Ai}) = k_c(C_A - C_{Ai}) \tag{9.11-1}$$

where N_A = the molar flux

k_G = the gas-film mass transfer coefficient per unit external surface area

p_A, p_{Ai} = the partial pressure of the component A in the bulk fluid and at the interface, respectively.

k_c, C_A, C_{Ai} = the corresponding terms on concentration basis

In case equimolar counterdiffusion may not occur, the driving force in the first term of Eq. (9.11-1) must be divided by the drift factor which is the logarithmic mean value of $P + \delta_A p_A$ and of $P + \delta_A p_{Ai}$ where P is the total pressure, p_A is the partial pressure of component A of the gas mixture, p_{Ai} is the value of p_A at the interface and δ_A is the net difference in number of moles of product and reactant per mole of component A. These mass-transfer coefficients are correlated in terms of Reynold's number $D_p G/\mu$, Schmidt number $N_{Sc} = \mu/(\rho D)$, Sherwood number $N_{Sh} = k_c D_p/D$, mass-transfer factor $j_D = (k_c \rho/G) N_{Sc}^{2/3}$ where G is the mass velocity based on empty reactor, μ is the fluid viscosity, ρ is the fluid density, D is the molecular diffusivity of the species being transferred in the system of interest, D_p is the equivalent diameter of the catalyst pellet. For a sphere, it is equal to the diameter of the sphere; for a cylinder, it is given by

$$4\left(\frac{D_p}{2}\right)^2 = 2r_c L_c + 2r_c^2$$

or

$$D_p = \sqrt{2r_c L_c + 2r_c^2} \qquad \textbf{(9.11-2)}$$

where r_c and L_c are the radius and length of the cylinder, respectively. The correlations are

9.11-1 Packed beds

1. Gas [Petrovic and Thodos (P1)]

$$\epsilon_B j_D = \frac{0.357}{N_{Re}^{0.359}} \qquad 3 < N_{Re} < 2000 \qquad \textbf{(9.11-3)}$$

2. Liquid [Wilson and Geankoplis (W1)]

$$\epsilon_B j_D = \frac{0.250}{N_{Re}^{0.31}} \qquad \begin{matrix} 55 < N_{Re} < 1500 \\ 0.35 < \epsilon_B < 0.75 \end{matrix} \qquad \textbf{(9.11-4)}$$

$$\epsilon_B j_D = \frac{1.09}{N_{Re}^{2/3}} \qquad 0.0016 < N_{Re} < 55 \qquad \textbf{(9.11-5)}$$

9.11-2 Fluidized beds

1. Chu et al. (C2)

$$j_D = 5.7(N'_{Re})^{-0.78} \qquad 1 < N'_{Re} < 30 \qquad \textbf{(9.11-6)}$$

$$j_D = 1.77(N'_{Re})^{-0.44} \qquad 30 < N'_{Re} < 10^4 \qquad \textbf{(9.11-7)}$$

2. Riccette and Thodos (R2)

$$j_D = \frac{1}{(N'_{Re})^{0.40} - 1.5} \qquad 100 < N'_{Re} < 7{,}000 \tag{9.11-8}$$

3. Sen Gupta and Thodos (S1)

$$\frac{\epsilon_B j_D}{\Omega} = \frac{0.300}{\left(\frac{\sqrt{A_p G}}{\mu}\right)^{0.35} - 1.90} \qquad \frac{\sqrt{A_p G}}{\mu} > 50 \tag{9.11-9}$$

where $N'_{Re} = \dfrac{D_p G}{\mu(1 - \epsilon_B)}$

G = superficial mass velocity based on empty reactor

A_p = surface area of a single pellet

ϵ_B = porosity

Ω = area availability factor (1.0 for spheres and 1.16 for cylinders in fluidized beds)

Analogous to the mass transfer, the rate of *heat transfer* of A from the bulk fluid to the interface is

$$q = h(T - T_i) \tag{9.11-10}$$

where q = the heat flux

h = the heat-transfer coefficient

T and T_i = the temperature of the bulk fluid and at the interface respectively

This heat-transfer coefficient is correlated in terms of N_{Re}, the Reynolds number, N_{St} the Stanton number, N_{Pr} the Prandtl number and $j_H = N_{St}N_{Pr}^{2/3}$ the heat-transfer factor, in which $N_{St} = h/(C_p G)$, $N_{Re} = DG/\mu$, $N_{Pr} = C_p\mu/k$ and h is the heat-transfer coefficient between the catalyst particle and the bulk fluid, C_p is the constant pressure heat capacity, k is the thermal conductivity of the fluid, μ is the viscosity of the fluid, and G is the superficial mass velocity of the fluid.

For packed beds, according to the Chilton-Colburn analogy,

$$j_H = j_D \tag{9.11-11}$$

This means that the previous correlations for mass transfer, Eqs. (9.11-3) to (9.11-5) can also be used to evaluate the heat-transfer coefficient. For fluidized beds, the correlation of Katten-ring which is valid in the range of N_{Re} between 9 and 55 is (K1)

$$\frac{hD_p}{k} = 0.0135\, N_{Re}^{1.3} \tag{9.11-12}$$

9.12 Overall Rate Expression

The seven steps in the phenomenon of a catalytic reaction have been discussed. It is now the time to combine all of these steps to obtain an overall rate expression for design purposes.

The intrinsic rate for adsorption, surface reaction, and desorption predicted from the previous section is

$$r_v = \frac{(\text{kinetic term})(\text{driving force})}{(\text{adsorption term})^n} \tag{9.12-1}$$

where r_v is the rate of reaction based on total volume. For catalytic reactor design, it is more convenient to express the rate in terms of the mass of the catalyst. This can be done by

$$r_v = r_m \rho_b \tag{9.12-2}$$

where r_m is the rate of reaction per unit mass of catalyst and ρ_b is the bulk density of the catalyst. Thus the rate of reaction is a function of concentrations and the rate constant. For isothermal condition,

$$r_m = f(C) \tag{9.12-3}$$

which means that the rate of reaction per unit mass of catalyst is a function of concentration. At the surface of the catalyst, this rate, because of internal diffusion, becomes

$$r_p = \eta f(C_\infty) \tag{9.12-4}$$

where η is the effectiveness factor and C_∞ is the concentration at the surface of the catalyst, and r_p is the rate of the reaction at the surface of the catalyst particle, which is also equal to the external mass transfer rate from the bulk fluid to the surface

$$N_i a_m = -\alpha_i r_m \tag{9.12-5}$$

where a_m is the external area per unit mass of catalyst, N_i is the molar flux of species i to the external surface of the catalyst, and α_i is the stoichiometric coefficient of the species i. The molar flux of species i relative to a fixed coordinate system is

$$N_i = y_i \left(\sum_{j=1}^{c} N_j \right) + k_{ci}(C_{ib} - C_{i\infty}) \tag{9.12-6}$$

where y_i is the mole fraction of the species, the subscripts b and ∞ refer to bulk and surface, respectively, and the summation involves all c components in the mixture. Combining Eqs. (9.12-5) and (9.12-6) with simplification gives

$$C_{ib} - C_{i\infty} = -\frac{\alpha_i r_m}{a_m k_{ci}}\left(1 - \frac{y_i}{\alpha_i}\sum_{j=1}^{c} \alpha_i\right) \tag{9.12-7}$$

The term within the parentheses is called the drift factor when there is no equal molal counterdiffusion. Assuming equal molal counterdiffusion, equating Eqs. (9.12-4) and (9.12-7) gives

$$\eta f(C_\infty) = k_c a_m (C_b - C_\infty) \tag{9.12-8}$$

Hence the surface concentration can be solved, and substituting back to Eq. (9.12-4) gives the overall or global rate equation r_p. For nonisothermal condition in which the temperature gradients are significant, the bulk fluid temperature is related to the surface temperature by

$$r_p(-\Delta H) = h a_m (T_\infty - T_b) \tag{9.12-9}$$

where ΔH is heat of reaction. The rate of reaction due to the internal diffusion is

$$r_p = \eta f(C_\infty) e^{-E/RT_\infty} \tag{9.12-10}$$

where E is the activation energy and R is the gas constant, and h is the heat-transfer coefficient. η is a function of ϕ, β, and γ as shown in the previous section. Hence for a given T_b and C_b, the global rate can be evaluated by the following numerical procedure:

1. Assume T_∞ and C_∞.
2. Calculate η from ϕ, β, and γ.
3. Obtain r_p from Eq. (9.12-10).
4. Calculate C_∞ from Eq. (9.12-8) and T_∞ from Eq. (9.12-9).
5. Repeat the procedure until the calculated C_∞ and T_∞ check the assumed values.

The isothermal case is illustrated by the following example.

Example 9.12-1 Global Reaction Rate Calculation

The intrinsic rate of the reversible reaction of the ortho and para hydrogen

$$o\text{-}H_2 \rightleftarrows p\text{-}H_2 \tag{A}$$

at $-196°C$, after simplification have been determined by Smith (S2):

$$r = \frac{k_1(K+1)}{K}(C_{eq} - C)_p \tag{B}$$

where $k_1 = \dfrac{1.1}{1 + 1.06 \times 10^3 C_t}$

K = equilibrium constant

$(C_{eq})_p$ = concentration of the para hydrogen at equilibrium

C_t = total concentration = $C_o + C_p$

C_o = concentration of the ortho hydrogen

C_p = concentration of the para hydrogen

The reaction is carried out in a tube of 0.5-in ID with $\frac{1}{8}$ in $\times$ $\frac{1}{8}$ in cylindrical catalyst pellets of Ni and Al_2O_3. Calculate the global reaction rate at a pressure of 400 psig at a location where the mole fraction of ortho hydrogen in the bulk gas stream is 0.60 with a superficial mass velocity of 14 lb/h·ft². The data are

Viscosity of H_2 at $-196°C = 348 \times 10^7$ P

Molecular diffusivity of H_2 at $-196°C$ and 400 psig = 0.00496 cm²/s.

$\mu/\rho D = 0.78$ at $-196°C$.

$\rho_p = 1.91$ g/cm³

$y_{eq} = 0.5026$

$\epsilon = 0.35$

Solution

The intrinsic rate equation is

$$r = \frac{k_1(K+1)}{K}(C_{eq} - C)_p \tag{A}$$

Because of internal diffusion, Eq. (A) becomes

$$r_p = \eta \frac{k_1(K+1)}{K}(C_{eq} - C_\infty)_p \tag{B}$$

The external mass transfer is

$$r_p = k_m a_m (C_\infty - C_b)_p \tag{C}$$

where k_m = the mass transfer coefficient on unit mass of catalyst basis

a_m = the external surface per unit mass

C_∞ = the surface concentration of the para hydrogen

C_b = the bulk concentration of the para hydrogen

Equating Eqs. (B) and (C) and solving for $C_{\infty,p}$ gives

$$C_{\infty,p} = \frac{\eta[k_1(K+1)/K]\,C_{\text{eq},p} + k_m a_m C_{b,p}}{\eta[k_1(K+1)/K] + k_m a_m} \tag{D}$$

Substituting $C_{\infty,p}$ into Eq. (C) yields the global rate of reaction as

$$r_p = \frac{(C_{\text{eq}} - C_b)_p}{K/\eta k_1(K+1) + 1/k_m a_m} \tag{E}$$

$$G = 14\left[\frac{454}{3600(12)^2(2.54)^2}\right] = 1.9 \times 10^{-3}\ \text{g/s}\cdot\text{cm}^2$$

$$N_{\text{Re}} = \frac{1/8(2.54)(1.9 \times 10^{-3})}{348 \times 10^{-7}} = 17.36$$

$$j_D = \frac{0.357}{\epsilon N_{\text{Re}}^{0.359}} = \frac{0.357}{0.35(17.36)^{0.359}} = 0.366$$

$$k_m = \frac{j_D G}{\rho}\left(\frac{\mu}{\rho D}\right)^{-2/3} = \frac{0.366(1.9 \times 10^{-3})}{8.95 \times 10^{-3}}(0.78)^{-2/3} = 0.0917\ \text{cm/s}$$

$$C_t = \frac{P}{RT} = \frac{114.7/14.7}{82(77)} = 1.235 \times 10^{-2}\ \text{g mol/cm}^3$$

$$k_1 = \frac{1.1}{1 + 1.06 \times 10^3(1.235 \times 10^{-3})} = 0.4762\ \text{cm}^3\text{/g catalyst per second}$$

$$D_{\text{eff}} = D\epsilon^2 = 0.00496(0.35)^2 = 6.076 \times 10^{-4}\ \text{cm}^2\text{/s} \quad \text{assuming random-pore model}$$

It has been shown by Smith (S2) that the Thiele modulus for a first-order reversible reaction is

$$\phi_s' = \frac{r}{3(2)}\sqrt{\frac{k_1(K+1)\,\rho_p}{KD_{\text{eff}}}}$$

$$= \frac{0.318}{3(2)}\sqrt{\frac{0.476(1.01+1)1.91}{1.01[6.076(0.0001)]}} = 2.89 \tag{F}$$

$$\eta = \frac{1}{\phi_s'}\left(\frac{1}{\tanh 3\phi_s'} - \frac{1}{3\phi_s'}\right) = \frac{1}{2.89}\left(\frac{1}{\tanh 8.67} - \frac{1}{8.67}\right) = 0.3059$$

$$(C_{eq} - C_b)_p = C_t(y_{eq} - y_b)_p = 1.235 \times 10^{-3}[0.5026 - (1. - 0.6)]$$
$$= 1.267 \times 10^{-4}$$

$$r_p = \frac{1.267 \times 10^{-4}}{\{1.01/0.3059[0.4762(1.01 + 1)]\} + 0.318(1.91)/6(0.0917)}$$
$$= 2.78 \times 10^{-5} \text{ g mol/g catalyst per second}$$

9.13 Generalized Mass Balance

Consider a fluid flowing through and undergoing a chemical reaction in a tube in which a mass balance of component j in the fluid in a volume element can be made as follows (see Fig. 9.13-1):

$$\underset{(1)}{\begin{bmatrix}\text{rate in}\\\text{by}\\\text{bulk flow}\end{bmatrix}} - \underset{(2)}{\begin{bmatrix}\text{rate out}\\\text{by}\\\text{bulk flow}\end{bmatrix}} + \underset{(3)}{\begin{bmatrix}\text{rate in}\\\text{by}\\\text{diffusion}\end{bmatrix}} - \underset{(4)}{\begin{bmatrix}\text{rate out}\\\text{by}\\\text{diffusion}\end{bmatrix}}$$

$$+ \underset{(5)}{\begin{bmatrix}\text{rate of}\\\text{generation}\end{bmatrix}} = \underset{(6)}{\begin{bmatrix}\text{rate of}\\\text{accumulation}\end{bmatrix}} \qquad \textbf{(9.13-1)}$$

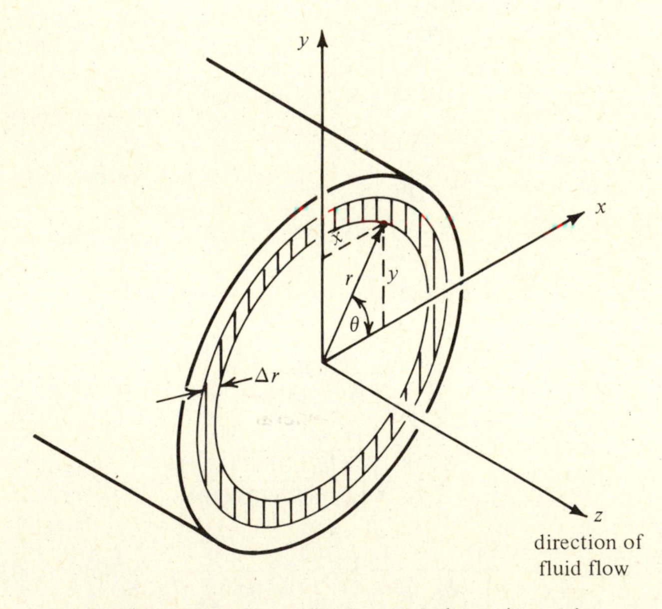

Figure 9.13-1 Generalized mass balance in a tube

Term 1: $[u_z C_j(2\pi r\,\Delta r)]_z$

Term 2: $[u_z C_j(2\pi r\,\Delta r)]_{z+\Delta z}$

Term 3: $\left[2\pi r\,\Delta z\left(-D_R\dfrac{\partial C_j}{\partial r}\right)\right]_r + \left[2\pi r\,\Delta r\left(-D_L\dfrac{\partial C_j}{\partial z}\right)\right]_z$

Term 4: $\left[2\pi r\,\Delta z\left(-D_R\dfrac{\partial C_j}{\partial r}\right)\right]_{r+\Delta r} + \left[2\pi r\,\Delta r\left(-D_L\dfrac{\partial C_j}{\partial z}\right)\right]_{z+\Delta z}$

Term 5: $\alpha r_v(2\pi r\,\Delta r\,\Delta z)$

Term 6: $2\pi r\,\Delta r\,\Delta z\dfrac{\Delta C_j}{\Delta t}$

Substituting these terms into Eq. (9.13-1), dividing through by $2\pi r\,\Delta r\,\Delta z$, and taking limits gives

$$D_L\frac{\partial^2 C_j}{\partial z^2} + \frac{D_R}{r}\frac{\partial}{\partial r}\left(r\frac{\partial C_j}{\partial r}\right) - \frac{\partial(uC_j)}{\partial z} + \alpha r_v = \frac{\partial C_j}{\partial t} \tag{9.13-2}$$

where D_L, D_R are the longitudinal and radial effective diffusivity, respectively, u is the superficial velocity based on empty tube, α is the stoichiometric coefficient of the j component in the reaction, and r_v is the reaction rate expressed in number of moles per unit time per unit volume.

9.14 Generalized Energy Balance

Similarly, an energy balance of the fluid flowing through the tube with chemical reaction can be made within the element volume in the previous section.

$$\underset{(1)}{\begin{bmatrix}\text{rate in}\\ \text{by}\\ \text{bulk flow}\end{bmatrix}} - \underset{(2)}{\begin{bmatrix}\text{rate out}\\ \text{by}\\ \text{bulk flow}\end{bmatrix}} + \underset{(3)}{\begin{bmatrix}\text{rate in}\\ \text{by}\\ \text{diffusion}\end{bmatrix}} - \underset{(4)}{\begin{bmatrix}\text{rate out}\\ \text{by}\\ \text{diffusion}\end{bmatrix}}$$

$$+ \underset{(5)}{\begin{bmatrix}\text{rate of}\\ \text{generation}\end{bmatrix}} = \underset{(6)}{\begin{bmatrix}\text{rate of}\\ \text{accumulation}\end{bmatrix}} \tag{9.14-1}$$

Term 1: $[u_z\rho C_p(T - T_R)2\pi r\,\Delta r]_z$

Term 2: $[u_z\rho C_p(T - T_R)2\pi r\,\Delta r]_{z+\Delta z}$

Term 3: $\left[2\pi r\,\Delta z\left(-k_R\dfrac{\partial T}{\partial r}\right)\right]_r + \left[2\pi r\,\Delta r\left(-k_L\dfrac{\partial T}{\partial z}\right)\right]_z$

Term 4: $\left[2\pi r\,\Delta z\left(-k_R\dfrac{\partial T}{\partial r}\right)\right]_{r+\Delta r}+\left[2\pi r\,\Delta r\left(-k_L\dfrac{\partial T}{\partial z}\right)\right]_{z+\Delta z}$

Term 5: $\alpha r_v(-\Delta H)2\pi r\,\Delta r\,\Delta z$

Term 6: $2\pi r\,\Delta r\,\Delta z\,\rho C_p\dfrac{\Delta T}{\Delta t}$

Substituting these terms into Eq. (9.14-1), dividing through by $2\pi r\,\Delta r\,\Delta z$, and taking limits gives

$$k_L^*\frac{\partial^2 T}{\partial z^2}+\frac{k_R^*}{r}\frac{\partial}{\partial r}\left(r\frac{\partial T}{\partial r}\right)-\frac{\partial(u_z\rho C_p T)}{\partial z}+\alpha r_v(+\Delta H)=\rho C_p\frac{\partial T}{\partial t} \qquad \textbf{(9.14-2)}$$

where ΔH is the heat of reaction, ρ is the density, C_p is the constant pressure heat capacity, T_R is the reference temperature, ϵ_T is the total porosity, k_L^* and k_R^* are the longitudinal and radial effective thermal conductivities, respectively.

9.15 Generalized Momentum Balance

Analogous to the mass and energy balances, a momentum balance can be made for the same fluid in previous sections as follows:

$$\underset{(1)}{\begin{bmatrix}\text{rate in}\\ \text{by}\\ \text{bulk flow}\end{bmatrix}}-\underset{(2)}{\begin{bmatrix}\text{rate out}\\ \text{by}\\ \text{bulk flow}\end{bmatrix}}+\underset{(3)}{\begin{bmatrix}\text{rate in}\\ \text{by}\\ \text{diffusion}\end{bmatrix}}-\underset{(4)}{\begin{bmatrix}\text{rate out}\\ \text{by}\\ \text{diffusion}\end{bmatrix}}$$

$$+\underset{(5)}{\begin{bmatrix}\text{rate of}\\ \text{generation}\end{bmatrix}}=\underset{(6)}{\begin{bmatrix}\text{rate of}\\ \text{accumulation}\end{bmatrix}} \qquad \textbf{(9.15-1)}$$

Term 1: $[u_z\rho u(2\pi r\,\Delta r)/g_c]_z$

Term 2: $[u_z\rho u(2\pi r\,\Delta r)/g_c]_{z+\Delta z}$

Term 3: $\left[2\pi r\,\Delta z\left(-\mu\dfrac{\partial u}{\partial r}\right)\Big/ g_c+P_z 2\pi r\,\Delta r\right]_r$

Term 4: $\left[2\pi r\,\Delta z\left(-\mu\dfrac{\partial u}{\partial r}\right)\Big/ g_c+P_z 2\pi r\,\Delta r\right]_{r+\Delta r,z+\Delta z}$

Term 5: $\rho g(2\pi r\,\Delta r\,\Delta z)/g_c$

Term 6: $2\pi r\,\Delta r\,\Delta z\,\rho\dfrac{\Delta u}{\Delta t}\Big/ g_c$

Substituting these terms into Eq. (9.15-1), dividing through by $2\pi r\,\Delta r\,\Delta z$, and taking limits gives

$$-\frac{\rho u}{g_c}\frac{\partial u}{\partial z}+\frac{u}{g_c r}\frac{\partial}{\partial r}\left(r\frac{\partial u}{\partial r}\right)-\frac{\partial P}{\partial z}+\frac{\rho g}{g_c}=\frac{\rho}{g_c}\frac{\partial u}{\partial t} \tag{9.15-2}$$

where g_c = the conversion factor in lb mass·ft/lb force·s²

P = the pressure of the fluid

μ = the viscosity of the fluid

u = the linear velocity of the fluid

9.16 Design of a Fixed-Bed Catalytic Reactor

In a fixed-bed catalytic reactor, the catalysts which fill the volume of the reactor are stationary. The gaseous reactants or sometimes liquid reactants entering the bottom flow through the void with an interstitial velocity u. Because it is hard to measure this velocity, the design is always based on the superficial velocity u_∞, which is defined as the velocity as if there were no catalyst in the reactor. Then $u_\infty = u\epsilon$. Because solid catalyst and the gaseous or liquid fluid are present in the reactor, the system is heterogeneous. The design of such a reactor could be equally or more difficult than the homogeneous one, depending on the model to be used. In general, there are two approaches for the design: simplified or sophisticated. The first approach is often called a *pseudo-homogeneous model,* in which the two phases are treated as one. That means that the concentrations and temperature on the solid phase are assumed to be the same as those in the fluid phase, $C_\infty = C$, $T_\infty = T$. This model can further be subdivided into (a) one-dimensional without axial mixing, (b) one-dimensional with axial mixing, (c) two-dimensional with radial mixing. The second approach is the heterogeneous model in which the concentrations and temperature in the solid phase are different from those in the fluid phase. This model can be subdivided into (a) one-dimensional with interfacial gradients, (b) one-dimensional with intraparticle gradients, (c) two-dimensional with radial mixing. All of them will be discussed below (F1).

9.17 One-Dimensional Without Mixing—Isothermal Pseudo-Homogeneous Model

In this case, plug flow is assumed, i.e., the velocity in the direction of the flow is assumed uniform in the radial direction. For isothermal case, the mass balance Eq. (9.13-2) becomes

$$\frac{\partial}{\partial z}(u_\infty C_j) = \alpha r_v \tag{9.17-1}$$

Here the superficial velocity u_∞ is used. Taking j as the limiting component, expanding Eq. (9.17-1), and multiplying both sides by the cross-sectional area of the tube A_c gives

$$u_\infty A_c \, dC_k = -r_v A_c \, dz \tag{9.17-2}$$

But

$$r_m \, \rho_b = r_v \tag{9.17-3}$$

where ρ_b is the bulk density of the catalyst, r_m is the rate of the reaction in moles of limiting component reacting per unit time per unit mass of catalyst, r_v is the rate of reaction of the limiting component reacting in moles per unit time per unit volume of the reactor. From the definition of fractional conversion f

$$f = \frac{C_0 - C}{C_0} \tag{9.17-4}$$

Differentiating gives

$$df = -\frac{dC_k}{C_0} \tag{9.17-5}$$

Combining Eqs. (9.17-2), (9.17-3), (9.17-4), and (9.17-5) yields

$$u_\infty A_c C_0 \, df = r_m \, \rho_b A_c \, dz \tag{9.17-6}$$

Since the feed rate of the limiting component F_0 and the mass of the catalyst W are

$$F_0 = u_\infty A_c C_0 \tag{9.17-7}$$

$$dW = \rho_b A_c \, dz \tag{9.17-8}$$

Eq. (9.17-6) becomes

$$F_0 \, df = r_m \, dW$$

or

$$W = \int_0^f \frac{F_0 \, df}{r_m} = F_0 \int_0^f \frac{1}{r_m} \, df \tag{9.17-9}$$

Example 9.17-1 Isothermal Pseudo-Homogeneous Model

Styrene is produced in an isothermally operated packed tubular reactor at a pressure of 1.2 bar by the catalytic dehydrogenation of ethyl benzene with steam. The rate of the steam is 10 times that of the ethyl benzene. The rate for the reaction at 600°C.

$$\underset{(E)}{C_6H_5C_2H_5} \rightarrow \underset{(S)}{C_2H_5CH{=}CH_2} + \underset{(H)}{H_2} \tag{A}$$

is

$$r_m = k\left(p_E - \frac{p_S p_H}{K}\right) \text{kmol/h} \cdot \text{kg catalyst} \tag{B}$$

where p_E, p_S, and p_H are the partial pressures of the ethyl benzene, the styrene, and the hydrogen, respectively, in bar, and the rate constant k, the equilibrium constant K relate with the absolute temperature T K as (W2)

$$k = 12{,}600e^{-11{,}000/T} \tag{C}$$

$$K = 0.027e^{0.021(T-773)} \tag{D}$$

Estimate the mass of catalyst required to produce 272 kmol styrene yearly at a conversion of 45 percent.

Solution

$$\text{Amount of ethyl benzene required} = \frac{272}{365(24)(0.45)} = 0.069 \text{ kmol/h}$$

$$\text{Amount of steam required} = (10)(0.069) = 0.69 \text{ kmol/h}$$

$$n_j = n_{jo} + \alpha\beta$$

For limiting reactant

$$n_k = n_{k0} + \alpha_k\beta = n_{k0}(1-f)$$

Thus

$$\beta = n_{k0}f$$

$$n_E = 0.069 - \beta = 0.069(1-f)$$

$$n_S = \beta = 0.069f$$

$$n_H = \beta = 0.069f$$

$$n_{H_2O} = 0.69$$

Hence

$$N = n_E + n_S + n_H + n_{H_2O} = 0.069(11+f)$$

Then

$$p_E = \frac{0.069(1-f)}{0.069(11+f)} 1.2 = \frac{1-f}{11+f}(1.2)$$

$$p_S = \frac{0.069f}{0.069(11+f)} 1.2 = \frac{f}{11+f}(1.2)$$

$$p_H = \frac{0.069f}{0.069(11+f)} 1.2 = \frac{f}{11+f}(1.2)$$

At 600°C = 873 K

$$k = 12{,}600e^{11{,}000/873} = 0.042475$$

$$\mathrm{K} = 0.027e^{0.021(873-773)} = 0.2205$$

$$r_m = 0.05097\left[\frac{1-f}{11+f} - \frac{1.2f^2}{0.2205(11+f)^2}\right]$$

Substituting into Eq. (9.17-9) and integrating numerically by Simpson's rule gives

$$W = 0.069\int_0^{0.45} \frac{1}{0.05097[(1-f)/(11+f) - 5.4422f^2/(11+f)^2]}\,df$$

$$= \frac{0.069}{0.05097}\left\{\frac{0.05}{3}[1(11) + 4(11.6466 + 13.2893 + 15.6300 + 19.1962) + 2(12.4000 + 14.3486 + 17.2084) + 1(21.7715)] + \frac{0.05}{2}(21.7715 + 25.2341)\right\} = 9.7072 \text{ kg}$$

9.18 One-Dimensional Without Mixing—Nonisothermal Pseudo-Homogeneous Model

For the nonisothermal case, an energy balance is needed. On account of the heterogeneous system, the accumulation term in the generalized equation (9.14-2) should be modified. The new energy equation for a catalytic reactor is

$$k_L^*\frac{\partial^2 T}{\partial z^2} + \frac{k_R^*\partial}{r\partial r}\left(r\frac{\partial T}{\partial r}\right) - \frac{\partial(u_z \rho C_p T)}{\partial z} + \alpha r_v(\Delta H) = [\epsilon_T \rho C_p + (1-\epsilon_T)\rho_s C_{ps}]\frac{\partial T}{\partial t} \quad \textbf{(9.18-1)}$$

where ρ, ρ_s, C_p, and C_{ps} are the densities and heat capacities of the fluid mixtures and the solid catalyst, respectively. The longitudinal and radial effective thermal conductivities k_L^* and k_R^* can be calculated by (S2)

$$k^* = \epsilon_B \left[k_g + \frac{D_p C_p G}{N_{\text{Pe},r} \epsilon_B} + \frac{4\sigma}{2-\sigma} D_p (0.173) \frac{T^3}{100^4} \right] + (1-\epsilon_B) \frac{h' k_s D_p}{2k_s + h' D_p} \quad \textbf{(9.18-2)}$$

where ϵ_B is the bed porosity, D_p is the particle diameter, C_p is the heat capacity, G is the mass velocity, T is the absolute temperature in degrees Rankine, σ is the emissivity, $N_{\text{Pe},r}$ is the radial Peclet number, which can be evaluated from Fig. 13-5 in Smith (S2) and h' is the heat-transfer coefficient which is the sum of convective h_c, radiative h_R, and particle h_p heat-transfer coefficients

$$h' = h_c + h_R + h_p \quad \textbf{(9.18-3)}$$

in which h_c can be calculated from the j_H factor (S2),

$$h_R = \left(\frac{2k_s + h' D_p}{D_p k_s} \right) \left(\frac{4\sigma}{2-\sigma} \right) D_p (0.173) \frac{T^3}{100^4} \quad \textbf{(9.18-4)}$$

$$h_p = \left(\frac{2k_s + h' D_p}{D_p k_s} \right) k_p \quad \textbf{(9.18-5)}$$

where

$$\log k_p = -1.76 + 0.0129 \frac{k_s}{\epsilon_B} \quad \textbf{(9.18-6)}$$

In addition to the heat transferred to the solid catalyst, there is some heat transferred to the wall. The heat-transfer coefficient at the wall is defined as

$$h_w = \frac{\dot{q}_w}{T_w - T_B} \quad \textbf{(9.18-7)}$$

where h_w is in Btu/h·ft²·°F, $\dot{q}_w$ is the heat flux at the wall, T_w and T_B are the temperatures at the wall and in the bulk stream, respectively. The h_w can be predicted by (C7)

$$\frac{h_w D_T}{k_g} = f^* D_T^{0.17} \left(\frac{D_p G}{\mu} \right)^{0.83} N_{\text{Pr}} \quad \textbf{(9.18-8)}$$

where D_T = the diameter of the tube in feet

D_p = the diameter of the particle in feet

k_g = the thermal conductivity of the fluid in Btu/h·ft²·°F

μ = the fluid viscosity in lb mol/h·ft

G = the superficial mass velocity in lb mole/h·ft²

f^* = a coefficient given by the following table in which D_p in inches and D_T in feet (J1)

D_p/D_T	0.04	0.06	0.08	0.12	0.19	0.295
f^*	0.145	0.175	0.207	0.225	0.235	0.195

Equation (9.18-1) can be transformed to account for the heat transfer from or to the surroundings. Multiplying each term of Eq. (9.18-1) by element area $2\pi r\, dr$, dividing by the cross-sectional area A_c, and integrating each term from 0 to R (the tube radius) gives

$$\frac{1}{A}\int_0^R \{(\epsilon_T\rho C_p)_{\text{gas}} + [(1-\epsilon_T)\rho C_p]_{\text{solid}}\}\frac{\partial T}{\partial t}\,2\pi r\,dr + \frac{1}{A_c}\int_0^R \frac{\partial(u_z\rho C_p T)}{\partial z}\,2\pi r\,dr$$
$$= \frac{k_L^*}{A_c}\int_0^R \frac{\partial^2 T}{\partial z^2}\,2\pi r\,dr + \frac{k_R^*}{A_c}\int_0^R \frac{1}{r}\frac{\partial}{\partial r}\left(r\frac{\partial T}{\partial r}\right)2\pi r\,dr + \frac{1}{A_c}\int_0^R \alpha r_v(\Delta H)2\pi r\,dr \quad \textbf{(9.18-9)}$$

From the definition of average values, we get

$$\{(\epsilon_T C_p)_{\text{gas}} + [(1-\epsilon_T)C_p]_{\text{solid}}\}\frac{\partial \mathbf{T}}{\partial t} + \frac{\partial(\overline{u_z C_p T})}{\partial z}$$
$$= k_L^*\frac{\partial^2 \mathbf{T}}{\partial z^2} + \overline{\alpha r_v(\Delta H)} + \frac{[2\pi r k_R^*(\partial T/\partial r)]_0^R}{A_c} \quad \textbf{(9.18-10)}$$

But

$$-k_R^*\left(\frac{\partial T}{\partial r}\right)_R = U(T - T^*) \quad \textbf{(9.18-11)}$$

where U is the overall heat-transfer coefficient from the surroundings to the reaction mixture and T^* is the wall temperature. Then the last term in Eq. (9.18-10) becomes

$$\frac{1}{A_c}\left[2\pi r k_R^*\left(\frac{\partial T}{\partial r}\right)\right]_0^R = -\frac{1}{A_c}[2\pi R U(T - T^*)] \quad \textbf{(9.18-12)}$$

Now under steady-state conditions with the assumption of one-dimensional plug flow, and neglecting the longitudinal dispersion, Eq. (9.18-10) and Eq. (9.18-12) become

$$u_\infty(\rho C_p)_{\text{fluid}}\frac{\partial \mathbf{T}}{\partial z} = \overline{\alpha r_v(\Delta H)} - \frac{1}{A_c}[2\pi R U(T - T^*)] \quad \textbf{(9.18-13)}$$

$$= \alpha r_m \rho_b(\Delta H) - \frac{4U(T - T^*)}{D_T} \quad \textbf{(9.18-14)}$$

Multiplying through by $A_c\ dz$ gives

$$(FC_p)_T\ dT = -r_m \Delta H\ dW - \frac{4U(T - T^*)}{D_T} A_c\ dz \tag{9.18-15}$$

But

$$F_k\ df = r_m\ dW \tag{9.18-16}$$

Hence

$$(FC_p)_T\ dT = F_k(-\Delta H)df - \pi D_T U(T - T^*)dz \tag{9.18-17}$$

where $(FC_p)_T$ is the product of the feed rate and the heat capacity of the mixture. F_k is the feed rate of the limiting reactant.

Example 9.18-1 Effective Thermal Conductivity

Predict the effective thermal conductivity for the fixed bed to carry out the oxidation of SO_2 from the following data:

Tube diameter 6 cm

Average bed temperature 460°C

Total pressure 800 mmHg

Superficial mass velocity 400 lb/h·ft²

Density of $\frac{1}{8}$ in. × $\frac{1}{8}$ in. cylinder of alumina as catalyst ρ_b = 64 lb/ft³

Porosity of the bed = 0.45

N_{Pr} of the mixture = 0.93

C_p = 1.07 J/g·K = 0.255 Btu/lb·°R

Thermal conductivity of solid k_s = 1.8 Btu/h·ft·°F

Thermal conductivity of fluid $k_g = 3.7 \times 10^{-4}$ J/cm·s·K = 0.0214 Btu/h·ft·°F

Solution

$$\frac{D_p}{D_T} = \frac{(\frac{1}{8})(2.54)}{6} = 0.0529$$

$$G = 400(1/3{,}600)(454)(1/144)(1/2.54)^2 = 0.05429 \text{ g/cm}^2\cdot\text{s}$$

$$N_{Re} = D_p G/\mu = \frac{(\frac{1}{8})(2.54)(0.05429)}{(3.2 \times 10^{-4})} = 53.87$$

From Fig. 13.5 in Smith (S2)

$$Pe_r = \frac{D_p}{D_r} = 9.0[1 + 19.4(0.0529)^2] = 9.4886$$

From Eq. (9.11-3)

$$j_D = \frac{0.357}{0.45(53.87)^{0.359}} = 0.1896$$

because

$$j_H = j_D$$

Therefore

$$h_c = 0.1896(1.07)(0.05429)(0.93)^{-2/3}$$
$$= 0.0115615 \text{ J/cm}^2\cdot\text{s}\cdot\text{K} = 20.46 \text{ Btu/h}\cdot\text{ft}^2\cdot{}^\circ\text{R}$$

From Eq. (9.18-6)

$$\log k_p = -1.76 + 0.0129(1.8/0.45) \qquad \text{or} \qquad k_p = 0.0195$$

Assume

$$h' = 40$$

From Eq. (9.18-5)

$$h_p = \frac{2(1.8) + 40(1/8)(1/12)}{(1/8)(1/12)(1.8)}(0.0195) = 4.177 \text{ Btu/h}\cdot\text{ft}^2\cdot{}^\circ\text{R}$$

From Eq. (9.18-4)

$$h_R = \frac{2(1.8) + 40(1/8)(1/12)}{(1/8)(1/12)(1.8)}\left(\frac{4(0.65)}{2-0.65}\right)\left(\frac{1}{8}\right)\left(\frac{1}{12}\right)(0.173)\left(\frac{1319.4^3}{100^4}\right)$$
$$= 17.0735 \text{ Btu/h}\cdot\text{ft}^2\cdot{}^\circ\text{R}$$

From Eq. (9.18-3), $h' = 20.46 + 4.177 + 17.07 = 41.707$ versus 40 assuming

$$\frac{D_pC_pG}{N_{\text{Pe},r}\epsilon_B} = \frac{(1/8)(1/12)(0.255)(400)}{9.4886(0.45)} = 0.2488 \text{ Btu/h}\cdot\text{ft}\cdot{}^\circ\text{F}$$

$$\frac{(1-\epsilon_B)h'k_sD_p}{2k_s + h'D_p} = \frac{(1-0.45)(41.707)(1.8)(1/8)(1/12)}{2(1.8) + 41.707(1/8)(1/12)} = 0.1066$$

From Eq. (9.18-2)

$$k^* = 0.45(0.0214 + 0.2488 + 0.0797) + 0.1066 = 0.264 \text{ Btu/h}\cdot\text{ft}\cdot{}^\circ\text{F}$$

Example 9.18-2 Adiabatic Pseudo-Homogeneous Model

What should be the inlet temperature of the gaseous mixture if the reactor in Example 9.17-1 is operated adiabatically with 9.7072 kg of the catalyst? The data are

ΔH of ethyl benzene = 140,000 kJ/kmol

C_p of the reaction mixture = 2.18 kJ/kg K

Solution

Since adiabatic, Eq. (9.18-17) becomes

$$(FC_p)_T dT = -F_k(\Delta H)df \tag{A}$$

$$[0.069(106)+0.69(18)](2.18)dT = -0.069(140{,}000)df$$

Integrating from T_o to T and 0 to f gives

$$T = T_o - 224.55f \tag{B}$$

But

$$k = 12{,}600e^{(-11{,}000/T)} \tag{C}$$

$$K = 0.027e^{0.021(T-773)} \tag{D}$$

$$r_m = \frac{1.2k}{11+f}\left[1-f-\frac{1.2f^2}{K(11+f)}\right] \tag{E}$$

$$W = \int_0^{0.45} \frac{0.069}{r_m}\,df = 0.069\int_0^{0.45} \frac{1}{r_m}\,df = 9.7072 \tag{F}$$

The calculation involves a trial-and-error solution. A T_o is assumed. Calculate T, k, K in sequence from Eqs. (B), (C), and (D). Then for each value of f, $1/r_m$ can be calculated. The integral can be calculated. Repeat the process until the result checks Eq. (F). The result is T_o, the inlet temperature of the reaction mixture, 941 K.

9.19 One-Dimensional Without Mixing with Pressure-Drop Consideration

With pressure-drop consideration, the generalized momentum balance equation (9.15-2) is used. Under steady-state conditions and neglecting the gravity term and the effect of the longitudinal direction to the velocity, Eq. (9.15-2) becomes

$$\frac{\mu}{g_c r}\frac{\partial}{\partial r}\left(r\frac{\partial u}{\partial r}\right) = \frac{\partial P}{\partial z} \tag{9.19-1}$$

However, the partial derivative of P with respect to z is simply the pressure drop in the flow direction for a length of L, or

$$\frac{\partial P}{\partial z} = -\frac{\Delta P}{L} \tag{9.19-2}$$

Thus Eq. (9.19-1) becomes

$$\frac{\mu}{g_c r}\frac{\partial}{\partial r}\left(r\frac{\partial u}{\partial r}\right) = -\frac{\Delta P}{L} \tag{9.19-3}$$

Integrating once yields

$$\frac{\mu}{g_c}\left(r\frac{\partial u}{\partial r}\right) = -\frac{\Delta P}{L}\frac{r^2}{2} + A \tag{9.19-4}$$

where A is an arbitrary constant. The boundary condition is

$$\frac{\partial u}{\partial r} = 0 \quad \text{at} \quad r = 0 \tag{9.19-5}$$

To satisfy this boundary condition, A must be zero. Therefore

$$\frac{\mu}{g_c}\frac{\partial u}{\partial r} = -\frac{\Delta P r}{L 2} \tag{9.19-6}$$

From transport phenomena, it is known that

$$\frac{\mu}{g_c}\frac{\partial u}{\partial r} = \tau_{rz} = \frac{\rho u^2}{2g_c} f_f \tag{9.19-7}$$

where f_f is the Fanning friction factor. Equating Eqs. (9.19-6) and (9.19-7) gives

$$-\frac{\Delta P r}{L 2} = \frac{\rho u^2}{2 g_c D} f_D \tag{9.19-8}$$

where D is the diameter of the tube and f_D is the Darcy friction factor, which is equal to $4f_f$. Equation (9.19-8) is good for the fluid flowing through an empty tube. For a packed bed, the interstitial velocity u is related to the superficial velocity u_∞ as

$$u = \frac{u_\infty}{\epsilon} \tag{9.19-9}$$

and the diameter D in Eq. (9.19-8) should be changed to the equivalent diameter

D_{eq} which is equal to four times the hydraulic radius R_h. But

$$R_h = \frac{\text{Cross-sectional flow area } A}{\text{Wetted perimeter } P}$$

$$= \frac{AL}{PL} = \frac{\text{Flow volume}}{\text{Total wetted surface}} = \frac{\text{Void volume/Total volume}}{\text{Total wetted surface/Total volume}} \tag{9.19-10}$$

$$= \frac{\epsilon}{a_p}$$

where a_p is the interstitial surface area. Because for spherical particle,

$$\frac{a_p}{1-\epsilon} = \frac{\pi D_p^2}{(\pi/6)D_p^3} \tag{9.19-11}$$

or

$$a_p = \frac{6(1-\epsilon)}{D_p} \tag{9.19-12}$$

Hence

$$R_h = \frac{\epsilon D_p}{6(1-\epsilon)} \tag{9.19-13}$$

and

$$D_{eq} = 4R_h = \frac{4\epsilon D_p}{6(1-\epsilon)} \tag{9.19-14}$$

With these modifications, the pressure drop for a fluid flowing through a packed bed is

$$-\frac{\Delta P}{L} = \frac{3}{2}\frac{f_D}{g_c D_p}\left(\frac{\rho u_\infty^2}{2}\right)\left(\frac{1-\epsilon}{\epsilon^3}\right) \tag{9.19-15}$$

Instead of using the Darcy friction factor, a friction factor for the particle f_p in the bed is correlated with the previous parameters as:

1. For flow in laminar, the modified

$$N_{Re} = \frac{u_\infty D_p \rho}{\mu(1-\epsilon)} < 10$$

$$\frac{3}{2} f_D = \frac{300}{N_{Re}} \quad \text{in Eq. (9.19-15)} \tag{9.19-16}$$

which is the Blake-Kozeny equation (K2).

2. For flow in turbulent, the modified

$$N_{Re} = \frac{u_\infty D_p \rho}{\mu(1-\epsilon)} > 1{,}000$$

$$\frac{3}{2} f_D = 3.5 \quad \text{in Eq. (9.19-15)} \tag{9.19-17}$$

which is the Burke-Plummer equation (B1).

3. For flow in transition region,

$$10 < N_{Re} = \frac{D_p u_\infty \rho}{\mu(1-\epsilon)} < 1000$$

$$-\frac{\Delta P}{L} = \left(3.5 + \frac{300}{N_{Re}}\right) \frac{\rho u_\infty^2 (1-\epsilon)}{2 g_c D_p \epsilon^3} \tag{9.19-18}$$

which is the Ergun equation (E1).

9.20 *One-Dimensional with Axial Mixing*

For one-dimensional with axial mixing under steady-state conditions, the generalized mass and energy balance equations (9.13-2) and (9.14-2) become

$$D_L \frac{\partial^2 C_j}{\partial z^2} - r_m \rho_b = \frac{u_\infty \partial C_j}{\partial z} \tag{9.20-1}$$

$$k_L^* \frac{\partial^2 T}{\partial z^2} - r_m \rho_b \Delta H - \frac{4U(T - T^*)}{D_T} = \frac{\partial(u_\infty \rho C_p T)}{\partial z} \tag{9.20-2}$$

The boundary conditions are:

$$\text{At } z = 0 \qquad u_\infty(C_{A0} - C_A) = -D_L \frac{dC_A}{dz} \tag{9.20-3}$$

$$\rho u_\infty C_p (T_0 - T) = -k_L^* \frac{dT}{dz} \tag{9.20-4}$$

$$\text{At } z = L \qquad \frac{dC_A}{dz} = 0 \tag{9.20-5}$$

$$\frac{dT}{dz} = 0 \tag{9.20-6}$$

A similar solution of the isothermal case, i.e., Eq. (9.19-19) with Eq. (9.19-21) and Eq. (9.19-23) has been presented in Chapter 8 in the section called "Dispersion

Model." For the nonisothermal case, the solution is better done by numerical method, which is a special case of the method in the following section.

9.21 Two-Dimensional Pseudo-Homogeneous Model Without Mixing

Under steady-state conditions, the generalized mass and energy balance equations (9.13-2) and (9.14-2) become

$$\frac{\partial(u_\infty C_A)}{\partial z} - \frac{D_R\partial}{r\,\partial r}\left(r\frac{\partial C_A}{\partial r}\right) + \rho_b r_m = 0 \quad \textbf{(9.21-1)}$$

$$\frac{\partial(u_\infty \rho C_p T)}{\partial z} - \frac{k_R^*\partial}{r\,\partial r}\left(r\frac{\partial T}{\partial r}\right) + \rho_b r_m \Delta H = 0 \quad \textbf{(9.21-2)}$$

The boundary conditions are:

$$\text{At } z=0 \qquad C_A = C_{A0} \qquad T = T_0 \quad \textbf{(9.21-3)}$$

$$\text{At } r=0 \qquad \frac{\partial C_A}{\partial r} = 0 \qquad \frac{\partial T}{\partial r} = 0 \quad \textbf{(9.21-4)}$$

$$\text{At } r=R \qquad \frac{\partial T}{\partial r} = -U\left(\frac{T-T^*}{k_R^*}\right) \quad \text{or others} \quad \textbf{(9.21-5)}$$

Assuming a first-order irreversible reaction and using the Arrhenius equation, Eq. (9.21-1) and Eq. (9.21-2) can be rearranged as functions of the fractional conversion f as

$$\frac{\partial f}{\partial z} - \frac{D_R}{u_\infty r}\frac{\partial}{\partial r}\left(r\frac{\partial f}{\partial r}\right) - \rho_b k_o \frac{e^{-E/RT}}{u_\infty}(1-f) = 0 \quad \textbf{(9.21-6)}$$

$$\frac{\partial T}{\partial z} - \frac{k_R}{\rho C_p u_\infty r}\frac{\partial}{\partial r}\left(r\frac{\partial T}{\partial r}\right) + \frac{\rho_b k_o e^{-E/RT}}{\rho C_p u_\infty} C_{A0}(1-f)\Delta H = 0 \quad \textbf{(9.21-7)}$$

The boundary conditions become

$$\text{At } z=0 \qquad f=0 \qquad T=T_0 \quad \textbf{(9.21-8)}$$

$$\text{At } r=0 \qquad \frac{\partial f}{\partial r} = 0 \qquad \frac{\partial T}{\partial r} = 0 \quad \textbf{(9.21-9)}$$

$$\text{At } r=R \qquad \frac{\partial f}{\partial r} = 0 \qquad \frac{\partial T}{\partial r} = -U\left(\frac{T-T^*}{k_R^*}\right) \quad \text{or others} \quad \textbf{(9.21-10)}$$

Example 9.21-1 Fixed-Bed Catalytic Reactor Design

Component A in a fluid undergoes a first-order irreversible reaction in a tubular fixed-bed catalytic reactor. The wall temperature maintains at 291°C and the entering conversion is everywhere zero. Using the finite-difference method, develop a computer program to calculate the steady-state temperature and conversion profile across and down the reactor for the following data:

Tube radius: 1.905 cm

Tube length: 9.144 cm

Average heat capacity of mixture = 0.2 cal/g·°C

Average density of mixture = 1.6×10^{-3} g/cm^3

Average velocity (plug flow) in reactor = 91,440 cm/h

Density of catalytic bed = 1.443 g/cm^3

Enthalpy of reaction = −16,651 cal/g A reacted

Average thermal conductivity of mixture = 0.2234 cal/h·cm·°C

Mass diffusion coefficient = 4645 cm^2/h

Inlet concentration of A = 4.00×10^{-5} g/cm^3

The rate of reaction is:

$$R_A = -r_A C_{A0} \rho_c \text{ g A reacted/h·cm}^3$$

where $r_A = 1{,}223{,}040 e^{-5{,}389/T}(1-f)$ cm^3/h·g of catalyst

$T = \text{K}$

$f =$ fractional conversion of A; grams A reacted per gram of A entering

$\rho_c =$ density of catalyst, g/cm^3

The entering temperature profile is presented below

Distance from center of tube, cm	Temperature, °C
0.0	316
0.381	316
0.762	316
1.143	316
1.524	316
1.905	291

Solution

Equations (9.21-6) and (9.21-7) can also be written as

$$\frac{\partial f}{\partial z} - \frac{D_R}{u_\infty}\left(\frac{\partial f}{r\,\partial r} + \frac{\partial^2 f}{\partial r^2}\right) - \frac{r_m \rho_b}{u_\infty} = 0 \quad \textbf{(A)}$$

$$\frac{\partial T}{\partial z} - \frac{k_R}{\rho C_p u_\infty}\left(\frac{\partial T}{r\,\partial r} + \frac{\partial^2 T}{\partial r^2}\right) + \frac{r_m C_{A0} \Delta H}{\rho C_p u_\infty} = 0 \quad \textbf{(B)}$$

where the rate of the reaction r_m is

$$r_m = k_o e^{-E/RT}(1 - f) \text{ in g}^3/\text{h}\cdot\text{g of catalyst}$$

The boundary conditions for Eq. (A) are

$$\text{At } z = 0, f = 0 \qquad \text{at } r = 0, \frac{\partial f}{\partial r} = 0 \qquad \text{at } r = R, \frac{\partial f}{\partial r} = 0$$

Those of Eq. (B) are

$$\text{At } z = 0, T = 316 \text{ K} \qquad \text{at } r = 0, \frac{\partial T}{\partial r} = 0 \qquad \text{at } r = R, T = 291 \text{ K}$$

Using the following finite difference approximations

$$\frac{\partial T}{\partial r} = \frac{\Delta T}{\Delta r} = \frac{T_{m+1,L} - T_{m,L}}{\Delta r} \tag{C}$$

$$\frac{\partial T}{\partial z} = \frac{\Delta T}{\Delta z} = \frac{T_{m,L+1} - T_{m,L}}{\Delta z} \tag{D}$$

$$\frac{\partial^2 T}{\partial r^2} = \frac{\Delta^2 T}{\Delta r^2} = \frac{(T_{m+1,L} - T_{m,L}) - (T_{m,L} - T_{m-1,L})}{\Delta r^2} \tag{E}$$

Equations (A) and (B) become

$$f_{m,L+1} = f_{m,L} + \frac{\Delta z}{(\Delta r)^2}\frac{D_R}{u_\infty}\left[\frac{1}{m}(f_{m+1,L} - f_{m,L}) + f_{m+1,L} - 2f_{m,L} + f_{m-1,L}\right] - \left[\frac{(r_m)_{av}\rho_b}{u_\infty}\right]\Delta z \tag{F}$$

$$T_{m,L+1} = T_{m,L} + \frac{\Delta z}{(\Delta r)^2}\frac{k_R}{\rho C_p u_\infty}\left[\frac{1}{m}(T_{m+1,L} - T_{m,L}) + T_{m+1,L} - 2T_{m,L} + T_{m-1,L}\right] - \left[\frac{(\Delta H)(r_m)_{av}C_{Ao}\rho_b}{\rho C_p u_\infty}\right]\Delta z \tag{G}$$

where $\Delta r/r$ has been replaced by $1/m$. At the center, L'Hospital's rule reduces the equations to

$$f_{0,L+1} = f_{0,L} + \frac{4\Delta z}{(\Delta r)^2}\frac{D_R}{u_\infty}(f_{1,L} - f_{0,L}) - \left[\frac{(r_m)_{av}\rho_b}{u_\infty}\right]\Delta z \tag{H}$$

$$T_{0,L+1} = T_{0,L} + \frac{4\Delta z\, k_R}{(\Delta r)^2 \rho C_p u_\infty}(T_{1,L} - T_{0,L}) - \left[\frac{(\Delta H)(r_m)_{av}C_{Ao}\rho_b}{\rho C_p u_\infty}\right]\Delta z \tag{I}$$

All of these equations are programmed in Fig. 9.21-1.

```
 81/12/23. 12.31.17.
PROGRAM   CATREC

00100 PROGRAM CATREC(INPUT,OUTPUT)
00110CGASRHO=GAS DENSITY,GM/CM**3
00120CCATRHO=CATALYST DENSITY,GM/CM**3
00130CSPHT=SPECIFIC HEAT,CAL/GM-DEGC
00140CVEL=VELOCITY,CM/HR
00150CTHCON=THERMAL CONDUCTIVITY,CAL/HR-CM-DEGC
00160CDIF=DIFFUSIVITY,CM**2/HR
00170CRAV=AVERAGE RATE OF REACTION,GM/HR-CM**3
00180CDH=HEAT OF REACTION,CAL/GM A REACTED
00190CYOA=MOL FRACTION OF A ENTERING
00200CDR=INCREMENT OF RADIUS,CM
00210CDZ=INCREMENT OF LENGTH,CM
00220CCDA=INITIAL CONCENTRATION OF A,GM/CM**3
00230CNR=NUMBER OF RADIUS INCREMENTS
00240CNZ=NUMBER OF LENGTH INCREMENTS
00250 DIMENSION T(7,8),X(7,8)
00260 READ*,GASRHO,CATRHO,SPHT,VEL,THCON,DIF
00270 READ*,RAV,DH,YOA,DR,DZ,CDA
00280 READ*,NR,NZ
00290 N1=NR+1
00300 DO 180 I=1,N1
00310 180 READ*,T(I,1),X(I,1)
00320 L=0
00330 PRINT 233,L
00340 DO 133 I=1,N1
00350 N2=I-1
00360 PRINT 22,N2,X(I,1),T(I,1)
00370 22 FORMAT(29X,I2,23X,E12.4,18X,E12.4)
00380 133 CONTINUE
00390 CONS1=(DZ*THCON)/(GASRHO*VEL*SPHT*(DR**2.))
00400 CONS2=(CATRHO*DZ)/(GASRHO*VEL*SPHT)
00410 CONS3=(DZ*DIF)/(VEL*(DR**2.))
00420 CONS4=(CATRHO*DZ)/VEL
00430 DO 300 L=1,N2
```

```
00440 DO 311 I=1,N1
00450 Z=I
00460 M=I+1
00470 K=I-1
00480 J=L+1
00490 99 IF(1-I) 199,111,111
00500 199 IF(I-N1) 100,700,100
00510 100 T(I,J)=T(I,L)+CONS1*((T(M,L)-T(I,L))/Z+T(M,L)-2.*T(I,L)+T(K,L))
00520+-CONS2*RAV*DH*CDA
00530 X(I,J)=X(I,L)+CONS3*((X(M,L)-X(I,L))/Z+X(M,L)-2.*X(I,L)+X(K,L))+RAV
00540+*CONS4
00550 GO TO 400
00560 111 T(I,J)=T(1,L)+2.*CONS1*(2.*T(2,L)-2.*T(1,L))-CONS2*RAV*DH*CDA
00570 X(I,J)=X(1,L)+2.*CONS3*(2.*X(2,L)-2.*X(1,L))+RAV*CONS4
00580 GO TO 400
00590 700 T(I,J) = 291.
00600 X(I,J)=X(I,L)+CONS3*((X(K,L)+X(I,L))/Z+X(K,L)-2.*X(I,L)+X(K,L))
00610++RAV*CONS4
00620 GO TO 311
00630 400 CALL RATE (T(I,J),X(I,J),R)
00640 CALL RATE(T(I,L),X(I,L),ROLD)
00650 IF(X(I,J)) 322,388,388
00660 388 IF(ABS(((ROLD+R)/2.-RAV)/RAV)-0.02) 311,322,322
00670 322 RAV=((ROLD+R)/2.+RAV)/2.
00680 GO TO 99
00690 311 CONTINUE
00700 PRINT 233,L
00710 DO 455 I=1,N1
00720 N2=I-1
00730 455 PRINT 22, N2,X(I,J),T(I,J)
00740 233 FORMAT(////48X,"LONGITUDINAL INCREMENT = ",I2//27X,"RADIAL"
00750+,22X,"CONVERSION",20X,"TEMPERATURE"/25X,"INCREMENT"//)
00760 300 CONTINUE
00770 STOP
00780 END
00790 SUBROUTINE RATE(TEMP,CON,R)
```

Figure 9.21-1

```
00800 R=1.223E+6*(EXP(-5389./(TEMP+273.)))*(1.-CON)
00810 RETURN
00820 END
READY.
RNH

? 1.6E-3,1.443,0.2,91440.,0.2234,4645.
? 3.205E-2,-16651.,0.1,0.381,1.524,4.E-5
? 5,6
? 316.,0.
? 316.,0.
? 316.,0.
? 316.,0.
? 316.,0.
? 291.,0.


                    LONGITUDINAL INCREMENT =  0

  RADIAL               CONVERSION                TEMPERATURE
INCREMENT

    0                     0.                     .3160E+03
    1                     0.                     .3160E+03
    2                     0.                     .3160E+03
    3                     0.                     .3160E+03
    4                     0.                     .3160E+03
    5                     0.                     .2910E+03
```

LONGITUDINAL INCREMENT = 1

RADIAL INCREMENT	CONVERSION	TEMPERATURE
0	.3257E-02	.3228E+03
1	.3257E-02	.3228E+03
2	.3257E-02	.3228E+03
3	.3257E-02	.3228E+03
4	.3257E-02	.3204E+03
5	.3257E-02	.2910E+03

LONGITUDINAL INCREMENT = 2

RADIAL INCREMENT	CONVERSION	TEMPERATURE
0	.6857E-02	.3303E+03
1	.6857E-02	.3303E+03
2	.6857E-02	.3303E+03
3	.6857E-02	.3300E+03
4	.6746E-02	.3250E+03
5	.7325E-02	.2910E+03

LONGITUDINAL INCREMENT = 3

RADIAL INCREMENT	CONVERSION	TEMPERATURE
0	.1091E-01	.3387E+03
1	.1091E-01	.3387E+03
2	.1091E-01	.3387E+03
3	.1084E-01	.3380E+03
4	.1092E-01	.3299E+03
5	.1170E-01	.2910E+03

Figure 9.21-1 (continued)

LONGITUDINAL INCREMENT = 4

RADIAL INCREMENT	CONVERSION	TEMPERATURE
0	.1549E-01	.3482E+03
1	.1549E-01	.3482E+03
2	.1543E-01	.3481E+03
3	.1551E-01	.3468E+03
4	.1538E-01	.3352E+03
5	.1689E-01	.2910E+03

LONGITUDINAL INCREMENT = 5

RADIAL INCREMENT	CONVERSION	TEMPERATURE
0	.2071E-01	.3591E+03
1	.2067E-01	.3591E+03
2	.2074E-01	.3589E+03
3	.2061E-01	.3566E+03
4	.2075E-01	.3409E+03
5	.2250E-01	.2910E+03

LONGITUDINAL INCREMENT = 6

RADIAL INCREMENT	CONVERSION	TEMPERATURE
0	.2676E-01	.3719E+03
1	.2689E-01	.3718E+03
2	.2675E-01	.3714E+03
3	.2682E-01	.3678E+03
4	.2648E-01	.3471E+03
5	.2917E-01	.2910E+03

Figure 9.21-1 (continued)

9.22 Heterogeneous Models—One-Dimensional Model with Interfacial Gradients

The following equations can also be derived from the generalized mass and energy balance equations (9.13-2) and (9.14-2):

Fluid:

$$-u_\infty \frac{dC}{dz} = k_g a_v (C - C_\infty) \tag{9.22-1}$$

$$u_\infty \rho_g C_p \frac{dT}{dz} = h_c a_v (T_\infty - T) - 4\frac{U}{D_T}(T - T_R) \tag{9.22-2}$$

Solid:

$$\rho_b r_m = k_g a_v (C - C_\infty) \tag{9.22-3}$$

$$-(\Delta H)\rho_b r_m = h_c a_v (T_\infty - T) \tag{9.22-4}$$

with boundary conditions

At $$z = 0 \qquad C = C_0 \qquad T = T_0 \tag{9.22-5}$$

where u_∞ = the superficial velocity
k_g = the mass transfer coefficient
C_∞ = the concentration at the surface of the catalyst
a_v = the surface area of the catalyst per unit volume
ρ_g = the density of the fluid mixture
C_p = the heat capacity of the fluid mixture
h_c = the heat-transfer coefficient
T_∞ = the surface temperature of the catalyst
T_R = the surrounding temperature
ρ_b = the bulk density of the catalyst
r_m = the rate of the reaction per unit mass of catalyst
ΔH = the heat of reaction

9.23 One-Dimensional Model with Interfacial and Intraparticle Gradients

The model equations are similarly derived from the generalized mass and energy equations (9.13-2) and (9.14-2) as follows:

Fluid:

$$-u_\infty \frac{dC}{dz} = k_g a_v (C - C_\infty) \tag{9.23-1}$$

$$u_\infty \rho_g C_p \frac{dT}{dz} = h_c a_v (T_\infty - T) - \frac{4U}{D_T}(T - T_R) \tag{9.23-2}$$

Solid:

$$\frac{D_e d}{\Omega^2\, d\Omega}\left(\Omega^2 \frac{dC_s}{d\Omega}\right) - \rho_s r_m(C_s, T_s) = 0 \tag{9.23-3}$$

$$\frac{k_{\text{eff}} d}{\Omega^2\, d\Omega}\left(\Omega^2 \frac{dT_s}{d\Omega}\right) + \rho_s(-\Delta H) r_m(C_s, T_s) = 0 \tag{9.23-4}$$

with boundary conditions:

$$\text{At } z = 0 \qquad C = C_0 \qquad T = T_0 \tag{9.23-5}$$

$$\text{At } \Omega = 0 \qquad \frac{dC_s}{d\Omega} = 0 \qquad \frac{dT_s}{d\Omega} = 0 \tag{9.23-6}$$

$$\text{At } \Omega = \frac{D_p}{2} \qquad k_g(C_\infty - C) = -D_{\text{eff}} \frac{dC_s}{d\Omega} \qquad h_c(T_\infty - T) = -k_{\text{eff}} \frac{dT_s}{d\Omega} \tag{9.23-7}$$

where D_e = the effective diffusivity

k_{eff} = the effective thermal conductivity

Ω = the radial coordinate of the particle

ρ_s = the solid density

C_s = the solid concentration

T_s = the solid temperature

If the effectiveness factor is used, the system reduces to Eqs. (9.23-1) and (9.23-2) with the boundary condition [Eq. (9.23-5)] and the following equations:

$$k_g a_v (C - C_\infty) = \eta \rho_b r_m(C_\infty, T_\infty) \tag{9.23-8}$$

$$h_c a_v (T_\infty - T) = \eta \rho_b (-\Delta H) r_m(C_\infty, T_\infty) \tag{9.23-9}$$

with

$$\eta = f(C_\infty, T_\infty) \tag{9.23-10}$$

$$= \left(\frac{3}{\phi^2}\right)(\phi \coth \phi - 1) \tag{9.23-11}$$

where

$$\phi = \left(\frac{D_p}{2}\right)\left[\frac{k(T_\infty)}{D_{\text{eff}}}\right]^{1/2} \tag{9.23-12}$$

If the effectiveness factor is expressed as a function of bulk fluid conditions

$$\eta_G = f(C,T) \tag{9.23-13}$$

$$= \frac{3\text{Sh}}{\phi^2}\left(\frac{\phi\cosh\phi - \sinh\phi}{\phi\cosh\phi + (\text{Sh}-1)\sinh\phi}\right) \tag{9.23-14}$$

where

$$\text{Sh} = \frac{k_g\delta}{D_{\text{eff}}} \quad \text{and} \quad \phi = \left(\frac{D_p}{2}\right)\left[\frac{k(T)}{D_e}\right]^{1/2} \tag{9.23-15}$$

and δ is the film thickness. The equations become

$$-u_\infty \frac{dC}{dz} = \eta_G \rho_b r_m(C,T) \tag{9.23-16}$$

$$u_\infty \rho_g C_p \frac{dT}{dz} = \eta_G \rho_b r_m(C,T) - \frac{4U}{D_T}(T - T_R) \tag{9.23-17}$$

9.24 Two-Dimensional Heterogeneous Models

The model equations are:

$$u_\infty \frac{\partial C}{\partial z} = D_e\left(\frac{\partial^2 C}{\partial r^2} + \frac{1\partial C}{r\partial r}\right) - k_g a_v(C - C_\infty) \tag{9.24-1}$$

$$u_\infty \rho_g C_p \frac{\partial T}{\partial z} = k_{er}^f\left(\frac{\partial^2 T}{\partial r^2} + \frac{1}{r}\frac{\partial T}{\partial r}\right) + h_c a_v(T_\infty - T) \tag{9.24-2}$$

$$k_g a_v(C - C_\infty) = \eta \rho_b r_m \tag{9.24-3}$$

$$h_c a_v(T_\infty - T) = \eta\rho_b(-\Delta H) r_m + k_{er}^s\left(\frac{\partial^2 T_s}{\partial r^2} + \frac{1}{r}\frac{\partial T_s}{\partial r}\right) \tag{9.24-4}$$

with boundary conditions

$$\text{At } z = 0 \qquad C = C_0 \qquad T = T_0 \qquad \frac{\partial C}{\partial r} = 0 \tag{9.24-5}$$

$$\text{At } r=0 \qquad \frac{\partial T}{\partial r}=\frac{\partial T_s}{\partial r}=0 \qquad \frac{\partial C}{\partial r}=0 \tag{9.24-6}$$

$$\text{At } r=R \qquad h_w^f(T_w - T)=k_{er}^f\frac{\partial T}{\partial r} \qquad h_w^s(T_w - T_s)=k_{er}^s\frac{\partial T_s}{\partial r} \tag{9.24-7}$$

where k_{er}^f = the effective thermal conductivity in the r direction for fluid
k_{er}^s = the effective thermal conductivity in the r direction for solid
h_w^f = the fluid convective heat-transfer coefficient in the vicinity of wall
h_w^s = the solid convective heat-transfer coefficient in the vicinity of wall
T_w = the temperature of the wall

9.25 Design of a Fluidized-Bed Reactor (S2)

The equipment to carry out a chemical reaction in a fluidized bed has been described previously. The essential feature of a fluidized bed is that the catalyst in the bed starts to move when the gas flowing through the bed reaches the minimum fluidization velocity. The voidage at minimum fluidization is calculated from

$$\Delta P_t = L_{\text{mf}}(1-\epsilon_{\text{mf}})(\rho_s - \rho_g) \tag{9.25-1}$$

where ΔP_t = the total pressure drop
L_{mf} = the height of fluidized bed at minimum fluidization
ρ_s = the density of the catalyst
ρ_g = the density of the gas

The minimum fluidization velocity is calculated from an empirical correlation by Leva (L1):

$$u_{\text{mf}} = 1.118\times 10^{-13}\frac{D_p^{1.82}(\rho_s-\rho_g)^{0.94}}{\rho_g^{0.06}\ \mu^{0.88}} \tag{9.25-2}$$

where u_{mf} = the minimum fluidization velocity in meters per second
D_p = the particle diameter in microns
μ = the viscosity of the fluid in Newton-seconds per square meter
ρ = the density of the fluid in kilograms per cubic meter

As the velocity is further increased, the bed may become less dense and finally the catalyst particles may be blown out. This maximum terminal velocity is calculated by

$$u_t = \sqrt{\frac{4gD_p(\rho_s - \rho_g)}{3\rho_g C_D}} \tag{9.25-3}$$

For sphere and laminar flow $C_D = 24/\text{Re}$ where $\text{Re} = D_p \rho_g u_t/\mu$. For Re > 0.4, C_D is obtained experimentally.

In a fluidized bed, there are three kinds of heat transfer worthy of discussion:

1. At internal surfaces, the heat-transfer coefficient h_i is calculated by the Wender and Cooper equation (W3)

$$\frac{h_i D_p/k_g}{1-\epsilon}\left(\frac{k_g}{C_g \rho_g}\right)^{0.43} = 0.033 C_R \left(\frac{D_p G}{\mu}\right)^{0.23}\left(\frac{C_s}{C_g}\right)^{0.80}\left(\frac{\rho_p}{\rho_g}\right)^{0.66} \tag{9.25-4}$$

where C_g = the heat capacity of the gas

C_s = the heat capacity of the solid

C_R = a correction factor for the displacement of the immersed tube from the axis of the vessel

r/R	0	0.1	0.2	0.3	0.4	0.5	0.6	0.68
C_R	0	1.50	1.62	1.70	1.72	1.70	1.63	1.60

r = the distance from the center of the vessel

R = the radius of the vessel

2. At external surfaces, the heat-transfer coefficient h_o is also calculated by the Wender and Cooper equation (W3):

$$\frac{h_o D_p}{k_g(1-\epsilon)(C_s \rho_p/C_g \rho_g)} = f(1 + 7.5e^{-0.44 L_H C_s/D_T C_g}) \tag{9.25-5}$$

where L_H = the bed depth

D_T = diameter of the vessel

f = a coefficient as a function of $D_p G/\mu$

D_pG/μ	0.04	0.1	0.4	1.0	4.0	10.0	40.0
$f(10^4)$	0.275	0.83	3.80	8.0	19.9	28.1	36.8

3. Particle-surface coefficient h_s is calculated by the Balakrishnan and Pei correlation (B2)

$$j_H = 0.043\left[\frac{D_p g(\rho_s - \rho_g)(1-\epsilon)^2}{u_\infty^2 \rho_g}\right]^{0.25} \tag{9.25-6}$$

In general, there are four important models: (1) May (M1) and Van Deemster (V1), (2) Davidson and Harrison (D1), (3) Rowe and Partridge (R3), and (4) Kunni and Levenspiel (K3). It is suggested that readers consult the references for details. However, a simplified model can be derived from the generalized mass equation (4.2-17) as follows. Imagine that there are two phases in a fluidized bed, a gas-bubble phase and a dense phase as shown in Fig. 9.25-1.

Bubble-gas phase

If there is no mixing, $D = 0$ in Eq. (4.2-17); if there is no reaction in this phase, $r = 0$; if the interstitial velocity in the direction of the flow z is constant, under steady-state conditions Eq. (4.2-17) becomes

$$u_b \frac{dC_{jb}}{dz} = -k_m a_m (C_{jb} - C_{jd}) = k_m a_m (C_{jd} - C_{jb}) \tag{9.25-7}$$

in which u_b is the interstitial velocity of the j component in the gas-bubble phase. However, it is more convenient to use the superficial velocity u_∞ than the interstitial velocity

$$u_b \epsilon = u_{b\infty} \tag{9.25-8}$$

where ϵ is the void volume. Equation (9.25-7) becomes

$$u_{b\infty} \frac{dC_{jb}}{dz} = k_m a_m \epsilon (C_{jd} - C_{jb}) \tag{9.25-9}$$

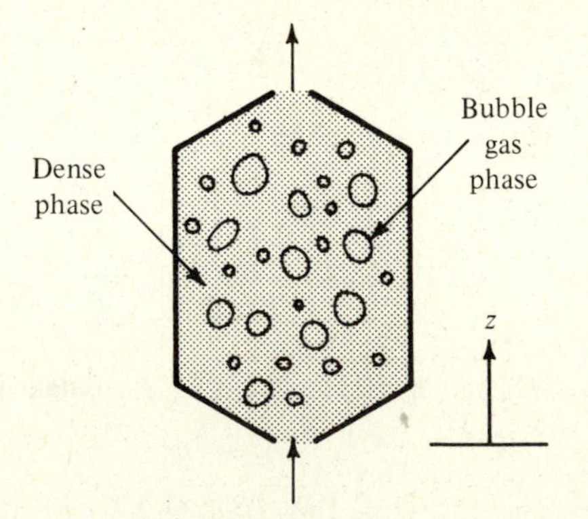

Figure 9.25-1 Two-phase model

where C_{jb} and C_{jd} are the concentrations of j component in the gas-bubble and dense phases, respectively. a_m is the mass-transfer area per unit mass between the gas-bubble phase and the dense phase. Hence the transfer area per unit volume a_V is

$$a_m \epsilon = a_v \tag{9.25-10}$$

Hence Eq. (9.25-9) can be further transformed to

$$u_{b\infty} \frac{dC_{jb}}{dz} = k_m a_v (C_{jd} - C_{jb}) \tag{9.25-11}$$

Dense phase

For a single reaction and steady-state conditions, Eq. (4.2-17) becomes

$$u_{d\infty} \frac{\partial C_{jd}}{\partial z} = \epsilon D \frac{\partial^2 C_{jd}}{\partial z^2} + \epsilon \alpha_j r_v + k_m a_v (C_{jb} - C_{jd}) \tag{9.25-12}$$

But

$$r_v = r_p \rho_B$$

where r_v is the reaction rate per unit volume, r_p is the intrinsic rate per unit mass of catalyst and ρ_B is the density of the catalyst in the dense phase. Then Eq. (9.25-12) becomes

$$-u_{d\infty} \frac{\partial C_{jd}}{\partial z} + \epsilon D \frac{\partial^2 C_{jd}}{\partial z^2} - \epsilon \rho_B r_p + k_m a_v (C_{jb} - C_{jd}) = 0 \tag{9.25-13}$$

where $u_{d\infty}$ is the superficial velocity in the dense phase and D is the diffusion coefficient in the dense phase. Equations (9.25-11) and (9.25-13) can be solved simultaneously with the following boundary conditions:

$$\text{At } z = 0 \qquad C_{jb} = (C_j)_i \tag{9.25-14}$$

$$-D \frac{dC_{jd}}{dz} = u_{d\infty}(C_{ji} - C_{jd}) \tag{9.25-15}$$

$$\text{At } z = L \qquad \frac{dC_{jd}}{dz} = 0 \tag{9.25-16}$$

For approximation, the first two terms in Eq. (9.25-13) can be neglected as suggested by Smith (S2). Then solving C_{jd} in terms of C_{jb} for a first-order reaction and

substituting into Eq. (9.25-11) gives

$$u_{b\infty}\frac{dC_{jb}}{dz} = -\left(\frac{1}{\epsilon k\rho_B} + \frac{1}{k_m a_v}\right)^{-1} C_{jb} \tag{9.25-17}$$

Integrating yields

$$1 - f = \frac{C_{jb}}{C_{ji}} = \exp\left[-\left(\frac{1}{\epsilon k\rho_B} + \frac{1}{k_m a_v}\right)^{-1} \frac{z}{u_{b\infty}}\right] \tag{9.25-18}$$

where f is the fractional conversion of the j component and C_{ji} is the concentration of the j component when $z = 0$.

If the operation is carried out in a CSTR, Eq. (9.25-11) can be modified by multiplying both sides by V

$$\iiint_V u_{b\infty}\frac{dC_{jb}}{dz}\, dV = k_m a_v V(C_{jd} - C_{jb}) \tag{9.25-19}$$

Using the divergence theorem, Eq. (9.25-19) can be further simplified to

$$(u_{b\infty}C_{jb}S)_2 - (u_{b\infty}C_{jb}S)_1 = k_m a_v V(C_{jd} - C_{jb}) \tag{9.25-20}$$

just as before, if the first two terms in Eq. (9.25-13) are neglected for the purpose of obtaining an approximate solution, C_{jd} can be solved in terms of C_{jb} and then substituted into Eq. (9.25-20) to yield

$$(u_{b\infty}C_{jb}S)_2 - (u_{b\infty}C_{jb}S)_1 = V\left(\frac{1}{k_m a_v} + \frac{1}{\epsilon\rho_B K}\right)^{-1} C_{jb} \tag{9.25-21}$$

Analogous to the tubular reactor mentioned before, the fractional conversion can be deduced as

$$1 - f_1 = \frac{C_{jb}}{C_{j1}} = \frac{1}{1 + [(1/k_m a_v) + (1/\epsilon\rho_B k)]\theta} \tag{9.25-22}$$

where C_{j1} is the concentration of the j component at the entrance to the CSTR. θ is the residence time which is equal to the volume of the CSTR divided by the volumetric flow rate of the j component at the entrance, i.e., $\theta = V/q = V/(u_{b\infty}S)$.

9.26 Design of Slurry Reactor (S2)

In this kind of reactor, gas is bubbled into and dissolved in the liquid in which catalyst particles are suspended (see below). Unlike the fluidized bed, there is

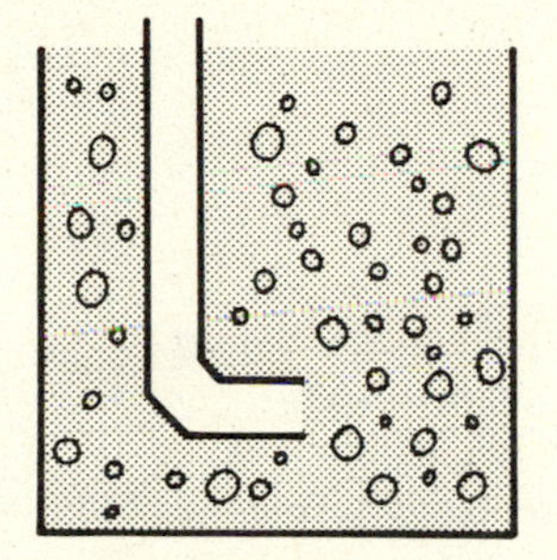

little relative movement between particles and liquid, even though the slurry is agitated. There are five processes in formulating the global rate:

1. Mass transfer from the bulk concentration in the gas-bubble to the bubble-liquid interface

$$r_v = k_g a_g (C_g - C_{ig}) \qquad \textbf{(9.26-1)}$$

2. Mass transfer from the bubble interface to the bulk-liquid phase

$$r_v = k_L a_g (C_{iL} - C_L) \qquad \textbf{(9.26-2)}$$

3. Mixing and diffusion in the liquid (because of agitation, this can be neglected).
4. Mass transfer to the external surface of the catalyst particles

$$r_v = k_c a_c (C_L - C_s) \qquad \textbf{(9.26-3)}$$

5. Reaction at the catalyst surface: Assuming first-order reaction,

$$r_v = k a_c C_s \qquad \textbf{(9.26-4)}$$

where a_c = the external area of catalyst particles per unit volume of slurry (bubble free)

a_g = the gas bubble–liquid interfacial area per unit volume of bubble-free slurry

c_s = the concentration of gas reactant at the outer surface of the catalyst particle

k_g, k_L, k_c = the appropriate mass-transfer coefficients

At equilibrium,

$$C_{ig} = HC_{iL} \tag{9.26-5}$$

where H is Henry's constant. Eliminating C_{ig}, C_{iL}, C_L, and C_s from these last five equations,

$$r_v = k_o a_c C_g \tag{9.26-6}$$

where

$$\frac{1}{k_o} = \frac{a_c}{a_g k_g} + \frac{a_c H}{a_g k_L} + H\left(\frac{1}{k_c} + \frac{1}{k}\right) \tag{9.26-7}$$

Since the resistance to diffusion from bulk-gas to bubble-liquid interface is insignificant, $C_g = C_{ig}$ and Eq. (9.26-7) becomes

$$\frac{1}{k_o H} = \frac{a_c}{a_g k_L} + \frac{1}{k_c} + \frac{1}{k} \tag{9.26-8}$$

Further simplification can be made by using the equilibrium value of C_L as

$$C_{ig} = C_g = H(C_L)_{\text{eq}}$$

Then Eq. (9.26-6) becomes

$$r_v = k_o a_c H(C_L)_{\text{eq}} \tag{9.26-9}$$

where $k_o H$ is given by Eq. (9.26-8). The mass-transfer coefficients k_L and k_c are evaluated by the following two empirical correlations (C3):

$$k_L\left(\frac{\mu_L}{\rho_L D}\right)^{2/3} = 0.31\left(\frac{\Delta\rho\mu_L g}{\rho_L^2}\right)^{1/3} \tag{9.26-10}$$

where μ_L = the viscosity of liquid phase

$\Delta\rho$ = the difference in density between liquid and gas phases

ρ_L = the density of the liquid phase

k_L = the mass transfer coefficient

g = the acceleration of gravity

and (C4)

$$(k_c)_{\text{grav}}\left(\frac{\mu_L}{\rho_L D}\right)^{2/3} = 0.34\left(\frac{\rho\mu_L g}{\rho_L^2}\right)^{1/3} \tag{9.26-11}$$

The design equation for the plug-flow assumption is

$$\frac{V}{F} = \int_0^f \frac{df}{r_v} \tag{9.26-12}$$

Substituting Eq. (9.26-6) into Eq. (9.26-12) yields

$$\frac{V}{F} = \int_0^f \frac{df}{k_o a_c C_g} \tag{9.26-13}$$

Example 9.26-1 Slurry Reactor

The reaction

$$C_2H_4(g) + H_2(g) \rightarrow C_2H_6(g)$$

is carried out in a slurry reactor in which an equimolal mixture of H_2 and C_2H_4 is bubbled through a slurry of nickel catalyst particles suspended in toluene. The operating temperature and pressure are 50°C and 10 atm, at which the rate is controlled by the rate of diffusion of hydrogen from the bubble interface to the bulk liquid. Find the volume of bubble-free slurry required to obtain a conversion of 40 percent H_2. The data are

Feed rate of hydrogen = 120 ft³ at 60°F and 1 atm/min

Henry's law constant = 9.4 g mol/cm³

$D = 1.1 \times 10^{-4}$ cm²/s

Density of toluene = 0.85 g/cm³

Viscosity of toluene = 0.45 cP

$a_g = 1.0$ cm²/cm³

Solution

$$\underset{(1)}{H_2(g)} + \underset{(2)}{C_2H_4(g)} \rightarrow \underset{(3)}{C_2H_6(g)}$$

$$N_j = N_{jo} + \alpha N_{ko} f/\alpha_k \tag{A}$$

For H_2 $\quad N_1 = 1 - f$

For C_2H_4 $\quad N_2 = 1 - f$

For C_2H_6 $\quad N_3 = f$

Hence $\quad N = 2 - f$

Assuming ideal $\quad V = (2 - f)\dfrac{RT}{P}$

$$C_1 = \frac{(1-f)P}{(2-f)RT} \tag{B}$$

Substituting Eq. (B) into Eq. (9.26-13) gives

$$\frac{V}{F} = \int_0^f \frac{df}{k_o a_c \dfrac{(1-f)P}{(2-f)RT}} \tag{C}$$

Integrating yields

$$\frac{V}{F} = \frac{RT}{k_o a_c P}[f - \ln(1-f)] \tag{D}$$

But

$$\frac{1}{k_o} = \frac{a_c H}{a_g k_L}$$

Thus

$$\frac{V}{F} = \frac{HRT}{a_g k_L P}[f - \ln(1-f)] \tag{E}$$

From Eq. (9.26-10)

$$k_L\left[\frac{0.0045}{0.85(0.00011)}\right]^{2/3} = 0.31\left[\frac{(0.85-0.0006)(0.0045)(32.2)}{0.85^2}\right]^{1/3}$$

$$k_L = 0.013 \text{ cm/s}$$

Substituting

$$\frac{V}{F} = \frac{9.4}{(1)(0.013)}\frac{82(273+50)}{10}[0.4 - \ln(1-0.4)]$$

$$= 1.74 \times 10^6 \text{ cm}^3/\text{g mol} \cdot \text{s}$$

Thus

$$V = \frac{120}{379}\left(\frac{454}{60}\right)(1.74)(10^6)(10^{-3}) = 4.169 \times 10^3 \text{ liters}$$

9.27 Design of a Gas-Solid Noncatalytic Reactor (*S2*)

In industry, there are many gas-solid reactions in which the solid is a reactant rather than a catalyst, such as the reduction of FeS_2

$$FeS_2(s) + H_2(g) \rightarrow FeS(s) + H_2S(g)$$

In general, when we have the following reaction

$$D(g) + bE(s) \rightarrow F(g) + H(s)$$

or similar, the reaction can be studied by the shrinking-core model as follows. In Fig. 9.27-1, solid reactant E is initially a sphere of radius R_s. This sphere is in contact with gas D whose bulk concentration is C_b. As reaction takes place, a layer of product F forms around the unreacted core of reactant E. Assuming this layer is porous, gas D can diffuse through this layer to react at the interface. Under pseudo-steady-state conditions, the three rate equations, expressed as moles of D disappearing per unit time per particle are

$$-\frac{dN_D}{dt} = 4\pi R_s^2 k_m (C_{Db} - C_{Ds}) \qquad \text{external diffusion} \tag{9.27-1}$$

$$-\frac{dN_D}{dt} = 4\pi R_c^2 D_e \left(\frac{dC_D}{dR}\right)_{R=R_c} \qquad \text{diffusion through product} \tag{9.27-2}$$

$$-\frac{dN_D}{dt} = 4\pi R_c^2 k (C_D)_c \qquad \text{reaction at } R_c \tag{9.27-3}$$

In order to obtain the gradient dC_D/dR in Eq. (9.27-2), a steady-state mass balance of D around a small element thickness ΔR in the product layer gives

$$-\left(\pi R^2 D_e \frac{dC_D}{dR}\right)_R - \left(\pi R^2 D_e \frac{dC_D}{dR}\right)_{R+\Delta R} = 0 \tag{9.27-4}$$

Taking the limit as $\Delta R \longrightarrow 0$ yields

$$\frac{d}{dR}\left(R^2 D_e \frac{dC_D}{dR}\right) = 0 \tag{9.27-5}$$

Integrating twice with the boundary conditions

$$\text{At } R = R_s \qquad C_D = C_{Ds}$$

$$\text{At } R = R_c \qquad C_D = C_{Dc}$$

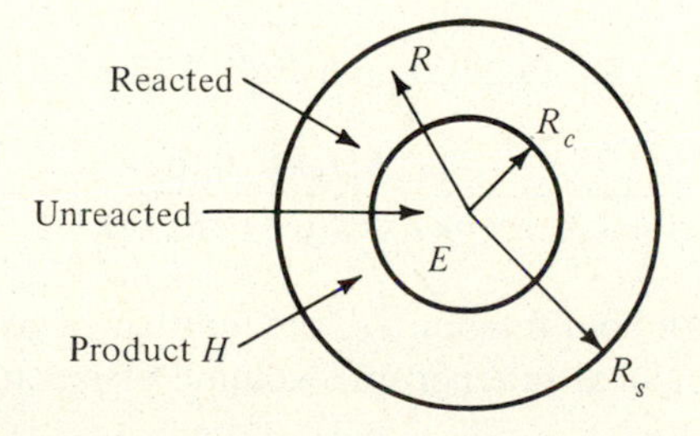

Figure 9.27-1 Gas-solid noncatalytic reactor

we obtain

$$C_D - C_{Dc} = (C_{Ds} - C_{Dc})\frac{1 - R_c/R}{1 - R_c/R_s} \tag{9.27-6}$$

Differentiating Eq. (9.27-6) with R and substituting into Eq. (9.27-2) yields

$$-\frac{dN_D}{dt} = 4\pi D_e \frac{C_{Ds} - C_{Dc}}{1 - R_c/R_s} \tag{9.27-7}$$

Now solving Eq. (9.27-1), Eq. (9.27-3), and Eq. (9.27-7) by eliminating dN_D/dt and C_{Ds} gives

$$C_{Dc} = \frac{C_{Db}}{1 + (R_c^2/R_s^2)(k/k_m) + (kR_c/D_e)(1 - R_c/R_s)} \tag{9.27-8}$$

Substituting C_{Dc} into Eq. (9.27-3) yields the global reaction rate per particle as

$$-\frac{dN_D}{dt} = \frac{4\pi R_c^2 C_{Db} k}{1 + (R_c^2/R_s^2)(k/k_m) + (kR_c/D_e)(1 - R_c/R_s)} \tag{9.27-9}$$

It is noted that R_c varies with time. Hence this relationship must be found before Eq. (9.27-9) can be used. The reaction of E in moles per unit time per particle is

$$\frac{dN_E}{dt} = \frac{\rho_E}{M_E}\frac{d}{dt}\left(\frac{4}{3}\pi R_c^3\right) = \frac{4\pi R_c^2 \rho_E}{M_E}\frac{dR_c}{dt} \tag{9.27-10}$$

Since the stoichiometry of the reaction shows

$$\frac{dN_D}{dt} = \frac{1}{b}\frac{dN_E}{dt} = \frac{4\pi R_c R_c \rho E}{bM_E}\frac{dR_c}{dt} \tag{9.27-11}$$

Combining Eq. (9.27-11), Eq. (9.27-3), and Eq. (9.27-8) yields

$$-\frac{dR_c}{dt} = \frac{bM_E k C_{Db}/\rho_E}{1 + (R_c^2/R_s^2)(k/k_m) + (kR_c/D_e)(1 - R_c/R_s)} \tag{9.27-12}$$

In a packed reactor with void fraction ϵ_B, the number of particle per unit volume is $3\epsilon/4\pi R_s^3$. Thus the global rate per unit volume of reactor is

$$r_p \rho_B = \frac{3\epsilon_B k (R_c/R_s)^2 C_{Db}}{R_s[1 + (R_c^2/R_s^2)(k/k_m) + (kR_c/D_e)(1 - R_c/R_s)]} \tag{9.27-13}$$

A mass balance for a fixed-bed factor is

$$-u\frac{\partial C_{Db}}{\partial z} = r_p\rho_B + \epsilon_B\frac{\partial C_{Db}}{\partial t} \tag{9.27-14}$$

Hence Eq. (9.27-14), Eq. (9.27-13), and Eq. (9.27-12) establish $C_D(t, z)_{Db}$ and $R_c(t, z)$. These equations may be solved numerically with the boundary conditions:

$$\text{At } z = 0 \text{ for } t > 0,\ C_{Db} = C_{D0}$$

$$\text{At } t = 0 \text{ for } z > 0,\quad R_c = R_s$$

In a moving-bed reactor, the residence time is

$$t = \theta = \frac{z\epsilon_s}{G_s/\rho_B} \tag{9.27-15}$$

$$dt = \frac{\epsilon_s\rho_B\, dz}{G_s} \tag{9.27-16}$$

where z is the reactor length, ϵ is the fraction of reactor volume occupied by solids, G_s is the mass velocity of the solids. Substituting Eq. (9.27-16) into Eq. (9.27-12) gives the change of R_c with z

$$\frac{dR_c}{dz} = -\frac{\epsilon_s bM_DkC_{Db}}{G_s[1 + (R_c^2/R_s^2)(k/k_m) + (kR_c/D_e)(1 - R_c/R_s)]} \tag{9.27-17}$$

Since

$$f_E = \frac{\text{Initial mass} - \text{mass at } t}{\text{Initial mass}} = \frac{(4/3)R_s^3\rho_E - (4/3)R_c^3\rho_E}{(4/3)R_s^3\rho_E} \tag{9.27-18}$$

$$= 1 - (R_c/R_s)^3$$

Equation (9.27-17) can be written as

$$\frac{df_E}{dz} = \frac{3\epsilon_s bM_EkC_{Db}(1 - f_E)^{2/3}}{G_sR_s\{1 + (k/k_m)(1 - f_E)^{2/3} + (kR_s/D_e)(1 - f_E)^{1/3}[1 - (1 - f_E)^{1/3}]\}} \tag{9.27-19}$$

which can be integrated because C_{Db} is related to f_E by taking a mass balance around the envelope in Fig. 9.27-2 as

$$u(C_{Df} - C_{Db}) = \frac{G_s}{(bM_E)(f_{E0} - f_E)} \tag{9.27-20}$$

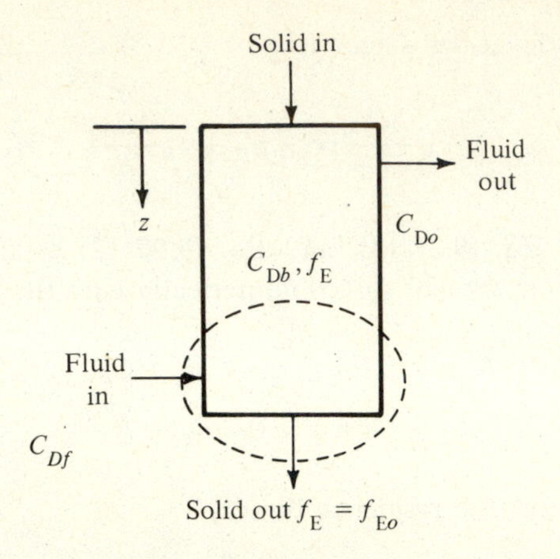

Figure 9.27-2 Counterflow moving-bed reactor

9.28 Design of a Gas-Liquid Noncatalytic Reactor (*J3*)

The typical example of a gas-liquid noncatalytic reaction in chemical engineering is absorption with chemical reaction. Two practical examples are given for illustration. The first one is the operation in a packed tower and the second in a plate tower.

9.28-1 Packed tower

A solute A in a gaseous mixture entering the bottom of a packed tower at a rate of G kmol/s is to be recovered by a liquid stream flowing countercurrently at a rate of L kmol/s. During the operation, A decomposes simultaneously by a first-order irreversible reaction. Find a relationship between the gas composition and the height of the tower. A solute balance around the envelope gives

Gas phase:

$$G\frac{dy}{dz} + K_g a A_c(y - y^*) = 0 \tag{9.28-1}$$

Liquid phase:

$$L\frac{dx}{dz} + K_g a A_c(y - y^*) - kA_c Hx = 0 \tag{9.28-2}$$

where x, y = mole fraction of the solute A in the liquid and gas phases, respectively

y^* = the equilibrium value of A in the gas phase (mole fraction)

A_c = the cross-sectional area of the tower

k = the first-order rate constant
K_g = the mass-transfer coefficient
a = the surface area per unit volume
H = the liquid hold-up in moles per unit volume

The equilibrium is

$$y^* = mx \tag{9.28-3}$$

Substituting y^* from Eq. (9.28-3) into Eq. (9.28-1) gives

$$G\frac{dy}{dz} + K_g a A_c (y - mx) = 0 \tag{9.28-4}$$

or

$$x = \frac{G}{mK_g a A_c}\frac{dy}{dz} + \frac{y}{m} \tag{9.28-5}$$

Differentiating x with respect to z yields

$$\frac{dx}{dz} = \frac{G}{mK_g a A_c}\frac{d^2y}{dz^2} + \frac{1}{m}\frac{dy}{dz} \tag{9.28-6}$$

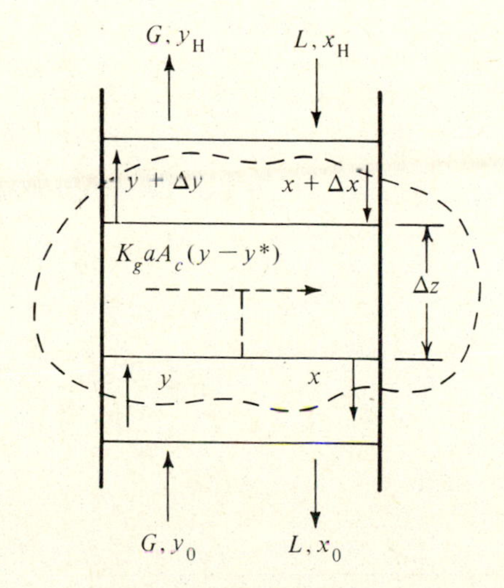

Figure 9.28-1 Packed column

Substituting Eqs. (9.28-5) and (9.28-6) into Eq. (9.28-2) results

$$\frac{L}{mK_gaA_c}\frac{d^2y}{dz^2}+\frac{L}{mG}\frac{dy}{dz}-\frac{kH}{mK_ga}\frac{dy}{dz}-\frac{kHA_c}{Gm}y=\frac{dy}{dz} \tag{9.28-7}$$

Let

$$H_G=\frac{G}{K_gaA_c} \qquad \lambda=\frac{mG}{L} \qquad \beta=\frac{kHG}{LK_ga}$$

Equation (9.28-7) becomes

$$\frac{H_G}{\lambda}\frac{d^2y}{dz^2}+\left(\frac{1}{\lambda}-1-\frac{\beta}{\lambda}\right)\frac{dy}{dz}-\frac{\beta}{H_G\lambda}y=0 \tag{9.28-8}$$

$$H_G\frac{d^2y}{dz^2}+(1-\lambda-\beta)\frac{dy}{dz}-\frac{\beta}{H_G}y=0 \tag{9.28-9}$$

Equation (9.28-9) is a second-order ordinary differential equation. Its solution is

$$y=Ae^{m_1z}+Be^{m_2z} \tag{9.28-10}$$

where m_1 and m_2 are roots of

$$am^2+bm+c=0 \tag{9.28-11}$$

where

$$a=H_G \qquad b=1-\lambda-\beta \qquad c=-\frac{\beta}{H_G}$$

A and B are integrating constants which can be determined by the following boundary conditions:

$$\text{At } z=0 \qquad y=y_0 \qquad \text{Thus } y_0=A+B \tag{9.28-12}$$

$$\text{At } z=H \qquad x=0 \qquad \text{Thus } \frac{dy}{dz}=-\frac{y}{H_G} \tag{9.28-13}$$

The complete solution is then

$$\frac{y}{y_0}=\frac{(m_1+1/H_G)e^{m_1H}e^{m_2z}-(m_2+1/H_G)e^{m_2H}e^{m_1z}}{(m_1+1/H_G)e^{m_1H}-(m_2+1/H_G)e^{m_2H}} \tag{9.28-14}$$

9.28-2 *Plate column*

Assuming first-order reaction, the rate is

$$r = kx_nH \tag{9.28-15}$$

where k is the rate constant and H is the moles of liquid hold-up per plate. A mass balance of A around plate n (see Fig. 9.28-2) is

$$G(y_{n-1} - y_n) - L(x_n - x_{n+1}) = kx_nH \tag{9.28-16}$$

The Murphree stage efficiency is

$$E = \frac{y_{n-1} - y_n}{y_{n-1} - y_n^*} \tag{9.28-17}$$

and the equilibrium is

$$y_n^* = mx_n \tag{9.28-18}$$

Combining Eqs. (9.28-16), (9.28-17), and (9.28-18) and rearranging yields the following second-order finite-difference equation

$$y_{n+1} + Ay_n + By_{n-1} = 0 \tag{9.28-19}$$

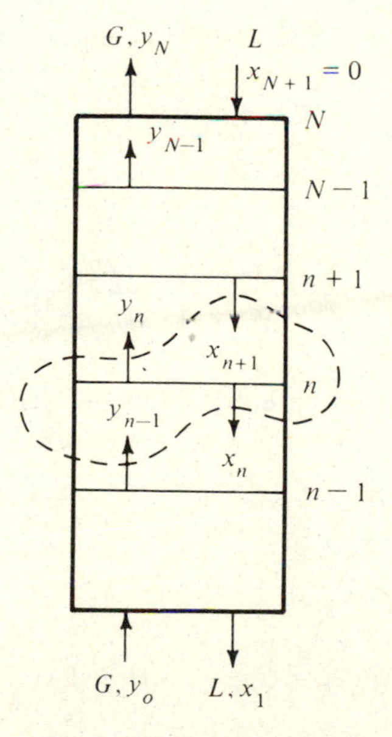

Figure 9.28-2 Plate column

where

$$A = [(E - 2) - \gamma(\delta + 1)] \tag{9.28-20}$$

$$B = [(1 - E)(1 + \gamma\delta) + \gamma] \tag{9.28-21}$$

and

$$\gamma = \frac{GmE}{L} \qquad \delta = \frac{Hk}{GmE}$$

Equation (9.28-19) can be solved with the boundary conditions:

$$\text{At } n = 0 \qquad y = y_0$$

$$\text{At } n = N - 1 \qquad y = y_{N-1}$$

The result is

$$y_n = \frac{(y_0 Z_2^{N-1} - y_{N-1})Z_1^n + (y_{N-1} - y_0 Z_1^{N-1})Z_2^N}{Z_2^{N-1} - Z_1^{N-1}} \tag{9.28-22}$$

or

$$y_N = \frac{(y_0 Z_2^{N-1} - y_{N-1})Z_1^N + (y_{N-1} - y_0 Z_1^{N-1})Z_2^N}{Z_2^{N-1} - Z_1^{N-1}} \tag{9.28-23}$$

where Z_1 and Z_2 are the roots of the auxiliary equation:

$$Z^2 + AZ + B = 0 \tag{9.28-24}$$

Simplifying Eq. (9.28-23) gives

$$\frac{1}{Z_1^N} = \frac{y_0(1/Z_1 - 1/Z_2)}{(y_N Z_1^{N-1}/Z_2^N - y_N/Z_2) - y_{N-1}[(Z_1/Z_2)^N - 1]} \tag{9.28-25}$$

Neglecting the terms in the first parentheses in the denominator, we get

$$N = \frac{\log [y_0(1/Z_1 - 1/Z_2)/(y_{N-1} - y_N/Z_2)]}{\log (1/Z_1)} \tag{9.28-26}$$

From the mass balance

$$Gy_{n-1} = Gy_N + (kH + L)x_N \tag{9.28-27}$$

But

$$x_N = \frac{(E-1)y_{N-1} + y_N}{mE}$$

Simplifying gives

$$y_{N-1} = \frac{y_N(mEG + KH + L)}{mEG - (KH + L)(E - 1)} \qquad \textbf{(9.28-28)}$$

9.29 *Design of a Liquid-Liquid Noncatalytic Reactor* (*J2*)

Analogous to the gas-liquid noncatalytic reactor, a liquid-liquid system can be developed. Similarly, two examples are given for packed tower and for plate column.

9.29-1 *Packed tower*

G kg/h of animal fat which will be hydrolyzed is fed to the base of a tower into which L kg/h of water is sprayed from the top. In an element of height Δz of the column in Fig. 9.29-1, the rate of mass transfer of glycerine from the fat to the water phase is

$$\text{Rate of mass transfer} = K_1 aS(x^* - x)\Delta z \qquad \textbf{(9.29-1)}$$

Assuming first-order reaction, the reaction rate is

$$\text{Rate of removal of fat} = kS\rho\sigma\ \Delta z \qquad \textbf{(9.29-2)}$$

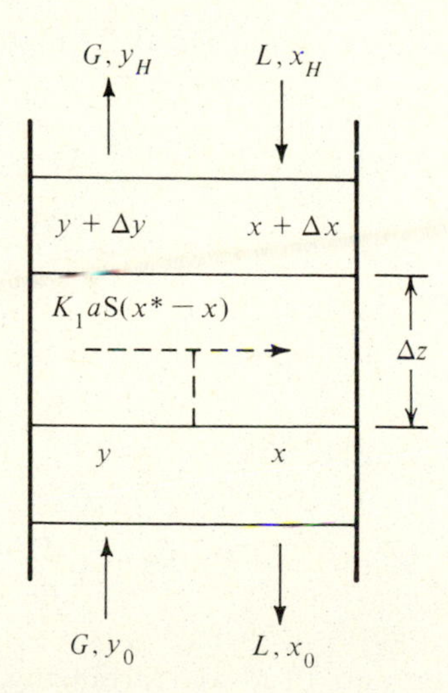

Figure 9.29-1 Liquid-liquid packed tower

where σ is the weight fraction of fat in the rising stream and ρ is the mass of fat per unit volume of column. If w is the weight of fat to produce unit mass of glycerine, the rate of production of glycerine is

$$\text{Rate of production of glycerine} = \frac{kS\rho\sigma\ \Delta z}{w} \tag{9.29-3}$$

A glycerine balance around the element height is

$$Gy + L\left(x + \frac{dx}{dz}\Delta z\right) + \frac{kS\rho\sigma\ \Delta z}{w} = G\left(y + \frac{dy}{dz}\Delta z\right) + Lx \tag{9.29-4}$$

which can be rearranged as

$$Gy + \frac{kS\rho\sigma\ \Delta z}{w} - G\left(y + \frac{dy}{dz}\Delta z\right) = Lx - L\left(x + \frac{dx}{dz}\Delta z\right) \tag{9.29-5}$$

$$= K_1 aS(x^* - x)\Delta z \tag{9.29-6}$$

An equivalent glycerine balance between the element and the bottom of the column is

$$\frac{G\sigma_0}{w} + Lx = G\left(y + \frac{\sigma}{w}\right) + Lx_0 \tag{9.29-7}$$

Substituting σ/w from Eq. (9.29-7) into Eq. (9.29-6) gives

$$S\rho k\left[\frac{\sigma_0}{w} + \frac{L}{G}(x - x_0) - y\right]\Delta z - G\frac{dy}{dz}\Delta z = K_1 aS(x^* - x)\Delta z \tag{9.29-8}$$

The equilibrium is

$$x^* = my \tag{9.29-9}$$

Equations (9.29-5) and (9.29-6) become

$$K_1 aSmy = K_1 aSx - L\frac{dx}{dz} \tag{9.29-10}$$

Substituting Eq. (9.29-10) into Eq. (9.29-8), rearranging, simplifying, and multiplying throughout by $K_1 aSm/LG$ yields

$$\frac{d^2x}{dz^2}-\frac{dx}{dz}\left(\frac{K_1aS}{L}-\frac{K_1aSm}{G}-\frac{S\rho k}{G}\right)$$
$$+\left(\frac{K_1aS^2m\rho k}{G^2}-\frac{K_1aS^2\rho k}{LG}\right)x=\frac{K_1aS^2m\rho k}{LG}\left(\frac{L}{G}x_0-\frac{\sigma_o}{w}\right) \quad (9.29\text{-}11)$$

Solving gives

$$x=Ae^{-\delta z}+Be^{-\nu z}+\frac{\gamma x_o-m\sigma_o/w}{\gamma-1} \quad (9.29\text{-}12)$$

where

$$\gamma=\frac{mL}{G} \qquad \delta=\frac{S\rho k}{G} \qquad \nu=\frac{K_1aS(\gamma-1)}{L}$$

The boundary conditions are

$$\text{At } z=0 \quad y=0 \qquad \text{At } z=H \quad x=0$$

Then the integrating constants A and B in Eq. (9.29-12) can be evaluated and the final solution is

$$x_0=\frac{m\sigma_o}{w(\gamma-e^{-\nu H})}\left[1+\left(\frac{\lambda-1}{\gamma-\lambda}\right)e^{-\nu H}-\left(\frac{\gamma-1}{\gamma-\lambda}\right)e^{-\delta H}\right] \quad (9.29\text{-}13)$$

where

$$\lambda=\frac{\nu-\delta+\gamma\delta}{\nu}=1+\frac{k\rho L}{K_1aG}$$

Finally substituting Eq. (9.29-13) into Eq. (9.29-12) yields

$$x=\frac{m\sigma_o}{w(\gamma-\lambda)}\left[e^{-\delta H}+\left(\frac{e^{-\delta H}-\lambda}{\gamma-e^{-\nu H}}\right)e^{-\nu H}\left(\frac{\lambda e^{-\nu H}-\gamma e^{-\delta H}}{\gamma-e^{-\nu H}}\right)\right] \quad (9.29\text{-}14)$$

9.29-2 Plate column

The same problem can be analyzed in a plate column. An equivalent glycerine balance between plate n and the bottom of the column is (see Fig. 9.29-2):

$$\frac{G\sigma_{N+1}}{w}+Lx_{n-1}=Lx_N+G\left(y_n+\frac{\sigma_n}{w}\right) \quad (9.29\text{-}15)$$

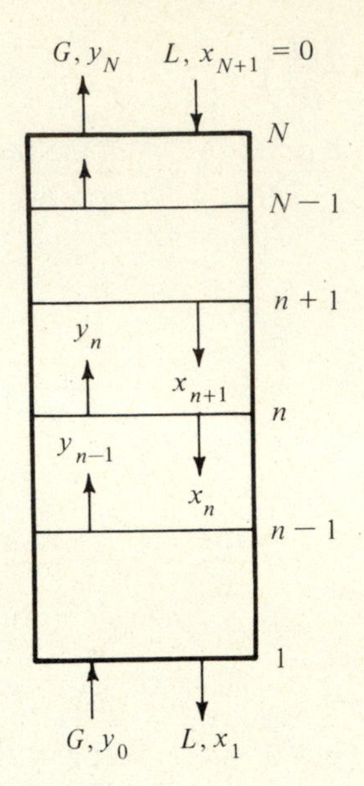

Figure 9.29-2 Plate column

A glycerine balance around nth plate is

$$Lx_{n-1} + Gy_{n+1} = Lx_n + Gy_n - \frac{kH_o \sigma_n}{w} \qquad \textbf{(9.29-16)}$$

where H_o is the fat hold-up per theoretical plate. Substituting σ_n/w from Eq. (9.29-15) into Eq. (9.29-16) with the use of the equilibrium relation

$$x_n = my_n$$

gives

$$\frac{mL}{G}(x_{n-1} - x_n) + x_{n+1} - x_n + \frac{kH_o}{G}\left(\frac{m\sigma_{N+1}}{w} + \frac{mL}{G}x_{n-1} - \frac{mL}{G}x_N - x_n\right) = 0 \qquad \textbf{(9.29-17)}$$

Let

$$\frac{mL}{G} = \gamma \qquad \text{and} \qquad \frac{kH_o}{G} = \delta'$$

Equation (9.29-17) becomes

$$x_{n+1} - x_n(\gamma + \delta' + 1) + x_{n-1}(\gamma + \gamma\delta') = \delta'\left(\gamma x_N - \frac{m\sigma_{N+1}}{w}\right) \quad \textbf{(9.29-18)}$$

or

$$x_{n+1} - Ax_n + Bx_{n-1} = C \quad \textbf{(9.29-19)}$$

where

$$A = \gamma + \delta' + 1 \qquad B = \gamma(1 + \delta') \qquad C = \gamma x_N - \frac{m\sigma_{N+1}}{w}$$

The boundary conditions are:

$$\text{At } n = 0 \qquad x = 0$$
$$\text{At } n = N + 1 \qquad y = 0$$

The complete solution of the second-order finite-difference equation (9.27-19) with the above boundary conditions is

$$x_N = \frac{m\sigma_{N+1}}{w(\gamma^{N+1} - 1)}\left[\gamma^N + \frac{(\gamma - 1)\gamma^N - \delta'(1 + \delta')^N}{(1 - \gamma + \delta')(1 + \delta')^N}\right] \quad \textbf{(9.29-20)}$$

9.30 Design of a Trickle-Bed Reactor (S2)

The trickle-bed is a three-phase reactor. Gas and liquid flow downward concurrently through a fixed bed of catalyst. A trickle bed is needed particularly when one reactant is too volatile to liquify and a second is too nonvolatile to vaporize.

To derive the design equations for operating this reactor, the following assumptions are made:

1. Isothermal and one dimensional
2. No radial gradients of concentration or velocity for the gas or liquid
3. Flowing liquid completely covers the particles
4. Reaction by reactant mass transfer through the liquid particle interface
5. Uniform distribution of gas and liquid phases
6. Negligible dispersion in the gas phase
7. Liquid reactant is nonvolatile

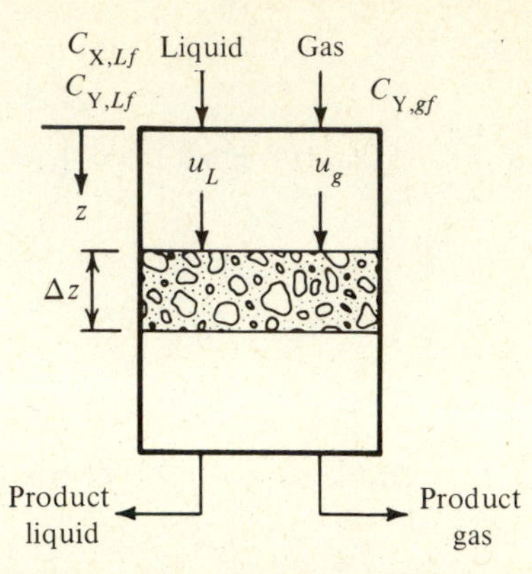

Figure 9.30-1 Trickle-bed reactor

Under these assumptions, let us consider the following reactions

$$a\text{Y(g)} + \text{X}(L) \rightarrow \text{Z(g or } L) + \text{W(g or } L)$$

Using the generalized mass balance equation, Eq. (4.2-17), the following equations can be obtained for different phases:

1. Reactant Y in gas phase

$$u_g \frac{dC_{Yg}}{dz} + (K_L a_g)_Y \left[\frac{C_{Yg}}{H_Y} - C_{YL}\right] = 0 \quad \textbf{(9.30-1)}$$

where u_g = the superficial velocity

C_{Yg} = the concentration of the gas Y in the gas phase

K_L = the overall mass-transfer coefficient between gas and liquid. It is related to the individual film coefficients k_L and k_g by

$$\frac{1}{K_L} = \frac{1}{Hk_g} + \frac{1}{k_L} \quad \textbf{(9.30-2)}$$

H = Henry's law constant

a_g = the mass-transfer area per unit volume

C_{Yg}/H_Y = the liquid-phase concentration in equilibrium with the bulk-gas concentration

C_{YL} = the concentration of the Y component in the liquid phase

2. Reactant Y in liquid phase

$$D_L \frac{d^2C_{YL}}{dz^2} - u_L \frac{dC_{YL}}{dz} + (K_L a_g)_Y \left[\frac{C_{Yg}}{H_Y} - C_{YL}\right] - (k_c a_c)_Y [C_{YL} - C_{Ys}] = 0 \quad \textbf{(9.30-3)}$$

where u_L = the superficial velocity of the liquid
D_L = the diffusion coefficient of the liquid
$k_c a_c$ = the liquid to particle mass-transfer coefficient
C_{Ys} = the concentration of the Y component in the interface

3. Reactant X in the liquid phase

$$D_{LX}\frac{d^2C_{XL}}{dz^2} - u_L\frac{dC_{XL}}{dz} - (k_c a_c)_X[C_{XL} - C_{Xs}] = 0 \qquad \textbf{(9.30-4)}$$

where D_{LX} = the diffusion coefficient of X in the liquid phase
C_{XL} = the concentration of the X component in the liquid phase
$(k_c a_c)_X$ = the mass transfer coefficient of the X component in the liquid phase
C_{Xs} = the concentration of the X-component in the interface

4. Reaction rate for component Y, r_Y

$$(k_c a_c)_Y(C_{YL} - C_{Ys}) = r_Y = \rho_B \eta f[C_{Ys}, C_{Xs}] \qquad \textbf{(9.30-5)}$$

where ρ_B = the bulk density of the catalyst in the bed
η = the effectiveness factor
$f(C_{Ys}, C_{Xs})$ = the intrinsic rate of reaction per unit mass of catalyst

5. Reaction rate for component X, r_X

$$(k_c a_c)_X(C_{XL} - C_{Xs}) = r_X = \frac{r_Y}{a} = \frac{\rho_B}{a}\eta f(C_{Ys}, C_{Xs}) \qquad \textbf{(9.30-6)}$$

With appropriate boundary conditions, the five concentrations C_{Yg}, C_{YL}, C_{XL}, C_{Ys}, and C_{Xs} in the five equations [(9.30-1) and (9.30-3) to (9.30-6)] can be solved as a function of the reactor bed depth. The mass transfer coefficients can be evaluated from the following correlations.

1. Gas to liquid, $k_L a_g$
(a) Correlations (R1, G1, C6) relating to pressure drop for two-phase flow in the reactor.
(b) Goto and Smith (G2) correlation

$$\frac{k_L a_g}{D_Y} = \alpha_L\left(\frac{G_L}{\mu_L}\right)^{\phi_L}\left(\frac{\mu_L}{\rho_L D_Y}\right)^{1/2} \qquad \textbf{(9.30-7)}$$

where D_Y = the molecular diffusivity of the Y component in square centimeters per second
G_L = the superficial mass velocity of the liquid in grams per square centimeter

$k_L a_g$ = the volumetric liquid side mass-transfer coefficient per second

α_L = about $7(\text{cm})^{\phi}L^{-2}$, and $\phi_L = 0.40$ for granular catalyst

μ_L = the liquid viscosity in gram per centimeter per second

2. Liquid to particle, $k_c a_c$
Dharwadkar and Sylvester correlation (D2)

$$\frac{k_c a_c}{u_L a_t}\left(\frac{\mu_L}{\rho_L D_Y}\right)^{2/3} = 1.64\left(\frac{d_p u_L \rho_L}{\mu_L}\right)^{-0.331} \quad (0.2 < Re_L < 2400) \qquad \textbf{(9.30-8)}$$

where u_L = the superficial velocity of liquid

a_t = the total external area of particles per unit volume of reactor

9.31 Optimization

Since heterogeneous reactors consist of many types, discussion of the optimization of each type within a single chapter is impossible. Hence attention is focused on the catalytic reactors, but even so, there are quite a few models of these. The simplest of these models, the pseudo-homogeneous model, is selected for discussion. By looking at the design equations for tubular and the pseudo-homogeneous models of the heterogeneous reactors, it is seen that if V, the volume of the reaction mixture, in Eq. (8.2-9) is replaced by the weight of the catalyst, w, this equation becomes identical to Eq. (9.17-9). This implies that if the V in the subsequent treatment for the optimization of the tubular reactor is replaced by w, the same equations can be applied to the optimization of the pseudo-homogeneous model of the heterogeneous reactor. Hence in order to minimize the catalyst used in a pseudo-homogeneous model of a heterogeneous reactor, the maximum rate of the reaction is also required. For constant conversion, if the derivative of the reaction rate with respect to temperature is equated to zero, the required maximum temperature can be determined. This yields the locus of temperatures at which the reaction rate is a maximum for a given conversion. The locus of these points passes through the maxima of curves of the conversion f as a function of T as contours of constant rate. Thus, to operate a series of adiabatic reactors along an optimum temperature path, hence minimizing the catalyst, the feed is heated to some point P and the reaction allowed to continue along an adiabatic reaction path until a point such as Q, in the vicinity of the optimum temperature curve, is reached (see Fig. 9.31-1). The products are then cooled to M before entering a second adiabatic reactor in which reaction proceeds to an extent indicated by point N again in the vicinity of the optimum temperature. For two reactors in this case, there will be four decisions to be made corresponding to the points P to N inclusive. These four decisions may be made in such a way to minimize the capital and operating costs of the system of reactors and heat exchangers. The following is one example of optimization.

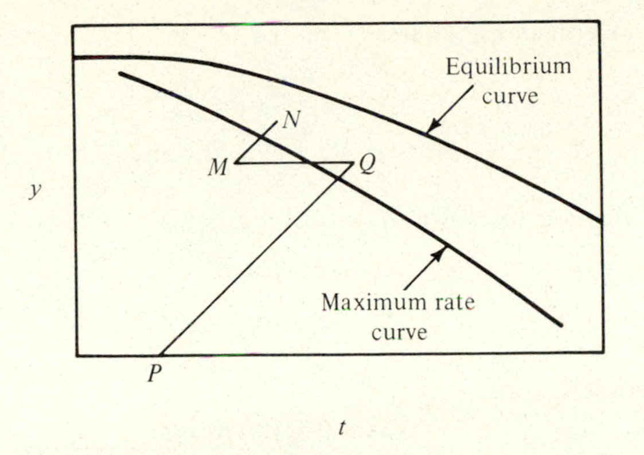

Figure 9.31-1 Optimum design of a two-staged adiabatic reactor

Example 9.31-1 Packed-Bed Reactor Optimization

The reaction rates for the following multiple reactions at 1,500°R

$$\text{Mesitylene} + H_2 \xrightarrow{k_1} m\text{-xylene} + \text{methane}$$
$$m\text{-xylene} + H_2 \xrightarrow{k_2} \text{toluene} + \text{methane}$$

are

$$r_1 = k_1 C_M C_H^{0.5}$$
$$r_2 = k_2 C_X C_H^{0.5}$$

where

$$k_1 = 3.83 \times 10^{-3}\ (\text{m}^3/\text{kmol})^{0.5}/\text{s}$$
$$k_2 = 2.09 \times 10^{-3}\ (\text{m}^3/\text{kmol})^{0.5}/\text{s}$$

and C_M, C_X, and C_H are the concentrations of mesitylene, *m*-xylene, and hydrogen, respectively. These reactions are carried out isothermally in a packed-bed catalytic reactor at 1,500°R and 40 atm. Find the volume required to obtain maximum concentration of *m*-xylene for a feed of 10 g mol H_2 and 5 g mol mesitylene per second.

Solution

$$\underset{(M)}{\text{Mesitylene}} + \underset{(H)}{H_2} \xrightarrow{k_1} \underset{(X)}{m\text{-xylene}} + \underset{(MN)}{\text{methane}}$$
$$\underset{(X)}{m\text{-xylene}} + \underset{(H)}{H_2} \xrightarrow{k_2} \underset{(T)}{\text{toluene}} + \underset{(MN)}{\text{methane}}$$

Using Eq. (1.4-47) and Eq. (1.4-32)

$$\begin{aligned} N_j &= N_{jo} + \Sigma \alpha_{ij} \beta_i \\ &= N_{jo} + \Sigma \alpha_{ij} N_{ko} f_i / \alpha_k \end{aligned}$$

For $j = \mathrm{H}$ (limiting reactant) $N_{\mathrm{H}} = 10(1 - f_1 - f_2)$

$j = \mathrm{M}$ $N_{\mathrm{M}} = 5 - 10 f_1$

$j = \mathrm{X}$ $N_{\mathrm{X}} = 0 + 10 f_1 - 10 f_2$

$j = \mathrm{T}$ $N_{\mathrm{T}} = 0 + 10 f_2$

$j = \mathrm{MN}$ $N_{\mathrm{MN}} = 0 + 10 f_1 + 10 f_2$

$$N_{\mathrm{H}} + N_{\mathrm{M}} + N_{\mathrm{X}} + N_{\mathrm{T}} + N_{\mathrm{MN}} = 10(1 - f_1 - f_2) + 5 - 10 f_1 + 10 f_1 - 10 f_2 + 10 f_2 + 10 f_1 + 10 f_2$$

or

$$\text{Total N} = \text{Total } N_o = 15$$

By ideal gas law

$$V = \frac{NRT}{P} = \frac{15(0.08206)(1500/1.8)}{40} = 25.64 \text{ liters}$$

Therefore, the concentrations of each component are

$$C_{\mathrm{H}} = \frac{10(1 - f_1 - f_2)}{25.64} = 0.39(1 - f_1 - f_2) \text{ g mol/liter}$$

$$C_{\mathrm{M}} = \frac{(5 - 10 f_1)}{25.64} = 0.39(0.5 - f_1) \text{ g mol/liter}$$

$$C_{\mathrm{X}} = \frac{(10 f_1 - 10 f_2)}{25.64} = 0.39(f_1 - f_2) \text{ g mol/liter}$$

$$C_{\mathrm{T}} = \frac{10 f_2}{25.64} = 0.39 f_2 \text{ g mol/liter}$$

$$C_{\mathrm{MN}} = \frac{(10 f_1 + 10 f_2)}{25.64} = 0.39(f_1 + f_2) \text{ g mol/liter}$$

The rates of the reactions are

$$\frac{dC_{\mathrm{M}}}{dt} = -k_1 C_{\mathrm{M}} C_{\mathrm{H}}^{0.5} \tag{A}$$

$$\frac{dC_{\mathrm{X}}}{dt} = k_1 C_{\mathrm{M}} C_{\mathrm{H}}^{0.5} - k_2 C_{\mathrm{X}} C_{\mathrm{H}}^{0.5} \tag{B}$$

In terms of fractional conversions f_1 and f_2, Eqs. (A) and (B) become

$$0.39\frac{df_1}{dt} = 0.39(0.5 - f_1)[0.39(1 - f_1 - f_2)]^{0.5}k_1 \tag{C}$$

$$0.39\frac{d(f_1 - f_2)}{dt} = 0.39(0.5 - f_1)[0.39(1 - f_1 - f_2)]^{0.5}k_1 - k_2(0.39)(f_1 - f_2)[0.39(1 - f_1 - f_2)]^{0.5} \tag{D}$$

Dividing Eq. (D) by Eq. (C) gives

$$1 - \frac{df_2}{df_1} = 1 - \frac{k_2(f_1 - f_2)}{k_1(0.5 - f_1)}$$

or

$$\frac{df_2}{df_1} = \delta\frac{f_1}{0.5 - f_1} - \delta\frac{f_2}{0.5 - f_1} \tag{E}$$

where

$$\delta = k_2/k_1 = 2.09 \times 10^{-3}/3.83 \times 10^{-3} = 0.5465$$

Equation (E) is a first-order linear ordinary differential equation with variable coefficients whose solution is

$$f_2 = \frac{(0.5 - \delta f_1) - 0.5(1 - 2f_1)^{\delta}}{1 - \delta}$$
$$= 1.1025[(1 - 1.093f_1) - (1 - 2f_1)^{0.5465}] \tag{F}$$

Moles of X formed per hour $= 10(f_1 - f_2)$

$$= 10\left[f_1 - \frac{(0.5 - \delta f_1) - 0.5(1 - 2f_1)^{\delta}}{1 - \delta}\right]$$

$$= 10\left[\frac{0.5(1 - 2f_1)^{\delta} - (0.5 - f_1)}{1 - \delta}\right]$$

In order to have maximum moles of X,

$$\frac{d}{df_1}[10(f_1 - f_2)] = \frac{d}{df_1}\left[10\,\frac{0.5(1 - 2f_1)^{\delta} - (0.5 - f_1)}{1 - \delta}\right] = 0$$

Solving for f_1 gives

$$f_{1,\text{opt}} = 0.368$$

Equation (A) can be transformed in terms of the fractional conversions f_1 and f_2 as

$$0.39\frac{df_1}{dt} = k_1(0.39)^{1.5}(0.5 - f_1)(1 - f_1 - f_2) \tag{G}$$

But $t = \tau = V/q_o$ and $q_o = 25.64$ liters/h. Thus Eq. (G) can be further simplified to

$$(0.39)(25.64)\frac{df_1}{dV} = k_1(0.39)^{1.5}(0.5 - f_1)(1 - f_1 - f_2) \tag{H}$$

or

$$dV = \frac{10}{3.83 \times 10^{-3}(0.39)^{1.5}} \int_0^{f_1} \frac{df_1}{(0.5 - f_1)(1 - f_1 - f_2)}$$

$$= 10.727 \int \frac{1}{(0.5 - f_1)(1 - f_1 - f_2)} df_1$$

The integral can be evaluated with Eq. (F) as follows:

f_1	f_2	$(0.5 - f_1)(1 - f_1 - f_2)^{0.5} = r$	$\frac{1}{r}$	*Integral*
0	0	0.5	2	0
0.05	0.0014	0.4383	2.2816	0.1070
0.10	0.0060	0.3782	2.6440	0.2301
0.15	0.0145	0.3199	3.1258	0.3744
0.20	0.0275	0.2637	3.7925	0.5474
0.25	0.0464	0.2097	4.7687	0.7614
0.30	0.0728	0.1584	6.3135	1.0384
0.35	0.1098	0.1102	9.0705	1.4231
0.368	0.1265	0.0939	10.6553	1.6006
0.40	0.1629	0.0661	15.1255	2.0131
0.45	0.2470	0.0275	36.6336	3.3071

Therefore the minimum volume at the optimal condition to produce maximum amount of *m*-xylene is

$$V = 10.727\ (1.6006) = 17.16\ \text{m}^3$$

Problems

1. A gaseous mixture of 11 percent SO_2, 10 percent O_2, and 79 percent N_2 at 400°C and 1 atm enters a packed adiabatic tubular reactor at a rate of 0.2 kmol/s. Find the amount of catalyst required for 60 percent conversion of the following reaction:

$$SO_2 + \frac{1}{2}O_2 \rightarrow SO_3$$

The data are

Specific heat in cal/g mol·K

$$SO_2\text{: } C_p = 6.157 + 1.384 \times 10^{-2}T - 0.9103 \times 10^{-5}T^2 + 2.057 \times 10^{-9}T^3$$

$$SO_3\text{: } C_p = 3.918 + 3.483 \times 10^{-2}T - 2.675 \times 10^{-5}T^2 + 7.744 \times 10^{-9}T^3$$

$$O_2\text{: } C_p = 6.732 + 0.1505 \times 10^{-2}T - 0.01791 \times 10^{-5}T^2$$

$$N_2\text{: } C_p = 6.529 + 0.1488 \times 10^{-2}T - 0.02771 \times 10^{-5}T^2$$

Heat of reaction in cal/g mol = −23,490 at 298 K

$$r_m = \frac{k_1 P_{SO_2} P_{O_2} - k_2 P_{SO_3} P_{O_2}^{1/2}}{P_{SO_2}^{1/2}} \tag{C5}$$

where $\ln k_1 = 12.07 - 31{,}000/RT$ where k_1 is expressed in moles per second per gram of catalyst per atm$^{3/2}$

$\ln k_2 = 22.75 - 53{,}600/RT$ where k_2 is expressed in moles per second per gram of catalyst per atmosphere

2. The oxidation rate in pound moles of NO converted to NO_2 per pound of catalyst per hour at 30°C and total pressure of 3 atm in the presence of active carbon is

$$r_m = \frac{p_{NO}^2 p_{O_2}}{1.619 \times 10^{-4} + 4.842 p_{NO}^2 + 0.001352 p_{NO_2}}$$

where p's are the partial pressures in atmospheres. Find the volume of the reactor for converting 40 tons per day of NO to NO_2 when using an air-NO mixture containing 2.0 mol percent NO and the conversion is 92 percent.

3. Predict the volume of reactor required to produce 14,000 kg styrene a day, using vertical tubes 1 m in diameter for an adiabatic conversion of 45 percent in Example 9.17-1. 1.7×10^{-3} kmol/s ethylbenzene and 34×10^{-3} kmol/s steam enter each reactor tube at a temperature of 625°C and a pressure of 1.2 atm. The data are

Bulk density of catalyst = 1,440 kg/m^3

Heat of reaction = 1.39×10^5 kJ/kmol

Surrounding temperature = 294 K

Heat capacity of the feed mixture = 0.52 cal/g K

4. What would be the conversion in Problem 3 if the reactor were operated nonadiabatically with an overall heat-transfer coefficient of 10 J/s·m^2 of inside tube area per Kelvin? Heat is lost to surroundings.

5. Determine the temperature at various depths of the bed for the hydrogenation of nitrobenzene. The data are

Rate $r_p = 5.79 \times 10^4 C^{0.578} e^{-2958/T}$ where r_p is in gram moles nitrobenzene reacting per cubic centimeter per hour expressed in terms of void volume; C is the concentration of nitrobenzene in gram moles per cubic centimeter

Reactor inside diameter = 3.5 cm

Feed rate = 70 g mol/h

Void fraction = 0.45

Entering temperature = 457 K

Heat capacity = 6.9 cal/g mol · °C

Heat-transfer coefficient = 8.67 cal/h · cm² · °C

Heat of reaction = −152,100 cal/g mol

Entering concentration of nitrobenzene = 5.0×10^{-7} g mol/cm³

6. The reaction rate of oxidation of naphthalene to phthalic anhydride is

$$r = 305 \times 10^5 p^{0.38} e^{-28{,}000/RT}$$

where r is in pound moles of naphthalene reacted to phthalic anhydride per hour per pound of catalyst, p is partial pressure of naphthalene in atmospheres, and T in Kelvin. The reaction is

$$C_{10}H_8 + 4.5O_2 \rightarrow C_8H_4O_3 + 2H_2O + 2CO_2$$

The heat of reaction is −6,300 Btu/lb of naphthalene, 0.2 mol percent of naphthalene and 99.8 percent air enter the catalytic reactor at 340°C and 1 atm. The data are

Tube size: 2.0 in. ID

Tube wall temperature = 340°C

Catalyst: 0.2 × 0.2 in cylinder

Bulk density = 50 lb/ft³

Superficial mass velocity = 420 lb/h · ft²

$C_p = 0.26$ Btu/lb · °F

Determine the height of the bed for 80 percent conversion under adiabatic conditions.

7. A fluidized catalyst bed operated at 300°C and 1 atm is used for the oxidation of ethylene to ethylene oxide with a feed gas containing 7 percent C_2H_4, 20 percent O_2, and 73 percent N_2. Assuming plug flow and neglecting pressure drop, estimate the minimum bed length required for 60 percent conversion of ethylene to ethylene oxide

$$C_2H_4 + \frac{1}{2}O_2 \rightarrow C_2H_4O$$

$$C_2H_4 + 3O_2 \rightarrow 2CO_2 + 2H_2O$$

The data are

Bulk density of catalyst = 20 lb/ft^3

Superficial velocity at 280°C = 2.2 ft/s

Selectivity of ethylene oxide with respect to CO_2 is 1.5

$r_p = kp_{C_2H_4}(p_{O_2})^{0.2}$ in pound moles of ethylene oxide produced per hour per pound catalyst where p, the partial pressures, are in atmospheres and k, the rate constant, is 0.12 at 300°C.

8. Find the exit conversion from a fluidized bed reactor to carry out a first-order gaseous reaction A → B at 480°F and 2 atm pressure. The data are

k = 0.06 ft^3/s·lb catalyst

Bulk density = 2.8 lb/ft^3

Superficial velocity = 0.16 lb/s·ft^2

Bed height = 12 ft

Molecular weight of A = 44

9. Calculate the conversion of A for 2A → B at 200°F and 1 atm in which

$$r_p = k_2 p_A^2$$

where k_2 is 5.0×10^{-6} lb mol/s·atm^2·lb catalyst. The linear velocity is 1.0 ft/s and the bulk density is 4.5 lb/ft^3. The bed height is 20 ft.

10. Remove ketene in a plate tower by absorption in glacial acetic acid while a first-order reaction is going on. 95 mol percent ketene is removed from a gaseous mixture containing 5.0 percent by volume of ketene entering the tower at a rate of 150 lb mol/h by 157 lb mol/h glacial acetic acid. The hold-up in the plate is 2.15 lb mol per plate. The equilibrium is $y^* = 2.0x$. The rate constant k is 0.075 s^{-1}. The plate efficiency is 0.38. Calculate the number of plates required.

11. Calculate the mass-transfer coefficient $K_E a$ in a liquid-liquid extraction tower from the following data:

m = 10.32; z_o/w = 0.0915; G = 0.5309 kg/s; L = 0.9678 kg/s; H = 22 m.

k = 0.0028 s^{-1}; ρ = 722.33 kg/m^3; y_o = 0.16; S = 0.3421 m^2; hold-up = 5,536.36 kg

References

(B1). Burke, S.P., and Plummer, W.B. *Ind. Eng. Chem.,* 20:1196 (1928).

(B2). Balakrishnan, A.R., and Pei, D.C.T. *Can. J. Chem. Eng.,* 53:231 (1975).

(C1). Chen, N.H., and Othmer, D.F. *J. Chem. Eng. Data,* 7:37 (1962).

(C2). Chu, J.C., Kalil, J., and Wetteroth, W.A. *Chem. Eng. Prog.,* 49:141 (1953).

(C3). Calderbank, P.H., and Moo-Young, M.B. *Chem. Eng. Sci.,* 16:39 (1961).

(C4). Calderbank, P.H., and Jones, S.J.R. *Trans. Inst. Chem. Engr.,* 39:363 (1961).

(C5). Calderbank, P.H. *Chem. Eng. Prog.,* 49:585 (1953).

(C6). Charpentier, J.C. *Chem. Eng. J.,* 11:161 (1976).

(C7). Colburn, A.P. *Ind. Eng. Chem.,* 23:910 (1931).

(C8). Coulson, J.M., Richardson, J.F., and Peacock, D.G. *Chemical Engineering,* Vol. III, 2nd ed. Elmsford, N.Y.: Pergamon Press, 1975.

(D1). Davidson, J.F., and Harrison, D. *Fluidized Particles.* New York: Cambridge Univ. Press, 1963.

(D2). Dharwadkar, A., and Sylvester, N.D. *AIChE J.,* 23:376 (1977).

(E1). Ergun, S. *Chem. Eng. Prog.,* 48:89 (1952).

(F1). Froment, G.F., and Bischoff, K.B. *Chemical Reactor Analysis and Design.* New York: John Wiley & Sons, Inc., 1979.

(G1). Gianettó, A., Specchia, V., and Baldi, G. *AIChE J.,* 19:916 (1973).

(G2). Goto, S., and Smith, J.M. *AIChE J.,* 21:706 (1975).

(H1). Hill, C.G. *Introduction to Chemical Engineering Kinetics and Reactor Design.* New York: John Wiley & Sons, Inc., 1977.

(J1). Jakob, M. *Heat Transfer.* New York: John Wiley & Sons, Inc., 1957.

(J2). Jeffreys, G.V., Jenson, V.G., and Miles, F.R. *Trans. Inst. Chem. Engr.,* 39:389 (1961).

(J3). Jeffreys, G.V., and Cooper, A.R. *Chemical Kinetics and Reactor Design.* Englewood Cliffs, N.J.: Prentice-Hall, Inc., 1973.

(K1). Kettenring, K.N., Manderfield, E.L., and Smith, J.M. *Chem. Eng. Prog.,* 46:139 (1950).

(K2). Kozeny, J. *Sitzber. Akad. Wiss. Wien, Math-naturw. kl. Abt. lla,* 136:271–306 (1927).

(K3). Kunni, D., and Levenspiel, O. *Fluidization Engineering.* New York: John Wiley & Sons, Inc., 1969.

(L1). Leva, M. *Fluidization.* New York: McGraw-Hill Book Company, 1960.

(M1). May, W.G. *Chem. Eng. Prog.,* 55:49 (1959).

(P1). Petrovic, L.J., and Thodos, G. *Ind. Eng. Chem. Fund.,* 7:274 (1968).

(R1). Reiss, L.P. *Ind. Eng. Chem. Proc. Des. Dev,* 6:486 (1967).

(R2). Riccetti, R.E., and Thodos, G. *AIChE J.,* 7:442 (1961).

(R3). Rowe, P.N., and Partridge, B.A. *Trans. Inst. Chem. Engr.,* 44:T335 (1966).

(S1). Sen Gupta, A., and Thodos, G. *Chem. Eng. Prog.,* 58:58 (1962).

(S2). Smith, J.M. *Chemical Engineering Kinetics,* 3rd ed. New York: McGraw-Hill Book Company, 1981.

(V1). Van Deemster, J.J. *Chem. Eng. Sci.,* 13:143 (1961).

(W1). Wilson, E.J., and Geankoplis, C.J. *Ind. Eng. Chem. Fund.,* 5:9 (1966).

(W2). Wenner, R.R., and Dybdal, E.C. *Chem. Eng. Prog.,* 44:275 (1948).

(W3). Wender, L., and Cooper, G.T. *AIChE J.,* 4:15 (1958).

CHAPTER TEN

Nonideal Chemical Reactors

10.1 Introduction

In previous chapters, we have discussed four types of reactors: batch, semibatch, CSTR, and tubular, all of which are ideal. The major equipment in the first three types of reactor is a vessel with small length-to-diameter ratio, usually a tank. The last type is a cylindrical vessel with large length-to-diameter ratio, usually a tube. The flow pattern for these two main types of vessel provides us two extreme cases. The first one, perfectly mixed or backmix flow, shows complete mixing of the feed with the reactor contents immediately on entry, due to vigorous agitation. The mixture is homogeneous down to a molecular scale. No differences at all exist between the various parts of the vessel. The outlet stream has the same properties as those fluids inside the vessel. At the other extreme, the tubular flow assumes the so-called plug or piston flow, in which the fluid velocity is uniform over the entire cross section of the vessel. Each element of fluid that enters the vessel moves forward through the vessel without intermingling with other fluid elements that enter earlier or later. These two cases are ideal. A real reactor lies between these extremes.

10.2 Nonideal Phenomena

If a tank reactor is poorly designed, pockets of stagnant fluid or "dead spots" may exist, such as the *S* regions in Fig. 10.2-1. The conversion in these regions will become very high but the fluid mixture will not leave the reactor. The overall conversion in the exit stream will be less because the remainder of the fluid would spend less time in the reactor. Dead spots can also occur if a baffled tank is poorly designed (see Fig. 10.2-2).

The other phenomenon which may occur in a nonideal tank reactor is the bypassing or short circuiting of the fluid as shown in Fig. 10.2-3. Here a part of the feed entering the reactor takes a short cut to the outlet and does not mix with the others. The conversion in the exit stream is therefore much lower than those in an ideal reactor.

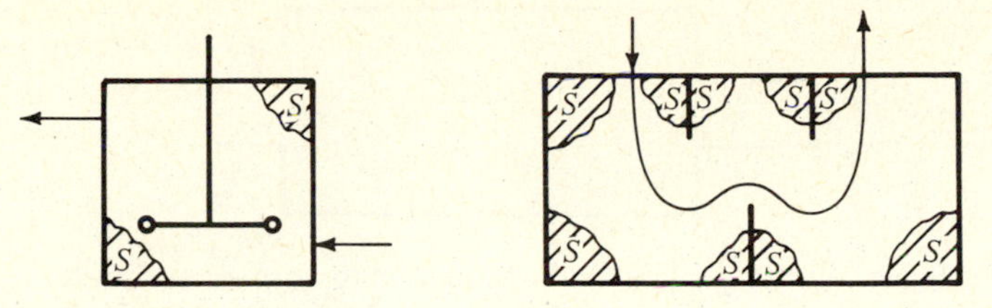

Figure 10.2-1 Dead spot

Figure 10.2-2 Another dead spot

In a tubular reactor, two kinds of deviations from ideal behavior are (1) some mixing in the longitudinal direction due to vortices and eddies (Fig. 10.2-4) and (2) incomplete mixing in the radial direction because of laminar flow (Fig. 10.2-5). The fluid forms a parabolic profile across the tube. Because of the relatively slow molecular diffusion process, the annular elements of the fluid mix only slightly in the radial direction. Also, the fluids near the wall will have a longer residence time and those in the center a shorter residence time. The net result again is the decrease of the conversion.

The third phenomenon is bypassing or short-circuiting in a fixed-bed catalytic reactor. The nonuniform packing arrangement causes some fluids to flow at a higher velocity than the others so that complete radial mixing cannot be realized (Fig. 10.2-6).

In conclusion, all these deviations can be classified into (1) segregated flow in which the elements of fluid do not mix but follow separate paths through the reactor so that they have different residence time and (2) micromixing in which the adjacent elements of fluid partially mix.

After these phenomena are identified, the next problem is to evaluate quantitatively the effects of these deviations on the conversion. If the velocity and local rate of mixing (micromixing) of every element of the fluid in the vessel were known and the differential mass balance could be integrated, an exact solution for the conversion could be obtained. Unfortunately, such complete information is unavailable in a real reactor. Experimentation to collect data becomes necessary. Since it is difficult to measure velocities and concentrations within a reactor, we

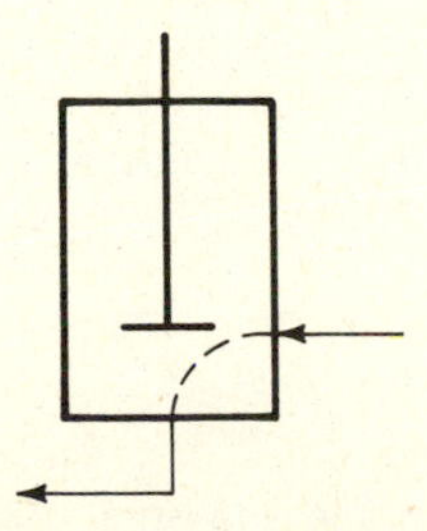

Figure 10.2-3 Bypassing

Figure 10.2-4 Longitudinal mixing due to vortices and turbulence

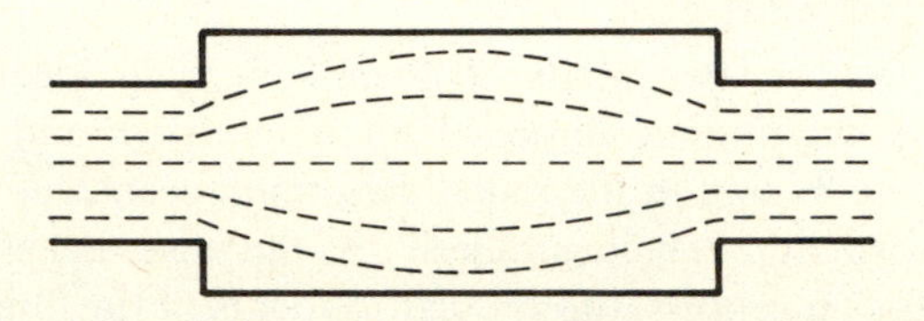

Figure 10.2-5 Poor radial mixing due to laminar flow

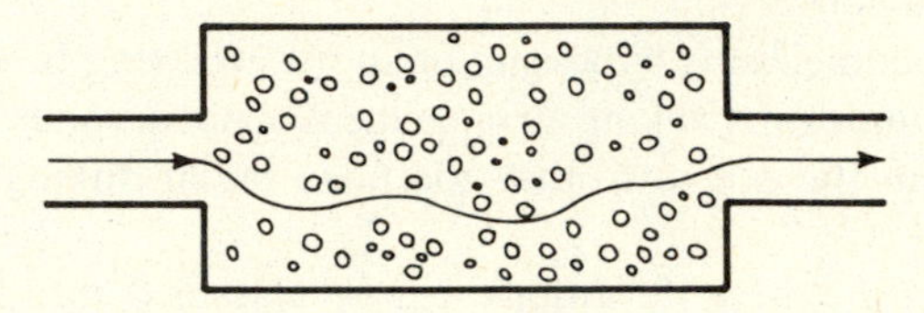

Figure 10.2-6 Bypassing in fixed-bed catalytic reactor due to nonuniform packing

often collect the data on the feed and effluent streams. Such end-effect or response data are obtained by measuring effects observed in the effluent stream when the concentration of an inert component in the feed is changed.

10.3 Terminology

1. Residence time of a fluid element is the time that elapses from the time the element enters the reactor to the time it leaves the reactor. For example, a fluid enters a tubular reactor of volume V at a volumetric flow rate q and also leaves at a rate of q, then the residence time τ is

$$\tau = \frac{V}{q} \tag{10.3-1}$$

2. The age of a fluid element at a given instant of time is the time that elapses between the element's entrance into the vessel and the given instant.

3. The internal age distribution $I(t)$. It is evident that the reactor contains fluid of varying ages. Let $I(t)$ be the internal age distribution frequency of fluid element in fraction of ages per unit time, then fraction of the fluid elements between t and $t + \Delta t$ in the reactor is given by $I(t)\Delta t$ (Fig. 10.3-1). Hence

$$\int_0^\infty I(t)\, dt = 1 \tag{10.3-2}$$

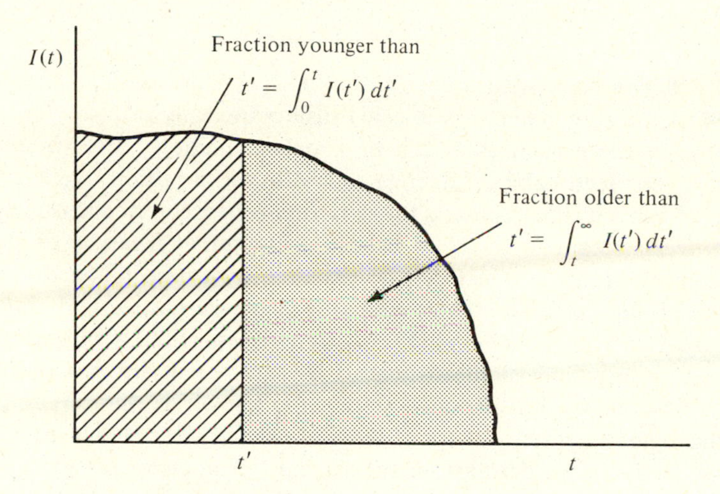

Figure 10.3-1 Internal age distribution

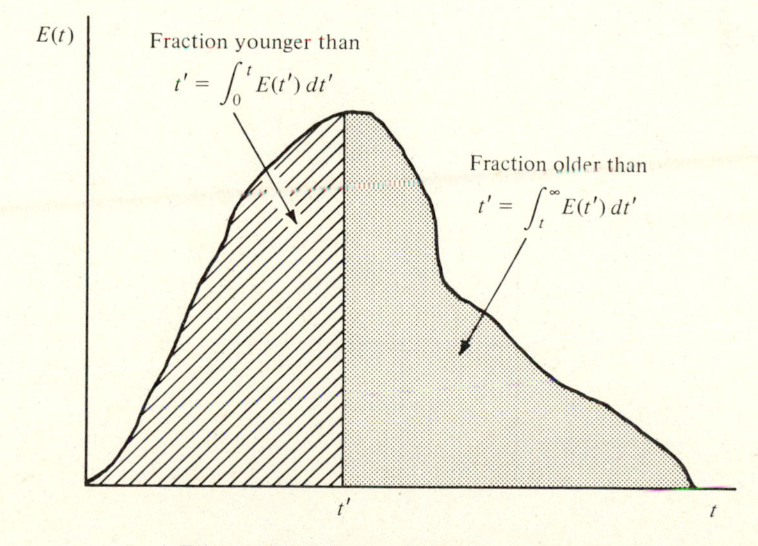

Figure 10.3-2 External age distribution

4. The external age distribution $E(t)$ or *RTD* (residence time distribution). The function $E(t)$ is the age-distribution frequency of the fluid elements leaving a vessel and has the units: fraction of ages per unit time. The fraction of exit ages itself is $E(t)\Delta t$ (see Fig. 10.3-2).

$$\int_0^\infty E(t)dt = 1 \tag{10.3-3}$$

10.4 Experimental Measurement

Consider a flow system with a constant volumetric flow rate q. At time $t = 0$, a mass m of a tracer material is injected instantaneously into the feed stream in some known fashion such as a step or a pulse function. The tracer can be a colored dye, an electrically conducting salt solution, a radioactive compound, or others. Preferably it should possess the same density, viscosity, and other properties as the process stream in order to be a representative measure of the mixing. The concentration of the tracer $c(t)$ is measured by a detector at the exit stream for a series of time intervals. These data are frequently plotted with dimensionless concentration $c(t)/c_o$ or $c(t)/(m/V)$ as ordinate where m is the mass and V is the volume against dimensionless time $t/\mathbf{t}$ or $t/(V/q)$ where q is the volumetric flow rate. The curves thus obtained are usually known as the response curve.

If the input is a step function, the response is also shown in Fig. 10.4-1. The shape of the curve depends on the system. This curve is called an F curve (Fig. 10.4-2). Applying the definition of $E(t)$ to the mass m of the tracer, a mass balance for a time interval Δt at the stream outlet gives

$$qc(t)\,\Delta t = mE(t)\,\Delta t \tag{10.4-1}$$

Hence

$$E(t) = \frac{q}{m}c(t) \tag{10.4-2}$$

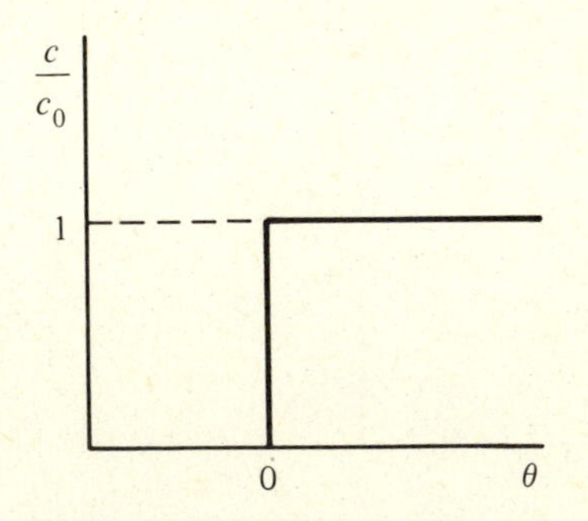

Figure 10.4-1 Step function input

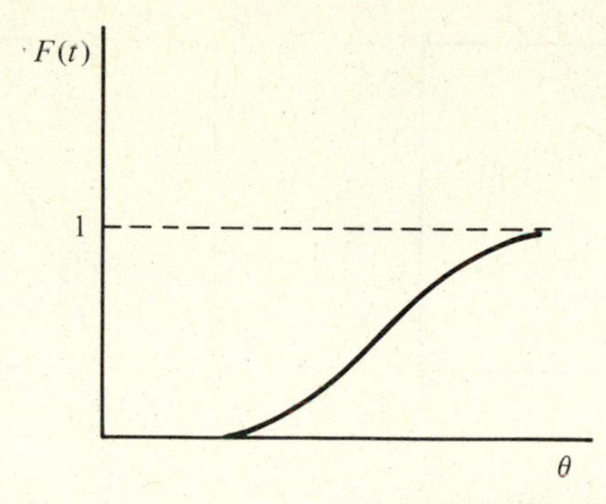

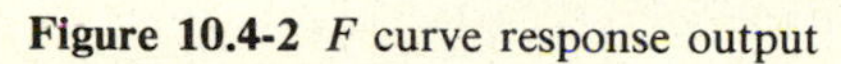

Figure 10.4-2 *F* curve response output

or

$$E(t)=\frac{q}{V}C(t) \tag{10.4-3}$$

where

$$C(t)=\frac{V}{m}c(t) \tag{10.4-1}$$

and the *F* curve is

$$F(t)=\frac{q}{m}\int_0^t c(t)dt \tag{10.4-5}$$

or

$$F(t)=\frac{q}{V}\int_0^t C(t)dt=\int_0^{qt/V} C(t)d(qt/V) \tag{10.4-6}$$

If the tracer input is a pulse function, or more frequently an impulse or Dirac delta function, the response is also shown in Fig. 10.4-3 as the *C* curve. It is seen that

$$\int_0^\infty C(\theta)d\theta=1=\int_0^\infty \frac{c(\theta)}{m/V}\,d\theta=1 \tag{10.4-7}$$

or

$$\frac{m}{V}=\int_0^\infty c(\theta)d\theta=\frac{1}{\mathrm{t}}\int_0^\infty c(t)dt \tag{10.4-8}$$

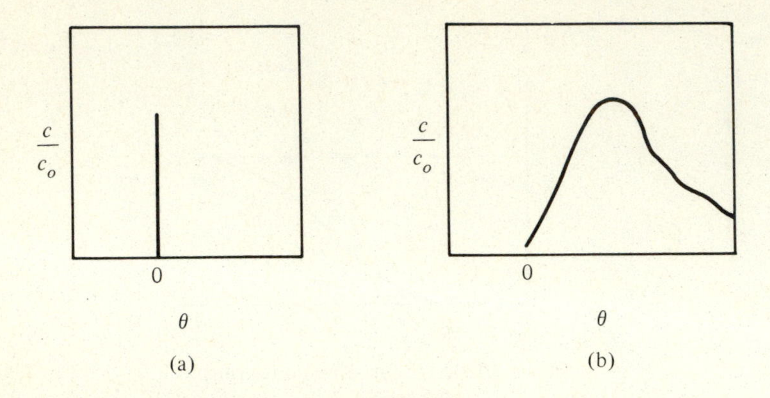

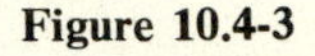
Figure 10.4-3

since

$$\boldsymbol{\theta} = \frac{t}{\mathbf{t}}$$

where

$$\mathbf{t} = \text{residence time} = \frac{V}{q}$$

Example 10.4-1 Residence Time Distribution (RTD)

A fluid flows at a rate of 3.776×10^{-3} m^3/s in a vessel of volume 3.3984 m^3. A solution containing 800 g of dye is added rapidly to the inlet stream. The dye concentration at the effluent stream was measured as follows:

Time, s	0	300	600	900	1200	1500	1800	2100
Dye conc., g/m^3	0	105.9322	176.5537	176.5537	141.2429	70.6215	35.3107	0

Develop the *C*, *E*, and *F* diagrams.

Solution

Check the total injected dye by Simpson's rule.

$$3.776 \times 10^{-3} \left\{ \frac{300}{3} \left[0 + 2(176.5537 + 141.2429 + 35.3107) + 4(105.9322 + 176.5537 + 70.6214) \right] \right\} = 800 \text{ g}$$

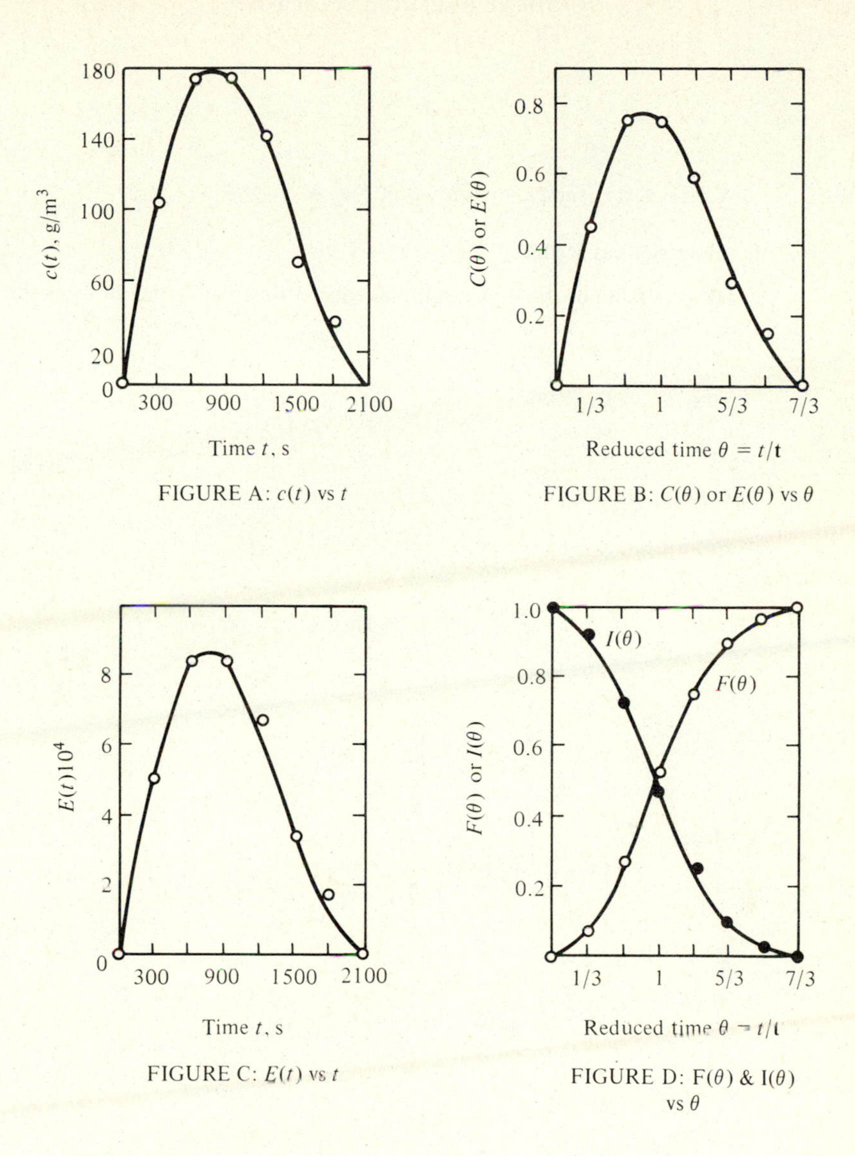

FIGURE A: $c(t)$ vs t

FIGURE B: $C(\theta)$ or $E(\theta)$ vs θ

FIGURE C: $E(t)$ vs t

FIGURE D: $F(\theta)$ & $I(\theta)$ vs θ

The residence time **t** is calculated by

$$\mathbf{t} = \frac{\Sigma(tc)}{\Sigma c} = \frac{300(105.9322) + 600(176.5537) + 900(176.5537) + 1{,}200(141.2429) + 1{,}500(70.6215) + 1{,}800(35.3107)}{105.9322 + 176.5537 + 176.5537 + 141.2429 + 70.6215 + 35.3107}$$

$$= \frac{6.355932 \times 10^5}{7.062147 \times 10^2} = 900 \text{ s}$$

or

$$\mathbf{t} = \frac{V}{q} = \frac{3.3984}{3.776 \times 10^{-3}} = 900 \text{ s} \qquad \text{check}$$

A plot of $c(t)$ versus t is shown in Fig. A on p. 415.

1. Dimensionless Reduced Time $\theta = t/\mathbf{t}$

Hence a table of the dimensionless reduced time can be made from the data as follows:

Time t	0	300	600	900	1,200	1,500	1,800	2,100
$\theta = t/\mathbf{t}$	0	1/3	2/3	1	4/3	5/3	2	7/3

2. C Diagram

From Eq. (10.4-8), we get

$$\frac{m}{V} = c_o = \int_0^\infty c(\theta) d\theta = \frac{1}{\mathbf{t}} \int_0^\infty c(t)\, dt$$

or

$$\frac{m}{V} = c_o = \frac{1}{\mathbf{t}} \Sigma c(t)\, \Delta t$$

Hence

$$c_o = \frac{7.062147 \times 10^2 (300)}{900} = 2.354049 \times 10^2$$

Then

$$C(t) = \frac{c(t)}{c_o} = \frac{c(t)}{2.354049 \times 10^2}$$

or by Eq. (10.4-4)

$$C(t) = \frac{V}{m} c(t) = \frac{3.3984}{800} c(t) \quad \text{for check}$$

By Eq. (10.4-7)

$$C(\theta) = \frac{c(t)}{c_o} = \frac{c(t)}{235.4049}$$

The C diagram can be made from the data as follows:

Time t	0	300	600	900	1,200	1,500	1,800	2,100
θ	0	1/3	2/3	1	4/3	5/3	2	7/3
$C(t)$	0	0.45	0.75	0.75	0.60	0.30	0.15	
$C(\theta)$	0	0.45	0.75	0.75	0.60	0.30	0.15	

A plot of $C(\theta)$ versus θ is shown in Fig. B on p. 415.

3. E Diagram

From the definition of $E(t)$, we can write

$$E(t) = \frac{c(t)}{\Sigma\, c(t)\, \Delta t} = \frac{c(t)}{706.2147(300)}$$

and

$$E(\theta) = \mathbf{t}E(t) = 900E(t)$$

or from Eq. (10.4-2)

$$E(t) = \frac{q}{m}\, c(t) = \frac{0.003776}{800}\, c(t)$$

from Eq. (10.4-3)

$$E(t) = \frac{q}{V}\, C(t) = \frac{0.003776}{3.3984}\, C(t) \quad \text{as a check}$$

Hence the E diagram can be made from the data as follows:

Time t	0	300	600	900	1,200	1,500	1,800	2,100
θ	0	1/3	2/3	1	4/3	5/3	2	7/3
$E(t)10^4$	0	5.0	8.333	8.333	6.667	3.333	1.667	
$E(\theta)$	0	0.45	0.75	0.75	0.60	0.30	0.15	

A plot of $E(\theta)$ versus θ is shown in Fig. B on p. 415 and a plot of $E(t)$ versus t is shown in Fig. C on p. 415.

4. F Diagram and I Diagram

The F diagram can be obtained from the E diagram by

$$F(t) = \int_0^t E(t)\, dt$$

or more conveniently by

$$F(\theta) = \int_0^\theta E(\theta)\, d\theta \qquad I(\theta) = 1 - F(\theta)$$

which can be evaluated by the trapezoidal rule as follows:

Time t	0	300	600	900	1,200	1,500	1,800	2,100
θ	0	1/3	2/3	1	4/3	5/3	2	7/3
$E(\theta)$	0	0.45	0.75	0.75	0.60	0.30	0.15	0
$\int_{\theta_1}^{\theta_2} E(\theta)\,d\theta$		0.075	0.200	0.250	0.225	0.150	0.075	0.025
$F(\theta)$		0.075	0.275	0.525	0.750	0.900	0.975	1.00
$I(\theta)$	1	0.925	0.725	0.475	0.250	0.100	0.025	0.00

A plot of $F(\theta)$ and $I(\theta)$ versus θ is shown in Fig. D on p. 415.

10.5 RTD for Ideal Reactors

For ideal reactors whose mixing characteristics are known, such as a plug-flow reactor, a single-tank reactor, or a tubular reactor with laminar flow, the response curve and the RTD can be predicted without experimentation. Each case is described below.

10.5-1 Plug-flow reactor

In plug-flow reactors, the velocity profile is uniform and there is no longitudinal mixing. This means that all fluid elements leaving the reactor have the same age. For a step input, the front or interface between the tracer and nontracer fluids proceeds down the vessel and comes out at the other end in a time equal to the mean residence time **t**. The input and response to a step function in a plug-flow reactor are shown in Figs. 10.5-1 and 10.5-2. Hence the phenomena can be presented by

$$\text{Input:}\quad c=\begin{cases}0 & \text{for } t<0\\ c_o & \text{for } t>0\end{cases}\quad\text{or}\quad \frac{c}{c_o}=\begin{cases}0 & \text{for } t<0\\ 1 & \text{for } t>0\end{cases}\tag{10.5-1}$$

$$\text{Response:}\quad F(\theta)=U(t-\mathbf{t})\quad\text{where } U(t-\mathbf{t})=\begin{cases}0 & \text{for } t<\mathbf{t}\\ 1 & \text{for } t>\mathbf{t}\end{cases}\tag{10.5-2}$$

$$E\text{ curve:}\quad E(t)=\frac{df}{dt}=\frac{d}{dt}\,U(t-\mathbf{t})=\delta(t-\mathbf{t})\tag{10.5-3}$$

$$E(\theta)=\delta(\theta-1)\tag{10.5-4}$$

$$I\text{ curve:}\quad I(\theta)=1-F(\theta)=1-U(\theta-1)\tag{10.5-5}$$

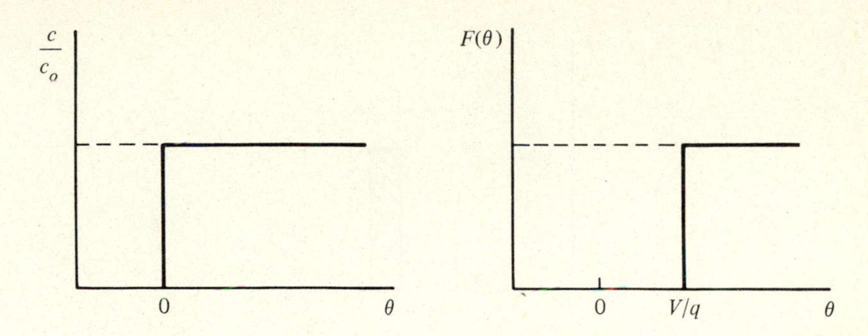

Figure 10.5-1 A step function input in a plug-flow reactor

Figure 10.5-2 Response in a plug-flow reactor

The input and response to a pulse function in a plug-flow reactor are shown in Figs. 10.5-3 and 10.5-4. The phenomena can be represented by:

$$\text{Input:}\quad c=\begin{cases}0 & \text{for}\quad t<0\\ c_o & \text{for}\quad 0<t<\Delta t_o\\ 0 & \text{for}\quad t>\Delta t_o\end{cases}\qquad \text{or}$$

$$\frac{c}{c_o}=\begin{cases}0 & \text{for}\quad t<0\\ 1 & \text{for}\quad 0<t<\Delta t_o\\ 0 & \text{for}\quad t>\Delta t_o\end{cases}\tag{10.5-6}$$

$$\text{Response:}\quad \frac{c}{c_o}=\begin{cases}0 & \text{for}\quad \mathbf{t}<0\\ 1 & \text{for}\quad 0<\mathbf{t}<\Delta t_o\\ 0 & \text{for}\quad \mathbf{t}>\Delta t_o\end{cases}\tag{10.5-7}$$

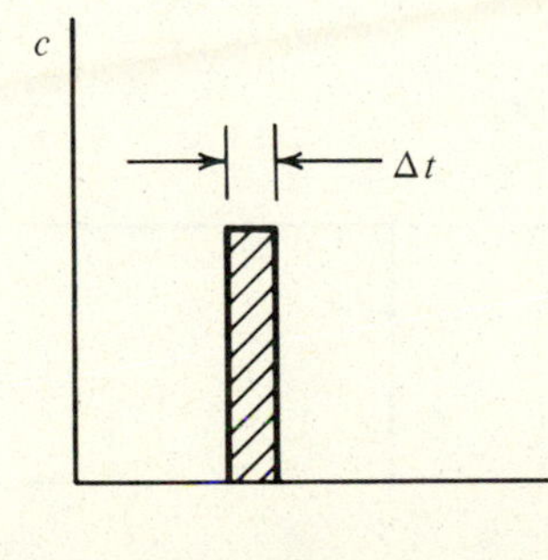

Figure 10.5-3 A pulse function input in a plug-flow reactor

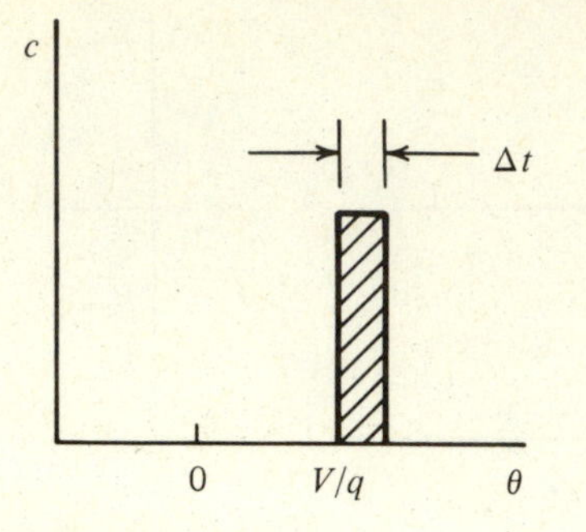

Figure 10.5-4 Response in a plug-flow reactor

10.5-2 Ideal stirred-tank reactor

The contents in an ideal stirred-tank reactor are perfectly homogeneous and have the same composition as the exit stream. If a step function input is applied to a perfectly mixed vessel, a mass balance gives

$$qc + V\frac{dc}{dt} = qc_o \tag{10.5-8}$$

or

$$c + \mathbf{t}\frac{tc}{dt} = c_o \tag{10.5-9}$$

where $\mathbf{t} = V/q$ is the residence time (see Figs. 10.5-5 and 10.5-6). Its solution for zero initial concentration is

$$\frac{c}{c_o} = 1 - e^{t/\mathbf{t}} = F(\theta) \tag{10.5-10}$$

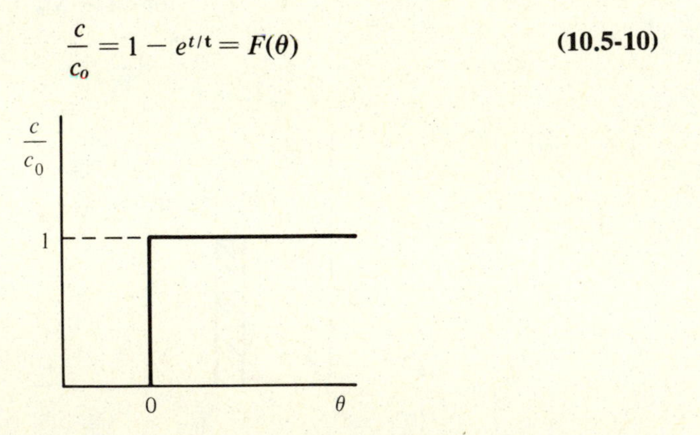

Figure 10.5-5 Step function input in a CSTR

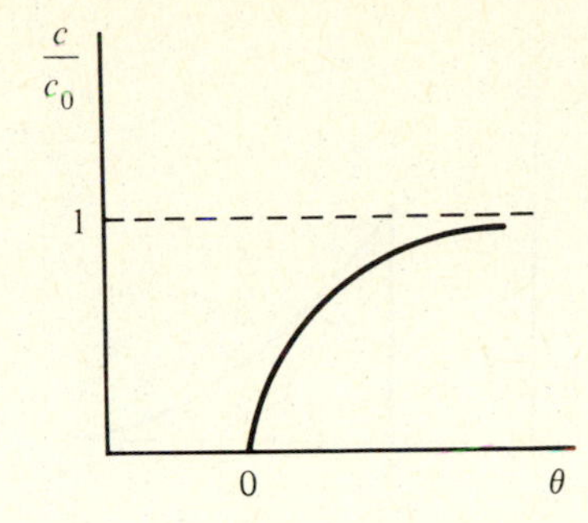

Figure 10.5-6 Response in a CSTR

Consequently,

$$E(t) = \frac{dF(\theta)}{dt} = \frac{1}{\mathbf{t}}\, e^{t/\mathbf{t}} \tag{10.5-11}$$

$$I(t) = \frac{1}{\mathbf{t}}\,[1 - F(\theta)] = \frac{1}{\mathbf{t}}\, e^{t/\mathbf{t}} \tag{10.5-12}$$

$$I(\theta) = e^{-\theta} = E(\theta) \tag{10.5-13}$$

If the input is a pulse function, the response in a CSTR is shown as in Figs. 10.5-7 and 10.5-8. If the input is an impulse function, the product qc_o in the following transient equation

$$qc + V\frac{dc}{dt} = qc_o \tag{10.5-14}$$

will become the mass m of tracer introduced as an impulse and will therefore have the transform m. Then for zero tracer in the vessel at time $t = 0$, the transform of the equation is

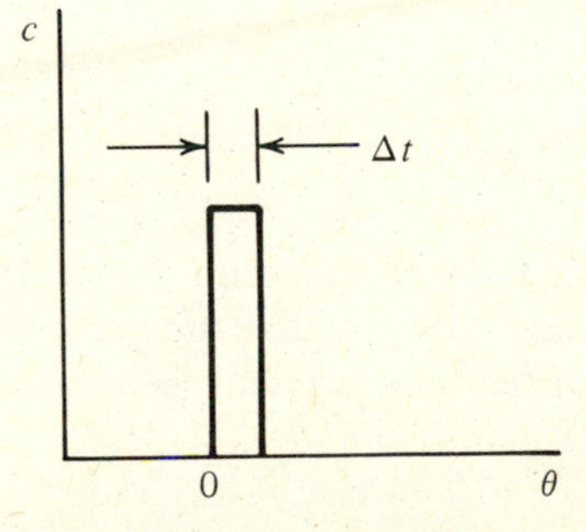

Figure 10.5-7 A pulse function in a CSTR

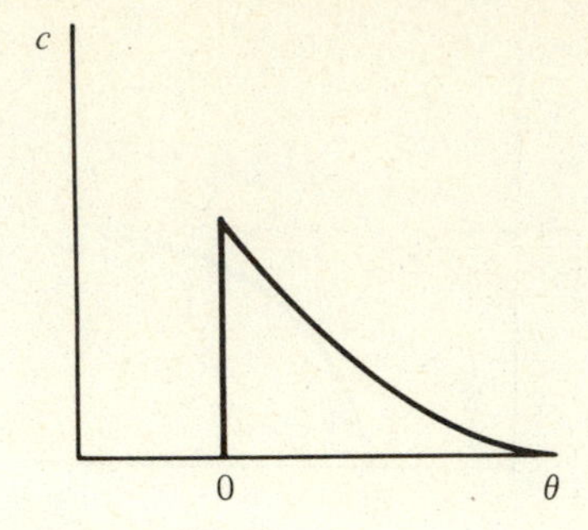

Figure 10.5-8 Response in a CSTR

$$q\mathbf{c}(s) + Vs\mathbf{c}(s) = m \tag{10.5-15}$$

Transposing and inverting gives

$$c(t) = \frac{m}{V} e^{-qt/V} \tag{10.5-16}$$

This gives the C diagram which is an exponentially decreasing function of time. Consequently the E- and F-diagrams become

$$E(t) = \frac{q}{V} e^{-qt/V} \tag{10.5-17}$$

$$F(t) = 1 - e^{-qt/V} \tag{10.5-18}$$

10.5-3 Laminar flow in a tubular reactor

The velocity profile in a tubular reactor with laminar flow in the absence of either radial or longitudinal diffusion is

$$u = u_o\left[1 - \left(\frac{r}{R}\right)^2\right] \quad \text{for} \quad 0 \le r \le R \tag{10.5-19}$$

where u_o is the velocity of the fluid at the center, r is the distance from the center of the tube, R is the inside radius of the tube. From fluid dynamics, it is seen that the average velocity of the fluid element is

$$\mathbf{u} = \frac{u_o}{2} \tag{10.5-20}$$

If L is the length of the tube, the time that it will take a fluid element to traverse the tube is

$$t = \frac{L}{u} = \frac{L}{u_0[1-(r/R)^2]} \qquad \textbf{(10.5-21)}$$

The average residence time $\mathbf{t}$ is

$$\mathbf{t} = \frac{L}{\mathbf{u}} = \frac{2L}{u_0} \qquad \textbf{(10.5-22)}$$

Combining gives

$$\frac{\mathbf{t}}{2t} = \left[1 - \left(\frac{r}{R}\right)^2\right] \qquad \textbf{(10.5-23)}$$

Since the fluid element at the center is moving the fastest, it will be the first to leave at a minimum time

$$t_{\text{min}} = \frac{L}{u_0} = \frac{\mathbf{t}}{2} \qquad \textbf{(10.5-24)}$$

Hence

$$F(t) = 0 \qquad \text{for} \qquad t < \frac{\mathbf{t}}{2} \qquad \textbf{(10.5-25)}$$

and when t is equal to the time necessary for a fluid element to traverse the tube length,

$$F(t) = \frac{\text{Volumetric flow rate between } r = 0 \text{ and } r = r}{\text{Total volumetric flow rate}} \qquad \textbf{(10.5-26)}$$

$$F(t) = \frac{\int_0^r u(r) 2\pi r \, dr}{\int_0^R u(r) 2\pi r \, dr} \qquad \textbf{(10.5-27)}$$

Substituting the velocity profile and integrating gives

$$F(t) = \left(\frac{r}{R}\right)^2 \left[2 - \left(\frac{r}{R}\right)^2\right] \qquad \textbf{(10.5-28)}$$

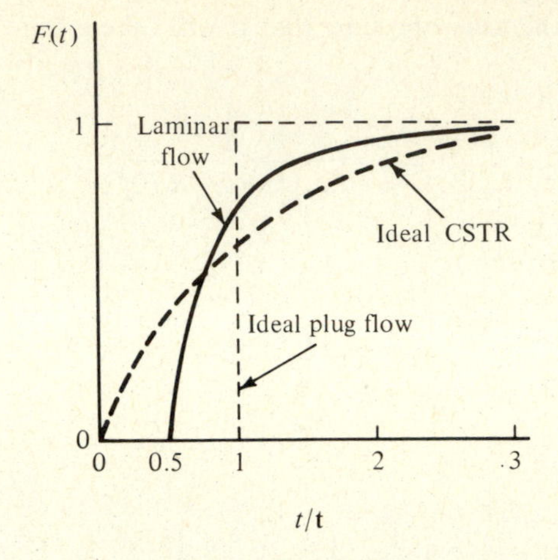

Figure 10.5-9 $F(t)$ for ideal reactors

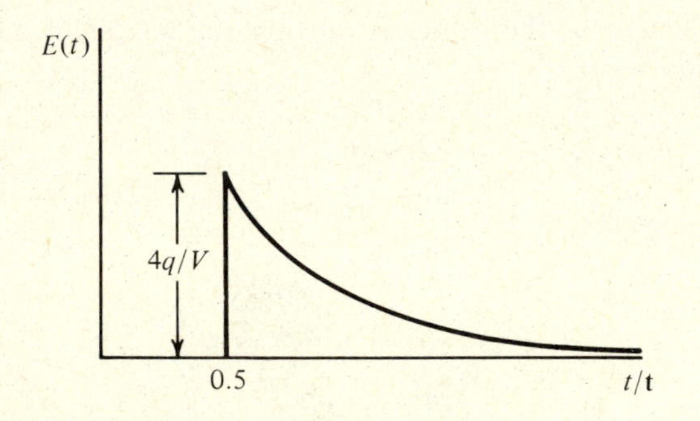

Figure 10.5-10 $E(t)$ for laminar flow

Eliminating $(r/R)^2$ by the previous expression gives

$$F(t) = 1 - \left(\frac{\mathrm{t}}{2t}\right)^2 \quad \textbf{(10.5-29)}$$

This response for a tubular reactor with laminar flow is compared with those in a plug flow and a CSTR reactor in Fig. 10.5-9. Consequently, the E diagram is (see Fig. 10.5-10)

$$E(t)dt = 0 \qquad \text{for } t < t_o \quad \textbf{(10.5-30)}$$

$$E(t)dt = 2t_o^2 \frac{dt}{t^3} \qquad \text{for } t > t_o \quad \textbf{(10.5-31)}$$

10.6 RTD for Nonideal Reactors

As mentioned earlier in this chapter, the deviations from the ideal flow conditions are due to (1) different fluid elements moving at different velocities through the reactor and (2) fluid elements with different ages mixing within the reactor. It is necessary to determine the RTD for these two types of nonideal reactors.

10.6-1 The axial dispersion model

In order to study the effect of the mixing in a reactor, let us assume that the mixing occurs only in the axial direction which can be accounted for by a term called *effective* or *apparent longitudinal diffusivity* D_L. The mixing in the radial direction is neglected. Further, the model also assumes that the fluid possesses constant velocity and constant concentration across the tube diameter. Within the reactor, there are no stagnant regions and no bypassing because of the magnitude of the dispersion. The differential equation including the dispersion term D_L can be easily formulated as

$$D_L \frac{\partial^2 c}{\partial z^2} - u \frac{\partial c}{\partial z} = \frac{\partial c}{\partial t} \qquad \textbf{(10.6-1)}$$

where z is the longitudinal direction and u is the linear velocity. Let us impose a step function of inert tracer concentration c_o into the feed at $\theta = 0$, then the initial condition is

$$c = \begin{cases} 0 & \text{at } z\ 0 \text{ for } \theta = 0 \\ c_o & \text{at } z\ 0 \text{ for } \theta = 0 \end{cases} \qquad \textbf{(10.6-2)}$$

and the boundary conditions are

$$-D_L\left(\frac{\partial c}{\partial z}\right) + uc = c_o \qquad \text{at } z = 0 \text{ for } \theta > 0 \qquad \textbf{(10.6-3)}$$

$$\frac{dc}{dz} = 0 \qquad \text{at } z = L \text{ for } \theta > 0 \qquad \textbf{(10.6-4)}$$

The solution of the differential equation with these boundary conditions is difficult. As a good approximation, let us further assume that D_L is small and the boundary conditions are changed to

$$c = \begin{cases} c_o & \text{at } z = -\infty \text{ for } \theta \geq 0 \\ 0 & \text{at } z = \infty \text{ for } \theta \geq 0 \end{cases} \qquad \textbf{(10.6-5)}$$

Then the solution can be accomplished by making the substitution

$$\beta = \frac{z - u\theta}{\sqrt{4D_L\theta}} \qquad \text{and} \qquad c^* = \frac{c}{c_o} \tag{10.6-6}$$

The differential equation becomes

$$\frac{d^2c^*}{d\beta^2} + 2\beta\frac{dc^*}{d\beta} = 0 \tag{10.6-7}$$

with the new boundary conditions:

$$c^* = \begin{cases} 1 & \text{for } \beta = -\infty \\ 0 & \text{for } \beta = \infty \end{cases} \tag{10.6-8}$$

Solving the equation and substituting back gives the response at the end of the reactor

$$\frac{c}{c_o} = \frac{1}{2}\left[1 - \text{erf}\left(\frac{1}{2}\sqrt{\frac{uL}{D_L}}\,\frac{1 - \theta/\boldsymbol{\theta}}{\sqrt{\theta/\boldsymbol{\theta}}}\right)\right]^* \tag{10.6-9}$$

where $\boldsymbol{\theta} = V/q = L/u$ is the mean residence time. This response is plotted in Fig. 10.6-1.

10.6-2 Series of stirred-tank model

Let us find the F and E diagrams in a series of well-mixed tanks. Just as in a single tank, the fluid elements in the tanks have different ages and have different velocities when they travel through the tanks. Exactly as in a single tank, at time $t = 0$ when all the tanks are full and the whole system is under steady state, a shot of N_o moles of inert tracer is injected into the feed. The tracer is assumed to be completely miscible with the liquid in the tanks. If V is the total volume of the n equally sized well-mixed reactors in series, the unsteady-state balance on the mth stage is

$$\frac{dN_m}{dt} = \frac{N_{m-1}}{V/n}\,q - \frac{N_m}{V/n}\,q \tag{10.6-10}$$

---

* Error function $\text{erf}\, x = \frac{2}{\sqrt{\pi}}\int_0^x e^{-u^2}\,du$.

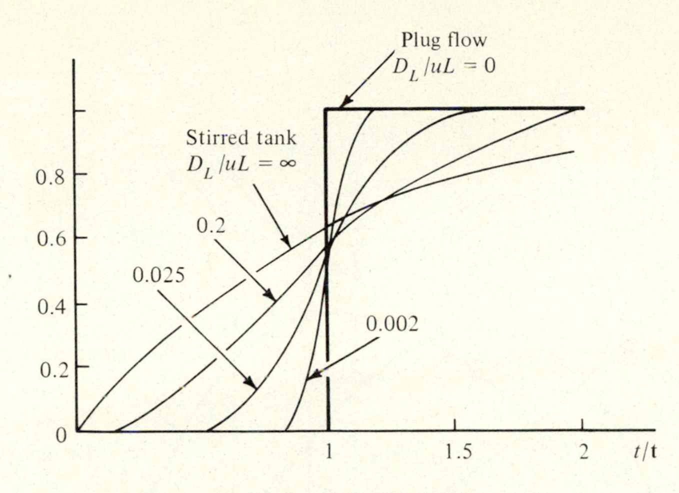

Figure 10.6-1 RTD for dispersion model

where q is the volumetric flow rate of the fluid. The solution for the first tank is

$$N_1 = N_o e^{-qnt/V} \tag{10.6-11}$$

Substituting for N_1 and integrating again gives

$$N_2 = N_o\left(\frac{qnt}{V}\right)e^{-qnt/V} \tag{10.6-12}$$

Continuing the process

$$N_3 = \frac{N_o}{2!}\left(\frac{qnt}{V}\right)^2 e^{-qnt/V} \tag{10.6-13}$$

$$N_r = \frac{N_o}{(r-1)!}\left(\frac{qnt}{V}\right)^{r-1} e^{-qnt/V} \tag{10.6-14}$$

If we plot c_i/c_o, where c_i is the concentration of the tracer leaving the ith tank and $c_o = N_o/V$ is the concentration of the tracer injected to the feed, against the reduced time (qt/V), we obtain the C diagram in Fig. 10.6-2. For the F diagram (Fig. 10.6-3), we note that any time t, the number of moles of the tracer remaining in the series of n tanks is

$$N_t = N_1 + N_2 + N_3 + \cdots + N_n \tag{10.6-15}$$

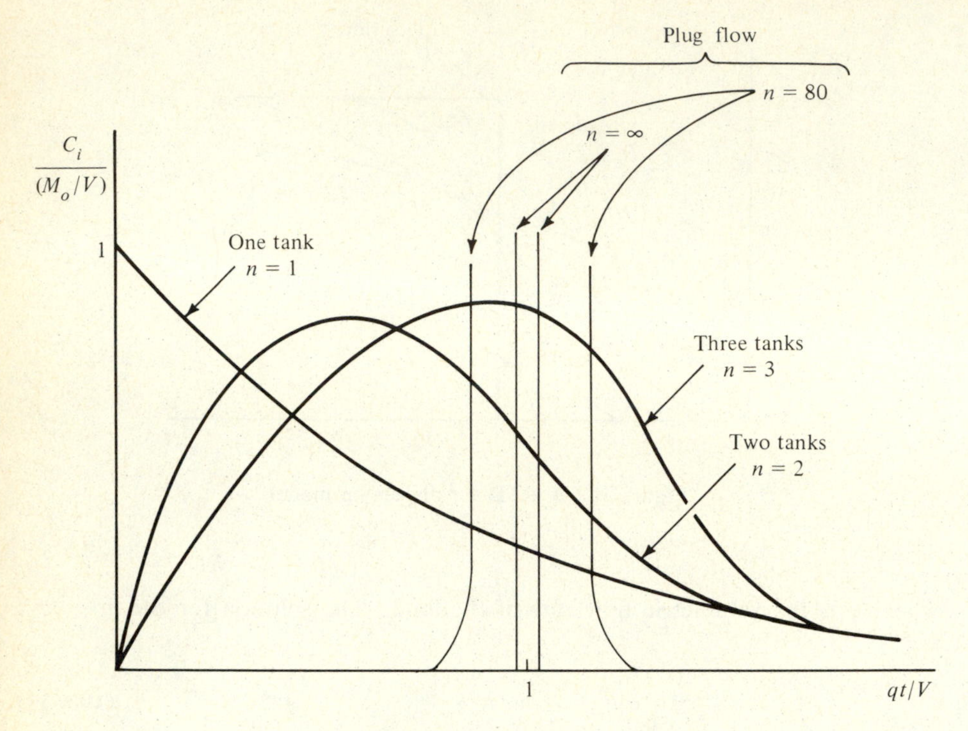

Figure 10.6-2 C diagram for a CSTR

The fraction of tracer F which has escaped completely after a time t is, therefore,

$$F = \frac{N_o - N_t}{N_o} = 1 - \frac{1}{N_o}(N_1 + N_2 + N_3 + \cdots + N_n) \qquad \textbf{(10.6-16)}$$

or

$$F = 1 - e^{-qnt/V}\left[1 + \frac{qnt}{V} + \frac{1}{2!}\left(\frac{qnt}{V}\right)^2 + \cdots + \frac{1}{(n-1)!}\left(\frac{qnt}{V}\right)^{n-1}\right] \qquad \textbf{(10.6-17)}$$

It is seen that as n approaches infinity, F approaches zero. This means that for an infinite number of tanks, the fraction of tracer that has escaped is zero for all times less than the residence time V/q. This is exactly the same for an ideal tubular plug-flow reactor. The E diagram can be obtained by taking the derivative of the $F(t)$ diagram with time t

$$E(t) = \frac{dF(t)}{dt} = \left(\frac{qnt}{V}\right)^{n-1}\left(\frac{nq}{V}\right)\frac{e^{-qnt/V}}{(n-1)!} \qquad \textbf{(10.6-18)}$$

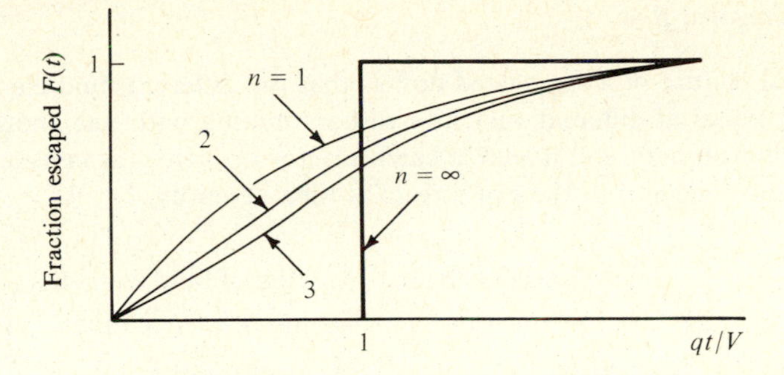

Figure 10.6-3 F diagram for a CSTR

10.7 Conversion in Nonideal Reactors

As mentioned before, there are two main types of phenomena for fluid flow: (1) segregated or no micromixing flow and (2) complete micromixing. The rate of reaction per unit volume for complete micromixing is

$$r_m = k\left(\frac{C_1 + C_2}{2}\right)^n \tag{10.7-1}$$

and for no micromixing, it is

$$r_s = \tfrac{1}{2}(kC_1^n + kC_2^n) \tag{10.7-2}$$

where n is the order of reaction, C_1, C_2 are the reactant concentrations for fluid elements one and two, respectively. For the same residence time, the ratio of the rates is equal to the ratio of the conversion f.

$$\frac{f_m}{f_s} = \frac{r_m}{r_s} = \frac{[(C_1 + C_2)/2]^n}{(C_1^n + C_2^n)/2} \tag{10.7-3}$$

Consequently, we may obtain

$$\frac{f_m}{f_s} = 1 \qquad \text{for } n = 1 \tag{10.7-4}$$

$$\frac{f_m}{f_s} > 1 \qquad \text{for } n < 1 \tag{10.7-5}$$

$$\frac{f_m}{f_s} < 1 \qquad \text{for } n > 1 \tag{10.7-6}$$

10.7-1 Segregated flow

The essential feature of a segregated flow is that the different fluid elements move through the vessel at different velocities without mixing with one another. Hence each fluid element acts as if it were a batch reactor operating at constant pressure. The average conversion in the exit stream is then given by

$$\mathbf{f}_A = \Sigma\left(\begin{array}{l}\text{fractional conversion} \\ \text{as a function of} \\ \text{residence time}\end{array} \times \begin{array}{l}\text{fraction of fluid elements} \\ \text{having residence times} \\ \text{between } t \text{ and } t+\Delta t\end{array}\right) \tag{10.7-7}$$

or in terms of integral

$$\mathbf{f}_A = \int_{F(t)=0}^{F(t)=1.0} f_A(t)\, dF(t) \tag{10.7-8}$$

Since

$$dF(t) = E(t)\, dt \tag{10.7-9}$$

$$\mathbf{f}_A = \int_0^{\infty} f_A(t)\, E(t)\, dt \tag{10.7-10}$$

For first-order kinetics,

$$\frac{dC_A}{dt} = -kC_A \tag{10.7-11}$$

or

$$\frac{df_A}{dt} = -k(1 - f_A) \tag{10.7-12}$$

Integrating and simplifying gives

$$f_A = 1 - e^{-kt} \tag{10.7-13}$$

Substituting into Eq. (10.7-10) with $E(t)$ obtained from Eq. (10.5-11) or Eq. (10.5-16) gives

$$\mathbf{f}_A = \int_0^{\infty} \left[(1 - e^{-kt})\left(\frac{1}{\mathbf{t}}\, e^{-t/\mathbf{t}}\right)\right] dt \tag{10.7-14}$$

Integrating gives

$$\mathbf{f}_A = \frac{k\mathbf{t}}{1 + k\mathbf{t}} \tag{10.7-15}$$

which is the same as Eq. (7.3-4) for complete mixing. Hence Eq. (10.7-4) is verified. That is, for first-order kinetics, the conversion is the same regardless of the extent of micromixing.

For second-order kinetics,

$$\frac{dC_A}{dt} = -kC_A^2 \tag{10.7-16}$$

or

$$\frac{df_A}{dt} = kC_{Ao}(1 - f_A)^2 \tag{10.7-17}$$

Integrating gives

$$f_A = 1 - \frac{1}{1 + kC_{Ao}t} \tag{10.7-18}$$

Substituting into Eq. (10.7-10)

$$\mathbf{f}_A = \int_0^\infty \left[\left(1 - \frac{1}{1 + kC_{Ao}t}\right)\left(\frac{1}{\mathbf{t}}\, e^{t/\mathbf{t}}\right) \right] dt \tag{10.7-19}$$

$$= 1 + \frac{1}{kC_{Ao}\mathbf{t}}\, e^{1/kC_{Ao}\mathbf{t}}\, Ei\left(\frac{-1}{kC_{Ao}\mathbf{t}}\right) \tag{10.7-20}$$

where Ei is the exponential integral. It is interesting to note that for complete micromixing under the same conditions, the average conversion can be obtained from Eq. (7.3-7) as

$$\mathbf{f}_A = 1 - \frac{C_{A2}}{C_{Ao}} = 1 + \frac{1}{2kC_{Ao}t}\,[1 - (1 + 4kC_{Ao}\mathbf{t})^{1/2}] \tag{10.7-21}$$

For half-order kinetics,

$$\frac{dC_A}{dt} = -kC_A^{1/2} \tag{10.7-22}$$

or

$$\frac{df_A}{dt} = \frac{k}{C_{Ao}^{1/2}}\,(1 - f_A)^{1/2} \tag{10.7-23}$$

Integrating gives

$$f_A = 1 - \left[1 - \frac{1}{2}\frac{kt}{C_{Ao}^{1/2}}\right]^2 \tag{10.7-24}$$

Substituting into Eq. (10.7-10) and integrating

$$\mathbf{f}_A = \int_0^\infty \left[1 - \left(1 - \frac{1}{2}\frac{kt}{C_{Ao}^{1/2}}\right)^2\right]\left(\frac{1}{\mathbf{t}}\, e^{(-t/\mathbf{t})}\right) dt \tag{10.7-25}$$

$$= \frac{k\mathbf{t}}{C_{Ao}^{1/2}}\left[1 - \frac{k\mathbf{t}}{2C_{Ao}^{1/2}}\,(1 - e^{-2C_{Ao}^{1/2}/k\mathbf{t}})\right] \tag{10.7-26}$$

which should be obtained from the integration from 0 to $t = 2C_{Ao}^{1/2}/k$. The corresponding equation for complete micromixing is

$$\mathbf{f}_A = \frac{(k\mathbf{t})^2}{2C_{Ao}}\,\{-1 + [1 + 4C_{Ao}/(k\mathbf{t})^2]^{1/2}\} \tag{10.7-27}$$

Example 10.7-1 Conversion for Segregated Flow

For $C_{Ao} = 1$, $\mathbf{t} = 10$ s, calculate the conversion for segregated flow in first-, second-, and half-order reactions. In each case, the rate constant is 0.1 units

Solution

First-order reaction:

Equation (10.7-15) for segregated flow
$$\mathbf{f}_A = \frac{k\mathbf{t}}{1 + k\mathbf{t}} = \frac{0.1(10)}{1 + 0.1(10)} = 0.5$$

Equation (7.3-4) for complete mixing
$$\mathbf{f}_A = \frac{k\mathbf{t}}{1 + k\mathbf{t}} = \frac{0.1(10)}{1 + 0.1(10)} = 0.5$$

Second-order reaction:

Equation (10.7-20) for segregated flow
$$\mathbf{f}_A = 1 + \frac{1}{kC_{Ao}\mathbf{t}}\, e^{1/kC_{Ao}\mathbf{t}} Ei(-1/kC_{Ao}\mathbf{t})$$

$$= 1 + \frac{1}{0.1(1)(10)}\, e^{1/(0.1\times1\times10)} Ei(-1)$$

$$= 1 + e^1(-0.21938) = 0.4036$$

Equation (10.7-21) for complete mixing

$$\mathbf{f}_A = 1 + \frac{1}{2kC_{Ao}\mathbf{t}}[1 - (1 + 4kC_{Ao}\mathbf{t})^{1/2}]$$

$$= 1 + \frac{1}{2(0.1)(1)(10)}\{1 - [1 + 4(0.1)(1)(10)]^{1/2}\}$$

$$= 0.382$$

Thus Eq. (10.7-6) is verified.

Half-order reaction:

Equation (10.7-26) for segregated flow

$$\mathbf{f}_A = \frac{k\mathbf{t}}{C_{Ao}^{1/2}}\left[1 - \frac{k\mathbf{t}}{2C_{Ao}^{1/2}}(1 - e^{-2C_{Ao}^{1/2}/k\mathbf{t}})\right]$$

$$= \frac{0.1(10)}{1}\left[1 - \frac{0.1(10)}{2}(1 - e^{-2/(0.1 \times 10)})\right]$$

$$= 0.5677$$

Equation (10.7-27) for complete mixing

$$\mathbf{f}_A = \frac{(k\mathbf{t})^2}{2C_{Ao}}\left\{-1 + \left[\frac{1 + 4C_{Ao}}{(k\mathbf{t})^2}\right]^{1/2}\right\} = \frac{1}{2}[-1 + (1 + 4)^{1/2}]$$

$$= 0.618$$

Thus Eq. (10.7-5) is verified.

Example 10.7-2 Conversion for Segregated Flow by Integration

Calculate the conversion for a first-order reaction ($k = 0.1$ s^{-1}, $\mathbf{t} = 10$ s) from the following RTD, assuming segregated flow. Compare the result with those for plug-flow, tubular-laminar-flow, and stirred-tank reactors.

$t/\mathbf{t}$	0	0.5	0.68	0.80	0.875	0.94	1.04	1.20	1.41	1.77	3.0
$F(t)$	0	0.1	0.2	0.3	0.4	0.5	0.6	0.7	0.8	0.9	1.0

Solution

1. Actual tubular reactor

The first-order kinetics as determined in Example 10.7-1 is

$$\mathbf{f}_A = 1 - e^{-0.1t}$$

Substituting into Eq. (10.7-8) gives

$$\mathbf{f}_A = \int_0^1 (1 - e^{-0.1t}\, dF(t) \qquad \textbf{(A)}$$

The value of the integral can be evaluated numerically by Simpson's rule as follows:

$F(t)$	$t/\mathbf{t}$	t	$1 - e^{-0.1t}$		
0	0	0	0	$\times 1$	$= 0$
0.1	0.5	5	0.3935	$\times 4$	$= 1.574$
0.2	0.68	6.8	0.4934	$\times 2$	$= 0.9868$
0.3	0.8	8.0	0.5507	$\times 4$	$= 2.2028$
0.4	0.875	8.75	0.5831	$\times 2$	$= 1.1662$
0.5	0.94	9.4	0.6094	$\times 4$	$= 2.4376$
0.6	1.04	10.4	0.6465	$\times 2$	$= 1.2930$
0.7	1.20	12.0	0.6988	$\times 4$	$= 2.7952$
0.8	1.41	14.1	0.7559	$\times 2$	$= 1.5198$
0.9	1.77	17.7	0.8297	$\times 4$	$= 3.3188$
1.0	3.00	30.0	0.9502	$\times 1$	$= 0.9502$
					18.2364

$$\mathbf{f}_A = 0.1(18.2364)/3 = 0.6079$$

2. Plug-flow reactor

First-order irreversible reaction, Eq. (8.2-9) becomes

$$\mathbf{f}_A = 1 - e^{k\mathbf{t}} = 1 - e^{(-0.1)(10)} = 0.6321$$

3. Tubular laminar-flow reactor

Substituting Eq. (10.7-13) and Eq. (10.5-28) into Eq. (10.7-8) gives

$$\mathbf{f}_A = \int_{1/2t}^{\infty} (1 - e^{-kt}) \frac{1}{2} \frac{\mathbf{t}^2}{t^3}\, dt$$

Inserting numerical values and integrating numerically gives

$$\mathbf{f}_A = 0.52$$

4. Stirred-tank reactor

$$\text{Equation (10.7-15)} \qquad \mathbf{f}_A = \frac{k\mathbf{t}}{1 + k\mathbf{t}} = \frac{(0.1)(10)}{1 + (0.1)(10)} = 0.50$$

10.7-2 Dispersion model

The conversion in a dispersion model has been described in Section 8.5. In order to use the design equation (8.5-7), the value of D/uL can be evaluated by actual response data. Then the conversion calculated from the design equation is the actual value for this nonideal reactor.

Example 10.7-3 Conversion in a Dispersion Reactor

Calculate the conversion for a first-order reaction $k = 0.1\ s^{-1}$, $\mathrm{t} = 10$ s in a dispersion reactor which has the same RTD as in the previous Example 10.7-2.

Solution

Using Eq. (10.6-9) to find out the best D_L/uL to fit the given RTD. The calculation is shown below:

t/t	0	0.5	0.68	0.80	0.875	0.94	1.04	1.20	1.41	1.77	3.0
$F(t)$ for $D_L/uL = 0.085$	0	0.04	0.17	0.29	0.37	0.44	0.54	0.67	0.80	0.92	0.99
$F(t)$ for $D_L/uL = 0.15$	0	0.15	0.24	0.34	0.40	0.46	0.53	0.63	0.74	0.85	0.98

Figure 10.7-1 shows that $D_L/uL = 0.085$ fits the upper part of the curve better, while $D_L/uL = 0.15$ is good for the lower part. Thus an average value of $D_L/uL = 0.1175$ is used. Hence from Eq. (8.5-7)

$$\beta = [1 + 4(0.1)(10)(0.1175)]^{1/2} = 1.2124$$

$$1 - f_A = \frac{4(1.2124)}{4.8947e^{0.9038} - 0.0451e^{-9.4144}}$$

$$= 0.4013$$

$$f_A = 0.5987$$

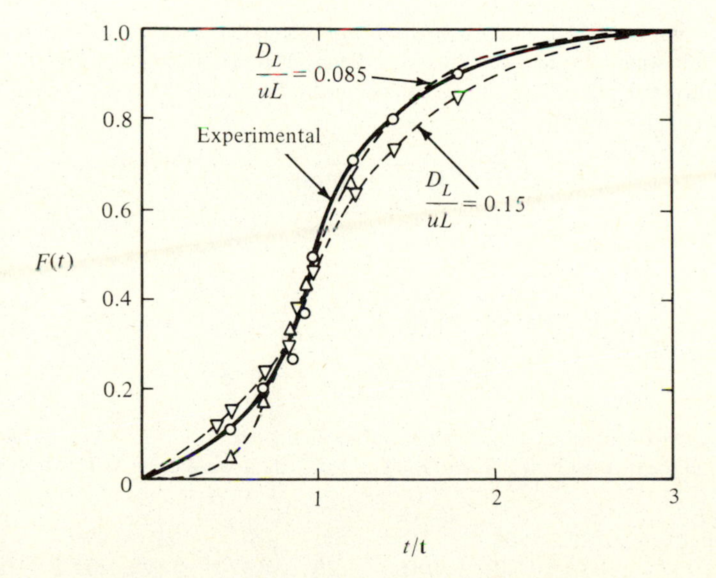

Figure 10.7-1 Fit of dispersion model to RTD data by Eq. (10.6-9)

10.7-3 CSTR model

The conversion in CSTR has been described in Section 7.8 Equations (7.8-10), (7.8-15), or (7.8-20) can be converted to

$$1 - f = \frac{1}{(1 + k\mathbf{t})^m} \tag{10.7-28}$$

The number of reactors m can be determined by fitting to the RTD data, using Eq. (10.6-17)

$$F = 1 - e^{-qnt/V}\left[1 + \frac{qnt}{V} + \frac{1}{2!}\left(\frac{qnt}{V}\right)^2 + \cdots \frac{1}{(n-1)}!\left(\frac{qnt}{V}\right)^{n-1}\right] \tag{10.6-17}$$

or

$$F = 1 - e^{-nt/\mathbf{t}}\left[1 + \frac{nt}{\mathbf{t}} + \frac{1}{2!}\left(\frac{nt}{\mathbf{t}}\right)^2 + \cdots \frac{1}{(n-1)!}\left(\frac{nt}{\mathbf{t}}\right)^{n-1}\right]$$

Example 10.7.4 Conversion in a CSTR

Calculate the conversion for a first-order reaction $k = 0.1\ \text{s}^{-1}$, $\mathbf{t} = 10$ s in a CSTR which has the same RTD as in Example 10.7-2.

Solution

Assuming $n = 5$, the values of the F are calculated from Eq. (10.6-17) as shown in the following table and the fitness to the given RTD is shown in Fig. 10.7-2. Consequently, the conversion can be calculated by Eq. (10.7-28) as

$$f = 1 - \frac{1}{[1 + 0.1(10)/5]^5} = 1 - 0.4 = 0.60$$

Problems

1. A first-order irreversible reaction

$$A \xrightarrow{k} B \qquad k = 0.05\ \text{s}^{-1}$$

is carried out in a flow reactor into which pure A enters. The effluent contains 10 percent A and 90 percent B. The RTD data were collected as follows:

Time t, s	25	31	46	55	61	67	91	180	360
$F(t)$	0	0.01	0.15	0.39	0.49	0.59	0.80	0.94	1.00

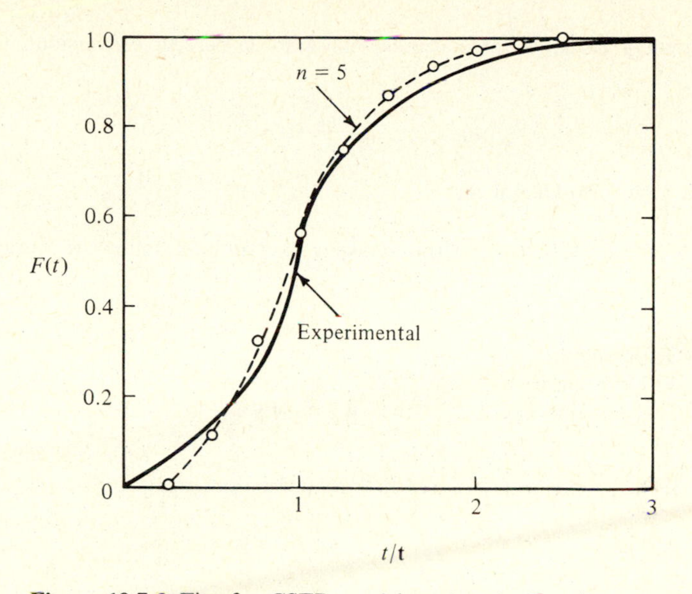

Figure 10.7-2 Fit of a CSTR model to data by Eq. (10.6-17)

(a) Find the average residence time.
(b) What are the values of the average residence time for a plug-flow reactor and for a CSTR with the same conversion?

2. Find the $F(t)$ curve for the following arrangement in two cases: first-order reaction and second-order reaction:
(a) A plug-flow reactor followed by a CSTR
(b) A CSTR followed by a plug-flow reactor

3. Determine the average residence time in a reactor where the RTD is

$$F(t) = 0 \qquad 0 \le t \le 0.42$$

$$F(t) = 1 - e^{1.2(t-0.42)} \qquad t > 0.42$$

What is the average fractional conversion for a first-order irreversible reaction ($k = 0.8 \text{ ks}^{-1}$), using a segregated model? If a space-time for a PFR is 0.5 ks and for a CSTR 0.8 ks, calculate the overall conversion for the arrangement in Problem 2.

4. Determine the conversion for a first-order reaction ($k = 0.4 \text{ ks}^{-1}$) to be carried out in two identical CSTR's in series using the $F(t)$ curve and segregated flow. The space-time for each reactor is 0.4 ks^{-1}. Check the results using an ideal CSTR.

5. Determine the volume of a PFR and a LFR required to take $0.5 \text{ m}^3/\text{ks}$ of feed containing 1.0 kmol/m³ of component A to 96 percent conversion. The reaction is

$$2A \xrightarrow{k} B + C$$

with $k = 5.5$ m³/kmol · ks.

6. (a) Find an $F(t)$ curve for the following data and determine the average residence time

Time t, min	0	3	6	9	12	15	18	21	24
Effluent tracer concentration	0	2.0	4.5	4.5	4.0	3.5	3.2	0.0	0.0

(b) If the reactor in (a) is used for first-order isomerization reaction

$$A \longrightarrow R$$

with a rate constant $k = 0.04$ min^{-1}, determine the average conversion.
(c) For the RTD in (a), find D_L/uL to fit the data and determine the conversion.
(d) Repeat (c) for a CSTR model.
(e) If the data were generated by an ideal CSTR followed by a PFR, what is the ratio of the reactor volume? What conversion can be expected?
(f) Summarize the previous results.

References

(H1). Hill, C.G. *Introduction to Chemical Engineering Kinetics and Reactor Design.* New York: John Wiley & Sons, Inc., 1977.

(L1). Levenspiel, O. *Chemical Reaction Engineering,* 2nd ed. New York: John Wiley & Sons, Inc., 1972.

(S1). Smith, J.M. *Chemical Engineering Kinetics,* 3rd ed. New York: McGraw-Hill Book Company, 1981.

CHAPTER ELEVEN

Design Considerations

11.1 Introduction

So far we have discussed the methods of designing three ideal basic types of reactors—batch, tubular, and continuous stirred tanks for homogeneous-phase and fixed-bed, fluidized-bed reactors for catalytic reactions, also gas-liquid and liquid-liquid reactors. Optimizations were also included and nonideal behavior mentioned. These are still many problems to be faced before designing or selecting a chemical reactor for a given chemical reaction. Some of the important problems are:

1. Is batch or continuous operation going to be used? What is the production rate?
2. What type of reactor, tank or tubular, do we have to design?
3. Are we going to use recycle or semibatch operation?
4. Is it advantageous to use reactants in excess?
5. If a well-mixed tank is adopted, do we use a single tank or multiple tanks in series? How many?
6. For multiple reactions, how can the yield of the product be increased? How can undesired byproducts be reduced?

This chapter will provide some of the answers.

11.2 Batch or Continuous Operation

For a heterogeneous reaction, particularly with catalysts, most of the commercial processes are continuous. For the homogeneous phase, both operations, batch or continuous, can be used, depending on the nature of the reaction. In general, batch operation is frequently favored for new and untried processes which are to be changed over to continuous operation at a more advanced stage of development. Batch reactors are often employed for liquid-phase reactions, particularly when the required production is small, the reaction time is long, and the reaction produces easy-fouling and contaminating materials which require frequent cleaning and sanitation. Consequently batch reactors are used extensively in the pharmaceutical and dyestuff industries. The merits of batch operation are the low capital investment and the ease in start-up, shut-down, and control—hence greater flexibility. The shortcomings are the high labor costs and high material-handling costs for charging, discharging, and cleaning, with the concomitant nonproductive period. Continuous reactors, such as tubular or well-mixed tanks in series, are better suited for large production. Gas-phase reactions often are carried out in tubular reactors; on the other hand, liquid-phase reactions are undertaken in well-mixed reactors in series. Despite the comparatively high investment, the operating and labor costs are lower. Furthermore, continuous reactors offer good quality control and facilitate automatic process control; they have wide application in the petroleum and petrochemical industries.

The difference between batch and continuous operations lies in the existence of a state of flow. Because of the flow, the yield and reaction rate in a continuous operation are quite different from those in a batch process, particularly for competing side reactions. The criteria for choosing the most suitable operation, batch or continuous, include many other factors as well (market conditions, labor conditions, economic analyses, and safety factors).

11.3 Economic Balance

The economic balance of either operation consists of:

1. Fixed costs, including depreciation, maintenance, and supervision are independent of annual production. They are expressed as a percentage of the total investment. Wages of operating personnel for the continuous plant are also included in fixed costs because the number of operators required does not depend on the annual production. In general, these fixed costs are relatively higher for continuous operation than for batch. Let us define these fixed costs per unit time F_c.
2. Operating costs: For single-unit production, these operating costs can be divided into two types of expenses: minimum expenses for raw materials, labor, power, steam, cooling water, and other utilities, laboratory service, etc., that remain constant and must be paid for each unit of production as long as any amount of material is produced; and extra expenses due to increased production, such as overloads on power lines, additional labor requirements, or a decreased efficiency of conversion. Let us denote these operating costs as $d + aP^b$, where P is the rate of production as total units of production per unit time, a, b, and d are empirical constants. They are generally higher for batch operations because of starting up and shutting down, alternate heating and cooling, and difficulties in recovery of heat.
3. Total product cost c_T per unit of production is then

$$c_T = d + aP^b + \frac{F_c}{P} \tag{11.3-1}$$

4. Total product cost per unit time C_T is

$$C_T = c_T P = \left(d + aP^b + \frac{F_c}{P} \right) P \tag{11.3-2}$$

5. Profit per unit of production e is, if s is the selling price per unit of production

$$e = s - c_T = s - d - aP^b - \frac{F_c}{P} \tag{11.3-3}$$

6. Profit per unit time E

$$E = eP = \left(s - d - aP^b - \frac{F_c}{P} \right) P \tag{11.3-4}$$

7. Optimum production rate for minimum cost per unit of production. Differentiating Eq. (11.3-1) with respect to P, equating to zero and solving for the optimum production rate P^*

$$P^* = \left(\frac{F_c}{ab} \right)^{1/(b+1)} \tag{11.3-5}$$

8. Optimum production rate for maximum total profit per unit time. Differentiating Eq. (11.3-4) with respect to P, equating to zero, and solving for P^* gives

$$P^* = \left[\frac{s - d}{a(b + 1)} \right]^{1/b} \tag{11.3-6}$$

9. Break-even point

At the break-even point, $E = 0$. Solving Eq. (11.3-4) for $P_{\text{break even}}$ gives

$$s - d - aP^b - \frac{F_c}{P} = 0 \tag{11.3-7}$$

which can be solved by trial and error or Newton's method.

Example 11.3-1 Determination of Profit at Optimum Production Rates

A plant produces urea at a rate of P tons per day. The operating costs per ton of urea have been determined to be \$50.00 + $0.1P^{1.25}$. The total fixed charges are \$2,000, and all other expenses are constant at \$8,000 per day. If the selling price per ton of urea is \$200, determine:

(a) The daily profit at a production schedule giving the minimum cost per ton of urea.
(b) The daily profit at a production schedule giving the maximum daily profit.
(c) The production schedule at the break-even point.

Solution

(a) Total cost per ton of urea $= c_T = 50 + 0.1P^{1.25} + \dfrac{(2{,}000 + 8{,}000)}{P}$

Production for minimum cost per ton of urea $= \dfrac{dc_T}{dP} = 0 = 0.125P^{*0.25} - \dfrac{10{,}000}{P^{*2}}$

Therefore

$$P^* = 151.06 \text{ tons per day}$$

$$\text{Daily profit} = \left[200 - 50 - 0.1(151)^{1.25} - \frac{10{,}000}{151}\right]151 = \$4{,}657.22$$

(b) Daily profit is

$$E = \left[200 - 50 - 0.1P^{1.25} - \frac{10{,}000}{P}\right]P$$

At production for maximum profit per day

$$\frac{dE}{dP} = 0 = 150 - 0.225P^{*1.25}$$

$$P^* = 181.60, \text{ rounded down to 181 tons per day}$$

$$\text{Daily profit} = \left[200 - 50 - 0.1(181)^{1.25} - \frac{10{,}000}{181}\right]181 = \$5{,}133.53$$

(c) At break-even point

$$\text{Total profit per day} = \left[200 - \left(50 + 0.1P^{1.25} + \frac{10{,}000}{P}\right)\right]P = 0$$

or

$$150 = 0.1P^{1.25} + \frac{10{,}000}{P}$$

Solving gives

$$P_{\text{break-even}} = 79 \text{ tons}$$

These solutions are shown in Fig. 11.3-1 in which the dollars per day is plotted against rate of production in tons per day. The horizontal dotted line is the total

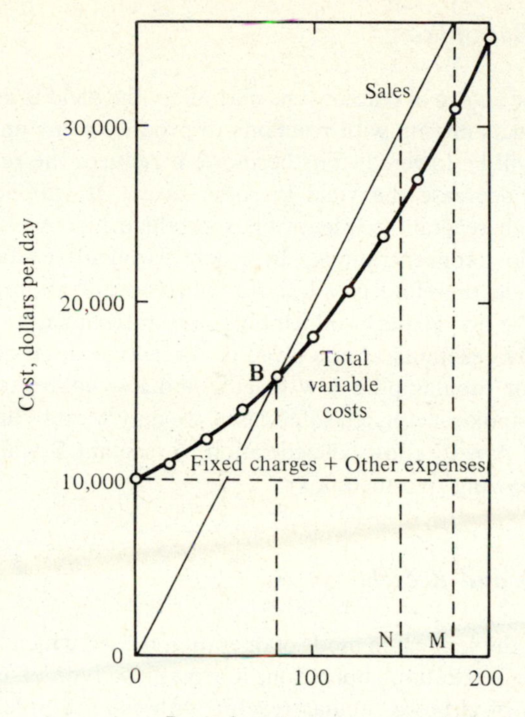

Figure 11.3-1 Economic analysis of a production plant

cost of the fixed charges and the other expenses per day. The sales revenue of the products is plotted as the straight line. The curve is the sum of the operating costs, fixed charges, and the other expenses per day, which may be defined as the total variable costs and are calculated by

$$\text{Total variable cost} = (\text{operating cost} + \text{fixed charges} + \text{expenses})P$$

$$= \left(50 + 0.1P^{1.25} + \frac{10{,}000}{P}\right)P$$

$$= 50P + 0.1P^{2.25} + 10{,}000$$

P	0	20	60	100	120	160	200
Total variable cost	10,000	11,084	14,001	18,162	20,766	27,104	35,042

The break-even point occurs at point B. The production for the maximum profit per day appears at X and the production for the minimum cost per ton of urea is at N.

11.4 Effect of Backmixing

The reason why the choice of reactor type may affect the yield is due to backmixing. It causes the products mixing with reactants to produce some unwanted products, so that the yield will be lowered. Furthermore, it reduces the reactant concentrations, which may decrease the yield in some cases. It will be shown that for some reactions high reactant concentrations produce high yields, whereas other reactions require low concentrations. In general, well-stirred tank reactors have complete mixing and therefore give low reactant concentrations, whereas tubular reactors which have no mixing produce high concentrations.

If two or more reactants are involved in a reaction, high concentrations can occur in a batch or tubular plug-flow reactor and low concentrations in a single continuous stirred-tank reactor. In some cases, we may want to have a high concentration of reactant A with a low concentration of reactant B which can be accomplished by the following two methods.

11.5 Semibatch and Recycle

In a batch reactor, the semibatch mode of operation is used which has been described in Chapter 6. For continuous operation, a cross-flow type of reactor (Fig. 11.5-1), which consists of either a tubular reactor with the multiple injection of x or a series of several stirred tanks (Fig. 11.5-2) with the feed of x divided between them may be employed.

Example 11.5

It is necessary to carry out the following reaction

$$X + Y \xrightarrow{k_1} Z \text{ (desired product)}$$

$$X + X \xrightarrow{k_2} P \text{ (unwanted product)}$$

in which X is expensive. Because of this, maintaining a high conversion of X, a

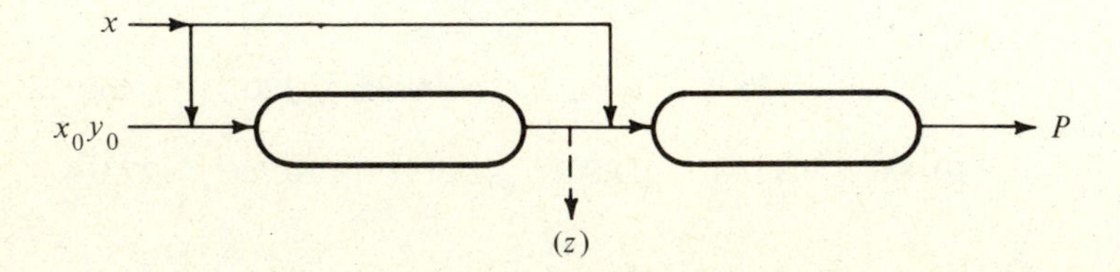

Figure 11.5-1 Cross-flow tubular reactor

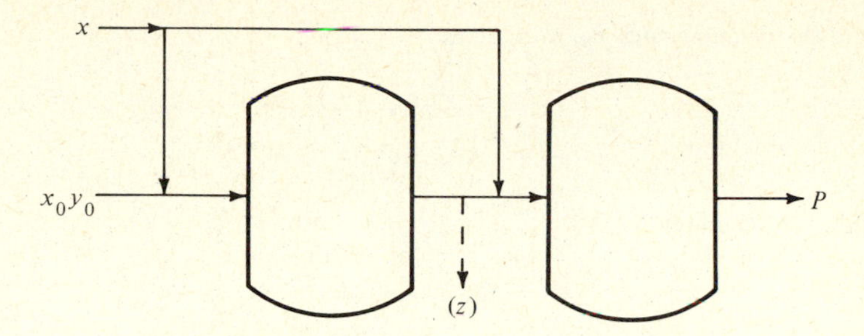

Figure 11.5-2 Cross-flow stirred-tank reactor

low concentration of X, and a high yield of Z are desired. The following cross-flow tubular reactor is proposed, in which X is fed along the length of the reactor and Y is fed at the inlet end only (see Fig. 11.5-3). It is required to determine the fractional conversion of the system.

Solution

Assumptions:

1. $C_X = C_{XL}$ over the entire reactor length
2. Total molar feed rate of X and Y are equal
3. $k_1 = k_2$
4. Density constant

A mass balance of X for a section dL is

$$-d(qC_X) + C_{Xw}\,dq + r_X S\,dL = 0 \tag{A}$$

where q is the volumetric flow rate, C_X and C_{Xw} are the concentrations of X at any position L and in the side injection, respectively. S is the cross-sectional area of the tubular reactor. A mass balance of Y for the same section dL is

$$-d(qC_Y) + r_Y S\,dL = 0 \tag{B}$$

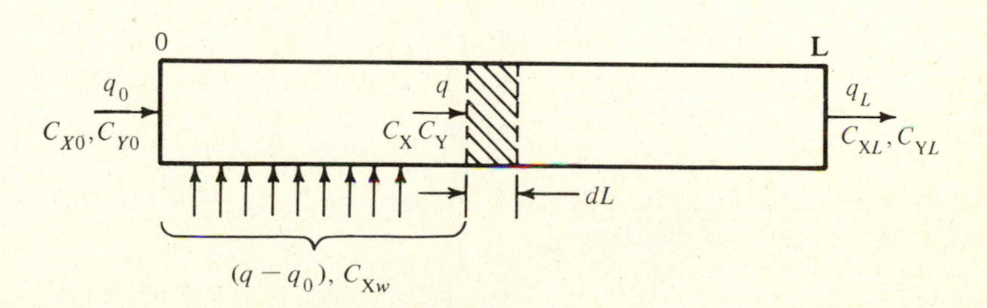

Figure 11.5-3

where the reaction rates r_X, *and* r_Y are

$$r_X = -k_1 C_X C_Y - 2(\tfrac{1}{2} k_2 C_X^2) \tag{C}$$

$$r_Y = -k_1 C_X C_Y \tag{D}$$

Because of assumption (1)

$$C_{X0} = C_X = C_{XL} \tag{E}$$

Because of assumption (2)

$$q_0 C_{Y0} = q_0 C_{X0} + (q_L - q_0) C_{Xw} \tag{F}$$

Eliminating $S\,dL$ from Eqs. (A) and (B) and substituting Eq. (C), (D), and from assumption (1)

$$C_{X0} = C_X = C_{XL}$$

we obtain

$$(C_{XL} - C_{Xw})dq = \left(1 + \frac{k_2 C_{XL}}{k_1 C_Y}\right) d(qC_Y) \approx \left[1 + q_L \frac{k_2 C_{XL}}{k_1 (qC_Y)}\right] d(qC_Y) \tag{G}$$

The latter approximation is only good if the variation in q is small. This means that the total flow of the side injection is relatively small, but its content in X is relatively high, i.e., $q_L - q_0 << q_0$ On this assumption, Eq. (G) can be integrated between $L = 0$ and $L = L$ and combined with the result from assumption (2), i.e.,

$$q_0 C_{Y0} = q_0 C_{X0} + (q_L - q_0) C_{Xw} \tag{H}$$

to give

$$\ln \frac{q_L C_{YL}}{q_0 C_{Y0}} = \frac{k_1}{k_2}\left(1 - \frac{C_{YL}}{C_{XL}}\right) \tag{I}$$

The fractional conversions of X and Y at the reactor outlet are

$$f_{XL} = 1 - \frac{q_L C_{XL}}{q_0 C_{X0}} = 1 - \frac{q_L C_{XL}}{q_0 C_{Y0}} \tag{J}$$

$$f_{YL} = 1 - \frac{q_L C_{YL}}{q_0 C_{Y0}} \tag{K}$$

because the molar feed rate of X and Y are equal, i.e., $q_0 C_{X0} = q_0 C_{Y0}$. Substituting Eqs. (J) and (K) into Eq. (I) gives

$$\ln (1 - f_{YL}) = \frac{k_1}{k_2} \left\{ \frac{f_{YL} - f_{XL}}{1 - f_{XL})} \right\} \tag{L}$$

For $k_1/k_2 = 1$, $f_X = 0.98$

$$\ln(1 - f_{YL}) = (1)\left(\frac{f_{YL} - 0.98}{1 - 0.98}\right) \tag{M}$$

Solving by trial and error gives

$$f_{YL} = 0.9275$$

It is interesting to compare the results from the performance of a batch reactor and from a continuous tank reactor under the same conditions. For the batch reactor

$$\frac{dc_X}{dt} = -k_1 C_X C_Y - k_2 C_X^2 \tag{N}$$

$$\frac{dC_Y}{dt} = -k_1 C_X C_Y \tag{O}$$

Eliminating the time by dividing Eq. (N) and Eq. (O) yields

$$\frac{dC_X}{dC_Y} = 1 + \frac{k_2}{k_1}\frac{C_X}{C_Y} \tag{P}$$

Solving for the initial conditions $C_X = C_{X0}$, $C_Y = C_{Y0}$ gives

$$\left(\frac{k_2}{k_1} - 1\right)\ln\frac{C_Y}{C_{Y0}} = \ln\left[\frac{1 + (k_2/k_1 - 1)C_X/C_Y}{1 + (k_2/k_1 - 1)C_{X0}/C_{Y0}}\right] \tag{Q}$$

In terms of fractional conversions, Eq. (Q) becomes when $C_{Xo} = C_{Yo}$

$$\left(\frac{k_2}{k_1} - 1\right)\ln(1 - f_Y) = \ln\left[1 + \left(\frac{k_2}{k_1} - 1\right)\frac{1 - f_X}{1 - f_Y}\right] - \ln\frac{k_2}{k_1} \tag{R}$$

For $k_2/k_1 = 1$, Eq. (R) can be reduced by L'Hospital's rule to

$$1 - f_X = (1 - f_Y)[1 + \ln(1 - f_Y)] \tag{S}$$

Then at $f_X = 0.98$, using trial and error gives

$$f_Y = 0.6125$$

For the continuous-tank reactor, the design equations become

$$C_{X0} - C_{X1} = k_1\theta C_{X1}C_{Y1} + k_2\theta C_{X1}^2 \tag{T}$$

$$C_{Y0} - C_{Y1} = k_1\theta C_{X1}C_{Y1} \tag{U}$$

Dividing Eq. (T) by Eq. (U) gives

$$\frac{C_{X0}-C_{X1}}{C_{Y0}-C_{Y1}}=1+\frac{k_2 C_{X1}}{k_1 C_{Y1}} \tag{V}$$

Because $C_{X0}=C_{Y0}$, Eq. (V) in terms of fractional conversions becomes

$$\frac{f_{X1}}{f_{Y1}}=1+\frac{k_2}{k_1}\left\{\frac{1-f_{X1}}{1-f_{Y1}}\right\} \tag{W}$$

For $f_{X1}=0.98$, solving Eq. (W) gives

$$f_{Y1}=0.8586$$

From these results, it is seen that the cross-flow reactor gives the highest yield of the desired product because of highest f_{YL} whereas the lowest f_Y in the batch or normal tubular reactor means that a relatively large portion of X is converted to the unwanted product.

If the cost of separating unreacted X is low, than a large excess of X can be maintained in the reactor. A single well-mixed tank reactor will give a low concentration of Y, while the large excess of X ensures a high concentration of X. Unreacted X is separated and recycled as shown in the following example.

Example 11.5-2

Let us consider a reversible reaction A ⇄ B to be carried out in a tubular reactor. The effluent from the reactor which consists of a mixture of product B and unreacted A enters a separating unit (such as distillation column) in which the product B leaves the top as overhead and the unreacted A is recovered from the bottom and recirculated back to the feed stream. It is necessary to calculate the volume of the reactor.

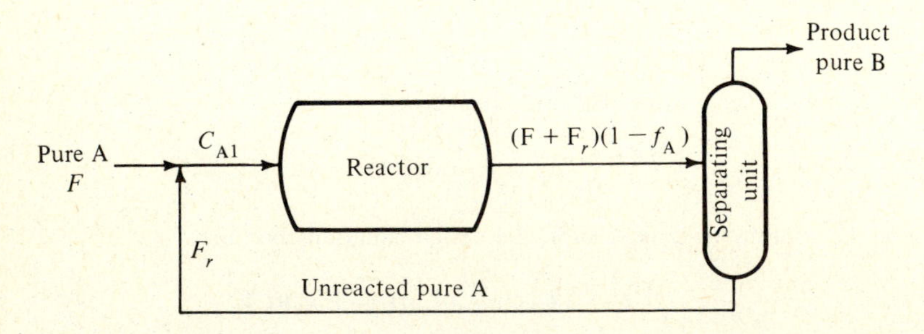

Figure 11.5-4

Solution

Let the molar flow rate of the fresh stream, and the recycle stream be F and F_r, respectively. Then the molar flow rate of the stream entering the tubular reactor is $F + F_r$ and that leaving the reactor is $(F + F_r)(1 - f_A)$ where f_A is the fractional conversion of A. A mass balance of A around the separating unit gives

$$(F + F_r)(1 - f_A) = F_r \tag{A}$$

The volume of the tubular reactor required is

$$V = (F + F_r)\int \frac{df}{r} \tag{B}$$

where

$$r = k_1 C_A - k_2 C_B \tag{C}$$

Since

$$C_A = C_{A1}(1 - f_A) \tag{D}$$

$$C_B = C_{A1} f_A \tag{E}$$

and

$$K = \frac{k_1}{k_2} \tag{F}$$

combining Eqs. (C), (D), (E), and (F) into Eq. (B) gives

$$V = (F + F_r)\int \frac{df_A}{k_1 C_{A1}[1 - (1 + 1/K)f_A]} \tag{G}$$

Integrating between 0 and f_{AL} gives

$$k_1 C_{A1} V = (F + F_r)\left\{\frac{1}{-(1 + 1/K)} \ln\left[1 - \left(1 + \frac{1}{K}\right) f_A\right]\right\} \tag{H}$$

Simplifying with Eq. (A) yields

$$\frac{k_1 C_{A1} V}{F} = \frac{1}{f_A}\left[-\frac{K}{K+1} \ln\left(1 - \frac{K+1}{K} f_A\right)\right] \tag{I}$$

From these equations, it is seen that if F, F_r, k_1, C_{A1}, K are specified, f_A can be evaluated from Eq. (A) and then the volume of the reactor is determined by Eq. (I). Since from Eq. (A) the ratio F_r/F decreases as f_A increase and from Eq. (I) $k_1 C_{A1} V/F$ increases as f_A increases, it is evident that an optimum value of f_A exists.

11.6 Excess in Reactants

Although theoretically it is advantageous to use high concentrations of the reactants in their stoichiometric ratio, it is sometimes rewarding to use the other, less expensive reactant in excess so as to increase the yield with respect to the more expensive reactant which may not be recovered too easily. Quite often reactants are used in excess to suppress unwanted side reactions or temperature changes such as by a large heat of reaction or a limited heat exchange capacity. Using excess reactant can reduce the required size of the reactor. This can be seen by the following illustration. Let us consider an irreversible reaction

$$\mathrm{X} + \mathrm{Y} \rightarrow \mathrm{Z}$$

to be carried out in a well-stirred reactor. The design equation is

$$qC_{X0} - qC_{X1} = rV \tag{11.6-1}$$

where q and V are the volumetric flow rate of X with concentration C_{X0} and Y with concentration C_{Y0} and volume of the reactor, respectively. C_{X1} is the concentration of X at the outlet and r is the rate of the reaction expressed as

$$\begin{aligned} r &= k_1 C_X C_Y \\ &= k_1 C_{X0}(1 - f_X)(C_{Y0} - C_{X0} f_X) \\ &= k_1 C_{X0}^2 (1 - f_X)(1 + \epsilon - f_X) \end{aligned} \tag{11.6-2}$$

where

$$\epsilon = \frac{C_{Y0}}{C_{X0}} - 1$$

Substituting Eq. (11.6-2) into Eq. (11.6-1) gives

$$V = \frac{q f_X}{k_1 C_{X0}(1 - f_X)(1 + \epsilon - f_X)} \tag{11.6-3}$$

It is seen that if q, k_1, C_{X0} and f_X are specified, V would decrease as ϵ increases. If V, q, and C_{X0} are specified, f_X would increase as ϵ increases. Hence there is an optimum value of ϵ. Sometimes excess of one of the reactants may shift the equilibrium.

11.7 Combination of Well-Mixed and Plug-Flow Reactors

Sometimes it is possible to arrange a reaction to be carried out in a combination of plug-flow and well-stirred reactors. Let us investigate the result with a simple irreversible second-order reaction to be carried out first in a plug-flow reactor and then in a well-mixed reactor (see Fig. 11.7-1). The design equation for the plug-flow reactor is

$$\theta_p = C_{A0} \int_0^{f_{A1}} \frac{df_A}{kC_{A0}^2(1-f_A)^2} \tag{11.7-1}$$

Integrating gives

$$kC_{A0}\theta_p = \frac{f_{A1}}{1-f_{A1}} \tag{11.7-2}$$

Solving for f_{A1}

$$f_{A1} = \frac{kC_{A0}\theta_p}{1+kC_{A0}\theta_p} \tag{11.7-3}$$

Then the design equation for the well-mixed reactor is

$$\theta_c = \frac{f_{A2}-f_{A1}}{kC_{A0}(1-f_{A2})^2} \tag{11.7-4}$$

Solving for f_{A2}

$$f_{A2} = 1 + \frac{1}{2kC_{A0}\theta_c}(1 \pm \sqrt{1+4kC_{A0}\theta_c(1-f_{A1})}) \tag{11.7-5}$$

Eliminating f_{A1} in Eq. (11.7-5) by Eq. (11.7-3) yields

$$f_{A2} = 1 + \frac{1}{2kC_{A0}\theta_c}\left(1 - \sqrt{1+\frac{4kC_{A0}\theta_c}{1+kC_{A0}\theta_p}}\right) \tag{11.7-6}$$

Now assume that the same reaction is carried out first in a well-mixed and then

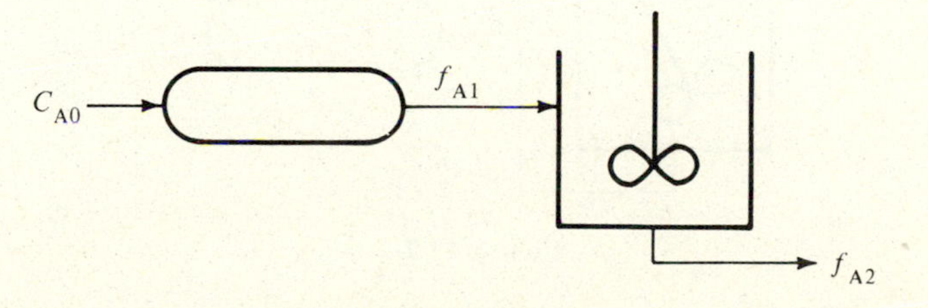

Figure 11.7-1

in a plug-flow reactor (see Fig. 11.7-2). The design equation for the well-mixed reactor is

$$\theta_c = \frac{f_{A1}}{kC_{A0}(1 - f_{A1})^2} \tag{11.7-7}$$

Solving for f_{A1}

$$f_{A1} = 1 + \frac{1}{2kC_{A0}\theta_c}(1 - \sqrt{1 + 4kC_{A0}\theta_c}) \tag{11.7-8}$$

Then the design equation for the plug-flow reactor is

$$\theta_p = C_{A0}\int_{f_{A1}}^{f_{A2}} \frac{df_A}{2C_{A0}^2(1 - f_A)^2} \tag{11.7-9}$$

Integrating and rearranging gives

$$kC_{A0}\theta_p = \frac{1}{1 - f_{A2}} - \frac{1}{1 - f_{A1}} \tag{11.7-10}$$

Eliminating f_{A1} by Eq. (11.7-8) and rearranging yields

$$f_{A2} = 1 + \frac{1 - \sqrt{1 + 4kC_{A0}\theta_c}}{2kC_{A0}\theta_c - kC_{A0}\theta_p(1 - \sqrt{1 + 4kC_{A0}\theta_c})} \tag{11.7-11}$$

When $\theta_c = \theta_p$ and $kC_{A0}\theta_p = 1$, f_{A2} by Eq. (11.7-6) gives 0.829 whereas f_{A2} by Eq. (11.7-11) gives 0.618. Hence the second case produces a much lower yield because mixing at a high concentration generates the greatest decrease in the average conversion rate.

For first-order irreversible reaction carried out isothermally, both reactor com-

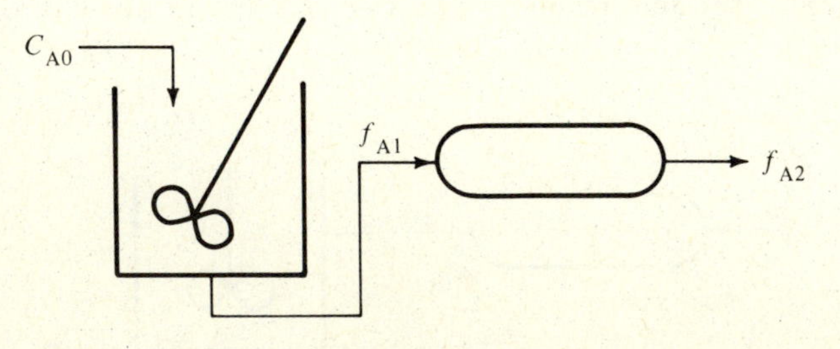

Figure 11.7-2

binations, having the same residence time distribution function, give rise to the same overall conversion as

$$f_{A2} = \frac{1 + k\theta_c - e^{-k\theta_p}}{1 + k\theta_c} \qquad \textbf{(11.7-12)}$$

which can be verified very easily.

11.8 Autothermal Operation

One of the important problems facing every design engineer is to utilize all the energy in the system. For in an exothermic reaction which involves a large amount of heat, it is always advantageous to feed back the reaction heat to preheat the incoming cold reactant stream. This process is called *autothermal operation* in which provision should be made for initiating the reaction. Once the reaction started, the reactor system can be self-supporting by the large amount of exothermic heat of reaction. The following are the two techniques employed to carry out this autothermal operation. In reactors operating under steady-state conditions, there may be heat exchange between the effluent stream or the reactor contents and the feed stream. In a well-mixed reactor, the fresh feed is rapidly mixed with the reactor contents to promote energy transfer. In a tubular reactor, a fraction of the reactor effluent may be recycled back for mixing with the fresh feed.

Example 11.8-1 Autothermal Operation in a CSTR Followed by PFR

Pure A enters to a well-mixed reactor ($V = 0.378$ m^3) at a rate of 4.2×10^{-6} m^3/s at 20°C followed by a PFR. Both are operated adiabatically. The overall fractional conversion at 163°C is 0.97. The reaction is

$$A \rightarrow B$$

The data are

$$k = 7.25 \times 10^{10} e^{(-14{,}570/T)} \text{ per second where } T \text{ is in Kelvin}$$

$$-\Delta H = 83 \text{ cal/g}$$

$$\mathbf{C}_p = 0.5 \text{ cal/g} \cdot {}^\circ\text{C}$$

It is necessary to determine the volume of the tubular reactor. For the well-mixed reactor, the mass balance equation for a first-order irreversible reaction is Eq. (7.3-4)

Solution

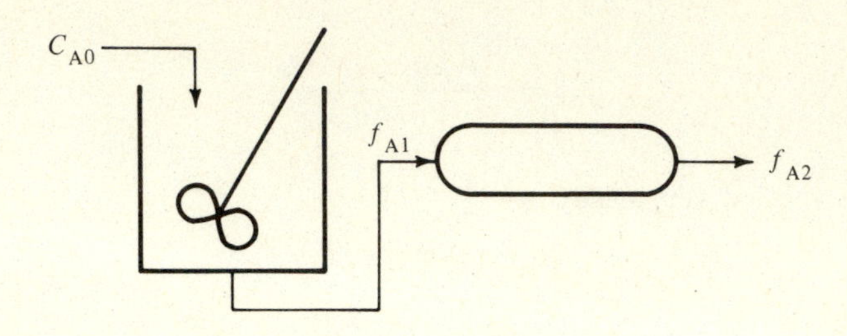

$$f_{A1} = \frac{k\theta}{1 + k\theta} \tag{7.3-4}$$

Since

$$\theta = \frac{V}{q} = \frac{0.378}{4.2 \times 10^{-6}} = 9 \times 10^4 \text{ s}$$

$$f_{A1} = \frac{9 \times 10^4 k}{1 + 90{,}000k} \tag{A}$$

where

$$k = 7.25 \times 10^{10} e^{-14{,}570/T_2} \tag{B}$$

The energy balance is obtained from Eq. (7.11-6). For a single and adiabatic reaction, Eq. (7.11-6) becomes

$$\rho \mathbf{C}_p q(T_1 - T_2) = \Delta H r V \tag{C}$$

Since

$$\rho q = F_A$$

and

$$rV = F_A - F_A(1 - f_{A1})$$

Eq. (C) becomes

$$\mathbf{C}_p(T_1 - T_2) = \Delta H f_{A1}$$

or

$$f_{A1} = \frac{0.5(T_2 - 293)}{83} \tag{D}$$

The three unknowns f_{A1}, k, and T_2 can be solved simultaneously from the three equations (A), (B), and (D) by trial and error. The result is

$$T_2 = 410 \text{ K} \qquad f_{A1} = 0.705 \qquad k = 2.67 \times 10^{-5} \text{ s}^{-1}$$

For the plug-flow reactor, the mass balance equation for a first-order irreversible reaction is Eq. (8.2-9)

$$\theta_p = \int_{f_{A1}}^{f_{A2}} \frac{df_A}{r} \tag{8.2-9}$$

or

$$= \int_{0.705}^{0.97} \frac{df_A}{k(1-f_A)} \tag{E}$$

The energy balance equation is obtained from Eq. (8.3-7). For an adiabatic reaction, Eq. (8.3-7) becomes

$$q\rho \mathbf{C}_p(T_1 - T_2) = \Delta H r V \tag{F}$$

Since

$$\rho q = F_A$$

and

$$rV = F_A \int_{f_{A1}}^{f_{A2}} df_A = F_A(f_{A2} - f_{A1})$$

Eq. (F) becomes

$$0.5(T_2 - 410) = 83(f_A - 0.705)$$

or

$$\begin{aligned} T_2 &= 410 + 166(f_A - 0.705) \\ &= 410 + 166(0.97 - 0.705) \\ &= 454 \text{ K} \end{aligned} \tag{G}$$

From Eq. (B)

$$k = 7.25 \times 10^{10} e^{-14{,}570/454} = 8.37 \times 10^{-4} \text{ s}^{-1}$$

From Eq. (E)

$$\begin{aligned} \theta_p &= \int_{0.705}^{0.97} \frac{df_A}{7.25 \times 10^{10} e^{-14{,}570/[410+166(f_A-0.705)]}(1-f_A)} \\ &= 13{,}400 \text{ s} \end{aligned}$$

Hence

$$V = 13{,}400\,(4.2 \times 10^{-6}) = 0.05628 \text{ m}^3$$

In a semibatch reactor, a cold feed may be heated by mixing with the reactor contents.

Example 11.8-2 Autothermal Operation in a Semibatch Reactor

At 165°C, the rate constant for the reaction A $\rightarrow$ B is $k = 0.85\ \text{h}^{-1}$. Into a well-mixed tank reactor which originally contains 727.27 kg B, a stream of A enters at the rate which varies with time as follows:

Time *t*, h	Feed rate of A, lb/h
0–2	150
2–5	250
5–6	300
6–7	350
7–10	420
10–11	350
11–12	300
12–13	250
13–14	200
14–15	150
15–16	125
16–17	0

Assuming constant temperature at 165°C, determine the total amount of A and B in the reactor as a function of time and also determine the direction and magnitude of the heat-transfer requirements if the average heat capacity per unit mass is 0.55 cal/g · °C and the heat of reaction is −85 cal/g. The initial temperature of B in the reactor is 165°C.

Solution

The mass balance of a semibatch reactor is indicated by Eq. (6.2–5) as

$$\frac{d(C_j V)}{dt} + (qC_j)_2 - (qC_j)_1 = \Sigma \alpha_{ij} r_i V \tag{6.2-5}$$

Since no product is withdrawn from this single reaction, and $n_j = C_j V$, Eq. (6.2-5) can be applied to the first-order irreversible reaction

$$\text{A} \rightarrow \text{B}$$

$$\frac{dn_A}{dt} = (qC_A)_1 - kC_A V \tag{A}$$

The units of n_A, q, C_A, k, t, and V are lb mole, ft³/h, lb mole/ft³, h⁻¹, h, and ft³, respectively. In terms of mass m_A and mass flow rate $\dot{m}_A$, Eq. (A) can be modified to

$$\frac{dm_A}{dt} = -km_A + \dot{m}_{A0} \tag{B}$$

where $\dot{m}_{A0}$ is the mass flow rate of A entering the reactor. Since there is no B in the inlet stream, the mass balance for B is

$$\frac{dm_B}{dt} = +km_A \qquad \text{(C)}$$

Combining Eqs. (B) and (C) gives

$$\frac{dm_B}{dt} = \dot{m}_{A0} - \frac{dm_A}{dt} \qquad \text{(D)}$$

Integrating Eq. (B) and Eq. (D) yields

$$m_A = m_{Ai}e^{-k(t-t_i)} + \left[\left(\frac{\dot{m}_{A0}}{k}\right)(1 - e^{-k(t-t_i)})\right] \qquad \text{(E)}$$

where m_{Ai} is the mass of A in the reactor at time t_i.

$$m_B = m_{B0} - m_A + \int_0^t \dot{m}_{A0}\, dt \qquad \text{(F)}$$

The overall heat requirement Q is

$$\begin{aligned} Q &= \begin{pmatrix}\text{heat required to raise the} \\ \text{mixture from 20° to 165°C}\end{pmatrix} + \begin{pmatrix}\text{heat evolved from} \\ \text{the reaction}\end{pmatrix} \\ &= (m_A + m_B - m_{B0})C_p(165 - 20) - 85(m_B - m_{B0}) \qquad \text{(G)} \\ &= 79.75m_A - 5.25(m_B - m_{B0}) \\ &= 143.55m_A - 9.45(m_B - m_{B0}) \qquad \text{where } Q \text{ is in Btu} \end{aligned}$$

The instantaneous heat q is obtained from Eq. (6.3-9). Because the temperature in the reactor is kept at constant 165°C and the single reaction is first-order, Eq. (6.3-9) becomes

$$\Delta HrV + (T_2 - T_1)\dot{m}_{A0}c_p = q \qquad \text{(H)}$$

Since $rV = km_A$ from Eq. (C), Eq. (H) in terms of mass becomes

$$q = -85(0.85)m_A + \dot{m}_{A0}(0.55)(165 - 20) = 79.75\dot{m}_{A0} - 72.25m_A$$

or

$$q = 143.55\dot{m}_{A0} - 130.05m_A \qquad \text{(I)}$$

where q is in calories per second, $\dot{m}_{A0}$ is the mass flow rate of A entering the reactor in grams per second, and m_A is the mass of A in the reactor at any time t. A computer program is provided in Fig. 11.8-1, in which the input data are time $TS(I)$ in hours, feed rate of A, $FA(I)$ in pounds per hour. All the others are in British units. However, the output results are kilograms, calories, and calories per second.

```
00100 PROGRAM SMB(INPUT,OUTPUT)
00110CPBO=INITIAL AMOUNT OF B,LB
00120CPAO=INITIAL AMOUNT OF A,LB
00130CNTOP=NUMBER OF TIME INCREMENT
00140CAK=RATE CONSTANT,HR**(-1)
00150 DIMENSION TS(20),FA(20)
00160 DATA(TS(I),I=1,12)/0.,2.,5.,6.,7.,10.,11.,12.,13.,14.,15.,16./
00170 DATA(FA(I),I=1,12)/150.,250.,300.,350.,420.,350.,300.,250.,200.,
00180+150.,125.,0./
00190 4 FORMAT(2X,"TIME",6X,"MASS A",3X,"MASS B",3X,"MAAS TOTAL",
00200+2X,"UNCONV A",2X,"Q TOTAL",3X,"Q-RATE",/3X,"HR",9X,"KG",
00210+7X,"KG",9X,"KG",18X,"CAL",5X,"CAL/SEC",/)
00220 2 FORMAT(3X,F3.0,6X,F6.2,3X,F7.2,4X,F7.2,4X,F6.4,1X,E10.4,
00230+2X,F7.0)
00240 READ*,PBO,PAO,NTOP,AK
00250 PAI=PAO
00260 PI=FA(1)
00270 PT=PAO+PBO
00280 TI=TS(1)
00290 U=PAO/(PAO-PBO)
00300 SPA=0.
00310 Q=0.
00320 QI=143.55*PI-130.05*PA
00330 K=2
00340 PRINT 4
00350 PAO=PAO/2.2
00360 PBO=PBO/2.2
00370 PT=PT/2.2
00380 Q=Q*252.
00390 QI=QI*252./3600.
00400 PRINT 2,TI,PAO,PBO,PT,U,Q,QI
00410 PAO=PAO*2.2
00420 PBO=PBO*2.2
00421 PT=PT*2.2
00430 Q=Q/252.
00440 QI=QI/252.*3600.
```

```
00460 DO 100 I=1,NTOP
00470 T=FLOAT(I)
00480 PA=PAI*EXP(-AK*(T-TI))+PI/AK*(1.-EXP(-AK*(T-TI)))
00490 SPA=SPA+PI
00500 PB=PBO+SPA-PA
00510 PT=PA+PB
00520 UFA=PA/(PT-PBO)
00530 IF(T.NE.TS(K)) GO TO 50
00540 PAI=PA
00550 PI=FA(K)
00560 TI=TS(K)
00570 K=K+1
00580 50 Q=143.55*PA-9.45*(PB-PBO)
00590 QI=143.55*PI-130.05*PA
00600 PA=PA/2.2
00610 PB=PB/2.2
00620 PT=PT/2.2
00630 Q=Q*252.
00640 QI=QI*252./3600.
00650 100 PRINT 2,T,PA,PB,PT,UFA,Q,QI
00660 STOP
00670 END
READY.
RNH
```

Figure 11.8-1

```
? 1600.,0.,20,0.85
 TIME       MASS A    MASS B   MAAS TOTAL  UNCONV A  Q TOTAL     Q-RATE
  HR          KG        KG        KG                   CAL       CAL/SEC

   0.        0.00    727.27     727.27    0.0000 0.                1507.
   1.       45.93    749.53     795.45     .6736  .3539E+07         587.
   2.       65.56    798.08     863.64     .4808  .4847E+07        1199.
   3.      104.57    872.70     977.27     .4183  .7560E+07         418.
   4.      121.24    969.67    1090.91     .3334  .8379E+07          84.
   5.      128.37   1076.18    1204.55     .2690  .8388E+07         444.
   6.      146.73   1194.18    1340.91     .2391  .9231E+07         578.
   7.      169.88   1330.12    1500.00     .2198  .1036E+08         818.
   8.      201.21   1489.70    1690.91     .2088  .1202E+08         191.
   9.      214.60   1667.22    1881.82     .1859  .1215E+08         -78.
  10.      220.33   1852.40    2072.73     .1638  .1164E+08        -896.
  11.      201.34   2030.48    2231.82     .1338  .9196E+07       -1018.
  12.      177.91   2190.27    2368.18     .1084  .6494E+07       -1051.
  13.      152.59   2329.23    2481.82     .0870  .3751E+07       -1046.
  14.      126.46   2446.27    2572.73     .0685  .1058E+07       -1025.
  15.       99.98   2540.93    2640.91     .0522 -.1545E+07        -746.
  16.       81.01   2616.72    2697.73     .0411 -.3452E+07       -1622.
  17.       34.62   2663.10    2697.73     .0176 -.7386E+07        -693.
  18.       14.80   2682.93    2697.73     .0075 -.9068E+07        -296.
  19.        6.33   2691.40    2697.73     .0032 -.9787E+07        -127.
  20.        2.70   2695.02    2697.73     .0014 -.1009E+08         -54.
```

Figure 11.8-1 (continued)

11.9 Comparison of Reactors

In order to compare the performance of different types of reactors, two criteria may be employed: reactor productivity and reactor selectivity. For a single reaction, we measure the reactor productivity, which is the output of product in relation to reactor size. It is more convenient to use the limiting reactant as a reference than to use the product formed. For this purpose, let us define a term, unit output H of a reactor system as the moles of limiting reactant A converted per unit time per unit volume of the reactor. From Chapter 1, the number of moles of limiting reactant A converted per unit volume is λ, defined by Eq. (1.4-31); so the unit output H is

$$\mathrm{H} = \frac{\lambda}{t} = \frac{C_{k0} f_k}{t} \tag{11.9-1}$$

H is the total moles of A converted by the whole reactor in unit time divided by the whole volume of the system. The unit output is therefore an average rate of reaction as a whole, while the specific rate r is the local rate of reaction.

The second criterion is the reactor selectivity, which is the extent to which formation of unwanted byproducts in multiple reactions can be suppressed. For multiple reactions, the reactor yield depends not only on the operating conditions but also on the type of reactor used. The types of multiple reactions have been mentioned in Chapter 3. There are two types of reactions, parallel and series, which we may further classify into desired and unwanted products. Our aim is to study the effect of the choice of reactor on the yield to these reactions.

When reactants are converted to desired, unwanted, and unreacted products, the amount of the desired product actually obtained is therefore smaller than the amount expected if all the reactant had been transformed into the desired product alone. Hence the reaction is said to produce a certain yield of the product. Relative yield ϕ_A is defined as

$$\phi_A = \frac{\text{moles of A converted to desired product}}{\text{total moles of A reacted}} \tag{11.9-2}$$

Operational yield Ω_A is

$$\Omega_A = \frac{\text{moles of A converted to desired product}}{\text{total moles of A entering reactor}} \tag{11.9-3}$$

Incidentally, another term, selectivity, is also useful in dealing with multiple reactions. Selectivity of the desired reaction is

$$S = \frac{\text{moles of A converted to desired product}}{\text{moles of A converted to unwanted product}} \tag{11.9-4}$$

11.10 Reactor Productivity

On the basis of the unit output, let us compare the performance of batch, tubular, and single and multiple stirred-tank reactors. The design equation for a batch reactor is

$$t = C_{Ao} \int_0^f \frac{df}{r} \tag{5.2-7}$$

and for a plug flow reactor it is

$$\frac{V}{q} = \tau = C_{Ao} \int_0^{f_1} \frac{df}{r} \tag{8.2-9}$$

If $1/r$ is plotted against f, the area under the curve is t/C_{Ao} for a batch reactor and τ/C_{Ao} for a plug-flow reactor, as seen from Figs. 11.10-1 and 11.10-2.

The design equation for a single-tank reactor is

$$\frac{V}{q} = \tau_1 = C_{A0} \frac{f_1}{r} \tag{7.2-8}$$

and for a multiple-tank reactor it is

$$\frac{V_m}{q} = \tau_m = \frac{C_{m-1} - C_m}{r_m} \tag{7.8-4}$$

$$= \frac{C_{A0}(f_m - f_{m-1})}{r_m} \tag{11.10-1}$$

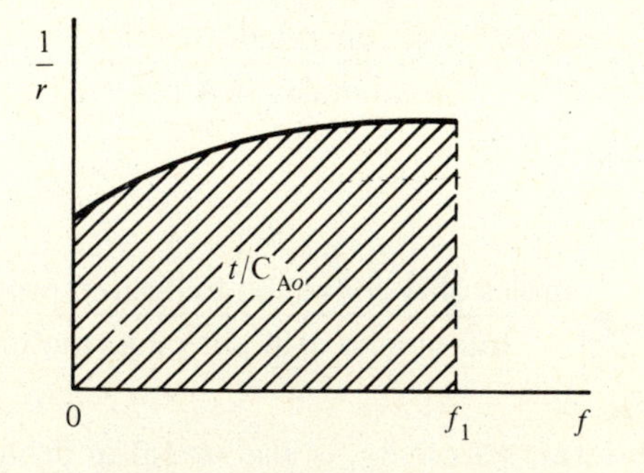

Figure 11.10-1 Batch reactor

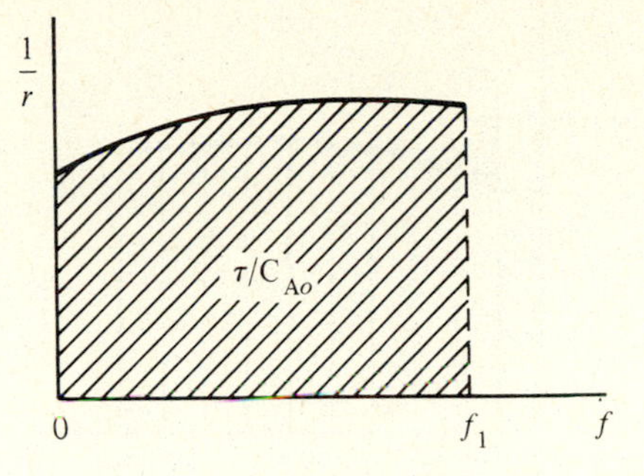

Figure 11.10-2 Plug-flow reactor

because

$$C_m = C_{A0}(1 - f_m)$$
$$C_{m-1} = C_{A0}(1 - f_{m-1})$$

where f_m and f_{m-1} are the overall fractional conversion of the limiting component A in tank m and $m - 1$, respectively. If $1/r$ is plotted against f, the area of the rectangle (not the area under the curve) is τ_1/C_{A0} for a single-tank reactor, $\sum_1^N \tau_i/C_{A0}$ for the multiple-tank reactor, as seen from Figs. 11.10-3 and 11.10-4. From these figures, we may conclude that:

1. If $\tau_{\text{plug flow}} = t_{\text{batch}}$, the area under the curve for the same rate of reaction and the same range of conversion is the same.
2. For the same rate of reaction and the same range of conversion, the area for a single-tank reactor is always greater than that of a batch or a plug-flow reactor.

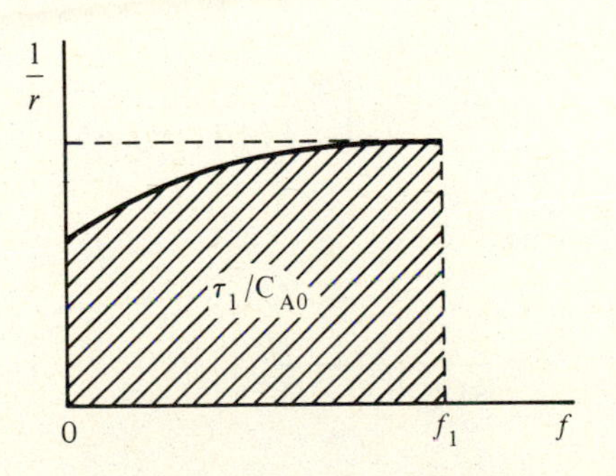

Figure 11.10-3 Single-tank reactor

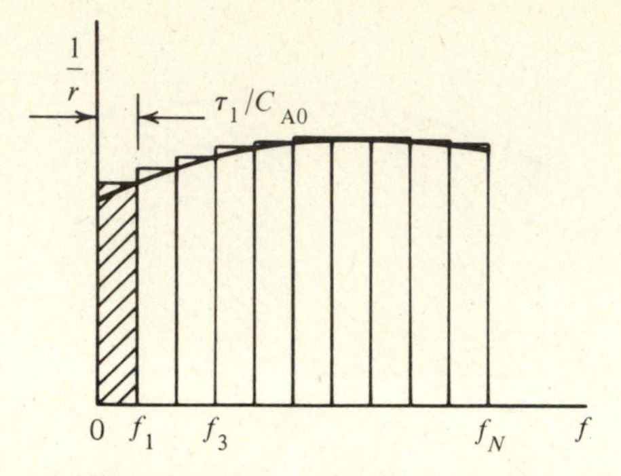

Figure 11.10-4 Multiple-tank reactor

This implies that the volume requirement for single-tank reactors is greater than the other two.

3. As the number of tanks in the multiple-tank-reactor cascade approaches infinity, the area in Fig. 11.10-4 becomes identical to that in Fig. 11.10-2.

From the above treatment, we see that although there are four types of reactors—batch, plug-flow, single-tank, and multiple-tank—we can compare only two, plug-flow and multiple-tank.

A plug-flow reactor is equivalent to a batch reactor if we treat $\tau = t$, i.e., the residence time in a plug-flow reactor is equal to the reaction time in a batch reactor and a multiple-tank reactor becomes a single-tank reactor if $N = 1$.

For a plug-flow reactor to carry out an nth-order reaction, the rate of the reaction is

$$r = kC_A^n = kC_{A0}^n(1 - f_A)^n \tag{11.10-2}$$

Substituting Eq. (11-10-2) into Eq. (8.2-9) gives

$$\tau = C_{A0}\int_0^{f_1} \frac{df_A}{kC_{A0}^n(1 - f_A)^n} \tag{11.10-3}$$

Integrating and simplifying

$$\tau = \frac{1}{kC_{A0}^{n-1}(n-1)}[(1-f)^{1-n} - 1] \tag{11.10-4}$$

For a multiple-tank reactor, Eq. (11.10-1) becomes

$$\tau_m = \frac{C_{A0}(f_m - f_{m-1})}{kC_{A0}^n(1 - f_m)^n} \tag{11.10-5}$$

The total residence time of the cascade is $\sum_{m=1}^{N} \tau_m$

$$\sum_{m=1}^{N} \tau_m = \frac{C_{A0}f_1}{kC_{A0}^n(1-f_1)^n} + \frac{C_{A0}(f_2 - f_1)}{kC_{A0}^n(1-f_2)^n} + \ldots + \frac{C_{A0}(f_N - f_{N-1})}{kC_{A0}^n(1-f_N)^n} \tag{11.10-6}$$

From the definition of the unit output, the ratio of the unit output of the plug-flow reactor to that of the tank cascade can be made

$$\frac{\eta_{PFR}}{\eta_{CSTR}} = \frac{(C_{k0}f_k/\tau)_{PFR}}{(C_{k0}f_k/\tau_i)_{CSTR}} = \frac{\tau_{CSTR}}{\tau_{PFR}} \tag{11.10-7}$$

Thus

$$\frac{\tau_{CSTR}}{\tau_{PFR}} = \frac{\sum_{m=1}^{N} [C_{A0}(f_m - f_{m-1})/kC_{A0}^n(1 - f_m)^n]}{\{[(1-f)^{1-n} - 1]/kC_{A0}^{n-1}(n-1)\}} \tag{11.10-8}$$

First let us compare the performance of a plug-flow reactor with a single CSTR under zero-, first-, second-, and higher-order reactions. In this case, Eq. (11.10-8) becomes

$$\frac{\tau_{CSTR}}{\tau_{PFT}} = \frac{C_{A0}f_1}{kC_{A0}^n(1-f_1)^n} \frac{kC_{A0}^{n-1}(n-1)}{[(1-f_1)^{1-n} - 1]} \tag{11.10-9}$$

Case (a): zero-order reaction, $n = 0$

Substituting $n = 0$ into Eq. (11.10-9) gives

$$\frac{\tau_{CSTR}}{\tau_{PFR}} = 1 \tag{11.10-10}$$

Case (b): first-order reaction, $n = 1$

Simplifying, Eq. (11.10-9) yields

$$\frac{\tau_{CSTR}}{\tau_{PFR}} = \frac{f_1}{(1-f_1)[1-(1-f_1)^{n-1}/(n-1)]} \tag{11.10-11}$$

For $n = 1$, the value within the bracket is indeterminate. It should be evaluated by L'Hospital's rule. The final result is

$$\frac{\tau_{\text{CSTR}}}{\tau_{\text{PFR}}} = \frac{f_1}{(f_1 - 1)\ln(1 - f_1)} = \frac{f_1}{(1 - f_1)\ln[1/(1 - f_1)]} \tag{11.10-12}$$

Case (c): nth-order reaction, $n = n$

$$\frac{\tau_{\text{CSTR}}}{\tau_{\text{PFR}}} = \frac{f_1(n - 1)}{(1 - f_1)[1 - (1 - f_1)^{n-1}]} \tag{11.10-13}$$

Next, we consider the CSTR cascade for the first-order reaction. Since the total residence time in the cascade is

$$\tau_{\text{CSTR}} = \sum_{m=1}^{N} \frac{C_{A0}(f_m - f_{m-1})}{kC_{A0}(1 - f_m)} \tag{11.10-14}$$

If we assume that each reactor has the same volume or the same residence time, then

$$\tau_{\text{CSTR}} = N\tau_m \qquad m = 1, 2, \cdots N \tag{11.10-15}$$

and

$$\tau_m = \frac{C_{A0}(f_m - f_{m-1})}{kC_{A0}(1 - f_m)} \tag{11.10-16}$$

Using concentration rather than fractional conversion

$$C_{Am} = C_{A0}(1 - f_m) \tag{11.10-17}$$

$$C_{A(m-1)} = C_{A0}(1 - f_{m-1}) \tag{11.10-18}$$

Eq. (11.10-16) becomes

$$\tau_m = \frac{C_{A(m-1)} - C_{Am}}{kC_{Am}} \tag{11.10-19}$$

Transposing

$$C_{Am} = \frac{C_{A(m-1)}}{1 + \tau_m k} \tag{11.10-20}$$

By induction

$$C_{A1} = \frac{C_{A0}}{1 + \tau_1 k} \quad \textbf{(11.10-21)}$$

$$C_{A2} = \frac{C_{A1}}{1 + \tau_2 k} \quad \textbf{(11.10-22)}$$

Substituting Eq. (11.10-21) into Eq. (11.10-22) and using $\tau_1 = \tau_2 = \cdots \tau_i$

$$C_{A2} = \frac{C_{A0}}{(1 + \tau_i k)^2} \quad \textbf{(11.10-23)}$$

For N reactors, Eq. (11.10-23) becomes

$$C_{AN} = \frac{C_{A0}}{(1 + \tau_i k)^N} \quad \textbf{(11.10-24)}$$

Using Eq. (11.10-15) and since $\tau_i = \tau_m$

$$C_{AN} = \frac{C_{A0}}{(1 + \tau_{CSTR} k/N)^N} \quad \textbf{(11.10-25)}$$

Solving for τ_{CSTR} gives

$$\tau_{CSTR} = \left[\left(\frac{C_{A0}}{C_{AN}} \right)^{1/N} - 1 \right] \frac{N}{k} \quad \textbf{(11.10-26)}$$

The residence time of a plug-flow reactor for a first-order reaction has been verified to be (see Eq. (11.10-12)

$$\tau_{PFR} = \frac{1}{k} \ln \frac{C_{A0}}{C_{AN}} \quad \textbf{(11.10-27)}$$

Thus, the ratio of the residence times for CSTR and for PFR is

$$\frac{\tau_{CSTR}}{\tau_{PFR}} = \frac{N[(C_{A0}/C_{AN})^{1/N} - 1]}{\ln C_{A0}/C_{AN}} \quad \textbf{(11.10-28)}$$

In terms of fractional conversion,

$$C_{AN} = C_{A0}(1 - f) \quad \textbf{(11.10-29)}$$

Eq. (11.10-28) becomes

$$\frac{\tau_{CSTR}}{\tau_{PFR}}=\frac{N\{[1/(1-f)]^{1/N}-1\}}{\ln[1/(1-f)]} \tag{11.10-30}$$

As N approaches infinity, the numerator is indeterminate, but it can be determined by L'Hospital's rule as

$$\frac{\tau_{CSTR}}{\tau_{PFR}}=\frac{-\ln(1-f)}{\ln[1/(1-f)]}=1 \tag{11.10-31}$$

Last, the ratio of the residence times of CSTR and PFR for second- or higher-order reactions is indicated by Eq. (11.10-8).

Example 11.10-1 Size Comparisons for First-Order Reactions

If the fractional conversion for first-order irreversible reaction is 0.95, the rate constant is $k = 1\ \text{ks}^{-1}$, the initial concentration of the limiting reactant $C_{Ao} = 1.0$ kmol/m³, determine the unit outputs for batch or plug-flow, single-tank, and two-tank reactors, and also evaluate their residence time ratios.

Solution

1. Batch or plug flow reactor

Equation (11.10-12)

$$t \text{ or } \tau=-\ln(1-f)=-\ln(1-0.95)=2.9957\ \text{ks}$$

Equation (11.9-1)

$$\eta=\frac{C_{A0}f_A}{t}=\frac{0.95}{2.9957}=0.3171\ \text{kmol/m}^3\cdot\text{ks}$$

2. Single-tank reactor

Equation (11.10-30)

$$\tau_{CSTR}=N\left[\left(\frac{1}{1-f}\right)^{1/N}-1\right]$$

$$=1\left[\frac{1}{1-0.95}-1\right]=19\ \text{ks}$$

$$\eta=\frac{C_{A0}f_A}{\tau}=\frac{0.95}{19}=0.05\ \text{kmol/m}^3\cdot\text{ks}$$

Equation (11.10-7)

$$\frac{V_{\text{CSTR}}}{V_{\text{PFR}}} = \frac{\tau_{\text{CSTR}}}{\tau_{\text{PFR}}} = \frac{\eta_{\text{PFR}}}{\eta_{\text{CSTR}}}$$

$$= \frac{0.3171}{0.05} = 6.34$$

3. Two-tank reactor

Equation (11.10-30)

$$\tau_{\text{CSTR}} = N\left[\left(\frac{1}{1-f}\right)^{1/N} - 1\right]$$

$$= 2\left[\left(\frac{1}{1-0.95}\right)^{1/2} - 1\right] = 6.9442 \text{ ks}$$

$$\eta = \frac{C_{\text{A0}} f_{\text{A}}}{\tau} = \frac{0.95}{6.9442} = 0.1368 \text{ kmol/m}^3 \cdot \text{ks}$$

Equation (11.10-7)

$$\frac{V_{\text{CSTR}}}{V_{\text{PFR}}} = \frac{0.3171}{0.1368} = 2.3179$$

Example 11.10-2 Size Comparisons for Second-Order Reaction

If the fractional conversion for second-order irreversible reaction is 0.875, the rate constant is $k = 9.92 \text{ m}^3/\text{kmol} \cdot \text{ks}$, the initial concentration of the limiting reactant $C_{\text{Ao}} = 0.08 \text{ kmol/m}^3$, determine residence time for batch or plug-flow, single-tank, two-tank, and three-tank reactors. Also evaluate their residence time ratios.

Solution

1. Batch or plug-flow reactor

Equation (11.10-4)

$$\tau = \frac{(1-f)^{1-n} - 1}{k C_{\text{A0}}^{n-1}(n-1)}$$

$$= \frac{(1-0.875)^{-1} - 1}{9.92(0.08)} = 8.8205 \text{ ks}$$

2. Single-tank reactor

Equation (11.10-5)

$$\tau_{\text{CSTR}} = \frac{C_{\text{A0}}(f-0)}{kC_{\text{A0}}^n(1-f)^n} = \frac{0.08(0.875)}{9.92(0.08)^2(0.125)^2} = 70.56 \text{ ks}$$

$$\frac{\tau_{\text{CSTR}}}{\tau_{\text{PFR}}} = \frac{70.56}{8.82} = 8.00$$

3. Two-tank reactors

Equation (11.10-6)

$$\tau_{\text{CSTR}} = \frac{C_{\text{A0}}f_1}{kC_{\text{A0}}^n(1-f_1)^n} + \frac{C_{\text{A0}}(f_2-f_1)}{kC_{\text{A0}}^n(1-f_2)^n}$$

Assuming equal volumes

$$\frac{C_{\text{A0}}f_1}{kC_{\text{A0}}^n(1-f_1)^n} = \frac{C_{\text{A0}}(f_2-f_1)}{kC_{\text{A0}}^n(1-f_2)^n}$$

or

$$\frac{f_1}{(1-f_1)^2} = \frac{f_2-f_1}{(1-f_2)^2}$$

$$\frac{f_1}{(1-f_1)^2} = \frac{0.875-f_1}{0.125^2} \quad \text{or} \quad f_1 = 0.725$$

Substituting

$$\tau_{\text{CSTR}} = \frac{0.08(0.725)}{9.92(0.08)^2(1-0.725)^2} + \frac{0.08(0.875-0.725)}{9.92(0.08)^2(1-0.875)^2}$$

$$= 24.18 \text{ ks}$$

$$\frac{\tau_{\text{CSTR}}}{\tau_{\text{PFR}}} = \frac{24.18}{8.82} = 2.74$$

4. Three-tank reactor

Equation (11.10-6)

$$\tau_{\text{CSTR}} = \frac{C_{\text{A0}}f_1}{kC_{\text{A0}}^n(1-f_1)^n} + \frac{C_{\text{A0}}(f_2-f_1)}{kC_{\text{A0}}^n(1-f_2)^n} + \frac{C_{\text{A0}}(f_3-f_2)}{kC_{\text{A0}}^n(1-f_3)^n}$$

Assuming equal volumes

$$\frac{0.875 - f_2}{(0.125)^2} = \frac{f_1}{(1 - f_1)^2}$$

$$\frac{f_2 - f_1}{(1 - f_2)^2} = \frac{f_1}{(1 - f_1)^2}$$

Solving these two equations gives $f_1 = 0.6285$ and $f_2 = 0.8038$. Substituting

$$\tau_{\text{CSTR}} = 5.75 + 5.75 + 5.75 = 17.26 \text{ ks}$$

$$\frac{\tau_{\text{CSTR}}}{\tau_{\text{PFR}}} = \frac{17.26}{8.82} = 1.96$$

For some other reactions, the residence time can be calculated by the appropriate equations and then the results can be compared.

Example 11.10-3 Size Comparisons for Reversible Reaction

A feed solution (density = 1020 kg/m^3) containing 23 percent by weight of acetic acid and 46 percent by weight of ethyl alcohol and no ester undergoes the following reversible reaction

$$CH_3COOH + C_2H_5OH \underset{k'}{\overset{k}{\rightleftharpoons}} CH_3COOC_2H_5 + H_2O$$

where $k = 7.93 \times 10^{-6}$ m^3/kmol · s and $k' = 2.7065 \times 10^{-6}$ m^3/kmol · s. For 35 percent fractional conversion, find: (1) reaction time in a batch reactor, (2) residence time in a plug-flow reactor, (3) residence time in a well-stirred reactor, (4) residence time in a cascade of three equal-size stirred reactors.

Solution

$$\underset{(1)}{CH_3COOH} + \underset{(2)}{C_2H_5OH} \underset{k'}{\overset{k}{\rightleftharpoons}} \underset{(3)}{CH_3COOC_2H_5} + \underset{(4)}{H_2O}$$

$$C_{1o} = \frac{1{,}020(0.23)}{60} = 3.91 \text{ kmol/m}^3$$

$$C_{2o} = \frac{1{,}020(0.46)}{46} = 10.2 \text{ kmol/m}^3$$

$$C_{3o} = 0$$

$$C_{4o} = \frac{1{,}020(0.31)}{18} = 17.56 \text{ kmol/m}^3$$

1. Batch reactor

Using Eq. (3.12-17)

$$D_1 = k - k' = 7.93 \times 10^{-6} - 2.71 \times 10^{-6} = 5.22 \times 10^{-6}$$

$$D_2 = -[k(C_{1o} + C_{2o}) + k'(C_{3o} + C_{4o})]$$

$$= -[7.93 \times 10^{-6}(3.91 + 10.2) + 2.71 \times 10^{-6}(17.56)]$$

$$= -1.5948 \times 10^{-4}$$

$$D_3 = kC_{1o}C_{2o} - k'C_{3o}C_{4o}$$

$$= 7.93 \times 10^{-6}[3.91(10.2) - 2.71 \times 10^{-6}(0)(17.56)]$$

$$= 3.1626 \times 10^{-4}$$

Thus

$$\lambda_e = \frac{-D_2 \pm \sqrt{D_2^2 - 4D_1D_3}}{2D_1}$$

$$= \frac{1.5948 \times 10^{-4} \pm \sqrt{(1.5948 \times 10^{-4})^2 - 4(5.22 \times 10^{-6})(3.1626 \times 10^{-4})}}{2(5.22 \times 10^{-6})}$$

$$= 28.42 \text{ and } 2.1318$$

$$\lambda = C_{1o}f = 3.91(0.35) = 1.3685$$

$$\sigma = \lambda_e + \frac{D_2}{D_1} = 28.42 + \frac{-1.5948 \times 10^{-4}}{5.22 \times 10^{-6}}$$

$$= -2.1317$$

and

$$\sigma = \lambda_e + \frac{D_2}{D_1} = 2.1318 + \frac{-1.5948 \times 10^{-4}}{5.22 \times 10^{-6}}$$

$$= -28.42$$

$$t = \frac{1}{(\lambda_e + \sigma)D_1} \ln \frac{\sigma(\lambda_e - \lambda)}{\lambda_e(\sigma + \lambda)}$$

$$= \frac{1}{(28.42 - 2.1317)(5.22 \times 10^{-6})} \ln \frac{(-2.1317)(28.42 - 1.3685)}{(28.42)(-2.1317 + 1.3685)}$$

$$= 7.1255588 \times 10^3 \text{ s}$$

$$= \frac{1}{(2.1318 - 28.42)(5.22 \times 10^{-6})} \ln \frac{(-28.42)(2.1318 - 1.3685)}{(2.1318)(-28.42 + 1.3685)}$$

$$= 7.124973 \times 10^3 \text{ s}$$

$$t_{av} = \frac{(7.1255588 + 7.124973)10^3}{2} = 7125 \text{ s}$$

$$\text{Unit output } \eta = \frac{C_{A0}f}{t} = \frac{3.91(0.35)}{7,125} = 1.92 \times 10^{-4} \text{ kmol/m}^3 \cdot \text{s}$$

If the time for the filling, emptying, and cleaning is 1 h, the unit output would become

$$\eta = \frac{3.91(0.35)}{7125 + 3600} = 1.276 \times 10^{-4} \text{ kmol/m}^3 \cdot \text{s}$$

2. Plug-flow reactor

Since the residence time in a plug-flow reactor τ is equal to the reaction time in a batch reactor t,

$$\tau = 7125 \text{ s}$$

and

$$\eta = 1.92 \times 10^{-4} \text{ kmol/m}^3 \cdot \text{s}$$

$$\frac{V_{\text{BR+cleaning}}}{V_{\text{PFR}}} = \frac{\tau_{\text{BR+cleaning}}}{\tau_{\text{PFR}}} = \frac{7,125 + 3,600}{7,125} = 1.5053$$

3. Single stirred-tank reactor

Using Eq. (7.8-4)

$$C_{km} - C_{k(m-1)} = -r_m \theta_m \qquad \textbf{(7.8-4)}$$

where

$$r_m = r = kC_{12}C_{22} - k'C_{32}C_{42}$$

and

$$\theta_m = \tau \qquad C_{km} = C_{12} \qquad C_{k(m-1)} = C_{11}$$

The concentrations of each species are

$$C_{12} = C_{11} - \lambda = 3.91 - 1.3685 = 2.5415$$

$$C_{22} = C_{21} - \lambda = 10.2 - 1.3685 = 8.8315$$

$$C_{32} = C_{31} + \lambda = 0 + 1.3685 = 1.3685$$

$$C_{42} = C_{41} + \lambda = 17.56 + 1.3685 = 18.9285$$

Substituting into Eq. (7.8-4)

$$2.5415 - 3.91 = -[7.93 \times 10^{-6}(2.5415)(8.8315) - 2.71 \times 10^{-6}(1.3685)(18.9285)]\tau$$

Solving for

$$\tau = 1.2695 \times 10^4 \text{ s}$$

$$\eta = \frac{(3.91)(0.35)}{1.2695 \times 10^4} = 1.078 \times 10^{-4} \text{ kmol/m}^3 \cdot \text{s}$$

$$\frac{\tau_{\text{CSTR}}}{\tau_{\text{PFR}}} = \frac{1.2695 \times 10^4}{7{,}125} = 1.78$$

4. Three equal-size stirred reactors

Using Eq. (7.8-4)

$$C_{k,m} + r_m \theta_m = C_{k(m-1)}$$

where

$$r_m = kC_1C_2 - k'C_3C_4$$

Since

$$C_1 = C_{10} - \lambda = 3.91 - \lambda$$

$$C_2 = C_{20} - \lambda = 10.2 - \lambda$$

$$C_3 = C_{30} + \lambda = \lambda$$

$$C_4 = C_{40} + \lambda = 17.56 + \lambda$$

Eliminating λ gives

$$C_2 = C_1 + 6.29$$

$$C_3 = 3.91 - C_1$$

$$C_4 = 21.47 - C_1$$

Substituting into the rate equation gives

$$r_m = kC_1(C_1 + 6.29) - k'(3.91 - C_1)(21.47 - C_1)$$

With $\theta_m = \tau_m$, and $k = 1$, Eq. (7.8-4) becomes

$$C_{1,m} + [kC_{1m}(C_{1m} + 6.29) - k'(3.91 - C_{1m})(21.47 - C_{1m})]\tau_m = C_{1(m-1)}$$

$$m = 1 \qquad C_{11} + [kC_{11}(C_{11} + 6.29) - k'(3.91 - C_{11})(21.47 - C_{11})]\tau_1 = C_{10}$$

$$m = 2 \qquad C_{12} + [kC_{12}(C_{12} + 6.29) - k'(3.91 - C_{12})(21.47 - C_{12})]\tau_2 = C_{11}$$

$$m = 3 \qquad C_{13} + [kC_{13}(C_{13} + 6.29) - k'(3.91 - C_{13})(21.47 - C_{13})]\tau_3 = C_{12}$$

With $C_{13} = 2.54$, $k = 7.93 \times 10^{-6}$, $k' = 2.71 \times 10^{-6}$, $C_{10} = 3.91$, and $\tau_1 = \tau_2 = \tau_3$, three unknowns τ, C_{11}, and C_{12} are left in these three equations. By trial

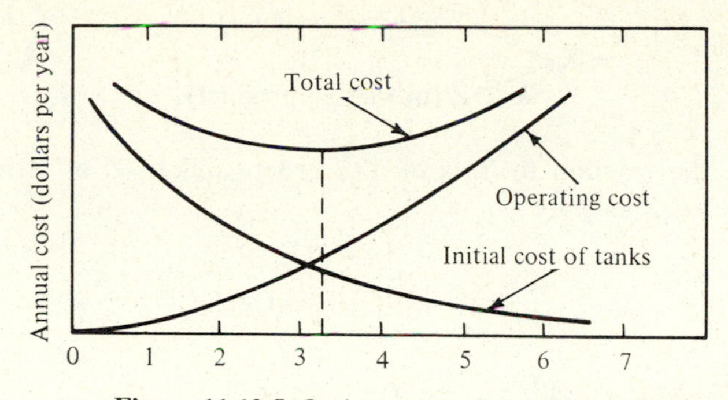

Figure 11.10-5 Optimum number of tanks

and error, it is found that $\tau = 2{,}865$ s, $C_{12} = 2.8482$, $C_{11} = 3.2680$ will give $C_{10} = 3.9128$ which is close to the given value of 3.91. Hence the unit output is

$$\eta = \frac{(3.91)(0.35)}{(2{,}865 + 2{,}865 + 2{,}865)} = 1.592 \times 10^{-4}$$

$$\frac{\tau_{\text{CSTR}}}{\tau_{\text{PFR}}} = \frac{2{,}865(3)}{7{,}125} = 1.21$$

From the previous three examples, we see that the total volume of the system reduces as the number of stirred tanks in the cascade increases. Hence the initial cost of the tanks decreases. However, the operating cost of more tanks and the cost of the auxiliary equipment increase as the number of tanks increases. There is an optimum number of tanks that make the total cost of the tanks and the operating cost a minimum (Fig. 11.10-5).

11.11 Reactor Selectivity

In Section 11.9, it was mentioned that one of the two criteria in comparing reactors is the reactor selectivity. Relative yield, operational yield, and selectivity have been defined in Eqs. (11.9-2), (11.9-3), and (11.9-4), respectively. Now we want to discuss the selectivity of multiple reactions: reactions in parallel and reactions in series.

11.11-1 Reactions in parallel

Let us first consider a single reactant to decompose to a desired product and an unwanted product

$$X \begin{array}{l} \xrightarrow{k_Y} Y \text{ (desired product)} \\ \xrightarrow{k_Z} Z \text{ (unwanted product)} \end{array}$$

Assuming that the reaction to Y is of Yth order, and to Z of Zth order, the component rate equations are

$$\frac{dC_X}{dt} = -(k_Y C_X^Y + k_Z C_X^Z) \tag{11.11-1}$$

$$\frac{dC_Y}{dt} = k_Y C_X^Y \tag{11.11-2}$$

$$\frac{dC_Z}{dt} = k_Z C_X^Z \tag{11.11-3}$$

Dividing Eq. (11.11-2) by Eq. (11.11-1) gives

$$\frac{dC_Y}{dC_X} = -\frac{1}{(1 + k_Z/k_Y)C_X^{(Z-Y)}} \tag{11.11-4}$$

The definition of relative yield ϕ_X in Eq. (11.9-2) can be written in differential form as

$$\phi_X = \frac{\text{Moles Y formed}}{\text{Moles X reacted}} = \frac{dC_Y}{-dC_X} \tag{11.11-5}$$

Combining with Eq. (11.11-4) yields

$$\phi_X = \frac{1}{[1 + (k_Z/k_Y)]C_X^{(Z-Y)}} \tag{11.11-6}$$

Similarly, the definition of selectivity in Eq. (11.9-4) can be written in differential form as

$$S = \frac{dC_Y}{dC_Z} = \frac{k_Y}{k_Z} C_X^{(Y-Z)} \tag{11.11-7}$$

Since ϕ_X is a function of C_X, the overall relative yield ϕ_{oX} which represents an average of the instantaneous value of ϕ_X over the whole concentration range can be developed from calculus as

$$\phi_{oX} = -\frac{1}{C_{Xo} - C_{Xf}} \int_{C_{Xo}}^{C_{Xf}} \phi_X \, dC_X \tag{11.11-8}$$

By similar reasoning, the overall relative yield in stirred tanks in series is

$$\phi_{o\mathrm{X}} = -\frac{1}{C_{\mathrm{X}o} - C_{\mathrm{X}f}} \Sigma\, \phi_{\mathrm{X}}\, \Delta C_{\mathrm{X}} \tag{11.11-9}$$

For a single stirred tank, because $\Delta C_{\mathrm{X}} = C_{\mathrm{X}f} - C_{\mathrm{X}o}$, Eq. (11.11-9) becomes

$$\phi_{o\mathrm{X}} = \phi_{\mathrm{X}} \tag{11.11-10}$$

which is justified because the concentration C_{X} in a single-tank reactor does not vary with time or position. Substituting Eq. (11.11-6) into Eq. (11.11-8) and Eq. (11.11-9) gives

$$\phi_{o\mathrm{X}} = -\frac{1}{C_{\mathrm{X}o} - C_{\mathrm{X}f}} \int_{C_{\mathrm{X}o}}^{C_{\mathrm{X}f}} \left[1 + \frac{k_{\mathrm{Z}}}{k_{\mathrm{Y}}} (C_{\mathrm{X}})^{Z-Y}\right]^{-1} dC_{\mathrm{X}} \tag{11.11-11}$$

and

$$\phi_{o\mathrm{X}} = -\frac{1}{C_{\mathrm{X}o} - C_{\mathrm{X}f}} \Sigma \left[1 + \frac{k_{\mathrm{Z}}}{k_{\mathrm{Y}}}(C_{\mathrm{X}})^{Z-Y}\right]^{-1} \Delta C_{\mathrm{X}} \tag{11.11-12}$$

Since the overall relative yield $\phi_{o\mathrm{X}}$ may also be defined as

$$\phi_{o\mathrm{X}} = \frac{\text{all Y formed}}{\text{all X reacted}} = \frac{C_{\mathrm{Y}f}}{C_{\mathrm{X}o} - C_{\mathrm{X}f}} \tag{11.11-13}$$

where $C_{\mathrm{Y}f}$ is the concentration of Y component at the exit. Thus

$$C_{\mathrm{Y}f} = \phi_{o\mathrm{X}}(C_{\mathrm{X}o} - C_{\mathrm{X}f}) \tag{11.11-14}$$

Substituting Eqs. (11.11-8), (11.11-9), and (11.11-10) gives

Plug-flow $$C_{\mathrm{Y}f} = \int_{C_{\mathrm{X}f}}^{C_{\mathrm{X}o}} \phi_{\mathrm{X}}\, dC_{\mathrm{X}} \tag{11.11-15}$$

Staged reactor $$C_{\mathrm{Y}f} = \Sigma \phi_{\mathrm{X}}(-\Delta C_{\mathrm{X}}) \tag{11.11-16}$$

Single reactor $$C_{\mathrm{Y}f} = \phi_{\mathrm{X}}(C_{\mathrm{X}o} - C_{\mathrm{X}f}) \tag{11.11-17}$$

Equations (11.11-15) to (11.11-17) are shown graphically in Figs. 11.11-1, 11.11-2, and 11.11-3. The requirements for high yield are:

1. Reaction concentration and reactor type. From Eq. (11.11-6), it is seen that if $Y > Z$, i.e., the order of the desired reaction is larger than the order of the undesired reaction, a high yield ϕ_{X} can be obtained when C_{X} is high. Because a batch or tubular reactor usually gives higher reactant concentration than the stirred-tank

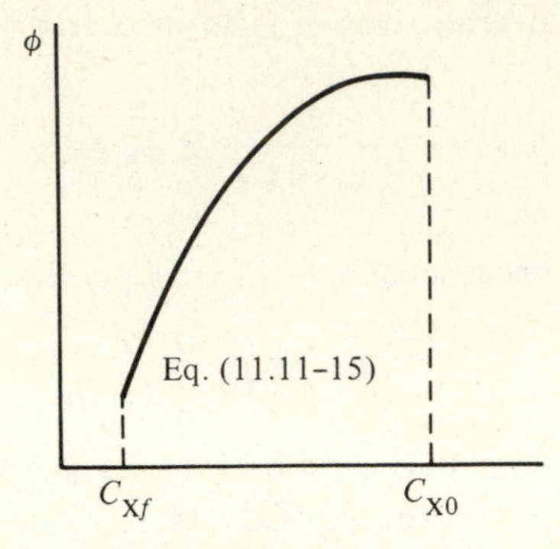

Figure 11.11-1 Plug flow

reactor, the former are thus preferred over the latter. However, if the latter is chosen for some other reasons, it should consist of several tanks in series without a recycling stream so as not to dilute the reactants. On the other hand, if $Y < Z$, because high yield is favored by a low reactant concentration, a single-tank reactor is the most suitable. In case that a batch or tubular reactor is used, dilution of the reactant by a recycle stream is preferred. Finally, if $Y = Z$, the yield is unaffected by reactant concentration.

2. Pressure in gas-phase reactions. If $Y > Z$, high pressure without inert gases is preferred. If $Y < Z$, low pressure is employed.
3. Temperature. The ratio k_Y/k_Z is affected by temperature according to the Arrhenius law

$$\frac{k_Y}{k_Z} = \frac{k_{oY}}{k_{oZ}} e^{-(E_Y - E_Z)/RT} \qquad \textbf{(11.11-18)}$$

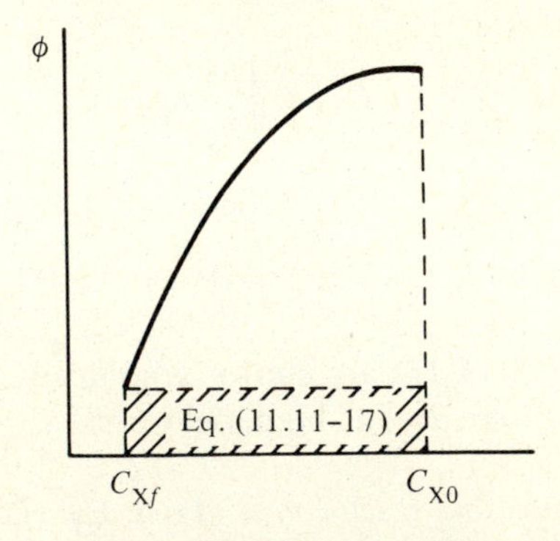

Figure 11.11-2 Single reactor

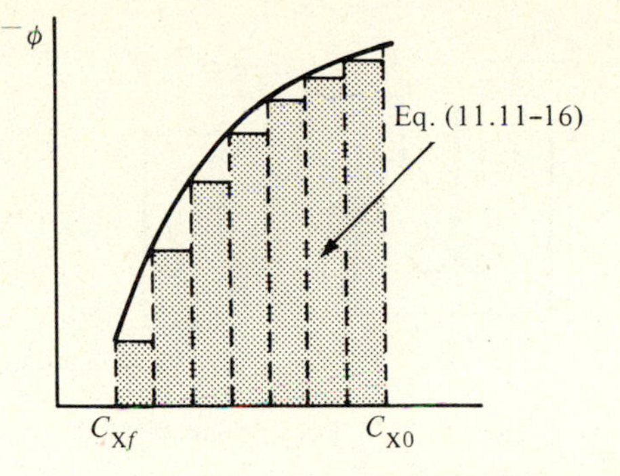

Figure 11.11-3 Staged reactor

in which k_{oY}, k_{oZ}, E_Y, E_Z are frequency factors and activation energies for XY and XZ reactions, respectively.

4. Choice of catalyst. If a suitable catalyst can be found to promote the desired reaction without producing an undesired reaction, the yield of the desired product will be increased.

5. Yield and reactor output. If a high yield is favored by a high reactant concentration, the reactor output will also be high. If by a low reactant concentration, the output will be low.

Next we consider reactions in parallel with two or more reactants. If a second reactant W is used with the first reactant X, the same principles apply to W as to X. There are three possible cases:

1. C_X, C_W are high. Either a batch or plug-flow reactor is employed.

2. C_X, C_W are low. A single stirred-tank reactor is used.

3. Either C_X or C_W is high and either C_W or C_X is low. A cross-flow reactor without recycle for continuous operation or a semibatch reactor for batch operation is employed. If the reactant required in high concentration can be easily recycled, a single-tank reactor is used (see Figs 11.11-4 and 11.11-5).

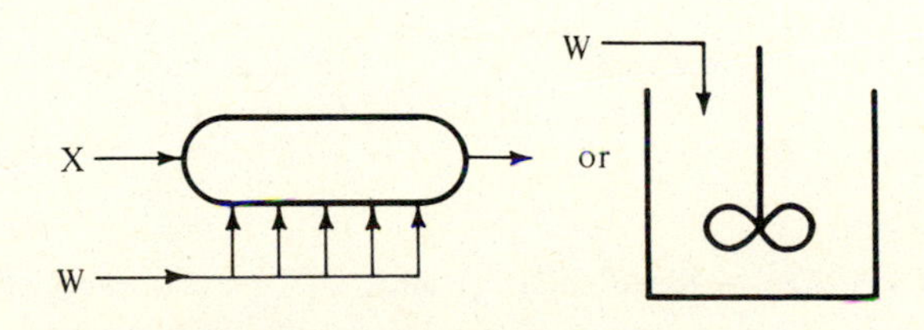

Figure 11.11-4 Recycle expensive

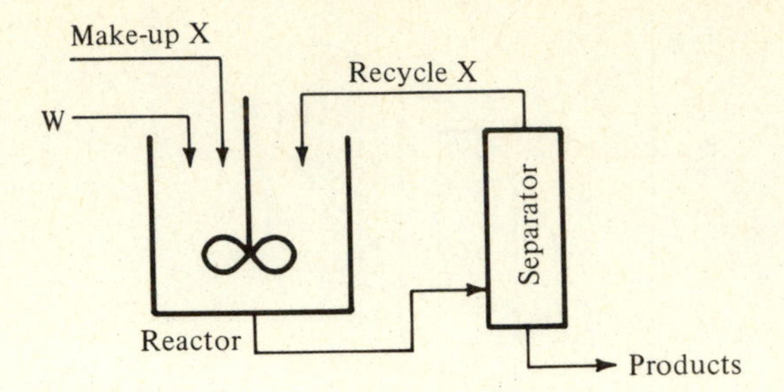

Figure 11.11-5 Recycle inexpensive

Example 11.11-1 Parallel Reactions

For the competitive reactions

$$X + W \xrightarrow{k_1} Y \quad \text{desired} \qquad \frac{dC_Y}{dt} = 1.0 C_X C_W^{0.3} \text{ mol/liter} \cdot \text{min}$$

$$X + W \xrightarrow{k_2} Z \quad \text{undesired} \qquad \frac{dC_Z}{dt} = 1.0 C_W^{1.8} C_X^{0.5} \text{ mol/liter} \cdot \text{min}$$

find the fraction of impurities in the effluent stream for 90 percent conversion of pure X and pure W (each has a density of 20 mol/liter) (1) for plug flow, (2) for single reactor, (3) for plug (X)-mixed(W) flow.

Solution

Since

$$-dC_X = dC_Y + dC_Z$$

$$\phi = \frac{dC_Y}{-dC_X} = \frac{dC_Y}{dC_Y + dC_Z} = \frac{k_1 C_X C_W^{0.3}}{k_1 C_X C_W^{0.3} + k_2 C_X^{0.5} C_W^{1.8}} = \frac{1}{1 + C_X^{-0.5} C_W^{1.5}}$$

1. Plug flow

$$C_{Xo} = C_{Wo} = 10 \text{ mol/liter} \qquad \text{and} \qquad C_X = C_W \text{ everywhere}$$

From Eq. (11.11-8)

$$\phi_{oX} = \frac{-1}{C_{Xo} - C_{Xf}} \int_{C_{Xo}}^{C_{Xf}} \phi_X \, dC_X = \frac{-1}{10-1} \int_{10}^{1} \frac{dC_X}{1 + C_X} = \frac{1}{9} \ln (1 + C_X) \Big|_{1}^{10} = +0.19$$

Therefore the percent of W in the product is $1 - 0.19 = 0.81$ or 81 percent

2. Single reactor

$C_{X0} = 10$

$C_{W0} = 10$

$C_{Xf} = 1$

$C_{Wf} = 1$

From Eq. (11.11-10)

$$\phi_{oX} = \phi_X = \frac{1}{1 + C_{Xf}} = \frac{1}{1+1} = 0.5$$

Thus the percent of W in the product is 50 percent.

3. Plug(X)-mixed(W) flow

$C_{Xo} = 19$ assume $C_W = 1$ everywhere

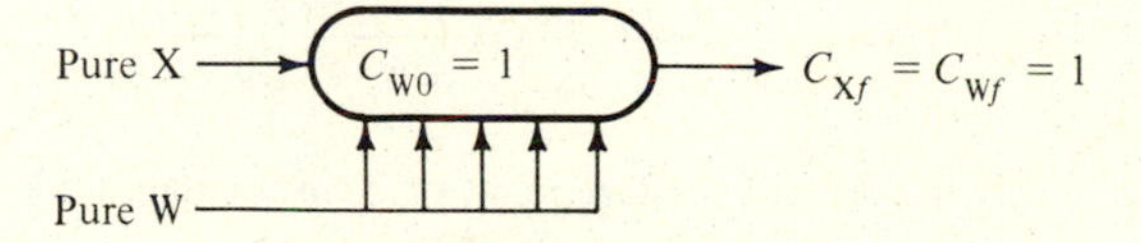

Assuming that W is introduced into the reactor in such a way that $C_W = 1$ mol/liter throughout the reactor, we can calculate

$$\phi = \frac{-1}{C_{Xo} - C_{Xf}} \int_{C_{Xo}}^{C_{Xf}} \phi \, dC_X = \frac{-1}{19-1} \int_{19}^{1} \frac{dC_X}{1 + C_X^{-0.5}(1)^{1.5}} = 0.741$$

Thus the percent of W in the effluent stream is 25.9 percent.

Example 11.11-2 Another Parallel Reaction

Find the overall relative yield in a plug-flow and a single-tank reactor for the following reactions:

$$X \xrightarrow{k_Y} Y$$

$$2X \xrightarrow{k_Z} Z$$

Solution

The component rate equations are

$$\frac{dC_X}{dt} = -(k_Y C_X + 2k_Z C_X^2) \tag{A}$$

$$\frac{dC_Y}{dt} = k_Y C_X \tag{B}$$

Dividing Eq. (A) by Eq. (B) gives

$$\phi_X = -\frac{dC_Y}{dC_X} = \frac{k_Y C_X}{k_Y C_X + 2k_Z C_X^2}$$

$$= \frac{1}{1 + (2k_Z/k_Y)C_X} \tag{C}$$

Substituting into Eq. (11.11-8) gives the overall relative yield of X in a plug-flow reactor

$$\phi_{oX} = \frac{1}{C_{Xo} - C_{Xf}} \int_{C_{Xf}}^{C_{Xo}} \frac{1}{1 + (2k_Z/k_Y)C_X} \, dC_X \tag{D}$$

$$= \frac{k_Y}{2k_Z(C_{Xo} - C_{Xf})} \ln \frac{1 + (2k_Z/k_Y)C_{Xo}}{1 + (2k_Z/k_Y)C_{Xf}} \tag{E}$$

From Eq. (11.11-10) the overall relative yield of X in a single-tank reactor is

$$\phi_{oX} = \phi_X = \frac{1}{1 + (2k_Z/k_Y)C_X} \tag{F}$$

11.11-2 Reactions in series

Let us consider

$$X \xrightarrow{k_Y} Y \xrightarrow{k_Z} Z$$

where Y is the desired product and Z is the undesired product. For simplicity, let us assume that both reactions are first-order. The component rate equations are

$$-\frac{dC_X}{dt} = k_Y C_X \qquad \textbf{(11.11-19)}$$

$$\frac{dC_Y}{dt} = k_Y C_X - k_Z C_Y \qquad \textbf{(11.11-20)}$$

$$\frac{dC_Z}{dt} = k_Z C_Y \qquad \textbf{(11.11-21)}$$

Dividing Eq. (11.11-20) by Eq. (11.11-19) gives

$$\frac{dC_Y}{dC_X} = \frac{k_Z C_Y}{k_Y C_X} - 1 \qquad \textbf{(11.11-21)}$$

which is a first-order linear differential equation. Its solution with initial conditions at $C_X = C_{Xo}$, $C_Y = 0$ is

$$C_Y = \frac{k_Y}{k_Y - k_Z}\left[-C_X + C_{Xo}\left(\frac{C_X}{C_{Xo}}\right)^{k_Z/k_Y}\right] \qquad \textbf{(11.11-22)}$$

From the definition of the relative yield

$$\phi_X = \frac{\text{moles Y formed}}{\text{moles X reacted}} = \frac{C_Y}{C_{Xo} - C_X}$$

$$= \frac{k_Y}{k_Y - k_Z}\left[\frac{-C_X}{C_{Xo} - C_X} + \frac{C_{Xo}}{C_{Xo} - C_X}\left(\frac{C_X}{C_{Xo}}\right)^{k_Z/k_Y}\right] \qquad \textbf{(11.11-23)}$$

$$= \frac{1}{1 - k_Z/k_Y}\left[\frac{1}{1 - C_{Xo}/C_X} + \frac{1}{1 - C_X/C_{Xo}}\left(\frac{C_X}{C_{Xo}}\right)^{k_Z/k_Y}\right] \qquad \textbf{(11.11-24)}$$

But

$$C_X = C_{Xo}(1 - f) \qquad \text{or} \qquad \frac{C_X}{C_{Xo}} = 1 - f; \qquad \frac{C_{Xo}}{C_X} = \frac{1}{1 - f}$$

Equation (11.11-24) becomes

$$\phi_X = \frac{1}{(k_Z/k_Y) - 1}\left[\frac{1 - f}{f} - \frac{(1 - f)^{k_Z/k_Y}}{f}\right] \qquad \textbf{(11.11-25)}$$

When $k_Z/k_Y = 1$, Eq. (11.11-25) can be evaluated by L'Hospital's rule as

$$\phi_X = -\frac{(1 - f)}{f}\ln(1 - f) \qquad \textbf{(11.11-26)}$$

Equations (11.11-25) and (11.11-26) are the instantaneous relative yield of X in a plug-flow reactor.

For a single stirred-tank reactor, Eq. (7.6-2) in this case becomes

$$C_{Y2} = \frac{k_Y C_{X0} \theta}{(1 + k_Y \theta)} \tag{11.11-27}$$

and

$$C_{X2} - C_{X0} = - k_Y C_{X2} \theta \tag{11.11-28}$$

Using

$$C_{X2} = C_{X0}(1 - f) \tag{11.11-29}$$

and solving Eq. (11.11-28) for θ yields

$$\theta = \frac{f}{k_Y(1 - f)} \tag{11.11-30}$$

Substituting Eq. (11.11.29) and Eq. (11.11-30) into Eq. (11.11-27) gives

$$C_{Y2} = \frac{C_{X0} f}{1 + k_Z f / k_Y(1 - f)} \tag{11.11-31}$$

Then from the definition of relative yield, we get

$$\phi_X = \frac{C_{Y2}}{C_{X0} - C_{X2}} = \frac{C_{X0} f}{(C_{X0} - C_{X2})[1 + k_Z f / k_Y(1 - f)]} \tag{11.11-32}$$

Simplifying Eq. (11.11-32) with Eq. (11.11-29) gives

$$\phi_X = \frac{1}{1 + k_Z f / k_Y(1 - f)} \tag{11.11-33}$$

Finally, the ratio of the overall relative yields of CSTR to PFR is

$$\frac{\phi_{CSTR}}{\phi_{PFR}} = \frac{\text{Eq. (11.11-33)}}{\text{Eq. (11.11-25)}} \qquad \text{for } \frac{k_Z}{k_Y} \neq 1 \tag{11.11-34}$$

$$= \frac{\text{Eq. (11.11-33)}}{\text{Eq. (11.11-26)}} \qquad \text{for } \frac{k_Z}{k_Y} = 1 \tag{11.11-35}$$

Using some of the results from these two equations, we may conclude:

1. Reactor type: For a high relative yield of Y, a batch or tubular plug-flow reactor should be used.
2. Conversion in reactor: The reactor should be designed for a low conversion of X per batch or per pass in a plug-flow reactor with separation of Y and recycling of the unused reactant.
3. Temperature: If $E_1 > E_2$, a high temperature should be chosen and conversely a low temperature if $E_1 < E_2$.
4. General: Remove Y continuously as soon as it is formed to avoid the degradation to the undesired product Z.

11.11-3 Series parallel reactions

The reactions

$$X + W \xrightarrow{k_Y} Y \qquad \text{desired reaction}$$

$$Y + W \xrightarrow{k_Z} Z \qquad \text{undesired reaction}$$

is equivalent to

$$X \xrightarrow{W} Y \xrightarrow{W} Z$$

The component rate equations are

$$\frac{dC_X}{dt} = -\, k_Y C_X C_W \tag{11.11-36}$$

$$\frac{dC_Y}{dt} = k_Y C_X C_W - k_Z C_Y C_W \tag{11.11-37}$$

$$\frac{dC_Y}{dt} = k_Z C_Y C_W \tag{11.11-38}$$

$$\frac{dC_W}{dt} = -\, k_Y C_X C_W - k_Z C_Y C_W \tag{11.11-39}$$

Dividing Eq. (11.11-37) by Eq. (11.11-36) gives the relative yield of X as

$$\phi = \frac{dC_Y}{-\, dC_X} = \frac{k_Y C_X C_W - k_Z C_Y C_W}{k_Y C_X C_W}$$

or

$$\frac{dC_Y}{dC_X} = -\, 1 + \frac{k_Z C_Y}{k_Y C_X} \tag{11.11-40}$$

which is identical to Eq. (11.11-21). Therefore the relative yield ϕ_X is also given by Eq. (11.11-25) for $k_Z/k_Y \neq 1$ and Eq. (11.11-26) for $k_Z/k_Y = 1$. Similarly, for a single stirred-tank reactor, the design equations are

$$C_{Y2} = k_Y C_{X2} C_{W2} \theta - k_Z C_{Y2} C_{W2} \theta \tag{11.11-41}$$

and

$$C_{X2} - C_{X0} = -k_Y C_{X2} C_{W2} \theta \tag{11.11-42}$$

Eliminating $C_W \theta$ gives

$$\frac{C_{Y2}}{C_{X2} - C_{X0}} = \frac{k_Z C_{Y2}}{k_Y C_{X2}} - 1 \tag{11.11-43}$$

which can be transformed into equations identical with those in Eq. (11.11-31) and Eq. (11.11-32). Hence the relative yield is also Eq. (11.11-33). The criteria in obtaining high yields are about the same. Since backmixing of products with reactants should be avoided, a plug-flow or a batch reactor is preferred. However, there is a difference between the series parallel reaction with the series reaction. This is in the stoichiometry of the reaction; the reaction cannot proceed completely to the product Z if less than 2 mol W per mole of X are supplied. Hence some control can be made by selecting the ratio of W to X.

11.12 Catalytic Reactor Selectivity

Similarly when multiple reactions occur in a catalytic reactor, unwanted reactions may be produced. Hence selectivity becomes important. In contrast to homogeneous reactors, two special phenomena, external and internal diffusions, exist in a catalytic reactor. Their effects on the reactor selectivity are discussed.

11.12-1 Effect of external diffusion and heat-transfer resistances

Suppose the following reactions

$$X \xrightarrow{k_1} Y$$

$$X \xrightarrow{k_2} Z$$

are taking place in a fixed-bed reactor through which fluid reactant X is passed. The desired product is Y. Assuming both reactions are first-order and irreversible, the global rates of reactions 1 and 2 are

$$r_{1p} = k_1 a_m C_s \tag{11.12-1}$$

and

$$r_{2p} = k_2 a_m C_s \tag{11.12-2}$$

in which a_m is the external surface area per unit mass of the pellet and C_s is the concentration of X at the surface. The corresponding rates due to external mass transfer are

$$r_{1p} = k_m a_m (C_b - C_s) \tag{11.12-3}$$

$$r_{2p} = k_m a_m (C_b - C_s) \tag{11.12-4}$$

where C_b is the concentration of X in the bulk gas. Eliminating C_s in Eqs. (11.12-1) and (11.12-3) or Eqs. (11.12-2) and (11.12-4) gives

$$C_s = \frac{k_m C_b}{k_1 + k_m} \tag{11.12-5}$$

and

$$C_s = \frac{k_m C_b}{k_2 + k_m} \tag{11.12-6}$$

Substituting C_s from Eqs. (11.12-5) and (11.12-6) into Eqs. (11.12-1) and (11.12-2) yields

$$r_{1p} = \frac{a_m C_b}{1/k_1 + 1/k_m} \tag{11.12-7}$$

and

$$r_{2p} = \frac{a_m C_b}{1/k_2 + 1/k_m} \tag{11.12-8}$$

Consequently the selectivity S which is defined as

$$S = \frac{r_{1p}}{r_{2p}} \tag{11.12-9}$$

becomes

$$S = \frac{1/k_2 + 1/k_m}{1/k_1 + 1/k_m} \tag{11.12-10}$$

For nonisothermal condition, heat transfer must be considered. When a fixed-bed reactor operates at steady state, the heat released by reaction on the catalyst pellet must be transferred to the bulk fluid. The reaction heat

$$Q_R = (-\Delta H) r_p = (-\Delta H) a_m A C_s e^{-E/RT_s} \tag{11.12-11}$$

where A is the frequency factor in the Arrhenius equation. However,

$$-\frac{E}{RT_s} = \frac{E}{RT_b}\left(\frac{T_s - T_b}{T_s} - 1\right) = -\frac{\alpha}{\theta + 1} \tag{11.12-12}$$

where $\alpha = E/RT_b$ and $\theta = (T_s - T_b)/T_b$, a dimensionless temperature. Hence Eq. (11.12-11) becomes

$$Q_R = (-\Delta H) a_m A C_s e^{-\alpha/(\theta+1)} \tag{11.12-13}$$

The heat transferred to the fluid Q is

$$Q = h a_m (T_s - T_b) = h a_m T_b \theta \tag{11.12-14}$$

Equating Eqs. (11.12-13) and (11.12-14) yields $T_s - T_b$ in terms of ΔH, A, E, h, and the unknown surface concentration C_s. Substituting C_s from Eqs. (11.12-5) or (11.12-6) into Eq. (11.12-13) gives

$$Q_R = \frac{(-\Delta H) a_m C_b}{(1/A) e^{\alpha/(\theta+1)} + 1/k_m} \tag{11.12-15}$$

For finite diffusion resistances, the ratio γ for first-order reaction

$$\gamma = \frac{r_p}{r_b} = \frac{a_m A C_s e^{-E/RT_s}}{a_m A C_b e^{-E/RT_b}}$$

Eliminating C_s yields

$$\gamma = \frac{e^{\alpha\theta/(\theta+1)}}{1 + (A/k_m) e^{-\alpha/(\theta+1)}} \tag{11.12-16}$$

Now the selectivity is

$$S = \frac{r_{1p}}{r_{2p}} = \frac{(r_p/r_b)_1}{(r_p/r_b)_2}\left(\frac{r_1}{r_2}\right)_b = \frac{\gamma_1}{\gamma_2} S_b \tag{11.12-17}$$

Using Eq. (11.12-16) gives

$$S = S_b e^{(\alpha_1-\alpha_2)\theta/(\theta+1)} \frac{1+(A_2/k_m)e^{-\alpha_2/(\theta+1)}}{1+(A_1/k_m)e^{-\alpha_1/(\theta+1)}} \tag{11.12-18}$$

11.12-2 Effect of internal diffusion

Isothermal Conditions Let us consider the following cases:

Case 1

$$\mathrm{C} \xrightarrow{k_1} \mathrm{D} \qquad \text{(desired product)}$$

$$\mathrm{Y} \xrightarrow{k_2} \mathrm{Z}$$

Assuming both reactions are first-order and neglecting longitudinal and radial dispersion effects, the ratio of the rates of these two reactions is

$$\frac{r_{v\mathrm{C}}}{r_{v\mathrm{Y}}} = \frac{k_1 C_\mathrm{C}}{k_2 C_\mathrm{Y}} \tag{11.12-19}$$

Due to internal diffusion, the rate of reaction for the slab model is

$$r_{v\mathrm{C}} = -A_c D_e \left(\frac{dC_\mathrm{C}}{dx}\right)_{x=\pm L} \tag{11.12-20}$$

The concentration profile of C through the slab is

$$C_\mathrm{C} = C_{\mathrm{C}\infty} \frac{\cosh(\lambda x)}{\cosh(\lambda L)} \tag{11.12-21}$$

Thus

$$r_{v\mathrm{C}} = \frac{-A_c D_e}{V_p} \left(\frac{dC_\mathrm{C}}{dx}\right)_{x=L} = \frac{-A_c D_e C_{\mathrm{C}\infty}}{L V_p} \phi_1 \tanh \phi_1 \tag{11.12-22}$$

where V_p is the geometric volume and $\phi_1 = L(k_1/D_e)^{1/2}$. Similarly,

$$r_{v\mathrm{Y}} = \frac{-A_c D_e C_{y\infty}}{L V_p} \phi_2 \tanh \phi_2 \tag{11.12-23}$$

where

$$\phi_2 = L\left(\frac{k_2}{D_e}\right)^{1/2}$$

For a general type of catalyst, the ratio of the rates of decomposition of C and Y then becomes

$$S = \frac{r_{vC}}{r_{vY}} = \frac{C_{C\infty}}{C_{Y\infty}} \frac{\phi_1}{\phi_2} \frac{\tanh \phi_1}{\tanh \phi_2} \tag{11.12-24}$$

where ϕ_1 and ϕ_2 are the Thiele moduli for the decomposition of C and Y, respectively.

Case 2

$$X \xrightarrow{k_1} Y \qquad \text{(desired product)}$$

$$X \xrightarrow{k_2} Z$$

In this case, the selectivity is, assuming equal diffusivity of Y and Z,

$$S = \frac{r_{vX}}{r_{vZ}} = \frac{(dC_Y/dx)_{x=L}}{(dC_Z/dx)_{x=L}} \tag{11.12-25}$$

The respective fluxes may be evaluated by writing the material balance equations for each component and solving the resulting simultaneous equations. If the two reactions are of the same order, then the selectivity is unaffected by mass transfer in pores. If not, the selectivity can be calculated.

Example 11.12-1

Derive an expression for the catalyst selectivity for the following concurrent irreversible reactions occuring isothermally in a flat slab-shaped porous catalyst pellet

$$X \xrightarrow{k_1} Y$$

$$X \xrightarrow{k_2} Z$$

The desired product Y is formed by a first-order reaction and the waste product Z is formed by a zero-order reaction.

Solution

Taking a material balance across the element Δx in Fig. 11.12-1. The flux in at $x + \Delta x$ minus the flux out at x is equal to the mass reacted in volume $2A_c\ \Delta x$ where A_c is the area of each of the faces. Thus

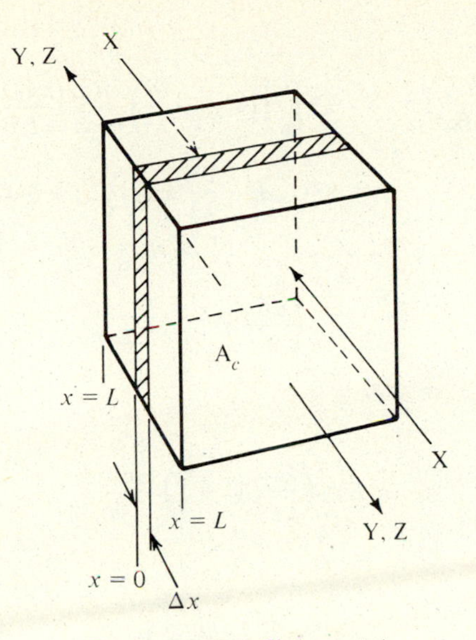

Figure 11.12-1 Slab model

$$\frac{d^2C_X}{dx^2} - \lambda^2 C_X = \theta^2 \tag{A}$$

$$\frac{d^2C_Y}{dx^2} + \lambda^2 C_X = 0 \tag{B}$$

$$\frac{d^2C_Z}{dx^2} = -\theta^2 \tag{C}$$

where $\lambda^2 = k_1/D_e$ and $\theta^2 = k_2/D_e$. The boundary conditions are

$$\text{At } x = \pm L \qquad C_X = C_{X0},\ C_Y = C_Z = 0$$

$$\text{At } x = 0 \qquad \frac{dC_X}{dx} = \frac{dC_Y}{dx} = \frac{dC_Z}{dx} = 0$$

The complete solution satisfying the boundary conditions is

$$C_X = \left(C_{X0} + \frac{\theta^2}{\lambda^2}\right)\frac{\cosh(\lambda x)}{\cosh(\lambda L)} - \frac{\theta^2}{\lambda^2} \tag{D}$$

Since the selectivity is

$$S = \frac{(dC_Y/dx)_{x=L}}{(dC_Z/dx)_{x=L}} \tag{E}$$

and

$$\left(\frac{dC_Y}{dx}\right)_{x=L} = -\lambda^2 \int_0^L \left\{\left(C_{X0} + \frac{\theta^2}{\lambda^2}\right)\frac{\cosh(\lambda x)}{\cosh(\lambda L)} - \frac{\theta^2}{\lambda^2}\right\} dx$$

$$= -\lambda^2 C_{X0} L - \frac{\theta^2}{\lambda^2}\tanh(\lambda L) + \theta^2 L \quad \text{(F)}$$

and

$$\left(\frac{dC_Z}{dx}\right)_{x=L} = -\theta^2 L \quad \text{(G)}$$

the final answer is

$$S = \left(\frac{k_1}{k_2} C_{X0} + 1\right)\frac{\tanh \phi}{\phi} - 1 \quad \text{(H)}$$

where

$$\phi = L\left(\frac{k_1}{D_e}\right)^{1/2}$$

Case 3

$$X \xrightarrow{k_1} Y \xrightarrow{k_2} Z$$

Assuming both reactions are first-order and using a slab model, a material balance for component Y is

$$D_e \frac{d^2 C_Y}{dx^2} = k_2 C_Y - k_1 C_X \quad \textbf{(11.12-26)}$$

and for component X, the equation is

$$D_e \frac{d^2 C_X}{dx^2} = k_1 C_X \quad \textbf{(11.12-27)}$$

The boundary conditions are

At $x = L$ $\quad C_X = C_{X\infty}$ and $C_Y = C_{Y\infty}$

At $x = 0$ $\quad \dfrac{dC_X}{dx} = \dfrac{dC_Y}{dx} = 0$

Solving Eqs. (11.12-26) and (11.12-27) simultaneously with the above boundary conditions leads to:

$$C_X = C_{X\infty}\frac{\cosh(\lambda_1 x)}{\cosh(\lambda_1 L)} \quad \textbf{(11.12-28)}$$

and

$$C_Y = C_{X\infty}\left(\frac{k_1}{k_1 - k_2}\right)\left\{\frac{\cosh(\lambda_2 x)}{\cosh(\lambda_2 L)} - \frac{\cosh(\lambda_1 x)}{\cosh(\lambda_1 L)}\right\} + C_{Y\infty}\frac{\cosh(\lambda_2 x)}{\cosh(\lambda_2 L)} \quad \textbf{(11.12-29)}$$

where

$$\lambda_1 = \left(\frac{k_1}{D_e}\right)^{1/2} \quad \text{and} \quad \lambda_2 = \left(\frac{k_2}{D_e}\right)^{1/2}$$

Since the selectivity is

$$S = -\frac{(dC_Y/dx)_{x=L}}{(dC_X/dx)_{x=L}} = \left(\frac{k_1}{k_1 - k_2}\right)\left(1 - \frac{\phi_2 \tanh \phi_2}{\phi_1 \tanh \phi_1}\right) - \frac{C_{Y\infty}\phi_2 \tanh \phi_2}{C_{X\infty}\phi_1 \tanh \phi_1} \quad \textbf{(11.12-30)}$$

which can be written in another form

$$-\frac{dC_Y}{dC_X} = \frac{S^{1/2}}{1 + S^{1/2}} - \frac{C_Y}{C_X S^{1/2}} \quad \textbf{(11.12-31)}$$

where S is the kinetic selectivity $= k_1/k_2$. Integrating Eq. (11.12-31) from the reactor inlet (where the concentration of X is C_{X0} and that of Y is zero) to any point in the reactor:

$$\frac{C_Y}{C_X} = \left(\frac{S}{S-1}\right)\left[\left(\frac{C_X}{C_{X0}}\right)^{(1-\sqrt{S})/\sqrt{S}} - 1\right] \quad \textbf{(11.12-32)}$$

When ϕ_1 and ϕ_2 are small, Eq. (11.12-30) becomes

$$-\frac{dC_Y}{dC_X} = 1 - \frac{C_Y}{SC_X} \quad \textbf{(11.12-33)}$$

which on integration gives

$$\frac{C_Y}{C_X} = \left(\frac{S}{S-1}\right)\left[\left(\frac{C_X}{C_{X0}}\right)^{(1-S)/S} - 1\right] \quad \textbf{(11.12-34)}$$

Nonisothermal Conditions As mentioned previously in the two concurrent first-order reactions

$$X \xrightarrow{k_1} Z$$

$$X \xrightarrow{k_2} Z$$

the selectivity is not affected by any limitations due to mass transfer. However, with heat transfer between the interior and exterior of the catalyst, the selectivity is changed. Assuming a slab model, the mass and heat balances of the two concurrent first-order reactions are:

$$\frac{d^2C_X}{dx^2}-\left(\frac{k_1+k_2}{D_e}\right)C_X=0 \qquad (11.12\text{-}35)$$

$$\frac{d^2C_Y}{dx^2}+\frac{k_1}{D_e}C_X=0 \qquad (11.12\text{-}36)$$

and

$$\frac{d^2T}{dx^2}+\frac{(-\Delta H_1)k_1+(-\Delta H_2)k_2}{k_e}C_X=0 \qquad (11.12\text{-}37)$$

where

$$k_i=A_ie^{-E_i/RT} \qquad i=1,2$$

and ΔH_1, ΔH_2 are heats of reaction of reactions 1 and 2. The boundary conditions are

$$\text{At } x=0 \qquad \frac{dC_X}{dx}=\frac{dC_Y}{dx}=\frac{dT}{dx}=0$$

$$\text{At } x=L \qquad C_X=C_{X\infty} \qquad C_Y=C_{Y\infty} \qquad T=T_\infty$$

This is a two-point boundary-value problem. Because of nonlinearity, no analytical solution is possible. However, if $E_1 = E_2$, the selectivity is the same as if there were no resistance to either heat or mass transfer. If $E_2/E_1 > 1$, the best selectivity is obtained for high values of the Thiele modulus. If $E_2/E_1 < 1$, a decrease in selectivity results.

Problems

1. 80 percent of a liquid reactant is converted in 15 min in an isothermal batch reactor. The reaction is a first-order irreversible. What space-time and space-velocity are required to have the same conversion in a plug-flow reactor and in a mixed reactor?

2. 50 percent of A is converted to R in a mixed reactor.

$$A \rightarrow R \qquad r=kC_A^2$$

What will be the conversion if the reaction is carried out in (a) one six times as large and (b) a plug-flow reactor, assuming all others unchanged.

3. Pure gas A at 5 atm and 350°C entering at a rate of 4 m^3/h to a reactor of 2.5 cm ID pipe 2.5 m long undergoes the following second-order reaction:

$$A \longrightarrow 3B$$

The conversion is 65 percent of the feed. A commercial plant is to handle 325 m^3/h of feed consisting of 50 percent A, 50 percent inert at 27 atm and 360°C to obtain 80 percent conversion.

(a) Determine the number of 2-m lengths of 2.5-cm ID pipe required for the conversion.

(b) Should they be placed in parallel or in series?

Assume plug-flow and ideal gas behavior with negligible pressure drop.

4. One mole per liter each of A and B reacts in a plug-flow reactor for the second-order irreversible reaction

$$A + B \longrightarrow R + S$$

The conversion is 60 percent.

(a) If a mixed reactor 10 times as large as the plug-flow reactor were connected in series with the existing unit, which unit should come first? By what fraction could production be increased?

(b) Does the feed concentration affect the answer?

5. Two mixed reactors of 1 m^3 and 2 m^3 are available for carrying out a first order irreversible reaction

$$A \xrightarrow{k} B \qquad k = 1 \text{ min}^{-1}$$

at constant temperature and density. The feed A of 1 g mol/liter enters the reactor at a rate of 1 m^3/h. Which of the following arrangements will be selected to give the highest production rate of B?

(a) The two reactors connected in parallel with equal residence times.

(b) The two reactors connected in parallel with different residence times.

(c) The two reactors connected in series with the feed entering the larger reactor.

(d) The two reactors connected in series with the feed entering the smaller reactor.

6. 50 percent of A (C_{Ao} = 4 kmol/m^3) undergoes the following elementary reaction:

$$A \xrightarrow{k_1} B \qquad k_1 = 10 \text{ ks}^{-1}$$

$$2A \xrightarrow{k_2} C \qquad k_2 = 1 \text{ m}^3/\text{kmol}\cdot\text{ks}$$

Find the effluent composition in (a) a CSTR (b) a PFR.

7. Consider the first-order reaction

$$X \xrightarrow{k_Y} Y \xrightarrow{k_Z} Z \qquad k_Y = k_Z = 0.5\ \text{h}^{-1}$$

where Y is the desired product. This liquid-phase reactions are to be carried out in a cascade of two equal volume CSTR in series. If the reactors are to be sized so as to maximize the concentration of Y in the effluent from the second reactor, determine the reactor volume necessary to process 400 gal/h of feed containing 5 mol/gal X only. What fraction of X is converted to Y?

8. For the following liquid reaction in a CSTR

$$X \xrightarrow{k_Y} Y + Z \xrightarrow{k_W} 2W$$

derive equations for the effluent concentrations of *Y, Z,* and W if no Y, Z, and W is present in the feed, the volumetric flow rate is q, and the reactor volume is V. If Y is the desired product, determine the space time for maximum production of Y.

9. A stream containing 2.5 kmol/m³ X and 0.4 kmol/m³ Y enters a CSTR at a rate of 0.8 m³/ks for the following reaction

$$X + Y \rightarrow Z \qquad r = 10\ \text{ks}^{-1} C_Y$$

$$Z + X \rightarrow W \qquad r = 4\ \text{ks}^{-1} C_Z$$

(a) Derive the equations for the effluent concentrations of X, Y, Z, and W.
(b) What should the reactor volume be for maximum concentration of Z in the effluent?

10. For the elementary reaction

$$\begin{array}{l} X \xrightarrow{k_Y} Y \xrightarrow{k_Z} Z \\ \downarrow k_W \\ W \end{array}$$

(a) Show for plug flow that

$$\frac{C_{Y,\max}}{C_{Xo}} = \frac{k_Y}{k_Y + K_W}\left(\frac{k_Z}{k_Y + k_W}\right)^{k_Z/(k_Y - k_Z + k_W)} \qquad \text{at } \tau_{\text{opt}} = \frac{\ln[(k_Y + k_W)/k_Z]}{k_Y + k_W - k_Z}$$

(b) Show for mixed flow that

$$\frac{C_{Y,\max}}{C_{Xo}} = \frac{k_Y}{(\sqrt{k_Y + k_W} + \sqrt{k_Z})^2} \qquad \text{at } \tau_{\text{opt}} = \frac{1}{\sqrt{k_Z(k_Y + k_W)}}$$

11. The following reactions are to be carried out in a 20-liter mixed reactor:

$$X \rightarrow Y \qquad r_Y = 3\ h^{-1} C_X$$

$$X \rightarrow Z \qquad r_Z = 0.5\ h^{-1} C_X$$

Determine the feed rate and conversion of reactant so as to maximize profits. What are these on an hourly basis? The data are feed material X costs \$1.20 per mole at $C_{Xo} = 1$ mol/liter, product Y sells for \$4.50 per mol, and Z has no value. The total operating cost of reactor and product separation equipment is \$2.6 per hour + \$1.3 per mole X fed to the reactor. Unconverted X is not recycled.

12. X and Y react as follows:

$$X + Y \rightarrow Z \qquad r_Z = 70\ \text{liter/h}\cdot\text{mol}\ C_X C_Y$$

$$2Y \rightarrow W \qquad r_W = 35\ \text{liter/h}\cdot\text{mol}\ C_Y^2$$

In this reaction 100 mol Z per hour are to be produced at minimum cost in a mixed reactor. Determine the feed rates of X and Y and the volume of the reactor. The data are the cost of the reactor = \$0.02 per hour-liter. Reactants are available in separate streams at $C_{Xo} = C_{Yo} = 0.1$ mol/liter and both cost \$0.55 per mole.

References

(H1). Hill, C.G. *Introduction to Chemical Engineering Kinetics and Reactor Design.* New York: John Wiley & Sons, Inc., 1977.

(L1). Levenspiel, O. *Chemical Reaction Engineering,* 2nd ed. New York: John Wiley & Sons, Inc., 1972.

(S1). Smith, J.M. *Chemical Engineering Kinetics,* 3rd ed. New York: McGraw-Hill Book Company, 1981.

APPENDIX ONE

Newton-Raphson Method

The Newton-Raphson method is an approximate and iterative method for solving nonlinear algebraic equations. It can be applied to a single variable equation or a system of equations for two or more variables.

A1.1 Single-Variable Equation

Let us find the root of $f(x) = 0$ by the Newton-Raphson method. First, expanding this function by Taylor's series gives

$$f(x_{n+1}) = f(x_n) + f'(x_n)h + \text{higher-order terms} \tag{A1.1}$$

where $h = x_{n+1} - x_n$. Neglecting higher-order terms and equating to zero leads to

$$f(x_{n+1}) = f(x_n) + f'(x_n)h = 0 \tag{A1.2}$$

Solving for h

$$h = -\frac{f(x_n)}{f'(x_n)}$$

or

$$x_{n+1} = x_n - \frac{f(x_n)}{f'(x_n)} \tag{A1.3}$$

which is the iterative equation to arrive at the root. That is to say, the next estimate of the root can be evaluated by Eq. (A1-3). The root is reached when the absolute value of the difference of x_{n+1} and x_n is smaller than a tolerance limit.

A1.2 System of Equations for Two or More Variables

For simplicity, let us consider equations for two variables. The same method can be extended to more variables. The problem is to solve the following two nonlinear or linear equations simultaneously:

$$\phi(x, y) = 0 \tag{A1.4}$$

$$\theta(x, y) = 0 \tag{A1.5}$$

Expanding Eqs. (A1-4) and (A1-5) by Taylor's series yields

$$\phi(x_o + h, y_o + k) = \phi(x_o, y_o) + h\phi'_{xo} + k\phi'_{yo} + \text{higher-order terms} \tag{A1.6}$$

$$\theta(x_o + h, y_o + k) = \theta(x_o, y_o) + h\theta'_{xo} + k\theta'_{yo} + \text{higher-order terms} \tag{A1.7}$$

Neglecting higher-order terms and equating to zero lead to

$$\phi(x_o, y_o) + h\phi'_{xo} + k\phi'_{yo} = 0 \tag{A1.8}$$

$$\theta(x_o, y_o) + h\theta'_{xo} + k\theta'_{yo} = 0 \tag{A1.9}$$

Solving Eqs. (A1.8) and (A1.9) simultaneously for h and k by Cramer's rule gives

$$h = \frac{\begin{vmatrix} -\phi(x_o, y_o) & \phi'_{yo} \\ -\theta(x_o, y_o) & \theta'_{yo} \end{vmatrix}}{D} \tag{A1.10}$$

$$k = \frac{\begin{vmatrix} \phi'_{xo} & -\phi(x_o, y_o) \\ \theta'_{xo} & -\theta(x_o, y_o) \end{vmatrix}}{D} \tag{A1.11}$$

where

$$D = \begin{vmatrix} \phi'_{xo} & \phi'_{yo} \\ \theta'_{xo} & \theta'_{yo} \end{vmatrix} \tag{A1.12}$$

Then the next values of x and y are

$$x_1 = x_o + h \tag{A1.13}$$

$$y_1 = y_o + k \tag{A1.14}$$

Repeat the process until the values of x and y are within a tolerance limit.

APPENDIX TWO

Numerical Differentiation

A2.1 Three-Point Formula

$$x_1' = \frac{-3x_1 + 4x_2 - x_3}{2h} \tag{A2.1}$$

$$x_2' = \frac{-x_1 + x_3}{2h} \tag{A2.2}$$

$$x_3 = \frac{x_1 - 4x_2 + 3x_3}{2h} \tag{A2.3}$$

A2.2 Four-Point Formula

$$x_1' = \frac{-11x_1 + 18x_2 - 9x_3 + 2x_4}{6h} \tag{A2.4}$$

$$x_2' = \frac{-2x_1 - 3x_2 + 6x_3 - x_4}{6h} \tag{A2.5}$$

$$x_3' = \frac{x_1 - 6x_2 + 3x_3 + 2x_4}{6h} \tag{A2.6}$$

$$x_4' = \frac{-2x_1 + 9x_2 - 18x_3 + 11x_4}{6h} \tag{A2.7}$$

A2.3 Five-Point Formula

$$x_1' = \frac{-25x_1 + 48x_2 - 36x_3 + 16x_4 - 3x_5}{12h} \tag{A2.8}$$

$$x_2' = \frac{-3x_1 - 10x_2 + 18x_3 - 6x_4 + x_5}{12h} \tag{A2.9}$$

$$x_3' = \frac{x_1 - 8x_2 + 8x_4 - x_5}{12h} \tag{A2.10}$$

$$x_4' = \frac{-x_1 + 6x_2 - 18x_3 + 10x_4 + 3x_5}{12h} \tag{A2.11}$$

$$x_5' = \frac{3x_1 - 16x_2 + 36x_3 - 48x_4 + 25x_5}{12h} \tag{A2.12}$$

A2.4 Seven-Point Formula

$$x_1' = \frac{1}{60h}(-147x_1 + 360x_2 - 450x_3 + 400x_4 - 225x_5 + 75x_6 - 10x_7) \tag{A2.13}$$

$$x_2' = \frac{1}{60h}(-10x_1 - 77x_2 + 150x_3 - 100x_4 + 50x_5 - 15x_6 + 2x_7) \tag{A2.14}$$

$$x_3' = \frac{1}{60h}(2x_1 - 24x_2 - 35x_3 + 80x_4 - 30x_5 + 8x_6 - x_7) \tag{A2.15}$$

$$x_4' = \frac{1}{60h}(-x_1 + 9x_2 - 45x_3 + 45x_5 - 9x_6 + x_7) \tag{A2.16}$$

$$x_5' = \frac{1}{60h}(x_1 - 8x_2 + 30x_3 - 80x_4 + 35x_5 + 24x_6 - 2x_7) \tag{A2.17}$$

$$x_6' = \frac{1}{60h}(-2x_1 + 15x_2 - 50x_3 + 100x_4 - 150x_5 + 77x_6 + 10x_7) \tag{A2.18}$$

$$x_7' = \frac{1}{60h}(10x_1 - 75x_2 + 225x_3 - 400x_4 + 450x_5 - 360x_6 + 147x_7) \tag{A2.19}$$

Reference

Milne, W.E. *Numerical Calculus.* Princeton, N.J.: Princeton University Press, 1949.

APPENDIX THREE

Analytical Solutions of Ordinary Differential Equations

A3.1 First-Order Linear Equation

$$P_1 y' + P_0 y = f(x) \tag{A3.1}$$

where P_1, P_0 are functions of x or constant. The solution is

$$y = \exp\left(-\int \frac{P_0}{P_1}\,dx\right)\left[\int \frac{1}{P_1} f(x) \exp\left(\int \frac{P_0}{P_1}\,dx\right) dx + C\right] \tag{A3.2}$$

where C is an integrating constant.

A3.2 Second-Order Linear Equation

$$P_2 y'' + P_1 y' + P_0 y = F(x) \tag{A3.3}$$

where P_2, P_1, P_0 are functions of x or constant. The solution is the sum of the complementary function and the particular solution.

A3.2-1 Complementary function

The auxiliary equation is

$$P_2 m^2 + P_1 m + P_0 = 0 \tag{A3.4}$$

whose solution is

$$m = \frac{-P_1 \pm \sqrt{P_1^2 - 4P_2P_0}}{2P_2} \tag{A3.5}$$

1. Distinct roots

$$y_c = C_1 e^{m_1 x} + C_2 e^{m_2 x} \qquad \text{for } m_1 \neq m_2 \tag{A3.6}$$

$$y_c = C_1 e^{mx} + C_2 e^{-mx} \qquad \text{for } m \text{ and } -m$$

$$= A \sinh mx + B \cosh mx \tag{A3.7}$$

2. Repeated roots, $m_1 = m_2 = m$

$$y_c = (C_1 + C_2 x)(\sinh mx + \cosh mx) \tag{A3.8}$$

3. Imaginary roots

$$y_c = e^{\gamma x}(A \cos \delta x + B \sin \delta x) \tag{A3.9}$$

where γ and δj are the real and imaginary parts of the roots.

A3.2-2 Particular solution

The particular solution can be obtained by the method of undetermined coefficients, the method of variation of parameter, etc. The method of undetermined coefficients consists of forming a linear combination of the different terms which occur in $F(x)$ in Eq. (A3.3) and their distinct first and second derivatives, then substituting into the complete differential equation and evaluating the coefficients. For details and the method of variation of parameters, consult textbooks on differential equations.

A3.3 Bessel Equations

$$x^2 y'' + xy' + (x^2 - n^2)y = 0 \tag{A3.10}$$

Its solution is:

$$y = C_1 J_n(x) + C_2 J_{-n}(x) \qquad \text{when } n \text{ is not an integer} \tag{A3.11}$$

$$y = C_1 J_n(x) + C_2 Y_n(x) \qquad \text{when } n = 0 \text{ or an integer} \tag{A3.12}$$

where $J_n(x)$ and $Y_n(x)$ are defined as the Bessel functions of the first and second kinds, respectively.

A3.4 Exponential Integral

It is defined as

$$Ei(x) = \int_{-\infty}^{x} \frac{e^t \, dt}{t} \tag{A3.13}$$

or

$$-Ei(-x) = \int_{x}^{\infty} \frac{e^{-t} \, dt}{t} \tag{A3.14}$$

in which the $Ei(x)$ can be evaluated by

$$Ei(x) = \gamma + \ln x + \sum_{n=1}^{\infty} \frac{x^n}{n \cdot n!} \qquad (x > 0) \tag{A3.15}$$

where $\gamma = 0.57721$ is Euler's constant. If $x > 15$,

$$Ei(x) \rightarrow \frac{e^x}{x} \tag{A3.16}$$

APPENDIX FOUR

Numerical Integration

A4.1 Trapezoidal Rule

$$I = \int_{x_0}^{x_1} y\,dx$$

Assume

$$y = a + bx$$

$$I = \int_{x_0}^{x_1} (a + bx)dx$$

$$= a(x_1 - x_0) + \frac{b}{2}(x_1^2 - x_0^2)$$

$$= (x_1 - x_0)\,a + \frac{b}{2}(x_1 + x_0) = \frac{(x_1 - x_0)}{2}[(a + bx_0) + (a + bx_1)]$$

At x_0

$$y_0 = a + bx_0$$

At x_1

$$y_1 = a + bx_1$$

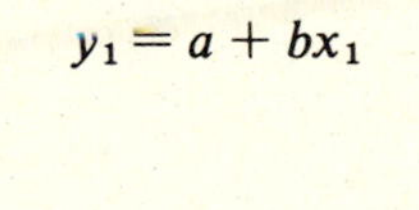

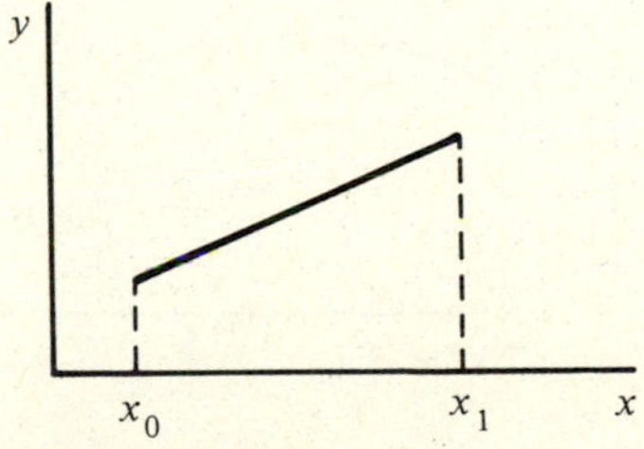

Figure A4.1 Trapezoidal rule

Thus

$$I = \frac{x_1 - x_0}{2}[y_0 + y_1]$$

In general

$$I = \int_{x_a}^{x_1} y\,dx + \int_{x_1}^{x_2} y\,dx + \cdots \int_{x_{N-1}}^{x_b} y\,dx$$

$$= \frac{x_b - x_a}{N}\left[y_a + 2\sum_{i=1}^{N-1} y_i + y_b\right] \tag{A4.1}$$

where N is the number of increments.

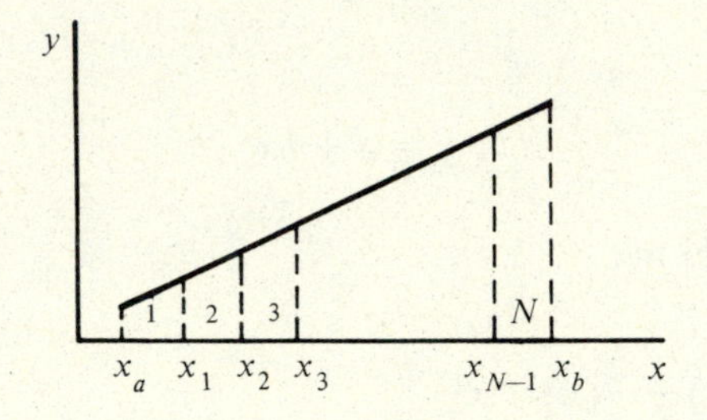

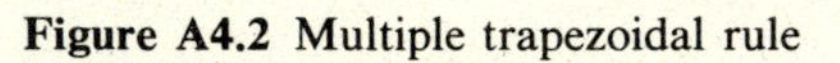

Figure A4.2 Multiple trapezoidal rule

A4.2 Simpson's Rule

$$I = \int_{x_1}^{x_3} y\,dx$$

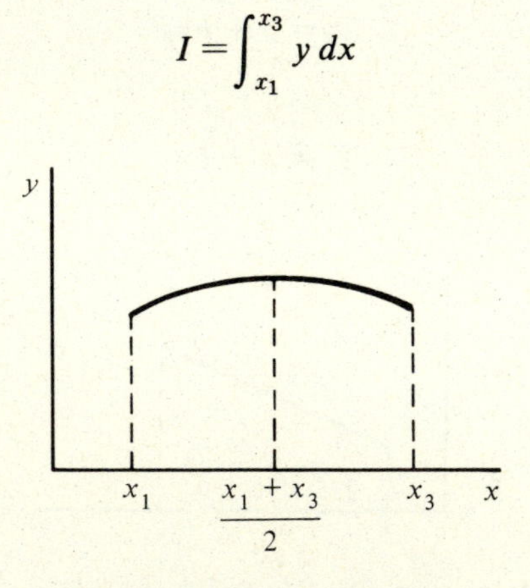

Figure A4.3 Simpson's rule.

Assume

$$y = a + bx + cx^2$$

$$I = \int_{x_1}^{x_3} (a + bx + cx^2)\, dx$$

Thus

$$I = a(x_3 - x_1) + \frac{b}{2}(x_3^2 - x_1^2) + \frac{c}{3}(x_3^3 - x_1^3)$$

$$= \frac{x_3 - x_1}{3}\left[c(x_3^2 + x_3x_1 + x_1^2) + \frac{3}{2}b(x_3 + x_1) + 3a\right]$$

At x_1

$$y_1 = a + bx_1 + cx_1^2$$

At $x_2 = (x_1 + x_3)/2$

$$y_2 = a + b\left(\frac{x_1 + x_3}{2}\right) + c\left(\frac{x_1 + x_3}{2}\right)^2$$

At x_3

$$y_3 = a + bx_3 + cx_3^2$$

Hence

$$y_1 + 4y_2 + y_3 = 2\left[c(x_1^2 + x_1x_3 + x_3) + \frac{3}{2}b(x_1 + x_3) + 3a\right]$$

Then

$$\mathrm{I} = \int_{x_1}^{x_3} y\, dx = \frac{x_3 - x_1}{6}(y_1 + 4y_2 + y_3)$$

In general

$$I = \int_{x_a}^{x_b} y\, dx = \int_{x_a}^{x_2} y\, dx + \int_{x_2}^{x_4} y\, dx + \cdots \int_{x_{N-2}}^{x_b} y\, dx$$

$$= \frac{h}{3}(y_a + 4y_1 + y_2) + \frac{h}{3}(y_2 + 4y_3 + y_4) + \frac{h}{3}(y_4 + 4y_5 + y_6) + \cdots \quad \textbf{(A4.2)}$$

$$= \frac{h}{3}\left(y_a + 2\sum_{i=1}^{n/2-1} y_{2i} + 4\sum_{i=1}^{n/2} y_{2i-1} + y_b\right)$$

where $h = (x_b - x_a)/N$ and N is the number of increments.

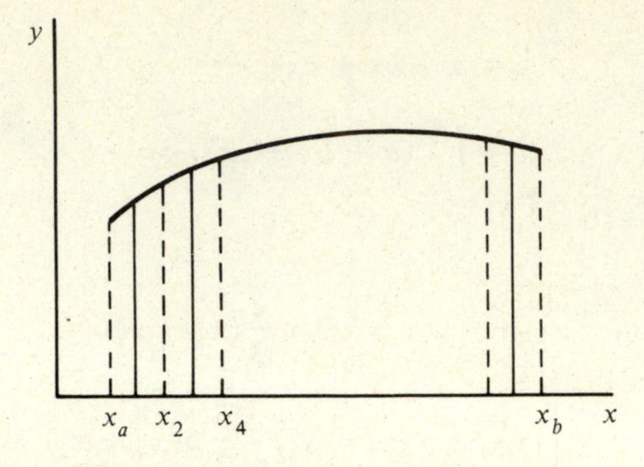

Figure A4.4 Multiple Simpson's rule

APPENDIX FIVE

Numerical Solution of Ordinary Differential Equations

A5.1 Runge-Kutta Method

To solve $y' = F(x, y)$

1. First-order

$$k_1 = hF(x_k, y_k) \qquad y_{k+1} = y_k + k_1 \tag{A5.1}$$

2. Second-order

$$k_1 = hF(x_k, y_k)$$

$$k_2 = hF\left(x_k + \frac{h}{2c}, y_k + \frac{h}{2c}k_1\right) \qquad y_{k+1} = y_k + (1-c)k_1 + ck_2$$

For $c = 1$

$$k_2 = hF\left(x_k + \frac{h}{2}, y_k + \frac{h}{2}k_1\right) \qquad y_{k+1} = y_k + k_2 \tag{A5.2}$$

which is the same as the modified Euler method.

For $c = \frac{1}{2}$

$$k_2 = hF(x_k + h, y_k + hk_1) \qquad y_{k+1} = y_k + \frac{1}{2}(k_1 + k_2) \tag{A5.3}$$

which is the same as the improved Euler method.

3. Third-order

$$k_1 = hF(x_k, y_k)$$

$$k_2 = hF\left(x_k + \frac{1}{2}h, y_k + \frac{1}{2}k_1\right)$$

$$k_3 = hF(x_k + h, y_k + 2k_2 - k_1) \qquad y_{k+1} = y_k + \frac{1}{6}(k_1 + 4k_2 + k_3) \qquad \textbf{(A5.4)}$$

4. Fourth-order

$$k_1 = hF(x_k, y_k)$$

$$k_2 = hF\left(x_k + \frac{1}{2}h, y_k + \frac{1}{2}k_1\right)$$

$$k_3 = hF\left(x_k + \frac{1}{2}h, y_k + \frac{1}{2}k_2\right)$$

$$k_4 = hF(x_k + h, y_k + k_3) \qquad y_{k+1} = y_k + \frac{1}{6}(k_1 + 2k_2 + 2k_3 + k_4) \qquad \textbf{(A5.5)}$$

A5.2 Runge-Kutta Fourth-Order Method

To solve

$$y' = f(x, y, z)$$

$$z' = g(x, y, z)$$

$$y_{k+1} = y_k + \frac{1}{6}(k_1 + 2k_2 + 2k_3 + k_4) \qquad \textbf{(A5.6)}$$

$$z_{k+1} = z_k + \frac{1}{6}(p_1 + 2p_2 + 2p_3 + p_4) \qquad \textbf{(A5.7)}$$

where

$$k_1 = hf(x_k, y_k, z_k) \qquad p_1 = hg(x_k, y_k, z_k)$$

$$k_2 = hf\left(x_k + \frac{h}{2}, y_k + \frac{k_1}{2}, z_k + \frac{p_1}{2}\right) \qquad p_2 = hg\left(x_k + \frac{h}{2}, y_k + \frac{k_1}{2}, z_k + \frac{p_1}{2}\right)$$

$$k_3 = hf\left(x_k + \frac{h}{2}, y_k + \frac{k_2}{2}, z_k + \frac{p_2}{2}\right) \qquad p_3 = hg\left(x_k + \frac{h}{2}, y_k + \frac{k_2}{2}, z_k + \frac{p_2}{2}\right)$$

$$k_4 = hf(x_k + h, y_k + k_3, z_k + p_3) \qquad p_4 = hg(x_k + h, y_k + k_3, z_k + p_3)$$

Extension of this method to a system of three or more differential equations is carried out in a manner analogous to that of going from one to two differential equations.

APPENDIX SIX

Matrices

A6.1 Matrix

A set of mn quantities, real or complex, arranged in a rectangular array of n columns and m rows is called a *matrix of order* (m, n) or $m \times n$. If $m = n$, the matrix is called a *square matrix.*

$$[A] = \begin{bmatrix} a_{11} & a_{12} & \dots & a_{1n} \\ a_{21} & a_{22} & \dots & a_{2n} \\ \dots & \dots & \dots & \dots \\ a_{m1} & a_{m2} & \dots & a_{mn} \end{bmatrix} \tag{A6.1}$$

A6.2 Row Vector

A set of n quantities arranged in a row which is a matrix of order $(1, n)$ is called a *row vector.*

$$\mathbf{a}^T = [a_1, a_2, a_3, \dots, a_n] \tag{A6.2}$$

A6.3 Column Vector

A set of n quantities arranged in a column which is a matrix of order $(n, 1)$ is called *column vector*

$$\mathbf{a} = \begin{bmatrix} a_1 \\ a_2 \\ a_3 \\ \vdots \\ a_n \end{bmatrix} \tag{A6.3}$$

A6.4 Scalar

A matrix of order (1, 1) is called a *scalar.*

A6.5 Diagonal Matrix

A square matrix in which all elements off the main diagonal are zero is called a *diagonal matrix.*

$$[A] = \begin{bmatrix} a_{11} & 0 & 0 \\ 0 & a_{22} & 0 \\ 0 & 0 & a_{33} \end{bmatrix} \tag{A6.4}$$

A6.6 Identity Matrix

A diagonal matrix having unit diagonal elements is called an *identity matrix.*

$$[I] = \begin{bmatrix} 1 & 0 & 0 \\ 0 & 1 & 0 \\ 0 & 0 & 1 \end{bmatrix} \tag{A6.5}$$

A6.7 Null or Zero Matrix

A matrix with all zero elements is called a *null* or *zero matrix.*

$$[0] = \begin{bmatrix} 0 & 0 & 0 \\ 0 & 0 & 0 \\ 0 & 0 & 0 \end{bmatrix} \tag{A6.6}$$

A6.8 Singular and Nonsingular Matrices

A square matrix in which the determinant of its elements, det $[A]$ is zero is called a *singular matrix.* If det $[A] \neq 0$, the matrix is *nonsingular.*

A6.9 Matrix Transpose

The *transpose* of a matrix $[A]$ is formed by the interchange of rows and columns of the matrix $[A]$.

$$[A]^T = \begin{bmatrix} a_{11} & a_{21} & \cdots & a_{m1} \\ a_{12} & a_{22} & \cdots & a_{m2} \\ a_{13} & & \cdots\cdots & \\ \vdots & & & \\ a_{1n} & a_{2n} & \cdots & a_{mn} \end{bmatrix} \tag{A6.7}$$

A6.10 Orthogonal Matrix

A matrix $[A]$ such that $[A]^T[A] = [A][A]^T = [I]$ is called an *orthogonal matrix.*

A6.11 Symmetric and Skew-Symmetric Matrices

A square and symmetrical about the main diagonal is called a symmetric matrix. A matrix such that $[A]^T = -[A]$ is called a *skew-symmetric matrix.*

$$[A]=\begin{bmatrix} 1 & 2 & 3 \\ 2 & 4 & -5 \\ 3 & -5 & 6 \end{bmatrix} \quad \text{symmetric matrix} \tag{A6.8}$$

$$[A]=\begin{bmatrix} 0 & -2 & 3 \\ 2 & 0 & 4 \\ -3 & -4 & 0 \end{bmatrix} \quad \text{skew-symmetric matrix} \tag{A6.9}$$

A6.12 Cofactor and Minor

The *cofactor* of an element in the ith row and jth column of a matrix is the value of the determinant formed by writing the elements of the matrix as a determinant, then deleting the ith row and jth column, and giving to it the sign $(-1)^{i+j}$. Without the sign, it is called the *minor* of a_{ij}.

$$[A]=\begin{bmatrix} a_{11} & a_{12} & a_{13} \\ a_{21} & a_{22} & a_{23} \\ a_{31} & a_{32} & a_{33} \end{bmatrix} \tag{A6.10}$$

The minors are

$$|M_{11}| = \begin{vmatrix} a_{22} & a_{23} \\ a_{32} & a_{33} \end{vmatrix}, \; |M_{12}| = \begin{vmatrix} a_{21} & a_{23} \\ a_{31} & a_{33} \end{vmatrix}, \; |M_{13}| = \begin{vmatrix} a_{21} & a_{22} \\ a_{31} & a_{32} \end{vmatrix} \tag{A6.11}$$

The cofactors are

$$\alpha_{11} = (-1)^{1+1}|M_{11}| = |M_{11}| \tag{A6.12}$$

$$\alpha_{12} = (-1)^{1+2}|M_{12}| = |M_{12}| \tag{A6.13}$$

$$\alpha_{13} = (-1)^{1+3}|M_{13}| = |M_{13}| \tag{A6.14}$$

The determinant of A is

$$|A| = \alpha_{11}|M_{11}| + a_{12}|M_{12}| + a_{13}|M_{13}| \quad \textbf{(A6.15)}$$
$$= a_{11}\,\alpha_{11} + a_{12}\,\alpha_{12} + a_{13}\,\alpha_{13}$$

A6.13 Adjoint Matrix

It is defined as the transpose of the matrix of the minors of the matrix A. For the matrix A in Section A6.12, the *adjoint matrix A* can be formed as:

$$[\text{adj } A] = \begin{bmatrix} |M_{11}| & |M_{12}| & |M_{13}| \\ |M_{21}| & |M_{22}| & |M_{23}| \\ |M_{31}| & |M_{32}| & |M_{33}| \end{bmatrix}^T = \begin{bmatrix} |M_{11}| & |M_{21}| & |M_{31}| \\ |M_{12}| & |M_{22}| & |M_{32}| \\ |M_{13}| & |M_{23}| & |M_{33}| \end{bmatrix} \quad \textbf{(A6.16)}$$

A6.14 Inverse of a Matrix

It is defined as the adjoint matrix divided by the determinant of the matrix.

$$[A]^{-1} = \frac{[\text{adj } A]}{|A|} \quad \textbf{(A6.17)}$$

A6.15 Characteristic Matrix, Characteristic Equation, and Eigenvalues

The *characteristic matrix* of a square matrix $[A]$ of order (n, n), and with constant elements, is

$$[C] = [\lambda I - A] \quad \textbf{(A6.18)}$$

The characteristic equation is

$$\det [C] = \det [\lambda I - A] = 0 \quad \textbf{(A6.19)}$$

The n roots of the characteristic equation $\det [\lambda I - A] = 0$ which will be of degree n in λ are called the *eigenvalues* of the matrix A.

A6.16 Matrix Addition and Subtraction

$$[A] + [B] = \begin{bmatrix} a_{11} & a_{12} \\ a_{21} & a_{22} \end{bmatrix} + \begin{bmatrix} b_{11} & b_{12} \\ b_{21} & b_{22} \end{bmatrix} = \begin{bmatrix} a_{11} + b_{11} & a_{12} + b_{12} \\ a_{21} + b_{21} & a_{22} + b_{22} \end{bmatrix} \quad \textbf{(A6.20)}$$

$$[A]-[B]=\begin{bmatrix} a_{11} & a_{12} \\ a_{21} & a_{22} \end{bmatrix}-\begin{bmatrix} b_{11} & b_{12} \\ b_{21} & b_{22} \end{bmatrix}=\begin{bmatrix} a_{11}-b_{11} & a_{12}-b_{12} \\ a_{21}-b_{21} & a_{22}-b_{22} \end{bmatrix} \quad \textbf{(A6.21)}$$

A6.17 Scalar Multiplication

$$k[A]=k\begin{bmatrix} a_{11} & a_{12} \\ a_{21} & a_{22} \end{bmatrix}=\begin{bmatrix} ka_{11} & ka_{12} \\ ka_{21} & ka_{22} \end{bmatrix} \quad \textbf{(A6.22)}$$

A6.18 Matrix Multiplication

$$[A][B]=[a_{ij}][b_{ij}]=[C_{ij}]=\left[\sum_{k=1}^{n} a_{ik}b_{kj}\right] \quad \textbf{(A6.23)}$$

$$[x_1 \quad x_2]\begin{bmatrix} y_1 \\ y_2 \end{bmatrix}=(x_1y_1+x_2y_2) \quad \textbf{(A6.24)}$$

$$\begin{bmatrix} x_1 \\ x_2 \end{bmatrix}[y_1 \quad y_2]=\begin{bmatrix} x_1y_1 \\ x_2y_2 \end{bmatrix} \quad \textbf{(A6.25)}$$

A6.19 Matrix Differentiation

$$\frac{d[A]}{dt}=\left[\frac{da_{ij}}{dt}\right] \quad \textbf{(A6.26)}$$

A6.20 Matrix Integration

$$\int_a^b [A]\,dx=\left[\int_a^b a_{ij}\,dx\right] \quad \textbf{(A6.27)}$$

A6.21 Matrix Diagonalization

For a nonsingular matrix A, find a matrix C that satisfies the following identity

$$[C]^{-1}[A][C]=[\lambda] \quad \textbf{(A6.28)}$$

which is equivalent to

$$[A][C]=[C][\lambda] \quad \textbf{(A6.29)}$$

or

$$([A]-[\lambda])[C]=0 \tag{A6.30}$$

which is a characteristic equation.

A6.22 Linear Form

$$\Sigma a_{ij}x_j=[A]^T[X]=[a_{11}\ a_{12}\ldots a_{1n}]\begin{bmatrix}x_1\\x_2\\\vdots\\x_n\end{bmatrix} \tag{A6.31}$$

A6.23 Bilinear Form

$$\sum_i^m \sum_j^n a_{ij}x_iy_j=[X]^T[A][Y]=[x_1\ x_2\ldots x_m]\begin{bmatrix}a_{11} & a_{12}\ldots & a_{1n}\\a_{21} & a_{22}\ldots & a_{2n}\\ \multicolumn{3}{c}{\ldots\ldots\ldots\ldots}\\a_{m1} & a_{m2}\ldots & a_{mn}\end{bmatrix}\begin{bmatrix}y_1\\y_2\\\vdots\\y_n\end{bmatrix} \tag{A6.32}$$

A6.24 Quadratic Form

$$\sum_i^n \sum_j^n a_{ij}x_ix_j=[X]^T[A][X]=[x_1\ x_2\ldots x_n]\begin{bmatrix}a_{11} & a_{12}\ldots & a_{1n}\\a_{21} & a_{22}\ldots & a_{2n}\\ \multicolumn{3}{c}{\ldots\ldots\ldots\ldots}\\a_{n1} & a_{n2}\ldots & a_{nn}\end{bmatrix}\begin{bmatrix}x_1\\x_2\\\vdots\\x_n\end{bmatrix} \tag{A6.33}$$

A6.25 Cayley-Hamilton Theorem

This theorem states that any square matrix satisfies its own characteristic equation.

A6.26 Sylvester's Theorem

For distinct root

$$P(A)=\sum_{r=1}^{n}\frac{P(\lambda_r)\,\mathrm{adj}(\lambda_r I-A)}{\det'(\lambda_r)} \tag{A6.34}$$

For multiple roots

$$P(A) = \frac{1}{(k-1)} \frac{d^{k-1}}{d\lambda^{k-1}} \left[\frac{P(\lambda) \operatorname{adj}(\lambda I - A)}{\det_s(\lambda)} \right] \bigg|_{\lambda=\lambda_s} \tag{A6.35}$$

where $\det_s(\lambda)$ is the determinant without the root λ_s.

A6.27 Linear Algebraic Equation

$$[A][X] = [B] \tag{A6.36}$$

Its solution is

$$[X] = [A]^{-1}[B] \tag{A6.37}$$

A6.28 Linear Differential Equation

$$\frac{d[X]}{dt} - [A][X] = [B] \tag{A6.38}$$

Its solution is

$$[X] = e^{[A]t}([X]_0 + [A]^{-1}[B]) - [A]^{-1}[B] \tag{A6.39}$$

A6.29 Linear Difference Equation

$$[A][Y(n+1)] + [B][Y(n)] = [f(n)] \tag{A6.40}$$

Its solution is

$$[Y(n+1)] = [D][Y(n)] + [F(n)] \tag{A6.41}$$

where

$$[D] = -[A]^{-1}[B] \qquad \text{and} \qquad [F(n)] = [A]^{-1}[f(n)]$$

APPENDIX SEVEN

The Laplace Transform

The Laplace transform is a useful method for handling differential equations though it can be applied with some difficulty to difference equations as well. It is defined as

$$F(s) = \int_0^\infty f(t)e^{-st}\,dt \tag{A7.1}$$

That means, we can transform a function of time $f(t)$ to another function of another domain $F(s)$ by multiplying with e^{-st} and then integrating from 0 to ∞. A large number of elementary functions, such as exponential, transcendental, and hyperbolic functions and some special functions, such as step, pulse, error, gamma, sine, cosine, and exponential integrals can all be transformed to functions of another domain *S*. From the properties of the Laplace transform, considerable functions can be extended for the transformation. Hence appreciable numbers of these transforms can be listed in a table for convenient use, particularly by using Laplace transform of derivatives

$$L\left(\frac{df(t)}{dt^n}\right) = s^nF(s) - s^{n-1}f(0) - s^{n-2}f'(0) - \cdots f^{n-1}(0) \tag{A7.2}$$

ordinary differential equations can be reduced to algebraic equations which are more easily solved. Hence in solving ordinary or partial differential equations, the first step is to transform the differential equation by looking up the result in very extensive tables or performing the integral operations indicated by Eq. (A7.1). Once the Laplace transform operation has been transformed and we have a relationship involving $F(s)$, it remains only to invert the process and transform $F(s)$ back to $f(t)$.

$$L^{-1}F(s) = f(t) = \frac{1}{2\pi j}\int_{Br} F(s)e^{st}\,dt \tag{A7.3}$$

This step of applying the inverse Laplace transform can be carried out in the following manner:

1. Look up $F(s)$ in the tables.
2. By partial fractions, break up the complicated $F(s)$ into simpler forms so that the tables can be used.
3. Use the Heaviside theorem.
4. Use the convolution theorem.
5. Use contour integration.

As to the details of these operations, consult a textbook on Laplace transform.

APPENDIX EIGHT

Generating Functions

Analogous to the Laplace transform which can reduce a differential equation to an algebraic equation, the generating function can transform a finite-difference equation to an algebraic equation. Although the Laplace transform can be used in handling a finite-difference equation, the generating function is easier. The main difference between these two types of equation is that the differential equation deals with continuous function, whereas the finite-difference equation treats discrete function. The generating function is defined as

$$G[f(n)] = F(s) = \sum_{n=0}^{\infty} f(n)s^n \tag{A8.1}$$

compared to the Laplace transform as

$$L[f(n)] = F(s) = \int_0^{\infty} f(n)e^{-sn}\,dn \tag{A8.2}$$

Similarly, we can take generating functions of the common functions such as unit step function, ramp function, reciprocal of factorial function, exponential function, sine and cosine functions, hyperbolic sine and cosine functions, and many others to form a table for convenient use. We can also extend the table by using the properties of generating functions such as shifting, scale change, complex differentiation or multiplication by n, complex integration or division by n, real convolution, initial- and final-value theorems. Particularly by using the shifting theorem

$$G[f(n+r)] = \frac{1}{s^r}\left[F(s) - \sum_{j=0}^{r-1} s^j f(j)\right] \tag{A8.3}$$

a finite-difference equation can be reduced to an algebraic equation. Thus in solving a finite-difference equation, the first step is to transform the difference equation by looking up the result in a very extensive table or by performing the operations indicated by Eq. (A8.1). Once the generating functions have been obtained and we have a relationship involving $F(s)$, it remains only to invert the process and transform $F(s)$ back to $f(n)$

$$G^{-1}F(s) = f(n) = \frac{1}{2\pi j}\int_{Br} \frac{F(s)}{s^{n+1}}\, ds \tag{A8.4}$$

This step of inverse generating functions can be carried out in the following manner:

1. Look up $F(s)$ in the table in N.H. Chen, *New Mathematics for Chemical Engineers* (1977), Hoover Book Co., P.O. Box 385, Lowell, MA 01853.
2. By partial fractions, break up the complicated $F(s)$ into simpler forms so that the tables can be used.
3. Use the convolution theorem.
4. Use the long-division theorem.
5. Use contour integration.

As to the details of these operations, consult N.H. Chen, *New Mathematics for Chemical Engineers* (1977), Hoover Book Co., P.O. Box 385, Lowell, MA 01853.

APPENDIX NINE

Maximum and Minimum Principles

A9.1 Continuous Function

$$\underset{u(t)}{\text{Minimize}}\ J = \theta(\vec{\mathbf{x}}, t)\Big|_{t_o}^{t_f} + \int_{t_o}^{t_f} \phi(\vec{x}, \vec{u}, t)\, dt \tag{A9.1}$$

$$\text{subject to}\ \frac{d\vec{x}}{dt} = \vec{f}(\vec{x}, \vec{u}, t) \tag{A9.2}$$

The first step is to couple the constraints

$$\underset{u(t)}{\text{Minimize}}\ J' = \theta(\vec{x}, t)\Big|_{t_o}^{t_f} + \int_{t_o}^{t_f} \{\phi(\vec{x}, \vec{u}, t) + \vec{\lambda}^T(t)[\vec{f}(\vec{x}, \vec{u}, t) - \dot{\vec{x}}]\}\, dt \tag{A9.3}$$

Define a Hamiltonian function H as

$$H(\vec{x}, \vec{u}, \vec{\lambda}, t) = \phi(\vec{x}, \vec{u}, t) + \vec{\lambda}^T(t)[\vec{f}(\vec{x}, \vec{u}, t)] \tag{A9.4}$$

The problem becomes

$$\text{Minimize}\ J' = \theta(\vec{x}, t)\Big|_{t_o}^{t_f} + \int_{t_o}^{t_f} [H(\vec{x}, \vec{u}, \vec{\lambda}, t) - \vec{\lambda}^T \dot{\vec{x}}\,]\, dt \tag{A9.5}$$

Since integrating by parts

$$\int_{t_o}^{t_f} \vec{\lambda}^T \dot{\vec{x}}\, dt = \vec{\lambda}^T(t)\vec{x}\,\Big|_{t_o}^{t_f} - \int_{t_o}^{t_f} \vec{\lambda}^T \vec{x}\, dt \tag{A9.6}$$

Equation (A9.5) can be rearranged to

$$\text{Minimize}\ J' = [\theta(\vec{x}, t) - \vec{\lambda}^T(t)\vec{x}(t)]\Big|_{t_o}^{t_f} + \int_{t_o}^{t_f} [H(\vec{x}, \vec{u}, \vec{\lambda}, t) + \vec{\lambda}^T\vec{x}(t)]\, dt \tag{A9.7}$$

The first variation of J' is

$$\delta J' = J('x + \delta x, u + \delta u, t) - J'(x, u, t) \tag{A9.8}$$

$$= [\theta(x + \delta x, t) - \lambda(t)(x + \delta x)] \Big|_{t_o}^{t_f} + \int_{t_o}^{t_f} \{H(x + \delta x, u + \delta u, \lambda, t) + \dot{\lambda}(x + \delta x)\}\, dt - \{\theta(x, t) - \lambda(t)x\} \Big|_{t_o}^{t_f} - \int_{t_o}^{t_f} \{H(x, u, \lambda, t) + \dot{\lambda} x\}\, dt \tag{A9.9}$$

Expanding Eq. (9.8-9) by Taylor's series, omitting higher-order terms, and simplifying gives

$$\delta J' = \left(\frac{\partial \theta}{\partial x} - \lambda\right) \delta x \Big|_{t_o}^{t_f} + \int_{t_o}^{t_f} \left[\left(\frac{\partial H}{\partial x} + \lambda\right) \delta x + \frac{\partial H}{\partial u} \delta u\right] dt = 0 \tag{A9.10}$$

At optimum, $\delta J' = 0$; therefore, the optimal conditions are

1. Transversality

$$\left[\frac{\partial \theta}{\partial x} - \lambda\right] \delta x = 0 \text{ for } t_o \text{ and } t_f \tag{A9.11}$$

2. Necessary

$$\text{A. } \dot{\lambda} = -\frac{\partial H}{\partial x} \tag{A9.12}$$

$$\text{B. } \dot{x} = f(x, u, t) = \frac{\partial H}{\partial x} \tag{A9.13}$$

$$\text{C. } \frac{\partial H}{\partial u} = 0 \tag{A9.14}$$

A9.2 Discrete Function

The problem is to minimize

$$\underset{u_n}{\text{Minimize}}\ J = \Big[\theta(x_n, n)\Big]_{n=n_o}^{n=n_f} + \sum_{n=n_o}^{n_f - 1} \emptyset(x_n, u_n, n) \tag{A9.15}$$

subject to

$$x_{n+1} = f(x_n, u_n, n) \tag{A9.16}$$

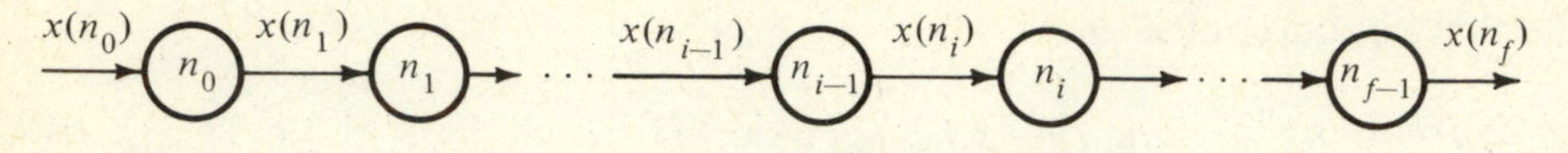

Figure A9.1 Cascade of stages

which is equivalent to

$$\underset{u_n}{\text{Minimize}}\ J' = [\theta(x_n, n)] \Big|_{n_o}^{t_f} + \sum_{n=n_o}^{n_f-1} \{\phi(x_n, u_n, n) - \lambda_{n+1}[x_{n+1} - f(x_n, u_n, n)]\} \quad \textbf{(A9.17)}$$

Now define the Hamiltonian function

$$H(x_n, u_n, \lambda_{n+1}, n) = H_n = \phi(x_n, u_n, n) + \lambda_{n+1} f(x_n, u_n, n) \quad \textbf{(A9.18)}$$

Eq. (A9.17) becomes

$$\underset{u_n}{\text{Minimize}}\ J' = [\theta(x_n, n)] \Big|_{n_o}^{n_f} + \sum_{n=n_o}^{n_f-1} [H_n - \lambda_{n+1} x_{n+1}] \quad \textbf{(A9.19)}$$

The first variation of J′ is then

$$\delta J' = J'(x_n + \delta x_n, u_n + \delta u_n, n) - J'(x_n, u_n, n) \quad \textbf{(A9.20)}$$

$$= [\theta(x_n + \delta x_n, n)] \Big|_{n_o}^{n_f} + \sum_{n=n_o}^{n_f-1} [H_{n+\delta n} - \lambda_{n+1}(x_{n+1} + \delta x_{n+1})] - [\theta(x_n, n)] \Big|_{n_o}^{n_f} + \sum_{n=n_o}^{n_f-1} [H_n - \lambda_{n+1} x_{n+1}] \quad \textbf{(A9.21)}$$

Expanding by Taylor's series, neglecting higher-order terms, and simplifying gives

$$\delta J' = \frac{\partial \theta}{\partial x_n} \delta x_n \Big|_{n_o}^{n_f} + \sum_{n=n_o}^{n_f-1} \frac{\partial H_n}{\partial x_n} \delta x_n + \sum_{n=n_o}^{n_f-1} \frac{\partial H_n}{\partial u_n} \delta u_n - \sum_{n=n_o}^{n_f-1} \lambda_{n+1} \delta x_{n+1} \quad \textbf{(A9.22)}$$

The last term can be written as

$$-\sum_{n=n_o}^{n_f-1} \lambda_{n+1} \delta x_{n+1} = -\sum_{n=n_o+1}^{n_f} \lambda_n \delta x_n = -\sum_{n=n_o}^{n_f-1} \lambda_n \delta x_n - \lambda_{n_f} \delta x_{n_f} + \lambda_{n_o} \delta x_{n_o} \quad \textbf{(A9.23)}$$

Substituting Eq. (A9.23) into Eq. (A9.22) and grouping yields

$$\left[\left(\frac{\partial\theta_{nf}}{\partial x_{nf}}\right)-\lambda_{nf}\right]\delta x_{nf}-\left[\left(\frac{\partial\theta_{no}}{\partial x_{no}}\right)-\lambda_{no}\right]\delta x_{no}+\sum_{n=n_o}^{n_f-1}\left[\left(\frac{\partial H_n}{\partial x_n}\right)-\lambda_n\right]\delta x_n + \sum_{n=n_o}^{n_f-1}\left(\frac{\partial H_n}{\partial u_n}\right)\delta u_n=0 \quad \text{(A9.24)}$$

Therefore the optimal conditions are:

1. Transversality:

(A) $$\left[\left(\frac{\partial\theta_{nf}}{\partial x_{nf}}\right)-\lambda_{nf}\right]\delta x_{nf}=0 \quad \text{(A9.25)}$$

(B) $$\left[\left(\frac{\partial\theta_{no}}{\partial x_{no}}\right)-\lambda_{no}\right]\delta x_{no}=0 \quad \text{(A9.26)}$$

2. Necessary:

(C) $$\frac{\partial H_n}{\partial x_n}=\lambda_n \quad \text{or} \quad \lambda_{n+1}=\left(\frac{\partial f}{\partial x_n}\right)^{-1}\left[\lambda_n-\frac{\partial\phi_n}{\partial x_n}\right] \quad \text{(A9.27)}$$

(D) $$\frac{\partial H_n}{\partial u_n}=0 \quad \text{or} \quad \frac{\partial\phi_n}{\partial u_n}+\left(\frac{\partial f}{\partial u_n}\right)\lambda_{n+1}=0 \quad \text{(A9.28)}$$

(E) $$\frac{\partial H_n}{\partial\lambda_{n+1}}=x_{n+1} \quad \text{or} \quad x_{n+1}=f(x_n, u_n, n) \quad \text{(A9.29)}$$

APPENDIX TEN

Notation

A	Surface area, frequency factor
A_c	Cross-sectional area of tube
A_m	Surface area per unit mass of particle
A_p	Surface area of a single pellet
a	Empirical constant in heat-capacity equation, activity, intercept in a straight-line equation, surface area per unit volume
a_m	External area per unit mass of catalyst
a_p	Interstitial surface area
B^0	Second virial coefficient
B^1	Second virial coefficient
b	Slope in a straight-line equation
C_0; C_o	Initial total concentration
C	Concentration
C^*	Concentration in the next phase
C_p	Heat capacity at constant pressure
C_s	Solid concentration
C_t	Total concentration
C_v	Heat capacity at constant volume
C_∞	Concentration at the surface
c	Constant exponentially related to the heats of adsorption of the first layer and the heat of liquefaction, empirical constant in heat-capacity equation
D	Diffusivity, diameter of tube

D_c	Combined diffusivity
D_e	Effective diffusivity
D_{eff}	Effective diffusivity
D_{eq}	Equivalent diameter
D_K	Knudsen diffusivity
D_T	Diameter of tube
d	Empirical constant in heat-capacity equation
d_c	Diameter of cylinder
E	Activation energy
E_D	Apparent activation energy
F	Feed rate
f	Fractional conversion, fugacity
f^*	Coefficient in the table on p. 356
f_D	Darcy friction factor
f_f	Fanning friction factor
G	Free energy, mass velocity, generation
$\mathbf{G}$	Partial molar free energy
g	Mass fraction
g_c	Conversion factor
(g)	Gas state
H	Henry's law constant
H'	Special function in Eq. (2-19)
$\hat{H}$	Enthalpy per unit mass
H_p	Enthalpy of product
H_f	Enthalpy of feed
ΔH_c	Heat of combustion
ΔH_f	Heat of formation

ΔH_r	Heat of reaction
h	Enthalpy
h'	Heat-transfer coefficient
h_c	Height of cylinder, convective heat-transfer coefficient
h_p	Particle heat-transfer coefficient
h_R	Radiative heat-transfer coefficient
h_w	Heat-transfer coefficient at the wall
h_w^f	Fluid convective heat-transfer coefficient in the vicinity of wall
h_w^s	Solid convective heat-transfer coefficient in the vicinity of wall
i	Number of reaction
J	$= -\Delta H/\Sigma(C_j C_{pj})$ in Eq. (5-35)
J_A	Mass flux due to diffusion
j	Species, subscript for component
j_D	Mass-transfer factor $= (k_c \rho/G) N_{Sc}^{2/3}$
j_h	Heat-transfer factor $= N_{St} N_{Pr}^{2/3}$
K	Equilibrium constant
K	Kinetic energy per unit mass
K_s	Surface reaction equilibrium constant
K_x	Adsorption equilibrium constant
K_Y	Reciprocal of the desorption equilibrium constant $= 1/K_d$
K_d	Desorption equilibrium constant
K_{bc}	Mass-transfer coefficient between bubble and clouds
K_{ce}	Mass-transfer coefficient between cloud and emulsion
k	Rate constant, thermal conductivity, subscript for limiting reactant
k'	Rate constant
k_c	Rate constant based on concentration, mass-transfer coefficient per unit external surface area based on concentration
k_f	Thermal conductivity of bulk fluid

k_g	Thermal conductivity of bulk fluid
k_G	Gas-film mass-transfer coefficient per unit external surface area
k_L^*	Effective thermal conductivity in the longitudinal direction
k_m	Interphase mass-transfer coefficient
k_p	Rate constant based on partial pressures, particle thermal conductivity
k_s	Surface reaction rate constant, solid thermal conductivity
k_R^*	Effective thermal conductivity in the radial direction
k_v	Rate constant based on total volume
k_{eff}	Effective thermal conductivity
k_{er}^f	Effective thermal conductivity in the r direction for fluid
k_{er}^s	Effective thermal conductivity in the r direction for solid
k_Y	Desorption rate constant
L	Length of tube
$\mathbf{L}$	Average pore length
L_{mf}	Height of the fluidized bed at minimum fluidization
(l)	Liquid state
M	Molecular weight; ratio of N_{jo}/N_{ko}
M_k	$=C_{ko}V$
M_T	Total moles $= \Sigma n_j$
m	Mass
m_k	Original mass of the limiting reactant k
m_p	Mass of particle
$\dot{m}_1$, $\dot{m}_2$	Mass flow rate at entrance and at exit, respectively
N_o	Initial total number of moles
N	Number of moles
N'	Molal flow rate
N_A, N_B	Molar flux of A and B, respectively

N_i	Molar flux of component i
n	nth order
$\mathbf{n}$	Unit normal
n_p	Number of pores per article; apparent order of reaction
$\dot{n}_p$	Total molar product rate
$\dot{n}_f$	Total molar feed rate
P	Perimeter, total pressure
p_j	Partial pressure of component j
P_o	Saturation pressure
Q	Heat transfer from surrounding
q	Volumetric flow rate, number of reaction
$\dot{q}_w$	Heat flux at the wall
R	Gas constant
R_h	Hydraulic radius
r	Rate of reaction
$\mathbf{r}$	Average pore radius
r_m	Reaction rate based on total mass of catalyst
r_o	Radius of cylinder or sphere
r_p	Reaction rate of the particle at the surface of the catalyst
r_v	Reaction rate based on total volume of catalyst
S	Chemical species, entropy, specific surface
S_g	Surface area per unit mass of particle
S_m	Variable in generating function
S_x	Geometric surface area
ss	Sum of squares
(s)	Solid state
T	Temperature

T^*	Surrounding temperature, temperature in next phase, wall temperature
T_B	Temperature in the bulk stream
T_c	Critical temperature
T_f	Reference temperature
T_R	Reference temperature
T_S	Solid temperature
T_w	Wall temperature
t	Time
U	Overall heat-transfer coefficient
$\hat{U}$	Internal energy per unit mass
u	Linear velocity
u_t	Terminal velocity
u_{br}	Rise velocity of a single bubble
V	Volume, velocity
V_b	Volume of bubble
V_c	Critical volume
V_g	Void volume per unit mass of particle
V_i	Interstitial velocity
V_m	Volume which would be adsorbed in a monolayer
V_∞	Superficial velocity
v	Volume of gas actually adsorbed
$\mathbf{v}$	Vector velocity
W	Mass of catalyst
W_s	Shaft work on surrounding
X_j	Chemical formula for component j
x	Mole fraction; normalized pressure; concentration
x_s	Surface concentration

$X_j(p)$	Physical state of component j
Y	Mole fraction
Z	Compressibility factor; distance

Greek Letters

α_j	Stoichiometric coefficient of j component, volume of wake/volume of bubble
$\boldsymbol{\alpha}$	Summation of $\alpha_j = \Sigma\alpha_j$
$\underline{\alpha}$	Order of reaction
β	Molar extent or degree of advancement of the reaction or reaction coordinate or unit conversion, energy generation function
τ	Space-time, momentum, tortuosity factor
ω	Distance
γ	λ/C_o, Arrhenius number
γ_b	Volume of solids dispersed in bubbles/ V_b
γ_c	Volume of solids within clouds and wakes/ V_b
γ_e	Volume of solids in emulsion/ V_b
δ	Bed fraction in bubbles, $= \beta/N_o$
ϵ	Void fraction, dummy variable in Eq. (6-20)
ϵ_p	Void fraction or porosity
ϵ_k	$= (\boldsymbol{\alpha} N_{k0})/N_0$
H	Solution conductivity, effectiveness factor, unit output
ϕ	Physical property, Thiele modulus, potential energy per unit mass, normalized temperature
η_f	Effective multiphase reaction-rate coefficient
μ	Viscosity
ν	Fugacity coefficient
Ω	Radial coordinate of the particle, area availability factor; $= \Sigma n_j C_{vj}$
Σ	Summation

σ	Absorption site, emissivity
Π	Product notation
θ	Residence time, fractional coverage
λ	Unit conversion per unit volume of reaction mixture, mean free path
$\mathfrak{z}_p$	$= \Sigma C_j C_{pj}$
$\mathfrak{z}$	$= C_{k0} \Sigma \Delta H_i / \mathfrak{z}_p$
ρ	Density
ρ_b	Bulk density, i.e., mass of catalyst per unit volume of bed
ρ_p	Particle density
ρ_s	Solid density, i.e., mass per unit volume of solid phase or skeletal density

Bibliography

1. ARIS, R. *Elementary Chemical Reactor Analysis.* Englewood Cliffs, N.J.: Prentice-Hall, Inc., 1965.

2. BOUDART, M. *Kinetics of Chemical Processes.* Englewood Cliffs, N.J.: Prentice-Hall, Inc., 1968.

3. BUTT, J.B. *Reaction Kinetics and Reactor Design.* Englewood Cliffs, N.J.: Prentice-Hall, Inc., 1980.

4. CARBERRY, J.J. *Chemical and Catalytic Reaction Engineering.* New York: McGraw-Hill Book Company, 1976.

5. CHEN, N.H. *New Mathematics for Chemical Engineers.* Lowell, MA: Hoover Book Co. (P.O. Box 385, Lowell, MA 01853), 1977.

6. COOPER, A.R., AND JEFFREYS, G.V. *Chemical Kinetics and Reactor Design.* Edinburgh: Oliver and Boyd Co., Inc., 1971.

7. COULSON, J.M., AND RICHARDSON, J.F. *Chemical Engineering, Volume III,* 2nd ed. Elmsford, N.Y.: Pergamon Press, 1975.

8. CREMER, H.W. *Chemical Engineering Practice, Volume VIII.* London: Butterworths, 1965.

9. DENBIGH, K.G., AND TURNER, J.C.R. *Chemical Reactor Theory,* 2nd ed. London: Cambridge University Press, 1971.

10. FAN, L.T. *The Continuous Maximum Principle.* New York: John Wiley & Sons, Inc., 1966.

11. FAN, L.T., AND WANG, C.S. *The Discrete Maximum Principle.* New York: John Wiley & Sons, Inc., 1964.

12. FOGLER, H.S. *The Elements of Chemical Kinetics and Reactor Calculations.* Englewood Cliffs, N.J.: Prentice-Hall, Inc., 1974.

13. FROMENT, G.F., AND BISCHOFF, K.B. *Chemical Reactor Analysis and Design.* New York: John Wiley & Sons, Inc., 1979.

14. FROST, A.A., AND PEARSON, R.G. *Kinetics and Mechanism,* 2nd ed. New York: John Wiley & Sons, Inc., 1961.

15. HILL, C.G. *Introduction to Chemical Engineering Kinetics and Reactor Design.* New York: John Wiley & Sons, Inc., 1977.

16. HOLLAND, C.D., AND ANTHONY, R.G. *Fundamentals of Chemical Reaction Engineering.* Englewood Cliffs, N.J.: Prentice-Hall, Inc., 1979.

17. KRAMERS, H., AND WESTERTERP, K.R. *Elements of Chemical Reactor Design and Operation.* New York: Academic Press, 1963.

18. LEVENSPIEL, O. *Chemical Reaction Engineering,* 2nd ed. New York: John Wiley & Sons, Inc., 1972.

19. MEARNS, A.M. *Chemical Engineering Process Analysis.* Edinburgh: Oliver and Boyd Co., Inc., 1973.

20. PETERSEN, E.E. *Chemical Reaction Analysis.* Englewood Cliffs, N.J.: Prentice-Hall, Inc., 1965.

21. RASE, H.F. *Chemical Reactor Design for Process Plants.* New York: John Wiley & Sons, Inc., 1977.

22. RUSSELL, T.W.F., AND DENN, M.M. *Introduction to Chemical Engineering Analysis.* New York: John Wiley & Sons, Inc., 1972.

23. SMITH J.M. *Chemical Engineering Kinetics,* 3rd ed., New York: McGraw-Hill Book Company, 1981.

24. WALAS, S.M. *Reaction Kinetics for Chemical Engineers.* New York: McGraw-Hill, Inc., 1959.

25. WILDE, D.J., AND BEIGHTLER, C.D. *Foundations of Optimization.* Englewood Cliffs, N.J.: Prentice-Hall, Inc., 1967.

26. WOJCIECHOWSKI, B.W. *Chemical Kinetics for Chemical Engineers.* Austin, Texas: Sterling Swift Co., 1975.

Index